# Cosmetic and Pharmaceutical Applications of Polymers

# Cosmetic and Pharmaceutical Applications of Polymers

Edited by

Charles G. Gebelein
Youngstown State University
Youngstown, Ohio

Tai C. Cheng
Advanced Cardiovascular Systems, Inc.
Santa Clara, California

and

Victor C. Yang
University of Michigan
Ann Arbor, Michigan

Springer Science+Business Media, LLC

Library of Congress Cataloging in Publication Data

American Chemical Society Symposium on Polymers for Cosmetic and Pharmaceutical Applications (1990: Washington, D.C.)
Cosmetic and pharmaceutical applications of polymers / edited by Charles G. Gebelein, Tai C. Cheng, and Victor C. Yang.
p. cm.
"Proceedings of an American Chemical Society Symposium on Polymers for Cosmetic and Pharmaceutical Applications, held August 26–31, 1990, in Washington, D.C."—T.p. verso.
Includes bibliographical references and index.
ISBN 978-1-4613-6722-2 ISBN 978-1-4615-3858-5 (eBook)
DOI 10.1007/978-1-4615-3858-5
1. Polymers in medicine—Congresses. 2. Polymeric drugs—Congresses. 3. Cosmetics—Congresses. I. Gebelein, Charles G. II. Cheng, Tai C. III. Yang, Victor Chi-Min, date. IV. Title.
[DNLM: 1. Chemistry, Pharmaceutical—congresses. 2. Cosmetics—congresses. 3. Polymers—congresses. QV 744 A5065c 1990]
R857.P6A43 1990
615'.19—dc20
DNLM/DLC 91-39442
for Library of Congress CIP

Proceedings of an American Chemical Society Symposium on Polymers for Cosmetic and Pharmaceutical Applications, held August 26-31, 1990, in Washington, D.C.

ISBN 978-1-4613-6722-2


Originally published by Plenum Press, New York in 1991
Softcover reprint of the hardcover 1st edition 1991

## PREFACE

Polymers continue to show almost amazing versatility. We have always known that polymers could be used for trinkets, toys and dishes. Now, however, we are no longer surprised to encounter these adaptable materials in almost every place we look. We find them in our cars, tools, electronic devices, building materials, etc. The use of polymeric materials in medicine is also well documented in previous books by one of the Editors (Gebelein) and by others. Likewise, the use of polymeric materials in pharmaceutical applications, especially in controlled release systems, is also well established.

Nevertheless, the use of these ubiquitous chemicals is far less obvious in the field of cosmetics, although modern cosmetic preparations rely heavily on polymers and this trend is certain to increase. This book brings together much of the basic information on polymers in cosmetics and compares this usage with similar applications in pharmaceutical and medical applications.

Cosmetics, like medicine and pharmacy, dates back to antiquity. We can find uses of perfumes, balms and ointments in various old books, such as the Bible. For example, the use of ointments and balms is noted more than thirty eight times, and perfumes and related materials are cited at least twenty nine times in the Bible.

Modern cosmetical formulations often utilize polymers because polymers impart special advantages to the end products, such as controlled release or surface modification. Interestingly enough, many of the same polymers are used in cosmetics that are used in pharmaceutical and medical applications, mainly because the end uses are somewhat similar in either requirements, usage or both. This book includes longer versions of most of the papers presented at the "First International Symposium on the Cosmetic and Pharmaceutical Applications of Polymers" which was held in Washington, DC in August, 1990 during the American Chemical Society National Meeting.

The sequence of the book is fairly simple. The first paper (Gebelein, Cheng and Yang) gives an overview and outlines the basic similarity in materials and applications. The next eight papers deal primarily with cosmetics and are followed by five other papers dealing fairly equally with cosmetics and pharmaceuticals. The following fourteen papers are mainly concerned with pharmaceutical use, and the last seven papers cover miscellaneous medical applications. Through this entire sequence, however, we find examples of the same polymers occurring in each area. Thus, we see the natural polymers hyaluronan and chitin (and derivatives) in a variety of cosmetic, pharmaceutical and medical uses. Likewise, we find synthetic polymers such as poly(silicones) in all three areas as well.

The first section (Cosmetic Applications) starts with Thomas outlining the governmental regulations. Conklin then describes lather enhancers, mainly using cellulosic materials. Next, two papers describe hair care polymers (Martino & Nowak; Tsushima & Kondo), followed by two papers describing various polymers used in cosmetics (Tazi, et al.; Floyd). This first section concludes with a pair of papers considering sunscreen systems (Wang, et al.; Nacht). In the latter paper, we see the natural polymer melanin used as a sunscreen.

The second section (Applications in Cosmetics and Pharmaceuticals) starts with four papers describing some important poly(saccharides), such as chitosan and hyaluronan, which are used in both fields (Hirano, et al.; Brode; Brode, et al.; Band, et al.). The remaining paper considers poly(vinyl pyrrolidone) polymers which are widely used in both cosmetics and pharmaceuticals.

The third section (Pharmaceutical Applications) continues the thrust of this book by considering hylan systems in drug delivery (Larsen, et al.), the basic design features of controlled release systems (Thombre), the use of a microsponge delivery system (Eury, et al.) and some bioelastic polymer derivatives of elastin as "smart" drug delivery systems (Urry). The next three papers are concerned with hydrogel systems (Bae & Kim; Kou, Fleisher & Amidon; Kim & Lee). This is followed by a pair of papers treating liposomes, vesicles and targeted delivery (Egbaria & Weiner; Kondo, et al.). Levy, et al., then discuss cardiovascular applications of controlled release systems, and Wang, et al., describe the release of vaccines from microspheres. Carraher, et al., discuss a new polymer system for treating yeast infections from *Candida albicans* and Stemberger, et al., cover the use of collagen as a drug carrier. This section concludes with a paper on anticancer polymers (Carraher & Williams).

The final section (Other Medical-Related Applications) starts with a review of the applications of hyaluronan (Balazs) and then turns to synthetic polymers which also can produce a lubricious surface (Fan). Kohn then describes some new biomaterials derived from amino acids. The next pair of papers are concerned with heparin therapy (Chen & Yang; Kim, Fu & Yang). Schrope, Shen and Wheatley then describe some polymers which enhance ultrasound medical analysis. Finally, Ngo discusses some polymers used in affinity chromatography and enzyme immobilization.

The Editors wish to thank the American Chemical Society Division of Polymeric Materials: Science and Engineering for sponsoring and supporting the Symposium from which this book is derived. We also wish to acknowledge the financial support of the following companies (in alphabetical order): Biomatrix, Inc., Dow Chemical, Felton Worldwide, FMC Corp., GAF Chemicals Corp., Goldschmidt Chemical Corp., B. F. Goodrich, Johnson & Johnson, Pfizer, Inc., Union Carbide Corp., and Van Dyke & Co. We especially wish to acknowledge and thank each author for their fine manuscripts. Typesetting was done by CG ENTERPRISES.

For the Editors:
Charles G. Gebelein
Youngstown State University

## CONTENTS

## *APPLICATIONS IN BOTH COSMETICS AND PHARMACEUTICALS*



## *PHARMACEUTICAL APPLICATIONS*

***OTHER MEDICAL-RELATED APPLICATIONS***

# COSMETIC, PHARMACEUTICAL AND MEDICAL POLYMER APPLICATIONS - THE SAME PLAYERS IN DIFFERENT GAMES

Charles G. Gebelein,[a] Tai Cheng[b] and Victor Yang[c]

(a) Dept. of Chemistry
Youngstown State University
Youngstown, OH 44555

(b) Advanced Polymer Systems
Redwood City, CA 94063

(c) College of Pharmacy
University of Michigan
Ann Arbor, MI, 48109

In recent years, the seemingly dissimilar applications of polymers in pharmaceutical, medical and cosmetic areas have drawn closer together, due partly to similar materials requirements and partly to similar chemistry in the end uses. This paper indicates how similar polymers can be used in these diverse fields, and how the basic technology is analogous in each case. Examples include poly(saccharides), poly(peptides), methacrylate derivatives and poly(silicones).

## INTRODUCTION

The extensive applications of polymers appear almost limitless. Part of this is no doubt due to the abundant selection of polymeric materials now accessible to the chemist and engineer, with a remarkable spectrum of physical and chemical properties. We are no longer surprised to see polymers used in automobiles, electronics, audio or visual recordings, photography, medicine, etc.

What may seem partially unanticipated, however, is to see the same polymers used in areas that sound reasonably distant from each other. This occurs in the seemingly disparate domains of medical, pharmaceutical and cosmetics polymer applications. Yet, this unusual situation can be elucidated if we merely examine the requirements in each of these unique domains. Often, these requirements are reasonably similar and can, therefore, be met by equivalent or identical polymers. In this paper, we'll overview some of the different polymers and applications in these three areas. We will examine the general application area of controlled release systems and then look at some specific classes of natural and synthetic polymers. In this manner, we will demonstrate that each of these areas actually uses the same types of polymeric systems, because the desired effects are essentially the same in all cases.

*Cosmetic and Pharmaceutical Applications of Polymers*
Edited by C.G. Gebelein *et al.*, Plenum Press, New York, 1991

## CONTROLLED RELEASE SYSTEMS

All three application areas use controlled release systems, although the nature of the chemicals released varies significantly. In spite of this stipulation, however, the primary technology remains the same whether we are considering a pharmaceutical, medical or cosmetic operation. Consequently, we encounter the same kinds of polymeric systems in all three cases. Thus, monolithic devices, reservoir devices, hydrogels, glassy polymers and microspheres are used in all three domains. In each of these cases, the object is to deliver a particular agent for a prolonged time period, and this goal can be achieved in nearly the same manner regardless of whether the agent is a perfume or a drug.

## POLY(SACCHARIDES)

Numerous medical or pharmaceutical agents have been used to treat abnormal skin conditions, but this same objective has been a mainstay of the cosmetic industry for decades. Recently some presumed inter-active cosmetics have appeared on the market, including derivatives of retinoic acid, which are claimed to reduce wrinkles. Even though retinoic acid is fairly low molecular weight, many polymer chemists wonder whether some polymeric material might achieve this same goal, possibly even more effectively. It is well known that several poly(saccharide) polymers have some emollient behavior. Perhaps with this in mind, the Japanese company Shiseido is currently marketing the sodium salt of hyaluronic acid, a natural poly(saccharide), Figure 1, as a skin treatment cosmetic.[1] This particular, potential cosmetic utility of hyaluronic acid probably arises from its hygroscopic nature, enabling it to moisturize the skin. Hyaluronic acid also possesses good adhesive qualities, which could aid in cosmetic usage.

Hyaluronic acid occurs naturally in the connective tissues of most vertebrate animals and can be isolated from such scrap materials as rooster combs. Medical uses of hyaluronic acid include the treatment of arthritis and eye surgery. In both cases, the poly(saccharide) achieves its special functions by producing a highly viscous solution. Hyaluronic acid has also been explored for replacement synovial fluid because it occurs naturally in this joint lubricating material. Hyaluronic acid has also been studied as a polymeric matrix in controlled release applications, where it behaves like a hydrogel.[2,3]

Chitin, a poly(saccharide) closely related to cellulose and shown in Figure 2, is being studied for a wide variety of biomedical, pharmaceutical, agricultural and cosmetic applications.[4-6] Chitin occurs mainly in insect and crustacean shells, and is the animal equivalent of cellulose.

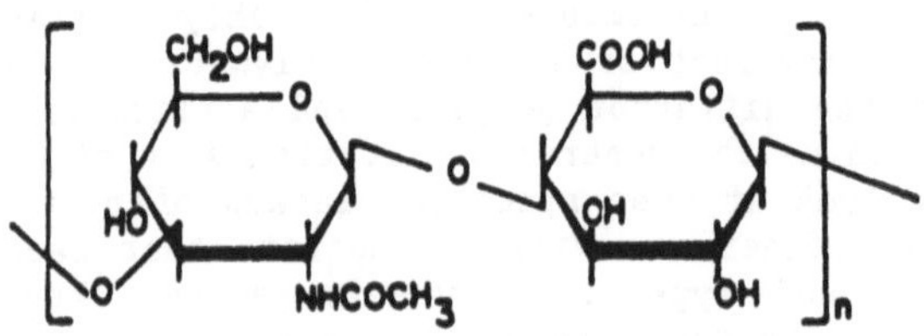


Figure 1. The structure of hyaluronic acid.

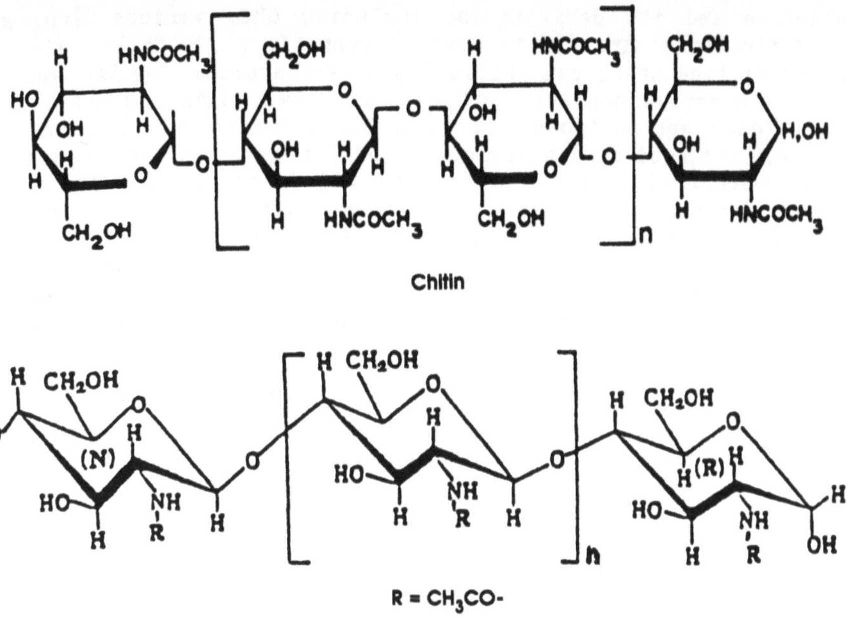


Figure 2. The structure of chitin.

Because chitin is insoluble in most solvents, most of the research centers on the deacetylated version, called chitosan, Figure 3.

Chitosan is now finding some new uses in the textile industry, waste water treatment and medicine.[7] While neither material is likely to be made *synthetically* on a commercial scale, both polymers are derived from formerly useless waste materials. Several chitin or chitosan derived materials are currently marketed in Japan. Cosmetic uses of chitin and chitosan, and derivatives are comparatively new, but are growing in importance. These include the use of chitin derivatives in various hair and skin care formulations.

Medically, various chitin derivatives have been studied as replacement blood vessels. This blood vessel use possibly stems from chitin's similarity to the structure of heparin, also a poly(saccharide). Carboxymethylchitin has been used to prepare vesicles which can serve as drug carriers or possibly as artificial red blood cells.[5] Chitin derivatives also are being used for artificial skin replacements (wound dressings).[8,9]

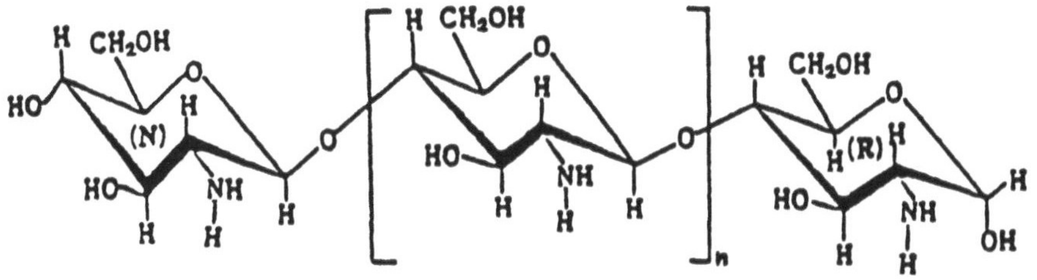

Figure 3. The structure of chitosan.

Cellulose and its derivatives, including CMC, various ethers, and other modifications, are widely used in cosmetic products.[10] Other poly(saccharides) find practical applications. For example, konjac and carrageenan have numerous cosmetic and medical uses. Konjac is used in controlled release applications and as a barrier film to protect wounds. Carrageenans are used in dentifrices, hand lotions, shaving creams, shampoos and controlled release systems.[11] Poly(saccharides) derived from fungi have been shown to possess anti-tumor properties,[12] and some recent research suggests curdlan sulfate may have antiviral activity, including potential anti-AIDS activity.[13] Some recent reports have also indicated that 5-fluorouracil can be attached to various poly(saccharides).[14] The topic of biologically active poly(saccharides) has been reviewed recently.[15]

Heparin is another poly(saccharide) which finds extensive biomedical usage, and it has been extensively reviewed recently,[16] and continues as the subject of much research, including its use in blood substitutes.[17]

## POLY(PEPTIDES)

Although it's an old poly(peptide), dozens of papers appear every year on potential biomedical uses of collagen, including skin and nerve regeneration[18-20] and artificial blood vessels.[21] Collagen has also been used to immobilize percutaneous implants.[22] Collagen has been used in controlled release applications as well.[23,24] For many decades, collagen has been used in cosmetic applications, such as treating the scars from small pox.

Poly(elastin), basically a synthetic poly(peptide) patterned after natural elastin, is currently being examined in both the medical and controlled release areas.[25,26] Although this material is relatively new, and of synthetic origin, it could probably find utility in cosmetics.[27]

Gelatin, also a poly(peptide), finds extensive use in the preparation of microcapsules for use in drug delivery and the encapsulation of various water-insoluble oils used in cosmetics. These gelatin systems are normally crosslinked to overcome the water solubility of the polymer.

Other important poly(peptide) applications include their use as drug carriers,[28] biodegradable drug delivery systems,[23] and pharmaceutical agents.[29] Enzymes are well known poly(peptides) with very high catalytic activity. Several groups are synthesizing modified proteins with "artificial" catalytic activity.[30] More recently, catalytic antibodies, which are also poly(peptide) in nature, have been developed.[31] In addition, the concept of totally artificial enzymes has been pursued vigorously by several groups; this has been reviewed recently.[32] Research also continues on immobilized enzyme systems.[33]

## ACRYLATES AND METHACRYLATES

Polymers and copolymers of acrylates and methacrylates have been used in medical, pharmaceutical and cosmetic applications for many years. Important medical uses include dental applications (e.g., fillings, dentures),[34] orthopedic (e.g., bone cement),[35] ophthalmic (contact lenses, intraocular lenses),[36-40] and other surgical applications (such as facial surgery).[41] Methacrylate copolymers also form the basic materials for glassy controlled release systems. Moreover, copolymers with hydroxyethyl methacrylate and/or methacrylic or acrylic acids form the basis for the hydrogel controlled release systems which are used in pharmaceutical applications, in addition to the soft contact lenses.[42] Copolymers with high levels or acrylic of methacrylic acid, crosslinked or uncrosslinked, are water soluble or water swellable polymers. These have found many cosmetic uses, including their use as thickeners.

Methacrylate derivatives have been synthesized with attached 5-fluorouracil and 6-methylthiopurine units. Copolymers prepared from these materials show the interesting, and potentially valuable, zero-order release of these antitumor drug agents.[43-46]

## SILICONE POLYMERS

Poly(dimethylsiloxane) has become a standby in the biomedical polymer field where it is used for a diversified assortment of applications, including catheters, implants, heart-lung machine membranes, and cosmetic surgery.[47] Likewise, silicone polymers are frequently used in reservoir and monolithic controlled release devices.[48] The water-repellent characteristics of silicones make them ideally suited for many cosmetic applications which require this property.[49]

## CONCLUSION

In conclusion, we note that the seemingly disparate use of polymers in the areas of pharmaceuticals, medicine and cosmetics are far less surprising than we might expect. In addition, these diverse fields have drawn closer together in recent years, partly because the material demands are similar and partly because the essential chemistry is parallel.

## REFERENCES:


1. Literature from Shiseido Company.
2. N. E. Larsen, E. A. Leshchiner, E. G. Parent and E. A. Balazs, Proc. Poly. Mater. Sci. Eng., **63**, 341-3 (1990).
3. E. A. Balazs, Proc. Poly. Mater. Sci. Eng., **63**, 689-91 (1990).
4. S. Hirano, Y. Noishiki, J. Kinugawa, H. Higashijima & T. Hayashi in: "*Advances in Biomedical Polymers,*" C. G. Gebelein, Ed., Plenum Press, New York, 1985, pp. 285-297.
5. A. Kato & T. Kondo in: "*Advances in Biomedical Polymers,*" C. G. Gebelein, Ed., Plenum Press, New York, 1985, pp. 299-310.
6. S. Hirano, M. Hayashi, K. Murae, H. Tsuchida & T. Nishida in: "*Applied Bioactive Polymeric Materials,*" C. G. Gebelein, C. E. Carraher, Jr. and V. R. Foster, Eds., Plenum Press, New York, 1988, pp. 45-59.
7. H. Struszczyk & O. Kivekäs, Brit. Polymer J., **23**, 261-5 (1990).
8. S. Hirano, H. Seino, Y. Akiyama and I. Nonaka, in: "*Progress in Biomedical Polymers,*" C. G. Gebelein & R. L. Dunn, Eds., Plenum, New York, 1990, pp. 283-90.
9. S. Hirano, K. Hirochi, K. Hayashi, T. Mikami and H. Tachibana, Proc. Poly. Mater. Sci. Eng., **63**, 699-703 (1990).
10. J. R. Conklin, Proc. Poly. Mater. Sci. Eng., **63**, 233-42 (1990).
11. R. J. Tye, Proc. Poly. Mater. Sci. Eng., **63**, 229-232 (1990).
12. K. Matsuzaki, I. Yamamoto, K. Enomoto, Y. Kaneko, T. Mimura & T. Shiio in: "*Applied Bioactive Polymeric Materials*", C. G. Gebelein, C.

E. Carraher, Jr. and V. R. Foster, Eds., Plenum Press, New York, 1988, pp. 165-174.
13. I. Yamamoto, K. Takayama, T. Gonda, K. Matsuzaki, K. Hatanaka, T. Yoshida, T. Uryu, O. Yoshida, H. Nakashima, N. Yamamoto, Y. Kaneko & T. Mimura, Brit. Polymer J., 23, 245-50(1990).
14. T. Ouchi, Poly. Mater. Sci. Eng., 62, 412-415 (1990).
15. C. Schuerch, in: "*Bioactive Polymeric Systems*," C. G. Gebelein & C. E. Carraher, Jr., Eds., Plenum Press, New York, 1985, pp. 365-386.
16. R. J. Linhardt and D. Loganathan, in: "*Biomimetic Polymers*," C. G. Gebelein, Ed., Plenum, New York, 1990, pp. 135-173.
17. F. Prouchayret, F. Bonneaux, M. Leonard, D. Sacco & E. Dellacherie, Brit. Polymer J., 23, 251-6 (1990).
18. I. V. Yannas, D. P. Orgill, J. Silver, T. V. Norregaard, N. T. Zervas & W. C. Schoene, in: "*Advances in Biomedical Polymers*", C. G. Gebelein, Ed., Plenum Press, New York, 1985, pp. 1-15.
19. I. V. Yannas, E. Lee & M. D. Bentz, in: "*Applied Bioactive Polymeric Materials*", C. G. Gebelein, C. E. Carraher, Jr. and V. R. Foster, Eds., Plenum Press, New York, 1988, pp. 313-318.
20. I. V. Yannas, J. F. Burke, D. P. Orgill & E. M. Skrabut in: "*Polymeric Materials and Artificial Organs*," C. G. Gebelein, Ed., ACS Symposium Series #256, Washington, DC, 1984, pp. 191-197.
21. S.-T. Li in: "*Advances in Biomedical Polymers*", C. G. Gebelein, Ed., Plenum Press, New York, 1985, pp. 171-183.
22. T. Okada and Y. Ikada, in: "*Progress in Biomedical Polymers*," C. G. Gebelein & R. L. Dunn, Eds., Plenum, New York, 1990, pp. 97-105.
23. R. V. Petersen in: "*Bioactive Polymeric Systems*," C. G. Gebelein & C. E. Carraher, Jr., Eds., Plenum Press, New York, 1985, pp. 151-177.
24. A. Stemberger, M. Unkauf, D. E. Arnold and G. Blümel, Proc. Poly. Mater. Sci. Eng., 63, 344-346 (1990).
25. D. W. Urry, R. D. Harris, H. Sugano, M. M. Long & K. U. Prasad in: "*Advances in Biomedical Polymers*", C. G. Gebelein, Ed., Plenum Press, New York, 1985, pp. 335-354.
26. D. W. Urry, Poly. Mater. Sci. Eng., 62, 587-593 (1990).
27. D. W. Urry, J. Jaggard, R. D. Haris, D. K. Chang and K. U. Prasad, in: "*Progress in Biomedical Polymers*," C. G. Gebelein & R. L. Dunn, Eds., Plenum, New York, 1990, pp. 171-178.
28. J. M. Whiteley in: "*Bioactive Polymeric Systems*," C. G. Gebelein & C. E. Carraher, Jr., Eds., Plenum Press, New York, 1985, pp. 345-363.
29. J. Samanen in: "*Bioactive Polymeric Systems*," C. G. Gebelein & C. E. Carraher, Jr., Eds., Plenum Press, New York, 1985, pp. 279-344.
30. M. H. Keyes and D. E. Albert, in: "*Biomimetic Polymers*," C. G. Gebelein, Ed., Plenum, New York, 1990, pp. 115-133.
31. D. Hilvert, in: "*Biomimetic Polymers*," C. G. Gebelein, Ed., Plenum, New York, 1990, pp. 95-113.
32. Y. Imanishi, in: *Bioactive Polymeric Systems*," C. G. Gebelein & C. E. Carraher, Jr., Eds., Plenum, New York, 1985, pp. 435-511.
33. M. K. Keyes & S. Sarawathi, in: *Bioactive Polymeric Systems*," C. G. Gebelein & C. E. Carraher, Jr., Eds., Plenum, New York, 1985, pp. 249-78.
34. J. F. Glenn in:"*Biomedical and Dental Applications of Polymers*," C. G. Gebelein & F. F. Koblitz, Eds., Plenum, New York, 1980, pp.317-335.
35. C. G. Gebelein, in: "*Kirk-Othmer: Encyclopedia of Chemical Technology*," John Wiley & Sons, New York, 1982, pp. 275-313.
36. T. B. Harvey, III, W. B. Meyers & L. M. Bowman, in: "*Progress in Biomedical Polymers*," C. G. Gebelein & R. L. Dunn, Eds., Plenum, New York, 1990, pp. 1-5.
37. L. M. Bowman, T. B. Harvey, III & W. B. Meyers, in: "*Progress in Biomedical Polymers*," C. G. Gebelein & R. L. Dunn, Eds., Plenum, New York, 1990, pp. 7-12.
38. W. B. Meyers, T. B. Harvey, III & L. M. Bowman, in: "*Progress in*

*Biomedical Polymers,"* C. G. Gebelein & R. L. Dunn, Eds., Plenum, New York, 1990, pp. 13-17.
39. E. P. Goldberg, M. Yalon & W. E. Longo, in: "*Progress in Biomedical Polymers,"* C. G. Gebelein & R. L. Dunn, Eds., Plenum, New York, 1990, pp. 19-26.
40. J. T. Jacob-LaBarre & D. R. Caldwell, in: "*Progress in Biomedical Polymers,"* C. G. Gebelein & R. L. Dunn, Eds., Plenum, New York, 1990, pp. 27-39.
41. A. N. Cranin, J. Dibling, A. Simons & M. Klein, in: "*Progress in Biomedical Polymers,"* C. G. Gebelein & R. L. Dunn, Eds., Plenum, New York, 1990, pp. 65-82.
42. S. W. Kim in: "*Bioactive Polymeric Systems,"* C. G. Gebelein & C. E. Carraher, Jr., Eds., Plenum Press, New York, 1985, pp. 143-150.
43. C. G. Gebelein, T. Mirza & R. R. Hartsough in: "*Controlled Release Technology*", P. I. Lee & W. R. Good, Eds., ACS Symposium Series 348, 1987, pp. 120-126.
44. C. G. Gebelein, M. Chapman & T. Mirza in: "*Applied Bioactive Polymeric Materials*", C. G. Gebelein, C. E. Carraher, Jr. and V. R. Foster, Eds., Plenum Press, New York, 1988, pp. 151-163.
45. C. G. Gebelein, M. Davison, T. Gober & M. Chapman, Polym. Mater. Sci. Eng. Proc., 59, 798-802, (1988).
46. C. G. Gebelein, in: *Bioactive Polymeric Systems,"* C. G. Gebelein & C. E. Carraher, Jr., Eds., Plenum, New York, 1985, pp. 1-15.
47. E. E. Frisch in: "*Polymeric Materials and Artificial Organs,"* C. G. Gebelein, Ed., ACS Symposium Series #256, Washington, DC, 1984, pp. 63-97.
48. D. L. Gardner & B. A. Metz in: "*Bioactive Polymeric Systems,"* C. G. Gebelein & C. E. Carraher, Jr., Eds., Plenum Press, New York, 1985, pp. 93-119.
49. D. T. Floyd, Polym. Mater. Sci. Eng., 63, 246-251 (1990).

# THE SUITABILITY AND REGULATION OF RAW MATERIALS FOR THE COSMETICS INDUSTRY

M. R. Thomas

Dow Chemical U. S. A.
2020 Dow Center
Midland, MI 48674

Cosmetic products in the major markets of the world are becoming increasingly regulated. As this progresses, the needs of the cosmetic industry increase for raw materials that are suitable and are made under manufacturing conditions that assure consistent purity, free from unexpected and harmful contaminants, even in minute quantities. This paper discusses the basics of regulation of the cosmetic industry in Japan, the EEC, and in the United States, in both formulated products and raw materials. Reference material comes from the Second International Conference on Cosmetic Regulation as well as a wealth of other sources.

## INTRODUCTION

Over the last five years, the increased concern for chance contaminants and for the harmful effects of various intentional ingredients in food and in drugs has also become manifested in cosmetic products. The public is asking for increased attention to safety by manufacturers and by regulatory groups in the three major market centers; The United States, Japan and the EEC and its 12-member states. Upwards of 8,000 ingredients are used in the manufacture of cosmetics today. Attention to the safety of pure materials and blends of ingredients used in the formulation cosmetics is, therefore, the topic of this paper.

## DEFINITIONS

The <u>cosmetic</u> product is any substance or preparation intended for contact with various external parts of the human body; for cleansing and promoting attractiveness, for perfuming or protecting these parts to keep them in good condition, to change their appearance or to correct body odors. No medical claim may be made for a defined cosmetic.

If a medical claim is made, i. e., if an article is intended to diagnose, cure, treat or prevent disease or affect the structure or any function of the body, then the article is regarded as a drug <u>under the law</u>.

*Cosmetic and Pharmaceutical Applications of Polymers*
Edited by C.G. Gebelein *et al.*, Plenum Press, New York, 1991

This has important implications because drug status subjects a product to proof of effectiveness before the product may be marketed and advertised with its medical claim.

Thus, ingredients used in an anti-perspirant, a lotion marketed to prevent or cure dry skin, a mouthwash that claims anti-plaque properties, a sunscreen product that prevents sunburn and subsequent skin disease, anti-dandruff shampoos and more than 25 other categories of products are regarded as drugs and are regulated as OTC drugs in the United States, as quasi-drugs in Japan and as drugs in the EEC. Any medical claim for a product makes it a drug subject to drug laws in composition formulation, shelflife, and proof of wholesomeness.

Thus, with this kind of product, there is no question of how the individual ingredient in each formulation must be screened by the formulator. He must list all ingredients in the label, the safety of each active ingredient must be determined and chance contaminants of unknown safety must be prevented.

But, what about cosmetics? What distinguishes the cosmetic product from the OTC drug category, other than the medical claim? The answer is increasingly less in the last two years and lies in the manner of registration with the various regulatory agencies in the three major markets for these products, the United States, the European Community, and Japan.

## THE REGULATORY RESPONSIBILITY

The U. S. Food and Drug Administration has regulatory authority for health problems and risks of food, drugs, biologics and medical devices as well as cosmetics. In Japan, the Ministry of Health and Welfare is the agency empowered by the Pharmaceutical Affairs Law which provides that a license must be obtained to manufacture or to import cosmetics and cosmetic raw materials. The ministry sets the rules for the owners of the license. In the EEC, consumer policy with respect to cosmetics is in development and has at this stage incorporated many of the policies of the twelve member-states. The 1976 Cosmetics Directive in the EEC was designed to "harmonize" the laws of member states relating to the composition, labeling, and packaging of cosmetic products. The apparent intention is to develop a series of lists as exist in Japan; an "approved" list, a "forbidden" list and a controlled list.

### 1. The Regulation of Cosmetics in Japan

As mentioned above, the Pharmaceutical Affairs Law regulates cosmetics by virtue of controlling manufacturing sites in Japan and an importing company's offices by license in a premarket clearance procedure.

The Standards of Cosmetics Ingredients includes standards of purity and assay for product categories. Specific quality standards of 588 ingredients have been published. The ministry does not grant a license to any cosmetic not meeting these standards. If the ingredient in the product to be imported is already on this list, or are those already used in Japan, it is considered that the safety is already confirmed. There is a reference guide titled "Japan Cosmetic Ingredient Dictionary" in which about 1800 ingredients are described. The list of prohibited ingredients for cosmetic formulations is extensive and includes vinyl chloride monomer, chloroform, mercury and bithional.

If a new product, or new ingredient for a product, is submitted for license, safety data is required for submission. The data include toxicity, sub-acute toxicity, chronic toxicity, effect on reproduction, skin sensitivity and photo-sensitivity, mutagenicity, carcinogenity, local irritation, absorption, metabolism and excretion. Yes, a great deal of data is necessary to license a new product or ingredient in Japan.

Even though ingredients may be considered safe, problems may arise. The Law requires the manufacturer or importer to forward to the Minister of Health and welfare within 30 days any report that they received that a cosmetic caused a harmful action. The Ministry may suspend the license based in that reporting.

In manufacturing ingredients for cosmetics used in Japan, therefore, the ingredient must be scrutinized for safety and must be manufactured free of chance contaminants.

## 2. The Regulation of Cosmetics in the European Community

The EEC is made up of 12 member states with nine official languages and many, quite different, legal systems needing to be accommodated. The EC Cosmetic Directive (76/768/EEC) regulates cosmetic labeling, colors, preservatives, sun screens, and restricted or prohibited ingredients. The 1976 Cosmetics Directive had two main objectives: to facilitate free circulation of goods and to safeguard the user of the products. In fact, the safeguard clause has, in some cases, caused restriction on products and it is hoped by the Commission of Protection and Promotion of Consumers Interests in EEC that the essential safety requirements and the development of specific lists with be forthcoming. The Initial focus is the implementation of "approved," "controlled" or "forbidden" ingredients for cosmetics as well as the standardization of safety testing. At present, the best guide is the specific requirements of the individual country of manufacture or importation.

## 3. The Regulation of Cosmetics in the United States

The FDA publishes requirements on cosmetics in the Federal Register and as a compilation, in the Code of the Federal Register, Title 21 and Sub-chapter G, Parts 700-799. The CFR is updated yearly with rules that have appeared during the year in the Federal Register. Sub-chapter G outlines definitions with regards to cosmetics, ingredients that are forbidden such as bithional, mercury, vinyl chloride monomer, chloroform, and certain chlorofluorocarbon propellants. It states a need for tamper-evident closures. Labeling requirements are also outlined. The Cosmetics, Toiletries and Fragrance Association has published the Cosmetic Ingredient Dictionary which has been adopted by the CFR as one guide on ingredient nomenclature. Other authorities are the United States Pharmacopoeia, the National Formulary and the Food Chemicals Codex.

Unlike food or drugs, there is no list of approved substances for use in cosmetics, but a judgment, by manufacturers, of appropriateness for use must be made and is a burden of the manufacturer. The fact that a product is listed in these references does not convey approval of safety by the FDA. Ingredients must be listed on the label in order of concentration to the 1% level, color additives must be listed as well. Changes in formulation for a product with an established name must be noted on label.

The FDA requests voluntary registration of cosmetic product establishments as well as cosmetic raw material manufacturing sites. In the registration of an ingredient, the ingredient is listed by its chemical name, or common name. A mixture of ingredients must be given by descending order of composition by weight. Requests for confidentiality of cosmetic ingredients submitted should contain a full statement of factual and legal grounds for the request. These grounds and petition forms are given in the CFR. A packet of instructions for the Voluntary Registration of Cosmetic Product Formulation and Cosmetic Raw Material Composition is available from the CTFA. These instructions and forms are in accordance with 21CFR 720. For packets or information call the CTFA at 202-331-1770.

A further FDA request is voluntary filing of cosmetic product experiences. Both positive and negative experiences are requested. Warning statements for labels are also spelled out and conditions for using in labeling product. A review of the Code of the Federal Register should be a part of the planning of the manufacture of a new cosmetic raw material or the formulation of a new cosmetic product.

## 4. The Suitability of Raw Materials for Cosmetics

The manufacture of a new raw material for cosmetics then includes a realization that the final ingredient must be free of unsuitable compounds, even in minute amounts. The final ingredient must present no hazard in the manner of its formulation and to the human that uses the final product. If the cosmetic product is used infrequently and removed completely afterward, there is less hazard than if the cosmetic product is applied everyday, remains on the body, and is only removed on cleansing or bathing and is reapplied immediately thereafter. Tests for suitability of use should reflect these varying conditions of use.

Each of the major markets and their regulatory agencies have lists of tests to which the new ingredient must or should be subjected. These have been under revision, both to predict the risk more clearly and to limit tests involving animals that have been important in setting guidelines. A test procedure for evaluating the irritancy potential of a chemical in animals was first published by Draize in 1944. When testing cosmetics, the nature of the procedure is such that errors in predicting toxicity in man is conservative. The irritant affects sometimes appear to be greater than actually observed in man. However, very permeable rabbit skin results in a test that will identify many substances that are irritants on human skin. To protect the consumer, the rabbit Draize test is required by law under provisions of the Federal Hazardous Substances Act.

The industry, in cooperation with the FDA, is attempting to devise non-animal tests or tests that utilize fewer animals which will reliably predict the irritancy, allergenicity and photo-sensitization by new chemical substances.

High standards of hygiene and microbiological safety guarantee that pure cosmetic products reach the customer. Once the formulation is developed, cosmetic purity is assured in production by complete plant hygiene, monitoring of new raw materials to detect microbiological contamination, and further testing of finished formulated products status for microbiological contamination by statistical sampling. Long-term storage tests are routinely conducted consisting of exposing finished products to a wide range of temperatures, relative humidity and atmospheric contamination.

The Cosmetic Ingredient Review (CIR) was established in 1976 by the Cosmetic, Toiletry and Fragrance Association (CTFA). This began as an effort to review and assess thoroughly the safety of ingredients used in cosmetics. A seven-member expert panel of scientists are nominated by consumers, scientific and clinical societies, government agencies as well as the cosmetic industry. The priority ingredient review list is selected by this panel based on indications of frequency of use and biological activity, the area of normal use, and estimate of penetration. Also, adding weight to the prioritization are: concentration of use, frequency of application, use by sensitive population subgroups, and consumer complaints about products containing these ingredients. It is useful to review these summary/monographs in the Cosmetic Ingredient Review to determine the kinds of tests that have been used to get baseline information on common cosmetic ingredients and to keep a file of all final reports on the various ingredients.

We are discussing here the use of new polymer systems for use in cosmetics. The processes to manufacture these polymers must be scrutinized for residuals in the final products; unreacted monomer, irritating solvents and/or competing side reactions. Batch to batch inconsistencies should be regarded as problematic in considering the polymer preparation for commercial use in cosmetics.

The interests of the industry are best served by voluntarily conducting appropriate safety testing, thereby keeping the need for a specific government imposed pre-market safety act a low priority. At this time, no pre-market notification is mandated by law.

It seems almost redundant to state that pre-introduction costs of a new skin care product or cosmetic are high. The development cost, distribution costs, testing costs and costs of advertising are made before the marketing of the product. These outlays will not be made if there is any question of safety of the ingredients or formulated product for its final use. From the Cosmetic Ingredient Review, to define the conditions of use, consider: (1) the amount of an ingredient used in a product, (2) the intended use, and reasonably foreseeable areas of use, (uses subject to ingestion, inhalation, contact with the mucous membranes or to the eye), (3) completeness of directions for use and, against misuse on the label, taking into consideration the formulation and its ingredients.

## SUMMARY

The cosmetic products in major markets of the world are becoming increasingly regulated. As this progresses, the needs of the cosmetic industry increase for ingredients that are in conformation with the safety consideration of their use.

## REFERENCES

1. Code of the Federal Register, Title 21, 700-799 (April, 1991).
2. Proceedings of the Second International Conference on Cosmetic Regulation sponsored by the Cosmetics, Toiletries, and Fragrances Association, the Japanese Cosmetic Industry Association and the Federation of National Trade Associations of the EEC in April 1988.
3. Drug Approval and Licensing Procedures in Japan, 1988 (Yakugyo Jiho Co., Ltd., Tokyo) Useful in determining what is a Quasi-Drug and procedures for licensing products, manufacturers and imports.

4. Cosmetic Science and Technology Series, Volume 5, Cosmetic Safety, J. H. Whittam, Ed., Volume 2, Cosmetic Industry, Scientific and Regulatory Formulations, N. F. Eptrin, Ed., Marcel Dekker, Inc., New York, 1987.
5. Handbook of Non-Prescription Drugs, published by the Americam Pharmaceutical Association, 9th edition, Washington, DC, 1990.
6. CTFA Cosmetic Ingredient Dictionary, published by the Cosmetic Toiletries and Fragrances Association, Washington, DC, 1985, and updates.
7. Instructions for the "Voluntary Registration of Cosmetic Product Formulations and Cosmetic Raw Material Composition" in accordance with CFR 21, Part 720, FDA Division of Colors and Cosmetics, May 1987, Forms and instructions are available from the CTFA (202) 331-1770.

# THE FORM AND FUNCTION OF LATHER ENHANCER CELLULOSICS IN PERSONAL CARE PRODUCTS

Jerry R. Conklin

Larkin Laboratory
The Dow Chemical Company
Midland, MI 48640

The objective of this paper is to present a novel technology for promoting foam in cleansing type consumer products. Specifically, the introduction of a new cellulose based polymeric surfactant which can be used to improve the properties of foaming products used in personal care. Such products include shampoo, bath and shower gels, facial cleansers and liquid hand-soap. Although many other uses are anticipated, this paper shall confine its discussion to personal care applications. In succession, the following aspects of this research will be discussed.

1. The need which is being addressed
2. Properties of the cellulosic polymer
3. Design of a lather generator
4. Lather enhancing shampoo studies
5. Conclusions

## MILDNESS

Formulators of personal care cleansing products have an obligation to minimize the extent to which their products may irritate the eyes and skin of consumers. Cleansing type products are particularly susceptible to this need because they contain surfactants as a primary ingredient. The tendency of some surface active agents to adsorb onto the skin and eyes and promote irritation is well known. The expert formulator is well aware of these properties and selects ingredients which accomplish the product's purpose while subjecting the user to a minimum of irritation.

However, products like shampoo and shower gels are primarily surfactants, and there is a conditioned expectation by the consumer to have them produce copious foam. Production of foam is a strong visual cue and is often used and relied upon by the consumer as an index of performance. The formulator must then satisfy this need by using an amount of surfactant which accomplishes this and also addresses the need for low irritation. A typical formulation makes use of raw materials like mild specialty surfactants and or fatty alkanolamides to boost foam volume while attempting to desirably control the irritancy properties.

*Cosmetic and Pharmaceutical Applications of Polymers*
Edited by C.G. Gebelein *et al.*, Plenum Press, New York, 1991

Some products, like those intended for use on infants and children, are formulated to contain especially mild surfactants or lower levels of conventional surfactants. Unfortunately, these formulations often display significantly reduced foaming capacities and subsequently are not readily accepted in the mass adult market.

Additionally, there is an increasing frequency of use of personal cleansing products which results in a heightened concern over mildness. The consumer that showered and shampooed two or three times a week a decade ago, may now be doing this daily. In many instances, sports active consumers are showering and shampooing twice a day. All of this points up to an increasing need for effective yet mild cleansers.

The key objective of this research is to address the evolving need for alternative, yet mild, lather enhancing agents. The novel cellulose based polymeric surfactant described here may represent such an alternative technology. It is a type of hydroxypropyl methylcellulose available as Methocel[R] 40-200.

## PORTRAIT OF A LATHER ENHANCER

Figure 1 depicts the concentration dependent surface tension of a cellulose ether lather enhancer at 25°C. This was obtained using dynamic du Nouy tensiometry. This provides very basic information regarding activity at the air/water interface.

However, in addition to interfacial activity, a lather enhancing additive should also provide the following multiple benefits.

1. Lather volume and stability
2. Lather lubricity
3. Viscosity building
4. Compatibility
5. Low irritation
6. Ease of addition

To examine whether our polymeric surfactants provided these additional and important benefits, a laboratory test was needed which would enable screening and evaluation of potential candidates.

## DEVELOPMENT OF A SEMI-AUTOMATIC LATHER GENERATOR

In the art, a variety of techniques have been reported to be of use in helping the formulator assess the contributions of foaming agents. Although the salon half-head test is widely relied upon as a definitive test, it is too costly and cumbersome to use as a screening tool. Nevertheless, a laboratory method which mimicked some of the essential elements of actual shampooing would be desirable. Therefore, the following semi-automated lather generator was designed and built. This technique uses four cells of a design described below in Figure 2.

This is a 1.3 liter cylindrical glass cell with graduations every 50 mL. The lower third of the cell has inward protruding "fingers" to provide mechanical shear in concert with the bottle brush bristles. Each finger projects inward 1.5 cm and is 0.4 cm in diameter. There are four circular rows of "fingers" with four finger per row (at ninety degree intervals). Each cell is jacketed for temperature control and connected

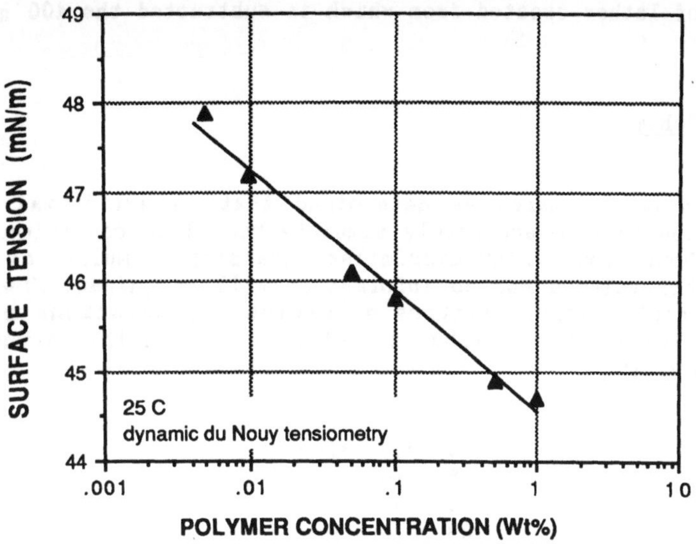


Figure 1. Surface tension of cellulose ether lather enhancer at 25°C.

to a constant temperature bath. The heavy duty beaker brush moves up and down within the cell on a 5 cm stroke with 120 strokes/min. A motor and speed control regulate the reciprocating brush action. The brush bristles are nylon and maintain integrity even after considerable use.

During use, 400 mL of test solution are poured into each lather generator cell. The reciprocating action of the lather generator brushes is maintained for two minutes. This is followed by an immediate reading of

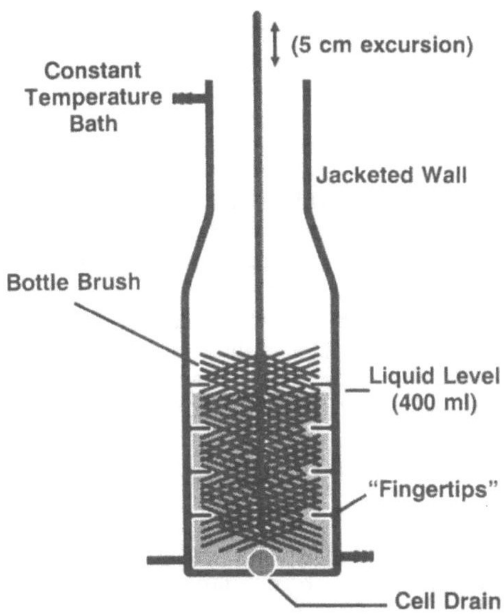


Figure 2. Semi-automated lather generator.

the volume of lather created from which is subtracted the 400 mL starting volume.

Artificial Sebum

Early in our research we determined that an artificial soil load should be used to more accurately simulate real life conditions of shampooing. Without sebum, many dissimilar cleansing formulations will give similar lather volumes and no information will be gained. However, with sebum, inherent susceptibilities of various formulations are easily determined. The artificial sebum formulation used in this study consists of the following:

| | |
|---|---|
| 96.0% | hair conditioner* |
| 1.6% | cetyl alcohol |
| 0.8% | triolein |
| 0.8% | squalene |
| 0.8% | cholesterol |
| 100.0% | |

*White Rain[R] Baby Conditioner (Gillete) contains: water, stearalkonium chloride, ceteth-2, dimethyl stearamine, phenoxyethanol, glyceryl stearate, citric acid, sodium chloride, stearyl alcohol, fragrance.

These ingredients were selected based upon literature references and discussions with others in the hair care industry. Hair conditioner was used as the primary base owing to its frequent use by consumers and the likelihood of encountering residual conditioning ingredients during actual shampooing. Conditioner ingredients such as fatty alcohols and quaternary ammonium compounds can function as antifoaming agents and therefore impact the lather generation of an applied shampoo.

Sixteen grams of the sebum emulsion is mixed with 384 g of the test solution for several minutes immediately prior to addition to the lather generator test cell. This ratio provides 0.31% active sebum solids in the test solution.

Lather Generator Results

In Figure 3 we see the relative lathering tendencies of two fatty alkanolamide foam boosters compared to our polymeric lather enhancer at 1.0% active in de-ionized water. Significantly, it was found that 0.16% sebum solids substantially defoamed all three nonionic additives. In Figures 4 and 5, the effect of added sebum is seen for two commercial U.S. shampoos. We have studied the sebum dependent lathering properties of many other commercial shampoos but selected these two to fairly represent the typical performance of shampoo formulations in the "gentle" to "normal" categories. These data were gathered at 40°C using 50x dilution of the shampoo in moderately hard water containing 180 ppm of calcium ion.

With the sebum susceptibility of lather from commercial shampoos established, it is of interest to compare the performance of prototype shampoos which use the cellulosic lather enhancer as a key component. Just like the two commercial shampoos, our prototypes were formulated

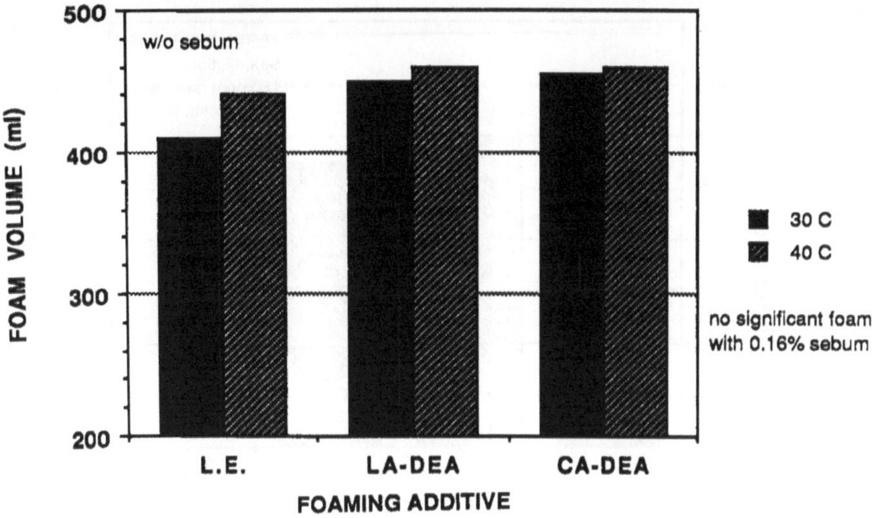


Figure 3. Foaming of polymeric surfactant vs fatty amide at 1% in D. I. water.

using ammonium lauryl sulfate (ALS) as the primary surfactant. The prototype formulae appear below:

Prototype Shampoo Formulations

| | |
|---|---|
| Surfactant* | 8, 10, or 14% ammonium Lauryl sulfate |
| Additive* | 0, 2, or 4%, either cellulosic lather enhancer (L.E.) or lauric diethanolamide (LA-DEA) |
| Electrolyte | as needed to effect thickening (0-2% NaCl) |
| Preservative | 0.10% Dowicil[R] 200 |
| Chelant | 0.05% Versene[R] 100 $Na_4EDTA$ |
| Citric acid | as needed to adjust to pH 6.0-6.5 |
| Fragrance | 0.15% (IFF, spice floral) |

*represents actual active concentration in the formulation
R is a trademark of the Dow Chemical Company

As is evident, most ingredients and proportions are fairly typical for U.S. shampoos in the "gentle" to "normal" categories. Experiments using the low value of 8% ALS were included to enable appraisal of an unusually low surfactant concentration.

In Figure 6, results are presented for prototype shampoos containing 8% ALS in combination with various concentrations of the cellulosic lather enhancer or lauric diethanolamide. The intent is to demonstrate that at such low primary surfactant levels, a conventional fatty amide "foam booster" may not in fact provide the impact on lather that is commonly assumed. On the other hand, the cellulose ether lather enhancer exhibits a large positive impact.

Figure 7 depicts similar trends for prototype shampoos based on 10% ALS. Of particular interest is the negative impact that the conventional fatty amide "foam booster" provides on lather generation. The cellulose ether, on the other hand, again contributes a substantial positive effect.

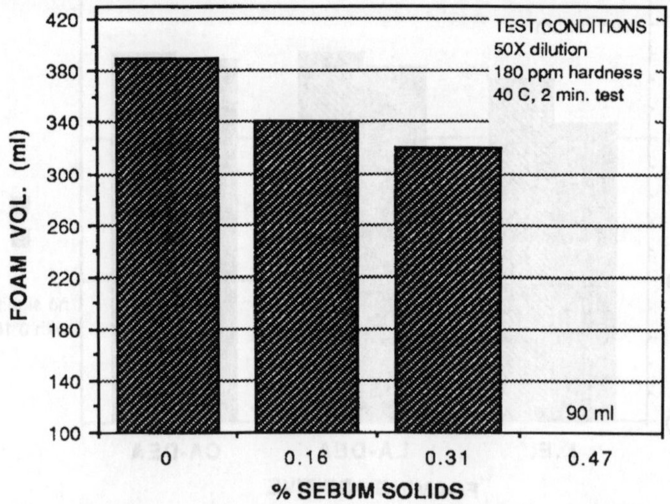


Figure 4. Foaming profile, Salon Selectives* shampoo with standard sebum.

The disparity between the performance of the two lather modifiers becomes minimal as the primary surfactant concentration is increased to 14% as shown in Figure 8. Note that the cellulosic can be seen to persist in offering the advantage in lather volume.

Figures 9 and 10 present this same data in another form so that the shampoo prototypes can be directly compared to existing commercial products. It is observed that those formulations containing the cellulosic lather enhancer do quite well in comparison to those prepared with a conventional fatty amide.

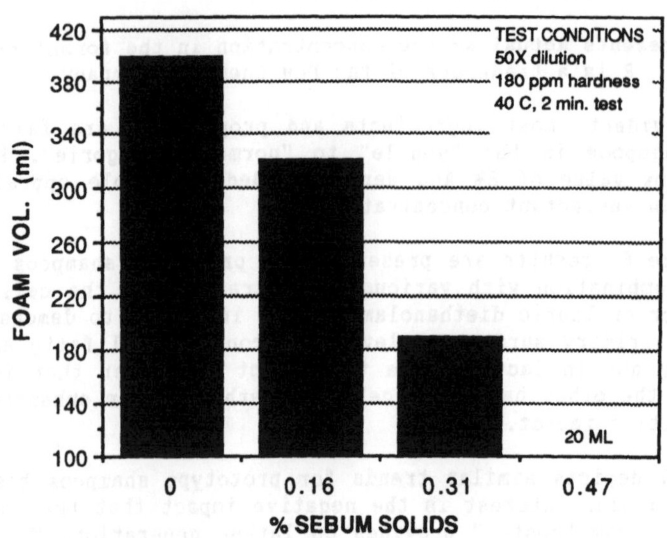


Figure 5. Foaming profile using Style* shampoo with standard sebum.

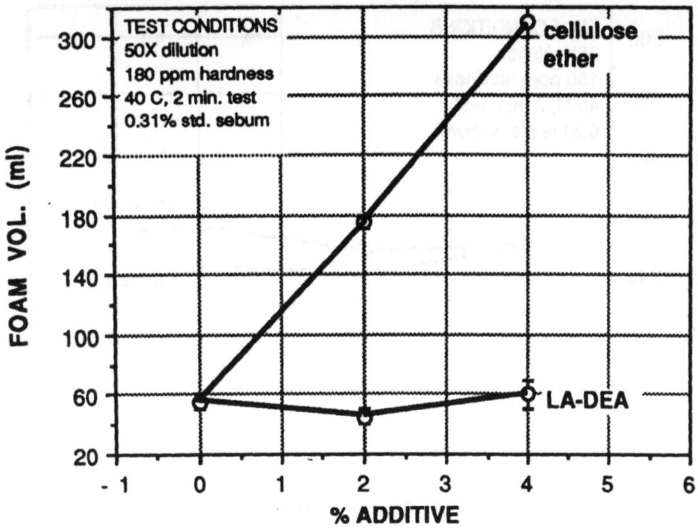


Figure 6. Foam volume of L. E. vs LA-DEA (8% ALS + additive).

We postulate three principal reasons for the exemplary performance of these lather enhancing cellulose ethers. The first is their rapid and efficient ability to emulsify the foam suppressing oleaginous components of the sebum. This would enable more primary surfactant to contribute to lather development.

The second reason lies in their inherent tendency toward formation of associative structures with many types of anionic surfactants. Formulation of such associative structures is demonstrated by the strong positive synergistic thickening effects observed when formulating. Viscosity also responds positively to electrolyte addition. This has carry-over implications as detailed below.

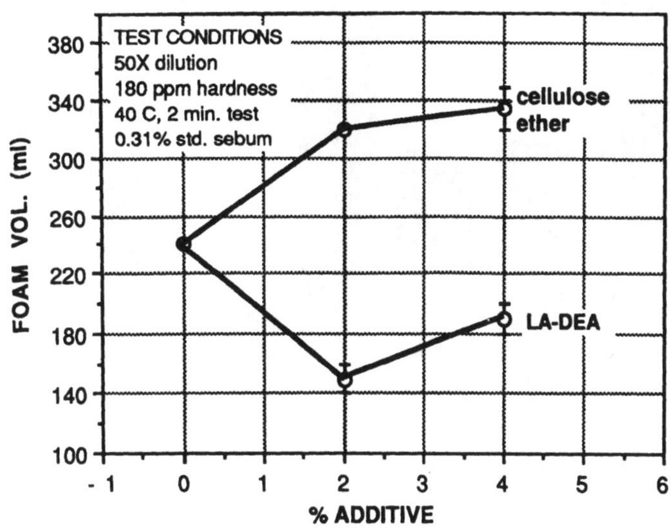


Figure 7. Foam volume of L. E. vs LA-DEA (10% ALS + additive).

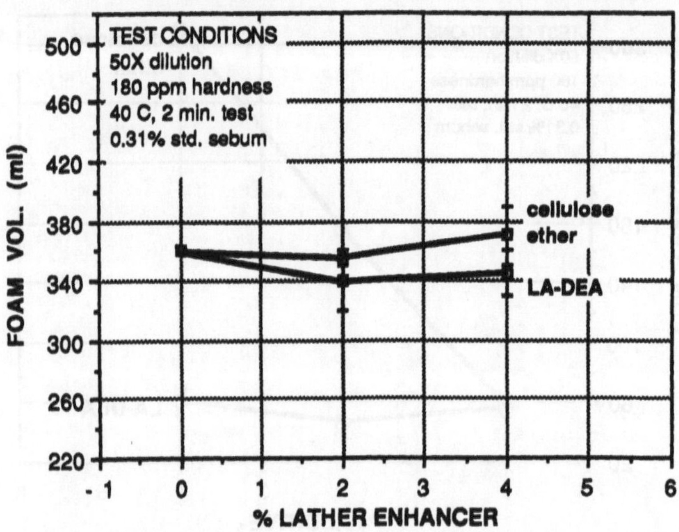


Figure 8. Foam volume of L. E. vs LA-DEA (14% ALS + additive).

The third reason is the presence of a property unique to this class of polymeric surfactants, that of interfacial gelation, Figure 11. Not only does the cellulosic lather enhancer rapidly sorb at many types of interfaces, but just as importantly, it's adsorbed multi-layers undergo an unusual sol-gel transition that is thermally driven and is operative over wide temperature ranges, especially the ambient to warm temperatures encountered in personal care applications.

These sorbed multi-layers likely incorporate anionic surfactants (when present) and critically alter bubble lamellae by imposing extremely

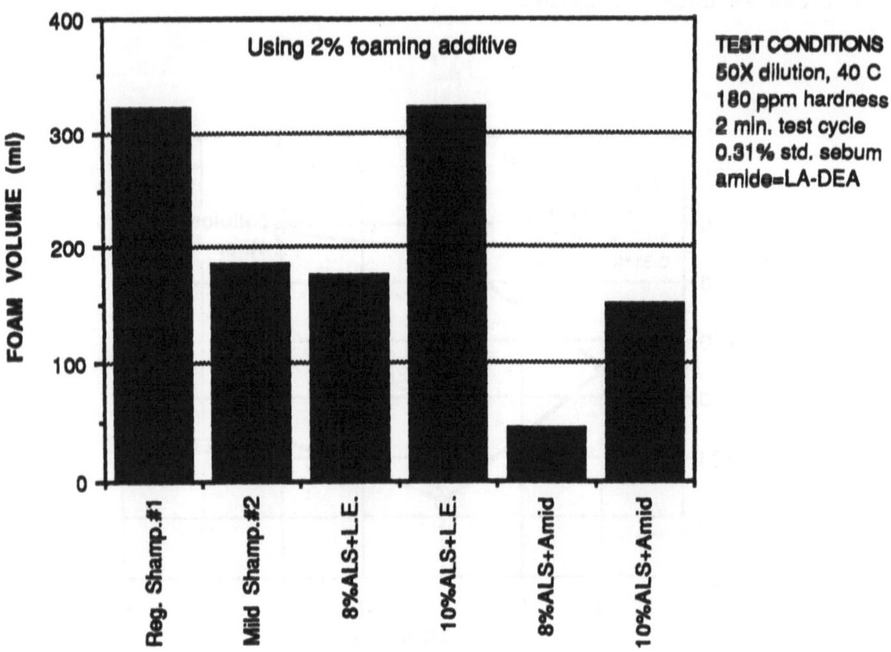


Figure 9. Comparative foaming of shampoos.

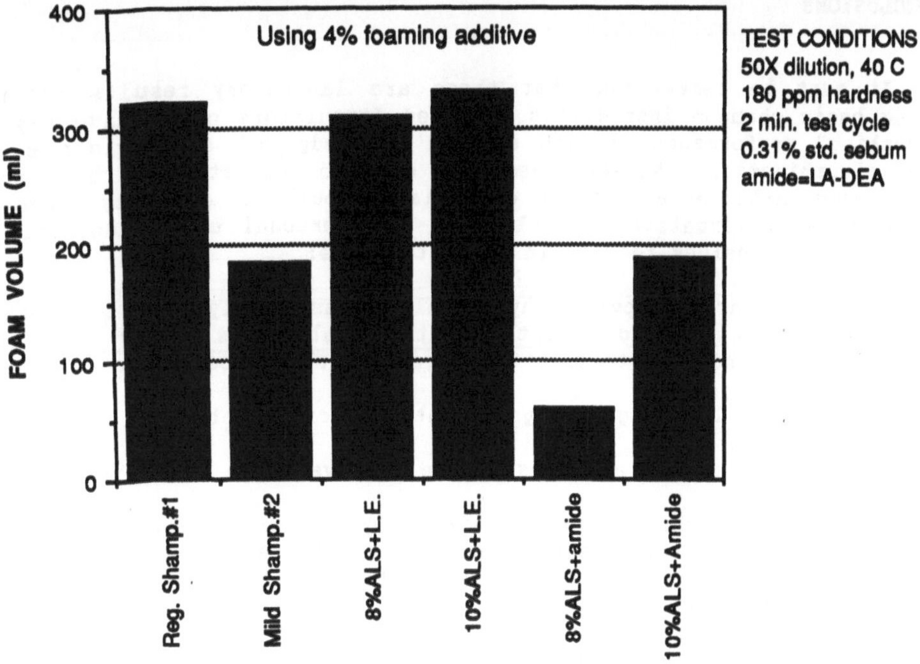


Figure 10. Comparative foaming of shampoos.

large surface viscosities and improved elasticity. Bubble lamellae appear much thicker and drain much more slowly. Interfacial gelation is likely the key operative force in retarding liquid drainage from the lather. It is undoubtedly the presence of the interfacial gel which accounts for the desirable lather lubricity, imparting an almost lotion-like feel. Presently, the unusual interfacial structural mechanical properties of this class of polymeric surfactants have been studied by only a few investigators.

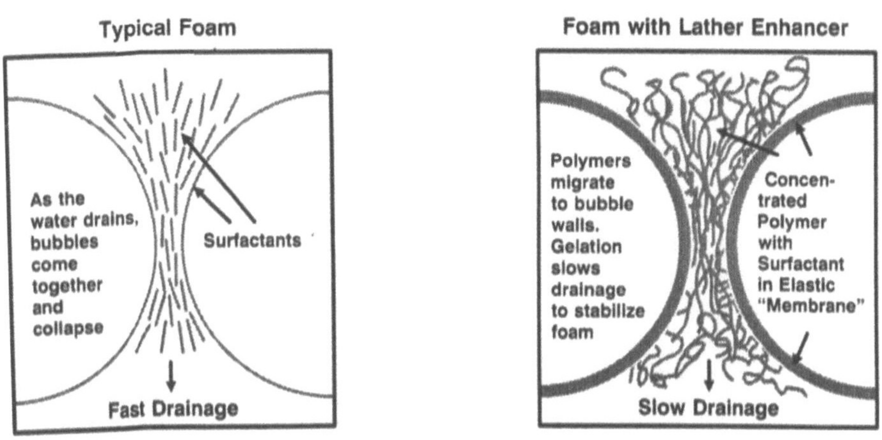


Figure 11. Interfacial gelation and lather development.

## CONCLUSIONS

It must be remembered that these are laboratory results obtained using basic shampoo ingredients. Skilled formulators often have ways of "tweaking" performance through special knowledge of appropriate synergistic combinations. Nevertheless, the data do suggest that these novel cellulosic lather enhancers can offer lather building advantages and may serve as an alternative technology in the personal care industry. The principal conclusions reached in this study are:

1. Novel hydroxypropyl methylcellulose based polymeric surfactants are efficient and effective lather enhancers. They are water soluble and easy to use. They are available from Dow as Methocel® 40-200.
2. Viscosity building is significant, especially at use levels of 3% or higher.
3. Lather lubricity and density are improved, the stability is also excellent.
4. Surfactant behavior coupled with unique interfacial gelatin formation of associative structures, and postulated rapid oil emulsification are important attributes of lather enhancement.
5. These polymeric surfactants are very mild to skin and eyes, even at high concentrations.
6. New and perhaps milder personal care cleansing products can be formulated. A new and alternative technology for lather enhancement is born and in development.

## REFERENCES


1. G. Barker, in: "*Surfactants in Cosmetics*," Surfactant Science Series, Vol. 16, Marcel Dekker, Inc., New York, 1985, Chapter 9. "Surfactants in Shampoos."
2. B. M. Abraham and K. Miyano, J. Colloid & Interface Sci., **107** (1), 264 (1985). "Static shear modulus of a methylcellulose solution and viscoelasticity of a polyvinyl alcohol solution at the air/water interface."

# APPLICATIONS OF POLYMERS IN HAIR CARE

Gary Martino and Frank Nowak

National Starch and Chemical Corporation
Bridgewater, New Jersey 08807

Over the past 5-10 years, the hair fixative industry has undergone somewhat of a metamorphosis due largely to changes in hair styles and a new surge in creativity in both the salon and consumer retail segments. Greater segmentation of the market is a continuing trend. New product forms and formulations have emerged. Products such as the "spritz" and "scrunching spray" were unknown until recent years. Superimposed upon this scenario is the impact of the ever strengthening movement toward "environmentally friendly" products. New polymers have emerged to provide features required of these new systems. Older polymers are also being widely used in a variety of new ways. Rapid market change continues. In this paper we review polymers and formulations in the context of newer product types and market trends.

## INTRODUCTION

Polymers are among the most highly, widely functional classes of raw materials available today to the cosmetic chemist. No single category of raw materials can perform so many functions in today's sophisticated, diverse hair care systems. Be they polyether polyols, silicone polymers, modified carbohydrates, vinyls or acrylics, they perform functions that no other class of raw materials can.

What is it that produces such versatility of polymers in hair care? There are two major factors. <u>First</u>, it's their rigidity due to molecular size and chemical composition providing hair fixing properties and conditioning, often without the greasiness associated with lower molecular weight raw materials. <u>Second</u>, a polymer's chemical functionality produces certain subjective effects. Also the ability of chemists to modify the molecular size, chemical composition, and functionality is the main reason for the extreme versatility of polymers in cosmetics. In many cases the polymer can be tailored to meet specific needs or applications.

## POLYMER FUNCTIONS IN COSMETICS

Polymers can be utilized for six general functions in hair care applications.

(1) Fixatives: in which their rigidity and moisture resistance hold hair and keep it held in place even under conditions of high humidity.

(2) Thickeners: in which high molecular weight and chemical functionality bind the vehicle used in the formulation into a highly viscous, three-dimensional matrix.

(3) Conditioners: in which a variety of different polymer types (silicones, modified proteins, and cationic, water soluble vinyl, acrylic and carbohydrate polymers) are placed on the hair from either a lotion or shampoo to impart any of a variety of subjective properties. Among the properties are ease of combing, gloss, body, manageability and static dissipation properties.

(4) Dye assists: in which water soluble, cationic polymers are used to form chemical bridges between anionic dyes and hair to facilitate deposition of the dye, enhance color uniformity, resist dry rub-off and improve color density.

(5) Opacifiers: in which certain hydrolytically stable latex polymers are utilized in aqueous systems to support the visual perception of richness or creaminess.

(6) Emulsifiers: in which surface active polymers are used to emulsify oils in creams and lotions without contributing the oiliness or greasiness of the stearate esters which are commonly used as emollients.

The three major functions of polymers are as fixatives, thickeners, and conditioners. Let's look at each major function individually, the applications, polymers used, and some of the reasons for their use.

## FIXATIVE POLYMERS

Fixatives represent the largest application, by far, for cosmetic polymers in hair care. Through a combination of film formation on the hair shaft and "seam welding" or "spot welding" the fibers to one another, fixatives hold hair in place. What are the properties of good hair fixatives?

### 1. Solubility

The polymer must be soluble in the solvent/propellant systems commonly used. Generally, this is not a problem when polar solvents and propellants are used, since most of the commonly used fixatives are themselves polar. In aerosol hair-sprays, non polar hydrocarbon propellants, however, will adversely effect low temperature stability for all commonly used resins.

## 2. Non-Toxic

Individual monomers selected for use should be relatively innocuous. Post polymerization treatment should reduce residual monomer levels to the parts per million range. A suitable health safety protocol should be performed.

## 3. Free of Objectionable Odor

The polymer should undergo post polymerization treatment to eliminate odoriferous low molecular weight fragments.

## 4. Additive Compatibility

The polymer must be compatible with a very broad variety of additives commonly used such as plasticizers, UV absorbers, thickeners, surfactants, and fragrances.

## 5. Optimum Molecular Weight

Molecular weight should be high enough for good holding. If too high, however it may not atomize well (hair spray), produce too dry a foam (mousse), or produce objectionable rheology (gel, lotion).

## 6. Drying Time

Drying time should be consistent with the application. Setting formulations require quick solvent release; styling formulations often require more "working" time. In both cases tackiness on drydown should be minimized.

## 7. Form Clear, Glossy Films

## 8. Form Hard Films

This is necessary for a stiff feel on the hair. Note that films can always be plasticized if softness is desired, however a soft polymer cannot be hardened easily. If the polymer is too hard however, the welds undergo brittle fracture from the enormous pressures exerted on these welds by movement of the hair. If the film is too soft ductible failure and a sticky feel will result.

## 9. Moisture Resistant

The polymer must remain non sticky and continue to hold hair well under high humidity conditions.

10. Adhesion to Hair

The polymer should not form unsightly flake when hair is combed or brushed.

11. Good Combing Properties

The polymer must provide good holding, yet allow the consumer to comb through and restyle.

12. Shampoo Removable

The polymer must be completely removable from hair even after heavy accumulation and subsequent use of shampoo.

**FIXATIVE POLYMERS - MAJOR APPLICATIONS**

Table 1 outlines the four major applications for fixative polymers and provides a list of many of the more commonly used polymers in each. The polymers used are largely acrylics, acetates, and modified carbohydrates. A common thread running through each is the requirement that the polymer be "stiff" or have a high $T_g$. The consumer believes that unless their hair is stiff, the product being used is not holding well. Silicones and the poly(ether polyol)s are not used because of softness and moisture sensitivity respectively.

1. Hairsprays

Well over 90% of the hairspray sold in the USA use one or more of the three polymers listed below. Although there are major differences between each polymer, they each provide the requisite properties of alcohol solubility, propellant compatibility and stiffness/humidity resistance.

Vinyl acetate/crotonic acid/vinyl neodecanoate copolymer
Octylacrylamide/acrylates/butylaminoethyl methacrylate copolymer.
Alkyl ester of PVM/MA copolymer.

A major challenge to the hairspray industry has arisen from the air quality focus of the California Air Resource Board (CARB). This and other regulatory bodies throughout the U. S. are investigating the feasibility of formulating hairsprays with greatly reduced levels of VOCs.

Many hair care chemists are evaluating water as a possible VOC replacement to comply with possible future restrictions. The ensuing problems associated with the use of water create a significant challenge to both hairspray marketer and polymer supplier alike.

Gels, Mousse, and a variety of lotions, under somewhat different designations, are all used to establish a coiffure. Note that several common resins are used throughout. Polymers are each water soluble. Delivery forms and polymer requirements are different in each case.

| Table 1. Fixative polymers major applications | |
|---|---|
| **Application** | **Polymers Commonly Used** |
| Hairspray | Vinyl Acetate/Crotonic Acid/Vinyl Neodecanoate Copolymer<br>Octylacrylamide/Acrylates/BAEMA Copolymer<br>Alkyl Esters of PVM/MA Copolymer |
| Gel | PVP and PVP/VA Copolymers<br>Polyquaternium- 4, 10, 11 |
| Mousse | Polyquaterium- 4, 10, 11<br>PVP/VA Copolymer<br>Octylacrylamide/Acrylates/BAEMA Copolymer<br>Alkyl Esters of PVM/MA Copolymer |
| Lotions | PVP and PVP/VA Copolymers<br>Polyquaternium- 4, 10, 11<br>Vinyl Acetate/Crotonic Acid Copolymer |

## 2. Gels

There are two main classes of gels: styling gels and conditioning gels. Styling gels are used for hair setting. A hard polymer which yields a stiff hold is required. PVP and PVP/VA have traditionally been used in these applications. Conditioning gels typically aid the mechanics of styling and relieve or prevent hair damage. The Polyquaterniums give good conditioning along with adequate hold.

Since most hair gels contain greater than 90% water the polymer must be very water soluble. Since the bulk of gels utilize carbomer as the thickener, the fixative polymer cannot either contain salt or be itself highly ionic. This maximizes thicker efficiency and solution clarity.

One disadvantage of the polymers traditionally used in gels is hygroscopicity due to inherent water solubility. Under high humidity conditions, the polymers pick up moisture and the hairstyle begins to sag. New gel formulations are being introduced which alleviate much of this problem. These use hard holding hairspray polymers as the main fixative.

## 3. Mousses

Mousses were initially largely based on the polyquaternium type polymers such as polyquaternium 4. Greater quantities of PVP types however, are now being utilized in combination with the polyquats to satisfy demands for greater stiffness.

## 4. Lotions

A variety of both "spray on" or manually applied lotions are available under many different designations. What is functionally a setting lotion may now be known also as a "glaze" or "spray gel. The trend in the

| Formula 1. An alcohol free spray gel | |
|---|---|
| **Ingredients** | **%** |
| Resyn® 28-1310 | 6.00 |
| AMP-95 | 0.65 |
| Glycerin | 0.10 |
| DC-190 silicone | 0.10 |
| Monamid® 716 | 0.20 |
| Uvinul MS-40 | 0.05 |
| Sodium bisulffite | 0.01 |
| Fragrance | 0.10 |
| Preservative | q.s. |
| Deionized water | 92.79 |
| | 100.00 |

US market is toward alcohol-free products. Lotions are formulated to provide good wet combing and be inherently very stiff and hard holding when applied to the hair. A typical spray gel formula is shown in Formula 1. In this case a very hard holding fixative resin (Vinyl acetate/Crotonic Acid Copolymer) provides not only stiff hold, but also wet combing ease-traits characteristic of the styling gel while providing the ease of application of a spray.

## CONDITIONING POLYMERS

The term "conditioning" relates to subjective functionality. Loosely described, a "conditioning agent" is one which improves some sensory appeal for the hair. It can make the hair appear fuller, shinier or healthier; feel silkier, smoother, and create more body. Moreover, deposition of a "conditioner" onto hair will facilitate wet and dry combing, all the while reducing static fly away upon brushing and combing. Further, some marketers make claims that their conditioner formulas can mend split ends. A well selected and well formulated conditioning polymer can easily accomplish each of these functions.

Table 2 describes the properties of a good conditioning polymer. Note that there is commonality between these and the properties of a fixative polymer. Both require good solubility and compatibility (although each property may be delivered from a different type of formulation). They should be non-toxic and of optimum molecular weight for best functionality. Both require clear, glossy films, adhesion to hair, and shampoo removability. There the similarity ends. Whereas a fixative polymer must be "hard," a conditioning polymer can be soft. Many of the silicone type polymers are in fact oils. Excellent "drag reducing" properties on hair are an absolute "must" for wet and dry combing. Further, to be successful, a conditioning polymer can provide "slip" to the hair but should never impart greasiness. This is the forte of a good conditioning polymer.

Table 3 outlines the major hair care applications for conditioning polymers. They are (a) shampoo in which the formula functions to not only cleanse the hair, but contributes a softness, shine, and controllability by depositing a functional conditioning polymer; (b) mousses; and (c) lotions or creams.

| Table 2. Desirable properties of conditioning polymers |
|---|
| Solubility |
| Non-toxic |
| Compatible with additives |
| Optimum molecular weight |
| Form clear glossy films |
| Adhesion to hair |
| Shampoo removable |
| Excellent combing properties |
| Can be soft or non-rigid |
| Non greasy |

Traditionally, water soluble cationics such as Polyquaternium-10 have been used in shampoos. In this a small quantity of the polymer-surfactant complex deposits on the hair during rinsing to achieve the desired conditioning effect. More recently, silicones (such as dimethicone in P&G's Pert Plus) are being used.

The original mousse products introduced to the US market in the early 1980s were termed "conditioning mousses" and were based on cationic water soluble polymers such as PQ-4. Many various brands have been brought to market since then. Some have responded to changing market needs by shifting focus. Many are now termed a "conditioning-fixative" mousse in which the fixative properties come from a cationic water soluble nonionic such as a PVP/VA copolymer.

Standard clear or opacified conditioning lotions which are applied to hair manually have been available for many years. Cationic, water soluble polymers, such as Polyquaternium 4, 7, 10 or 11 provide the necessary wet combing effects while adding "body" and a degree of set holding to the formula. Silicones such as dimethicone copolyol can augment the combing effects and gloss.

## THICKENING POLYMERS

The functionality of thickening polymers needs little description. Desirable properties (Table 4) are remarkably similar to those of fixative polymers. Primarily, they must thicken the solvent matrix in which they are placed, provide clear solutions, and impart the desired rheology or feel to the consumer on application to the hair.

Table 5 outlines the applications and lists some commonly used poly-

Table 3. Major applications of conditioning polymers

| Application | Polymers commonly used |
|---|---|
| Shampoo | Polyquaternium-4, 10, 11 Silicones |
| Mousse | Polyquaterium- 4, 11 |
| Lotion & Cream | Polyquaternium-4, 7, 10, 11 Silicones |

| Table 4. Desirable properties of a thickening polymer |
|---|
| Soluble |
| Non-toxic |
| Compatible with additives |
| Optimum molecular weight |
| Forms clear, glossy films |
| Adhesion to hair |
| Shampoo removable |
| Efficient thickening of solvent matrix |
| Form clear solution |
| Provide desired rheology and "skin feel" |

Table 5. Major applications for thickening polymers

| Application | Polymers Commonly Used |
|---|---|
| Gel | Carbomer types<br>Polyquaternium-10<br>Modified cellulose types<br>Gums |
| Lotions | Carbomer types<br>Polyquaternium-10<br>Modified cellulose types<br>Gums |

mers. A good thickener obviously must be capable of rheology modification (i.e., slight "bodying" of lotions and glazes, etc.) as well as thickening gel formulas all the way up to the 30,000 - 50,000 cps desired. The Carbomer types are the most widely used; they offer the property of being salt sensitive. Gel structure "breaks" on application to the hand such that the resulting lotion is easily spread into the hair. Other thickeners include the modified cellulosics, gums, and high viscosity grade of Polyquaternium-10. These appear to be used largely in salon type gels.

Thus, polymers are multi-faceted, and they offer the formulating hair care chemist the possibility of use in a broad variety of applications. New polymers are continually being introduced to the market. Most will be the driving force behind new hair care products and concepts. Conversely, new product concepts may be the driving force behind the synthesis of a specialized new polymer. Either way, the polymer suppliers and hair care product marketers must work together to bring truly unique products to fruition.

**REFERENCES**

1. Robert Y. Lochhead, Cosmetics and Toiletries, 103 (12), 23 (1988).

# POLYMERS FOR HAIR CARE PRODUCTS

Rikio Tsushima and Akihiro Kondo

Wakayama Res. Lab.
Kao Corp.
1334, Minato, Wakayama

Polymers, which are adsorbed or deposited on hair during shampooing and conditioning processes to improve the properties of hair, are discussed. (1) A water soluble wool-keratin hydrolyzate was prepared, and this oligomer permeates into the hair and reinforces it. (2) A polystyrene latex of 0.05 μ diameter was prepared, which is easily adsorbed on hair and gives body-feeling to the treated hair by raising the static friction coefficient. (3) An amphoteric polymer is soluble in aqueous salt solution and precipitates by dilution with water. This phenomenon was applied to new type conditioner with hair-styling ability.

## INTRODUCTION

In hair care products, it is an important technology to change the properties of hair fibers by utilizing polymers which would penetrate into the hair or would be deposited on the hair during the shampooing and conditioning processes.[1]

In this paper, we will report on the following three examples:

(1) A hydrolyzed wool-keratin which penetrates into damaged hair fibers and reinforces them.

(2) A polystyrene micro-latex which adsorbs on the surface of hair fibers and gives body-feeling to the hair by enhancing the static friction coefficient.

(3) An amphoteric polymer which shows a salting-in phenomenon. It is applied to a dilution-deposition type conditioner, which has hair-styling ability.

*Cosmetic and Pharmaceutical Applications of Polymers*
Edited by C.G. Gebelein *et al.*, Plenum Press, New York, 1991

## EXPERIMENTAL

### 1. Materials

#### 1A. Hydrolyzed wool-keratin:

Wool fibers were heated with an aqueous solution of phosphoric acid. After neutralization with calcium hydroxide, the unreacted wool and calcium phosphate were removed by filtration to give an aqueous solution of hydrolyzed wool-keratin. The solution was treated with activated charcoal and was freeze-dried to give a pale yellow powder.

#### 1B. Polystyrene micro-latex:

The emulsion polymerization of styrene was carried out by the use of nonionic emulsifiers and ammonium persulfate as the initiator.[2] The particle size of the latex was varied by the kind and concentration of emulsifier.

#### 1C. Amphoteric polymer:

An aqueous solution of 2-acrylamido-2-methylpropane sulfonic acid, dimethylaminoethyl methacrylate, and 2,2-azobis-(amidinopropane)-dihydrochloride was suspended in n-hexane containing sorbitan monostearate, and the polymerization was carried out under reflux.

### 2. Measurements

The diameter of the hair fiber was measured by a ANRITU SLB DIA MEASURING SYSTEM M551A. The elastic modulus of hair fiber was measured by a TENSILON UIM-III-100. The particle size of latex was determined by a COLTER N4. The friction coefficient of each hair fiber was measured by a NIPPON RHEOLOGY TYPE NRF-II. HPLC was carried out using a HITACH 655.

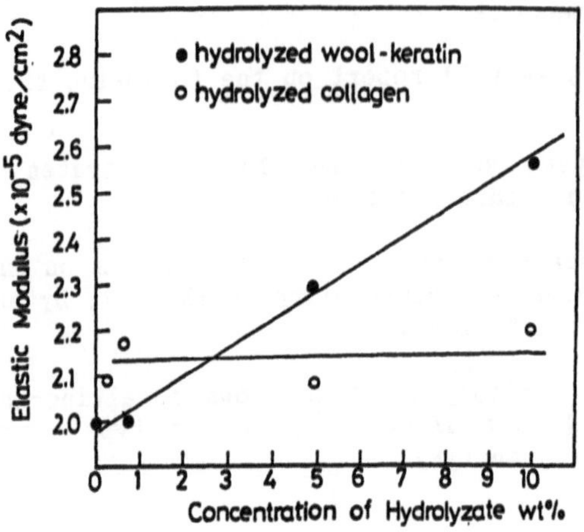


Figure 1. Effect of hydrolyzates on the elastic modulus of hair.

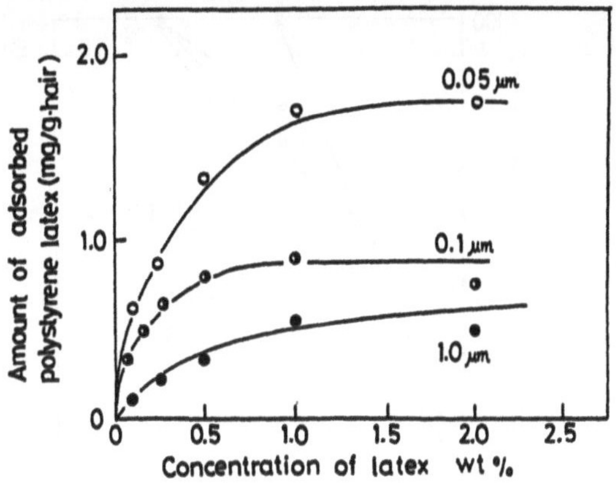


Figure 2. The adsorption behavior of polystyrene latex on hair.

## RESULT AND DISCUSSION

The adsorption of hydrolyzed wool-keratin to permanent waved hair and the cosmetic effects were studied. Figure 1 shows the elastic modulus of hair fiber treated with an aqueous solution of hydrolyzed wool-keratin and collagen. The molecular weight of the hydrolyzed wool-keratin was *ca.* 2000 and was almost the same as that of the hydrolyzed collagen. The elastic modulus of the hair fiber treated with hydrolyzed wool-keratin increased with an increase in the hydrolyzate concentration, although the elastic modulus of the hair fiber treated with the hydrolyzed collagen did not increase.

The swelling degree of hair fibers immersed in water was compared with that of the hair after treatment with hydrolyzed wool-keratin. The ratio of the diameter of swelling hair/diameter of dry hair for the treated hair was 1.23, while that for the untreated hair was 1.31. These results suggest that the hydrolyzed wool-keratin permeates into the hair and reinforces it.

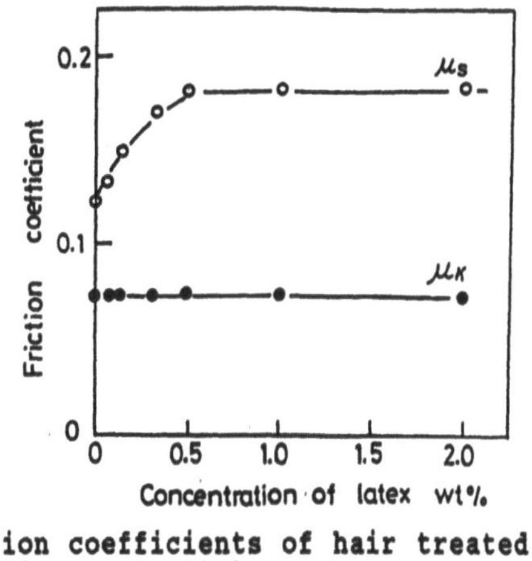


Figure 3. Friction coefficients of hair treated with a model conditioner. Conditions: temperature = 20°C; humidity = 50% R.H.; rotation of nylon pulley = 5 rpm; weight = 3.76 g.

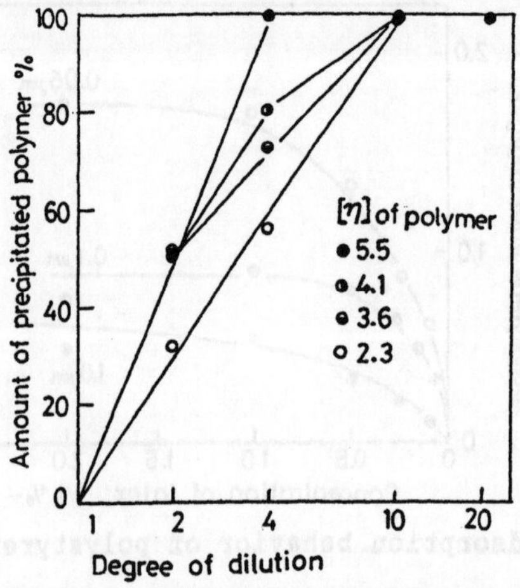


Figure 4. Precipitation behavior of the amphoteric polymer. Polymer solution before dilution: polymer = 1.5%, NaCl = 5.0%.

The adsorption of polystyrene latex on the hair and its effect on the friction coefficient of the hair was investigated. Figure 2 shows the adsorption behavior of a polystyrene latex added to a model conditioner, a mixed solution of a cationic surfactant (2.0%), cety alcohol (3.0%) and propylene glycol (2.0%). The amount of adsorbed polystyrene latex increased with a decrease in the particle size of the emulsion.

Figure 3 shows the dynamic and static friction coefficients of a hair fiber treated with the model conditioner containing 0.5% (by weight) of the latex of 0.05 μ diameter. The static friction coefficient increased with an increase in the concentration of latex, while the dynamic one remained constant. This latex could be applied to a new conditioner which gives body-feeling to the hair and retains hair style with staying at low combing force level.

The amphoteric polymer is generally soluble in an aqueous salt solution and it's not soluble in pure water, so precipitation of amphoteric polymer occurs when diluting the aqueous salt solution of the amphoteric polymer with water.[3,4]

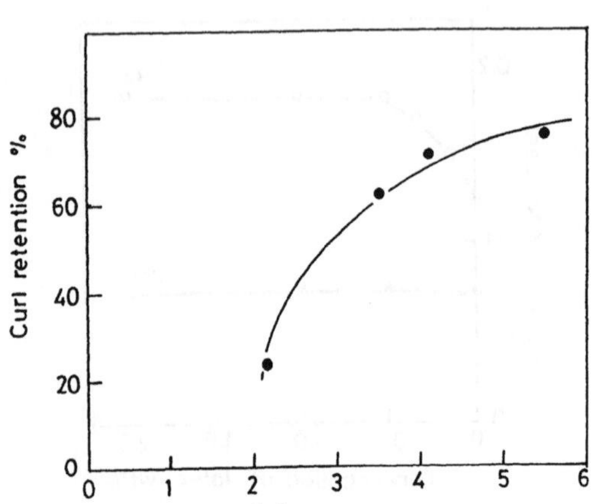


Figure 5. Effect of the molecular weight on curl retention at 95% R.H. For [η] solvent is 1M NaCl, 25°C.

The precipitation behavior of an amphoteric polymer and its applications to hair conditioner were investigated. Figure 4 shows the precipitation of copolymer of 2-acrylamido-2-methylpropane sulfonic acid and dimethylaminoethyl methacrylate dissolved in aqueous sodium chloride solution by diluting with water. With an increase in the molecular weight of the copolymer, it precipitated easier.

The curl retention of waving hair was measured after treating the hair with a mixed solution of sodium chloride (5%) and the copolymer (1.5%), followed by rinsing with water and drying. Figure 5 shows the effect of polymer molecular weight on the curl retention. Curl retention increased with an increase in molecular weight of amphoteric polymer. This salt-type polymer could be applied for dilution-deposition conditioner, which enhances the retention of hair-style.

## REFERENCES


1. R. Y. Lochhead, Cosmetics & Toiletries, **103**, 23 (1988).
2. G. G. Greth and J. E. Wilson, J. Appl. Polym. Sci., **5**, 135 (1961).
3. R. Hart and D. Timmerman, J. Polymer Sci., **28**, 638 (1958).
4. J. C. Salamone, W. Volkson, A. P. Olson and S. C. Israel, Polymer, **19**, 1157 (1978).

# ACETYLENE-DERIVED POLYMERS FOR PERSONAL CARE APPLICATIONS

Mohammed Tazi, Robert Login, Stephen Kopolow

GAF Chemical Corporation
1361 Alps Road
Wayne, NJ 07470 USA

Acetylene is a versatile intermediate in the synthesis of polymers with end-uses in the cosmetic and toiletries industry. As examples, two new products derived from N-vinyl pyrrolidone and methylvinylether are presented. N-vinyl pyrrolidone-co-methacrylamidopropyl triethylammonium chloride is a new cationic conditioning resin with a unique combination of substantive and film forming properties. The half ethyl ester of a maleic anhydride/methylvinylether copolymer is a new product for aerosol and non-aerosol hair sprays. It is engineered to meet the growing need for stiffer, harder, holding products.

## INTRODUCTION

In the 1930s, a chemist by the name of Reppe developed techniques for controlling the reaction of acetylene, a highly reactive compound which was difficult to handle. From acetylene, alkyl vinyl ethers and vinyl pyrrolidone were about to become the key monomers in the polymerization and copolymerization of products useful in a wide range of applications including cosmetic and pharmaceutical uses.

### Vinyl Pyrrolidone Based Polymers

In this class of polymers, acetylene is reacted with formaldehyde to form 2-butyne-1,4-diol. The hydrogenation of butynediol leads to 1,4-butanediol which, in turn, forms $\gamma$-butyrolactone via cyclization. The reaction of this lactone with ammonia gives 2-pyrrolidone. Further reaction of acetylene with 2-pyrrolidone leads to the formation of N-vinyl pyrrolidone. A simplified version of the flow chart is shown in Figure 1. All these compounds find important industrial applications as solvents and intermediates.

N-vinyl pyrrolidone is a versatile vinyl monomer which may be readily homopolymerized or copolymerized with many other monomers, in virtually any proportions and under a wide variety of conditions, to give resins

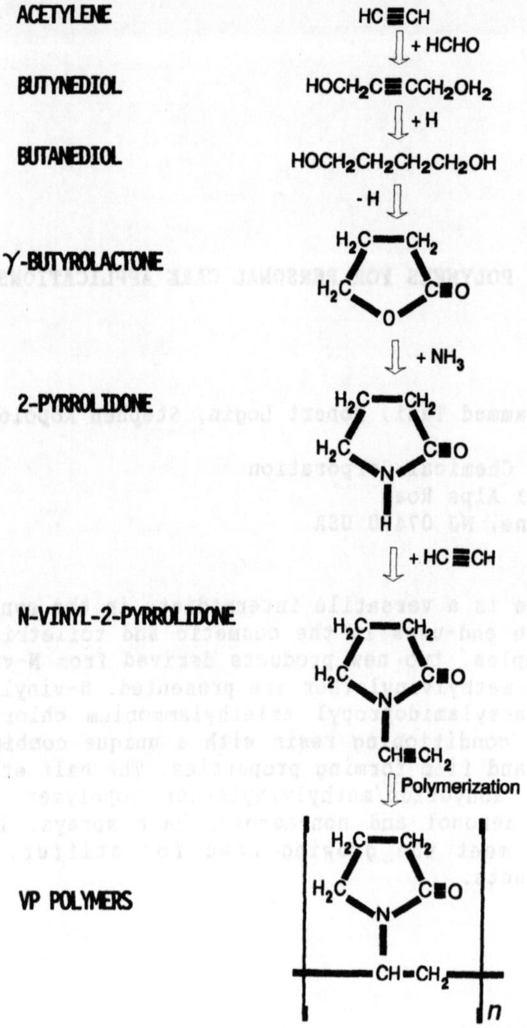


**Figure 1. Chemical derivation of VP polymers simplified reaction flow chart for products of VP polymer from acetylenes.**

**with unique properties. The homopolymer is soluble in water and in a wide variety of organic solvents, including alcohols and chlorinated solvents.**

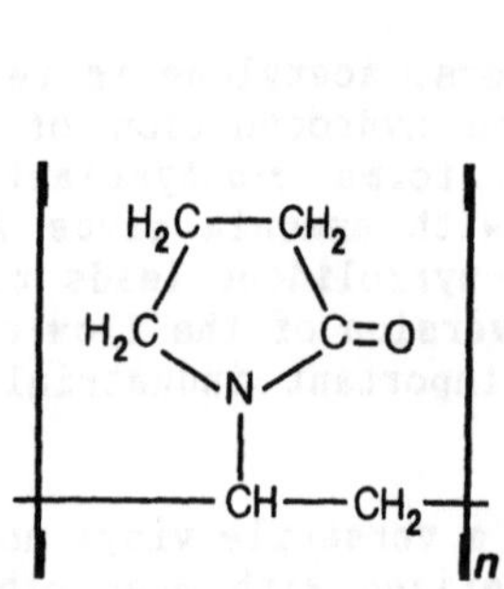

**Structure 1. Poly(vinyl pyrrolidone).**

In view of its unique chemical and physical properties, PVP has found significant uses in pharmaceuticals, cosmetics and toiletries, as well as in textiles, paper, beverages, detergents, soaps, adhesives, etc.

Because of an exceptionally low animal oral toxicity, and very high parenteral lethal dose ($LD_{50}$), PVP has found extensive use in the pharmaceutical industry. Over 8,000 papers have been published on the use of PVP in pharmacy and medicine. The major use of PVP in pharmacy is as an excipient in oral dosage forms. PVP has been included under the name of Povidone in the United States Pharmacopia XIX, as well as numerous other national compendia.

In the early 1950s, PVP was the first synthetic polymer to be used in hairsprays. Substantivity to hair, the formation of rewettable, transparent films and contributions to luster and smoothness were reasons for the shift to the whole series of PVP-based hair grooming products, shampoos, wave sets and creams, hair tints, and shaving products. Moreover, the broad spectrum of solubility of PVP in propellants used in pressurized products, and its anti-corrosive nature, made it a natural for the rapidly developing aerosol industry. PVP improves physical properties and contributes to performance in skin and eye make-up, lipsticks, deodorants, suntan lotions, and dental hygiene, in addition to hair preparations. The polymer promotes emolliency, lubricity, and a moisture balance in skin cosmetics, It stabilizes creams, lotions, and foam products, disperses pigments, and aids in the reduction of toxicity and sensitivity problems.

An interesting aspect of vinyl pyrrolidone copolymers is that their potential applications are so varied and are not limited to aqueous systems. Table 1 summarizes the vinyl pyrrolidone based polymers used in cosmetic applications.

As an example of cosmetic polymers, a new copolymer of vinyl pyrrolidone with methacrylamidopropyl trimethylammoniumchloride (MAPTAC) is presented. This monomer, like vinyl pyrrolidone, is a water soluble monomer. The cationic character of the copolymer provides substantivity to

Table 1. Cosmetic polymer based on vinylpyrrolidone

| Chemical name | CTFA adopted name |
|---|---|
| Polyvinylpyrrolidone | PVP |
| Vinylpyrrolidone/vinyl acetate copolymers | PVP/VA copolymer |
| Vinylpyrrolidone/dimethylaminoethyl methacrylate copolymers: | |
| unquaternized | PVP/dimethylaminoethyl methacrylate copolymer |
| quaternized | Polyquaternium-11 |
| Vinylpyrrolidone/methacrylamidopropyltrimethylammonium chloride | Polyquaternium-28 |
| Vinylcaprolactum/vinylpyrrolidone/dimethylaminoethyl methacrylate copolymer | Vinylcaprolactum/PVP/dimethyl methacrylate copolymer |
| Alkylated vinylpyrrolidone polymers | Butylated PVP<br>PVP/Hexadecene copolymer<br>PVP/eicosene copolymer |
| Vinylpyrrolidone/styrene copolymer | Styrene/PVP copolymer |

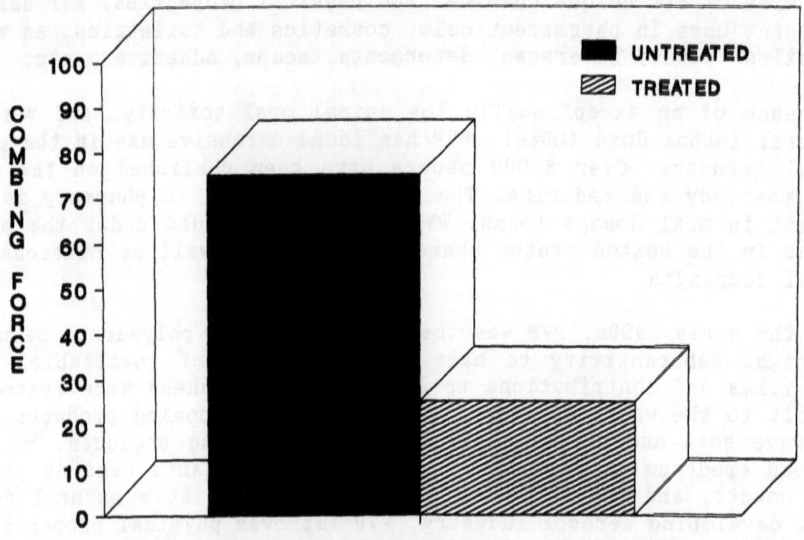


Figure 2. Wet combing force measurements. Hair tresses treated with polyquaternium-28.

both hair and skin. The substantivity translates into conditioning benefits as demonstrated by combing force data using the Instron test. Hair tresses were immersed in 2% active solutions of the resin, rinsed, and combing force measurement performed according to the method of Carcia, et al. Figure 2 shows significant improvement in wet combing compared to control tresses immersed in water.

Vinyl pyrrolidone/MAPTAC copolymer is compatible with anionic surfactants. Clear solutions were obtained with a 2% resin and various concentrations of anionic surfactants such as ammonium lauryl sulfate, sodium lauryl sulfate, sodium laureth-3 sulfate and TEA lauryl sulfate. The

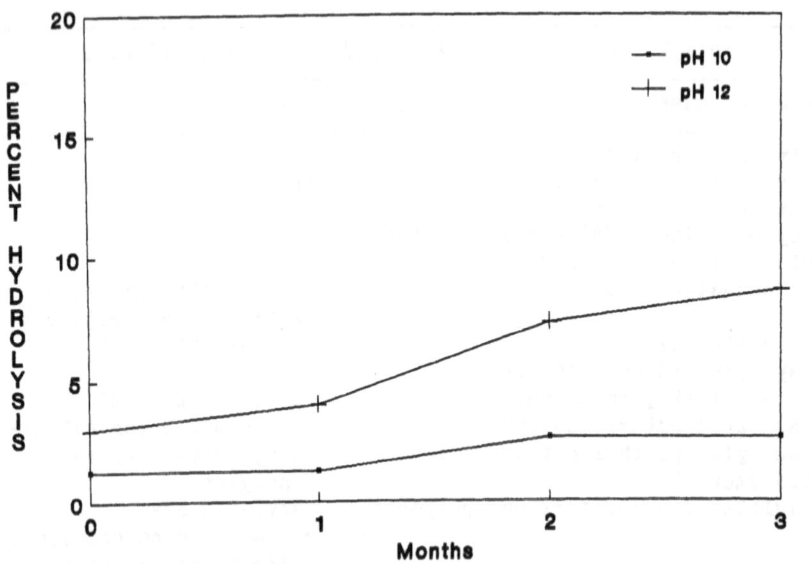


Figure 3. Hydrolytic stability of polyquaternium-28, pH 10 and pH 12 at 45°C.

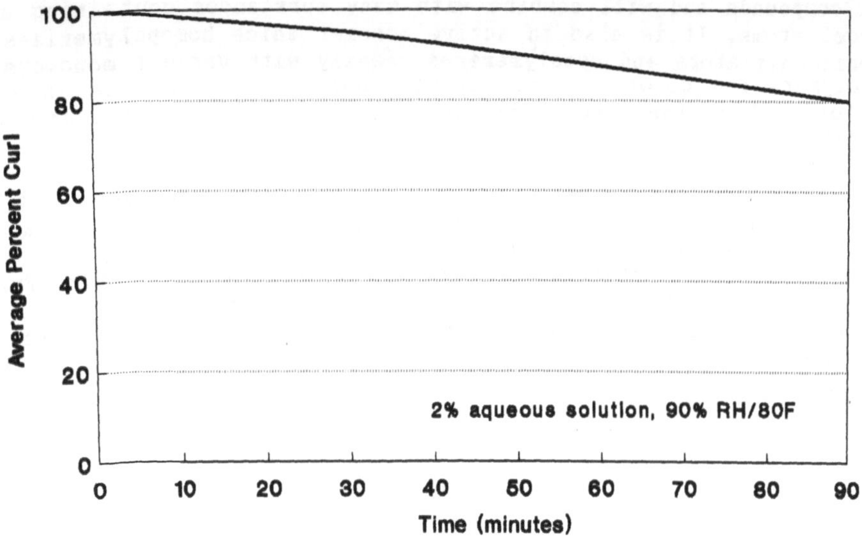


Figure 4. Polyquaternium-28, curl retention.

resin was found to give no build-up, a concern associated with cationic polymers.

The most significant properties of this copolymer is the fact that the comonomer of vinyl pyrrolidone is an amide which is more stable than an acrylate. The hydrolytic stability of vinyl pyrrolidone/MAPTAC makes the resin an ideal choice for a conditioning agent in permanent waving systems. Figure 3 shows the hydrolytic stability of the resin measured as the % cationic monomer hydrolyzed over time at 45°C for pH = 10 and pH = 12. Similar hydrolytic stability is expected at extreme acid pH.

Vinyl pyrrolidone/MAPTAC resin possesses film properties that support use in fixative products such as mousses and gels. The films are clear, glossy, and flexible. Our retention studies at 2% solids demonstrate its excellent hair holding performance when applied from a leave-on styling product (Figure 4). Table 2 summarizes the potential applications of hydrolytically stable vinyl pyrrolidone/MAPTAC copolymer.

## Alkyl Vinyl Ether Polymers

The second class of acetylene-derived cosmetic polymers consists of copolymers of methyl vinyl ether. Acetylene undergoes a vinylation reaction in the presence of alcohols. With methanol, the result is methyvinyl ether. This compound undergoes reactions typical of unsaturated

Table 2. Potential uses for the hydrolytically stable vinyl pyrrolidone /MAPTAC copolymer

| | |
|---|---|
| Shampoos | Gels |
| Conditioners | Glazes |
| Permanent waving systems | Hair and body lotions |
| Mousses | Moisturizing creams |

vinyl compounds and will combine with many substances containing active hydrogel atoms. It is also an active monomer which homopolymerizes with cationic initiators and copolymerizes readily with various monomers. The homopolymers and copolymers of methyl vinyl ether find uses in a wide range of applications including cosmetic, pharmaceutical, and agricultural.

As an acetylenic cosmetic polymer, a new resin designed for high solids in hair spray is discussed. Figure 5 shows the reaction of acetylene with methanol to form methyl vinyl ether which, in turn, is copolymerized with maleic anhydride to give an alternating copolymer. Further reaction of the resin with ethanol leads to a half ethyl ester of methyl vinyl ether/maleic anhydride. The copolymerization is carried out in a controlled way to achieve a molecular weight and molecular weight distribution providing the following properties to the resin when formulated in a hair spray.

stiffness
hard hold
increase hold by increasing solids content
excellent spray pattern
quick drying time of solids

In addition, the other qualities of PVP/MA copolymer that are expected in today's fixative products include:

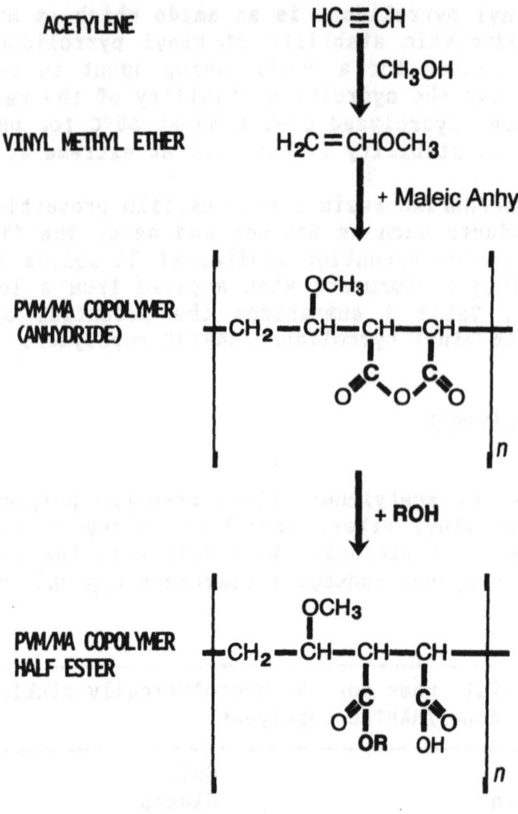


Figure 5. Chemical derivation of PVM/MA copolymer, simplified reaction flow chart for production of half esters of PVM/MA.

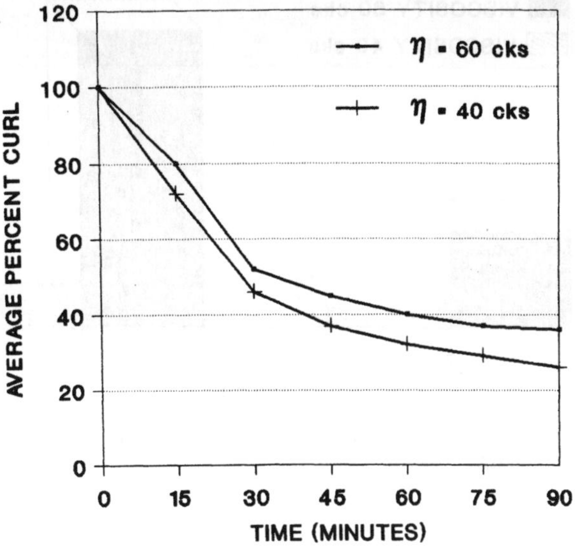


Figure 6. Ethyl ester of PMVE/MA at 5% solids, effect of specific viscosity.

high humidity resistance
negligible tack
no flaking
high gloss/sheen
natural feel
good compatibility
easily removed by shampoo
compatibility with additives

Figures 6 and 7 show the effect of specific viscosity of the polymer on the curl retention at 5%. Figure 8 shows the effect of specific viscosity of the polymer on the drying time at 5% and 8%.

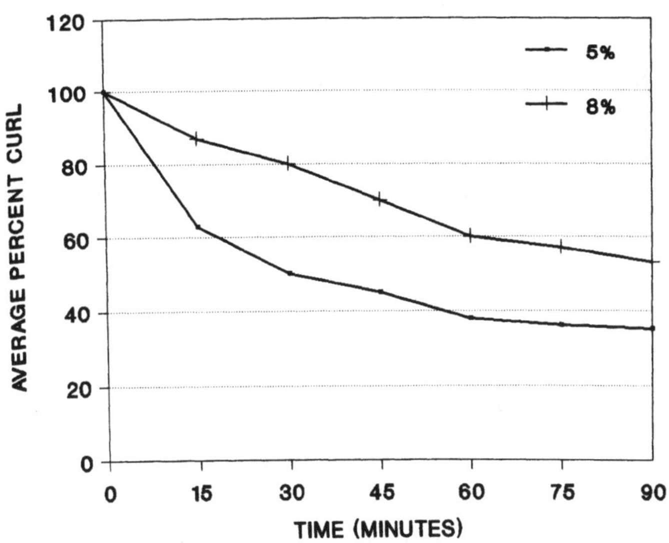


Figure 7. Curl retention of ethyl ester of PMVE/MA, at 5% and 8% solids.

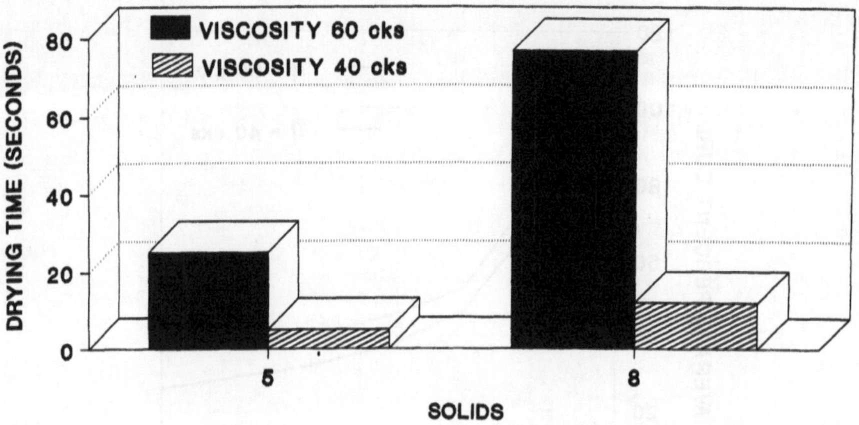


**Figure 8.** Drying times vs. per cent solids, ethyl ester of PMVE/MA.

## REFERENCES


1. W. Reppe, C. Schuster and A. Hartmann, U.S. Patent 2,265,450. "Polymeric N-vinyl lactams and process of producing same."
2. W. Reppe, H. Krzikalla, O. Dornheim and R. Sauerbier, U.S. Patent 2,317,804. "N-vinyl lactams."
3. C. Schuster, R. Sauerbier and H. Fikentscher, U.S. Patent 2,335,454. "Polymerization of N-vinyl lactams."
4. L. Blecher, D. H. Lorenz, H. L. Lowd, A. S. Wood and D. P. Wyman, "*Handbook of Water-Soluble Gums and Resins*," R. L. Davidson, Ed., McGraw-Hill, New York, 1980, Chapter 21. "Polyvinylpyrrolidone."
5. M. Bergmann, I. Flance and H. T. Blumenthal, New Engl. J. Med., 258, 471-476 (1958). "Thesaurosis following inhalation of hair spray."
6. M. Bergmann, I. J. Flance, P. T. Cruz, N. Klam, P. R. Aronson, R. A. Joshi and H. T. Blumenthal, New Engl. J. Med., 266, 750-755 (1962). "Thesaurosis due to inhalation of hair spray, Report of twelve new cases including three autopsies."
7. "PVP, a novel approach to the reduction of toxicity and sensitivity problems inherent in other chemical compounds," Technical Bulletin 9630-006, GAF Corporation, New York.
8. R. L. Goldemberg and L. Safrin, J. Soc. Cosmet. Chem., 28, 667-679 (1977). "Reduction of topical irritation."
9. R. L. Goldemberg, J. Soc. Cosmet. Chem., 30, 415-427 (1979). "Anti-irritants."
10. H. A. Shelanski, M. V. Shelanski and A. Cantor, J. Soc. Cosmet. Chem., 5, 129-131 (1954). "Polyvinylpyrrolidone (PVP)-A useful adjunct in cosmetics."
11. M. F. Nelson, U. S. Patent 2,783,182. "Non-staining chlorophyll preparations."
12. P. Molyneux in: "*Proceedings of the International Symposium on Povidone,*" G. A. Digenis and J. Ansell, Eds., University of Kentucky, Lexington, USA, April 1983, pp. 1-19. "The physical chemistry and pharmaceutical applications of polyvinylpyrrolidone."
13. Cosmetic, Toiletry and Fragrance Association, J. Am. Coll. Toxicol., 2(5), 141-159 (1983). "Final report on the safety assessment of polyvinylpyrrolidone/vinyl acetate copolymer."
14. G. T. Hewitt, U.S. Patent 3,769,398. "Polyethyleneimine shampoo composition."
15. R. J. Crawford and C. R. Robbins, J. Soc. Cosmet. Chem., 31, 273-278 (1980). "A replacement for Rubine dye for detecting cationic on keratin."

16. P. J. Petter, Aerosol Rep., **23**(3), 124-132 (1984). "Aerosol foam mousse products for hair care."
17. P. J. Petter, Soap Perfumery Cosmet., **57**(9), 450-451 (1984). "Formulating hair mousse products."
18. E. J. Murphy, Soap, Cosmet. Chem. Spec., **56**(2), 34, 39, 40, 82 (1980). "Cationic polymer conditioner evaluation."
19. P. Flemming, Parfuem. Kosmet., **61**(9), 332-336 (1980). "Foenfrisuren-foenwellmittel."
20. "*GAFQUAT quaternary polymers for the cosmetic industry*," Technical Bulletin, 2550-017, GAF Corporation, New York.
21. Y. K. Kamath, C. J. Dansizer and H. D. Weigmann, J. Appl. Polym. Sci., **29**, 1011-1026 (1984). "Surface wettability of human hair, I, Effect of deposition of polymers and surfactants."
22. Cosmetic, Toiletry and Fragrance Assoc., J. Am. Coll. Toxicol., **2**(5), 161-178 (1983). "Final report on the safety of polyquaternium-II."
23. S. C. Johnson, Cosmet. Toiletries, **99**(6), 77-81 (1984). "Acetylene-derived polymers."
24. "*Copolymer VC-713*," Technical Bulletin 2550-047, GAF Corporation, Wayne, NJ.
25. P. J. Petter, Seifen, Oele, Fette, Wachse, **112**(4), 103-104 (1986). "Modern haircare aerosols-hairsprays and mousses."
26. "*Antaron V Polymers*," Technical Bulletin 2550-052, GAF Corporation, Wayne, NJ.
27. R. Oteri, S. Johnson and S. Dastis, Cosmet. Toiletries, **102**(3), 107-109 (1987). "A new waterproofing agent for sunscreen products."
28. H. Hoeffkes, Ger. Offen., DE3 34-365. "Conditioning agent for oily hair."
29. J. Petter, Manuf. Chem., **53**(11), 34-35, 37, 39 (1982). "Hairsprays-the role of resins."
30. M. Hoyt, Cosmet. Perfumery, **88**(4), 53-55 (1973). "New concepts in hair products."
31. P. J. Petter, Aerosol Rep., **24**(9), 479-480, 482-484 (1985). "Aerosol formulations for the new look in hair fashions."
32. P. J. Petter, Aerosol Rep., **21**(9), 369, 372-375 (1982). "Comparison of the hydrocarbon compatibility of hairspray polymers."

## ORGANO-MODIFIED SILICONE COPOLYMERS FOR COSMETIC USE

David T. Floyd

Goldschmidt Chemical Corporation
914 E. Randolph Road
Hopewell, VA 23860

Silicones have been used in the cosmetic industry for over 30 years, the most common being the poly(dimethylsiloxane)s. Their use is limited by their poor compatibility with cosmetic oils and emollients. The hydrophobic nature of these silicones renders them unsuitable for aqueous formulations. Organic modifications of these poly(dimethylsiloxane)s are made by replacing one or more of the methyl groups with lipophilic or hydrophilic moieties to form new polymers. Examples of polymers include: poly[alkyl- poly(dimethyl-) and poly(methylsiloxane)]s, poly[ether poly(dimethylsiloxane)]s and poly[alkyl poly(ether) poly(dimethylsiloxane)]s. Ionic silicone surfactants can be produced by way of an epoxy reaction on the poly(dimethylsiloxane). Both quaternized and amphoteric poly(siloxane)s have been made. The reactions to produce these chemicals will be discussed as well as their application. The silicone polymers have been extensively tested, and these versatile polymers are recognized as safe for use in personal care products and are biodegradable.

### SUMMARY

Organo-modified silicones were structurally derived from poly(dimethylsiloxane)s by replacing one or more of the methyl groups with lipophilic or hydrophilic organic moieties to produce several new types of polymers, which have been shown to be major interest to cosmetic formulators.

To produce organo-modified silicones, three basic organic modifications were used to provide the base building blocks for the polymers. These modifications or adducts are alkyl or alkoxy substitutions, alcohol additions and epoxide intermediates. From these basic modifications, silicone based nonionic, amphoteric and cationic surfactants were synthesized, along with silicone waxes and polymers which combine both alkyl or wax modifications and surfactant properties. These latter types of polymers are powerful emulsifiers and stabilizers for emulsions.

These highly stable polymers give the cosmetic and pharmaceutical

*Cosmetic and Pharmaceutical Applications of Polymers*
Edited by C.G. Gebelein *et al.*, Plenum Press, New York, 1991

formulator powerful new tools to develop more effective formulations. These polymers contain both silicone and organic properties in the same molecule. The contributions of silicone, such as their hydrolytic stability, liquidity at high molecular weights, lubricity, gloss, surface spreading, and low toxicity profiles, can be combined with the substantivity, the compatibility - aqueous and/or organic, and the conditioning and after-feel properties of the organic moiety. This improved functionality gives increased performance and is easily compounded into formulations.

These new silicone copolymers are highly versatile polymers, are recognized as safe for use in personal care products, and are biodegradable in the environment.

## INTRODUCTION

Since the first silicone containing skin lotion was introduced over 30 years ago, silicones have grown increasingly important in skin care formulations.[1-3] Today, with advances in their chemistry, silicones can be used in specialized skin care products that are applied to protect, lubricate and clean the skin, such as skin creams and lotions, cleansing creams, suntan lotions, and sunscreens. They also are useful in formulating antiperspirants, hair care preparations, and make-up. Their use in pharmaceutical products is also increasing.

As formulation aids, silicones help to detackify other ingredients. They serve as emulsifying agents or as carriers that deliver active ingredients to the skin,[4] then quickly spread, leaving little residue. Silicones impart a characteristic soft feel, provide a protective barrier to irritants, and contribute lubricity and emollient benefits to skin care products. They are essentially physiologically and chemically inert, showing a low order of oral toxicity, and are nonirritating and nonsensitizing when applied to the skin. Presently, the most widely used silicones in personal care products are the poly(dimethylsiloxane)s or dimethicones.

The dimethicones have valuable properties in that they can enhance water repellency, improve lubricity, and exhibit a controlled foaming behavior. However, the range of application of these dimethicones is limited because they also demonstrate some properties which limit their use in personal care products. In particular, these poly(dimethysiloxane)s are not soluble in water, alcohol, mineral oil and other esters and solvents used in cosmetic products. They also tend to leave oily residues and they have the ability to suppress foaming, making them undesirable in products such as shampoos and shaving creams. Dimethicones, when used in w/o or o/w emulsions, do improve the rub out or spreading and reduce tackiness. In hair care products, they function as glossing agents and, in skin care products, they impart moisture retention and softness.[2]

Organo-modified silicone copolymers are of increasing interest in the sciences as well as for industrial applications. The cosmetic and pharmaceutical industries have shown an increased interest in these polymers. Their unique properties, which are quite different to those of conventional hydrocarbon polymers, have led them to widespread applications in these industries. The reason for these different properties arise mainly from the nature of the base dimethylsiloxane structure.[5-8] This structure gives these polymers flexibility and good flow properties; even at low temperatures, good thermal and oxidative resistance, low surface tensions and either hydrophobicity or lipophobicity; depending on the modification, a high permeability to many gases and biological inertness. With organo-modified silicones, we can maintain the desired properties of Dimethicone and reduce the negative attributes.

These organic modifications of the poly(dimethylsiloxane)s are made by replacing one or more of the methyl groups with lipophilic or hydrophilic moieties to form new polymers.[5,9] Some examples of these polymers include: poly[(alkyl) poly(dimethylsiloxane)]s, poly[(alkyl) poly(methylsiloxane)]s, poly[(ether) poly(dimethylsiloxane)]s, and poly[(alkyl) poly(ether) poly(dimethylsiloxane)]s. Ionic silicone surfactants can be produced by epoxide reactions on the poly(dimethylsiloxane). Both cationic and amphoteric surfactants have been produced. The reactions to produce these polymers will be reviewed and their applications in personal care products will be defined. The silicone polymers were extensively tested, and these versatile polymers were found to be safe for use in personal care products.

## Silicone Copolymers

Prior to the introduction of organo-modified siloxane copolymers, it may be helpful to look first at poly(methylsiloxane)s and poly(hydrocarbon)s in general. The basic structure of these compounds are shown in Figure 1.

The flexibility of the silicone-oxygen linkage allows poly(methylsiloxane)s to remain liquid even at high molecular weights, in contrast to polymeric hydrocarbons which become increasingly viscous with the lengthening of the organic chain. Generally, hydrocarbons have a clearly defined (fairly narrow) molecular weight; whereas dimethicones usually exhibit a broad statistical molecular weight distribution. The unique feature of the poly(dimethylsiloxane)s is their high surface activity. Pure poly(dimethylsiloxane)s exhibit surface tension values of 21 dynes/cm, which are exceeded only by fluorocarbons. The prevailing explanation for this phenomenon is the dense packing of methyl groups at the surface of the poly(dimethylsiloxane)s. The low surface tension of poly(dimethylsiloxane)s gives them surface-active properties both in aqueous and organic solutions. The practical use of these properties is limited by the insufficient compatibility of the pure poly(dimethylsiloxane)s in water and polar organic solvents.

## Silicone Chemistry

Silicones can be viewed as organic derivatives of $SiO_2$, with some of

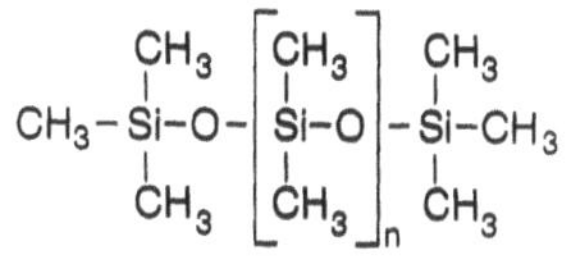

$CH_3-(CH_2)_n-CH_3$

Figure 1. Structures of poly(methylsiloxane) and poly(hydrocarbon)s.

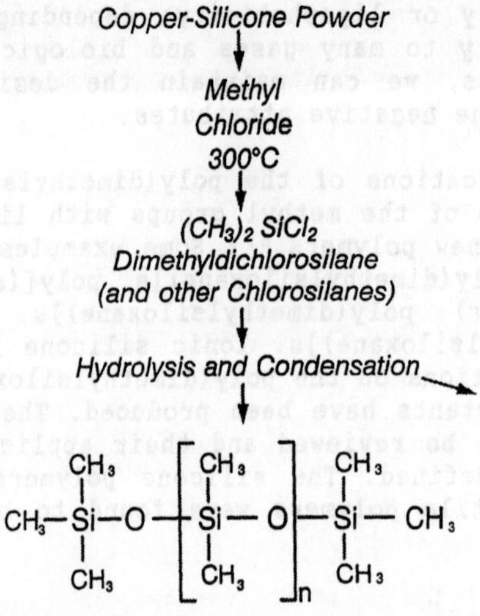


Figure 2. Synthesis of the base silicone polymer.

the oxygen atoms in the high polymer $SiO_2$ molecule being replaced by organic groups.

Production of poly(methylsiloxane)s begins with a mixture of elemental silicon powder and a copper catalyst which is heated with methyl chloride to produce dimethyldichlorosilane, $(CH_3)_2SiCl_2$, as well as lesser amounts of other chlorosilanes having one to four chlorines. These methylchlorosilanes are hydrolyzed to methyl substituted silicic acids or silanols. Condensation occurs as the silanols are formed; and as the elements of water are split out, the basic dimethylsiloxy segments of the silicone polymer are formed (Figure 2).

The resulting silicone fluids vary in viscosity depending on the molecular weight of the polymer, a property directly related to the length of the silicone-oxygen central chain.

The term "silicones" is used for a variety of technical grade products, the major property determining compounds of which are poly(organosiloxane)s. The bulk of commercial silicones are poly(dimethylsiloxane)s, which consist of siloxane units arranged in the form of chains or rings.

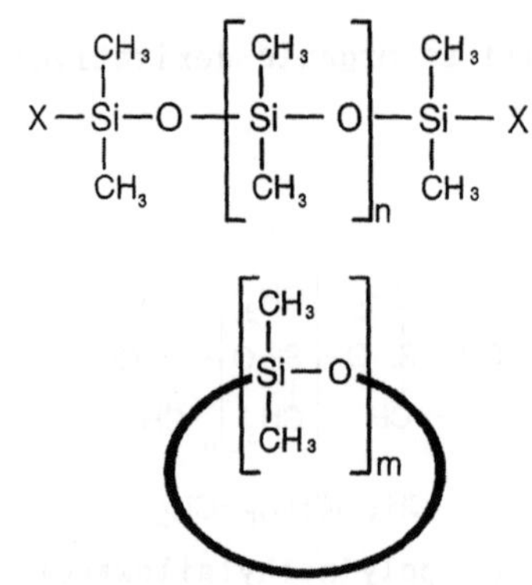


Figure 3. Forms of poly(dimethylsiloxane)s.

This class of substances is described by the following formulas (Figure 3).

As a result of this molecular structure, the siloxanes occupy an intermediate position between inorganic and organic compounds, in particular between silicates and organic polymers. The uses of the linear and cyclic poly(dimethylsiloxane)s in personal care products has been widely discussed and their uses are widely known.[10]

## Synthesis of Organo-Modified Silicone Polymers

The synthesis of organo-poly(siloxane) copolymers can be achieved by several synthetic routes. First, reactive poly(siloxane)s are synthesized as precursors. These may contain Si-Cl, Si-OR or Si-H groups. This leads to an active site as shown in the structure below (Figure 4).

Secondly, organic groups are built in by way of further reaction. Two main types of reaction have gained importance in the industry and are widely used (Figure 5).[11] (a) Esterification of reactive chlorine-groups containing poly(siloxane)s or transesterifications of alkoxy-groups containing poly(siloxane)s with OH-functional organic compounds. This process leads to products which contain Si-O-C linkages. (b) Another type of reaction is the addition of terminal olefinic groups containing compounds to poly(methylsiloxane)s with reactive Si-H groups in the presence of a platinum catalyst. The resulting products are connected with Si-C linkages.

The organic groups can be attached to the poly(methylsiloxane) variety at the terminal Si-atom of the molecule and/or in side-branch positions (Figure 6).

For applications in personal care products, especially those containing aqueous solutions, the copolymers linked with silicone to carbon bonds (Si-C) are preferred because of their increased hydrolytic stability.[1,4,10,11] The range of organic groups which may be attached to the poly(methylsiloxane) moiety varies widely in their chemical composition and may include nonionic, cationic, amphoteric and anionic groups. Examples of the more common organic moieties are as follows:

| Nonionic | Ionic |
|---|---|
| $-C_6H_5$ | $-(CH_2)_nNR_3{}^+X^-$ |
| $-(CH_2)_nCH_3$ | $-(CH_2)_3-O-CH_2-CH(OH)-CH_2NR_2{}^+X^-$ |
| $-(CH_2)_nNR_2$ | |
| $-(CH_2)_n-C(=O)-N(H)-(CH_2)_mNR_2$ | |
| $-(CH_2)_3-O-(C_nH_{2n}O)_xR$ | $-(CH_2)_3-O-CH_2-CH(OH)-CH_2SO_3{}^-Na^+$ |

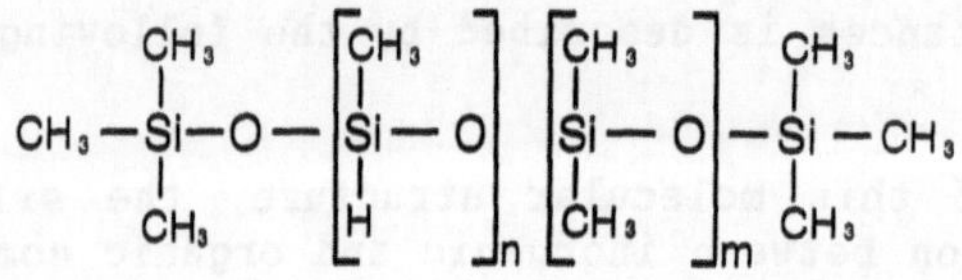

**Figure 4. Typical reactive poly(siloxane).**

From these basic types of moieties a number of types of organo-modified silicones are derived.

__Poly(ether)-Poly(siloxane) Copolymers__

Among the nonionic modified polymers, the products containing one or several $-(CH_2)_3-O-(C_nH_{2n}O)_xR^1$ pendent groups have gained considerable significance in cosmetic formulations.[1] These compounds are generally described as poly(ether)-poly(siloxane) copolymers. Their CTFA name is Dimethicone Copolyol (Figure 7).

The polyether part in Dimethicone Copolyols can consist of homopolymeric chains of either ethylene or propylene oxide or can contain varying proportions of both ethylene and propylene oxide. Through the chemical variations which can be realized in the composition of Dimethicone Copolyols, these products can provide unique surfactant properties. They offer significant advantages as surface tension depressants, wetting agents, emulsifiers and foam builders in hair treatment and shaving preparations.[12-15]

Of particular interest, in terms of compatibility, is the orientation of poly[(ether) poly(siloxane])s in aqueous solutions. The dimethylsiloxane moiety is hydrophobic and, therefore, aligns itself at the surface of the aqueous medium. The ethylene oxide moiety is hydrophilic and attaches itself into the aqueous medium.

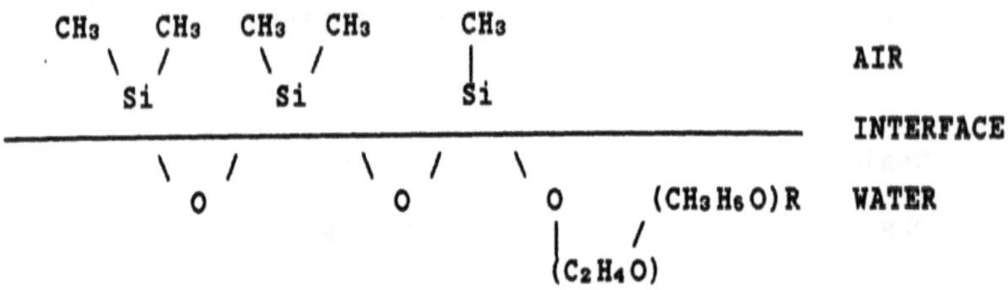


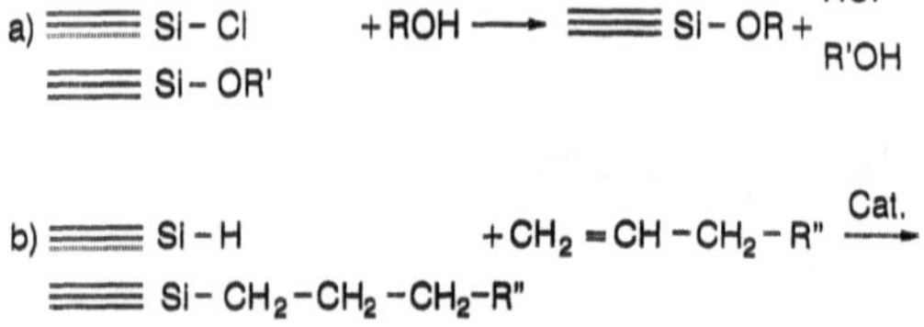

**Figure 5. Types of reaction.**

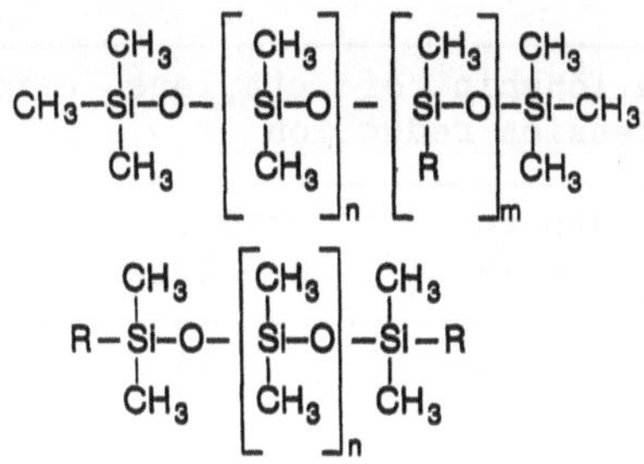

*R = Non-Ionic or Ionic Organic Group*

Figure 6. Organo-modified poly(siloxane)s.

The physical properties pertinent to aqueous systems include surface tension, cloud point, foam height, wetting, solubility and pH. The parameters influencing these properties are as follows:

1. Aqueous surface tension lowering:

(a) Result of tight packing of dimethyl siloxane units at aqueous interface;
(b) Short Si chains provide better lowering;
(c) The EO chain length (not % EO content) is important (a short chain is desirable) (Table 1);
(d) In mixed EO/PO systems, the surface tension is generally raised due to the higher surface tension of the propylene oxide moiety (approximately 32 dynes/cm);
(e) The lower limit of aqueous surface tension is 20 dynes/cm.

2. Cloud point: the temperature at which the hydrogen bonding between the ethylene oxide moiety and the aqueous phase is cleaved causing the poly(ether) poly(siloxane) to drop out of solution, resulting in a turbid or "cloudy" liquid.

(a) Cloud point is an inverse temperature/solubility phenomenon;
(b) Generally, the chain length of the silicone backbone is not a determining factor;
(c) EO content is of minimal influence;
(d) EO chain length is the predominant determining factor with longer chains providing greater solubility and higher cloud points (Table 2);

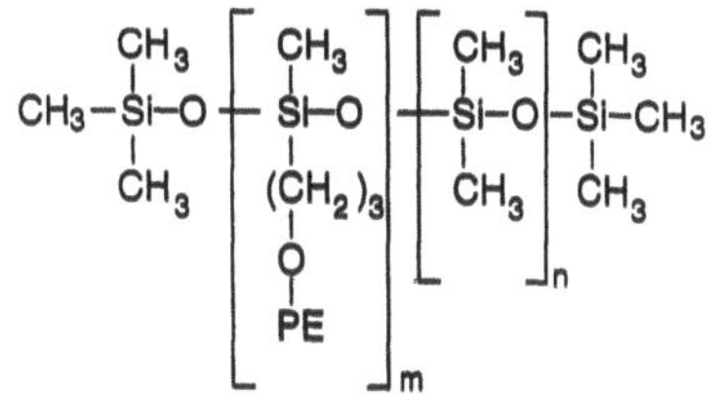

$PE = (-C_2H_4O)_x(-C_3H_6O)_y-H$

*CTFA: Dimethicone Copolyol*

Figure 7. Poly(siloxane)-, poly(ether)-copolymers.

Table 1. The relationship of ethylene oxide content to surface tension reduction

| Degree of Polymerization | #EO in Chain | %EO in Copolymer | 1.0% Aqueous Surface Tension |
|---|---|---|---|
| 5 | 7.5 | 43 | 21.2 |
| 5 | 7.5 | 54 | 21.5 |
| 5 | 17.5 | 64 | 25.7 |
| 5 | 17.5 | 74 | 25.4 |
| 10 | 7.5 | 43 | 23.2 |
| 10 | 7.5 | 55 | 23.0 |
| 20 | 7.5 | 43 | 24.9 |
| 20 | 7.5 | 55 | 25.3 |

(e) Below the cloud point, poly(ether) poly(siloxane)s tend to act as foam stabilizers. Above the cloud point they become defoamers.

While the cloud point of silicone copolymers in general should have little or no effect on the physical performance of a cosmetic or personal care formulation at the point of application, it may be used with advantage by the cosmetic formulator. For example, above their cloud point, these copolymers may function as antifoams. Because the copolymers exist as discrete aggregates or droplets (above the cloud point) which spread rapidly on the surface of a liquid film or bubble, they tend to drag away the original foaming liquid and replace it with the copolymer. As the copolymer is not capable of sustaining a film, the bubble ruptures, resulting in antifoam characteristics. In effect, the droplet of copolymer may be thought of as a discontinuity in the surface tension profile of the foaming liquid. The discontinuity allows for rapid bubble rupture - hence antifoaming behavior. This allows the copolymers to be used as processing aids where foaming hampers filling operations.

For example, suppose copolymer A exhibits a cloud point of 40°C in formulation X. Bottles normally are filled with this formulation at about 35°C, but severe foaming greatly reduces line speed. By adding a small amount of copolymer A to the formulation and raising the filling temperature to 42°C (just above the cloud point), the foam collapses rapidly, allowing a significant increase in line speeds. Below the cloud point,

Table 2. The relationship of ethylene oxide chain length and content to cloud points

| Degree of Polymerization | #EO in Chain | %EO in Copolymer | 0.1% Aqueous Cloud Point °C |
|---|---|---|---|
| 5 | 7.5 | 43 | 56 |
| 5 | 7.5 | 54 | 53 |
| 5 | 17.5 | 64 | 88 |
| 5 | 17.5 | 74 | 93 |
| 10 | 7.5 | 43 | 54 |
| 10 | 7.5 | 55 | 56 |
| 40 | 17.5 | 64 | 89 |
| 40 | 17.5 | 49 | 88 |
| 40 | 17.5 | 72 | 97 |

the copolymer rapidly redissolves and does not interfere with foaming properties up to the cloud point temperature. Thus, the copolymers are "transient" antifoams and avoid some problems caused by the more conventional antifoams which are completely insoluble in the foaming medium.

The cloud points may be used to advantage where a swirling, opalescent or opaque delivery is desired. For example, a clear bath oil containing a Dimethicone Copolyol of suitable cloud point will cause a bathtub of hot water to become cloudy. The bather will perceive a benefit due to the "milk bath" effect. At the same time, the bather will perceive a softening effect on the skin due to the surface tension reduction of the silicone copolymer at the skin/water interface. Finally, the bather should not feel an oily nature to the skin, because if properly selected, the copolymer should be water soluble at body temperatures. As the bather steps from the hot tub water, the skin surface water would cool rapidly to a point below the copolymer cloud point. Any copolymer deposited in bulk on the skin would redissolve and be toweled away. Because of its nature, some copolymer would remain in intimate contact with the skin resulting in a softening/smoothing effect without being oily or greasy. Similar effects would be noted also for hair care products especially shampoos.

In addition to functioning in this application due to the cloud point phenomenon, the subject copolymers may function as emulsifying agents to help disperse additional, but water insoluble, emollients. The nature of the hydrophobic and hydrophilic portions of these copolymers determines their emulsifying ability as well as their cloud points. Like traditional organic surface active agents, the copolymers exhibit a wide variety of emulsifying characteristics. However, unlike the organic materials, the behavior of the Dimethicone Copolymers cannot be predicted easily.

3. Wetting:

(a) Rapid wetting is best achieved with a very short silicone backbone;
(b) Long chain EO provides better solubility but poorer wetting characteristics compared to short chain EO;
(c) As with cloud point, EO chain length rather than %EO content is the determining factor.

Dimethicone Copolyols are well noted for their wetting properties in aqueous solutions. A variety of methods for determining wetting abilities are available including poly(ethylene) wet-out, teflon wet-out, contact angle measurements and the like. The degree of polymerization (silicone chain length) is a dominant factor. The shortest wetting times correspond to the shortest silicone chains. As the degree of polymerization increases, wetting time rapidly increases at first then levels out and begins to decrease at higher silicone chain lengths. It is interesting to note that, although the lower degree of polymerization copolymers provide for the lowest surface tensions along a homologous series, low surface tensions themselves do not insure the most rapid wetting. It appears, however, that packing efficiency plays a major role in wetting as it does in surface tension reduction. It is sufficient to state that the smaller, more easily and efficiently packed copolymers (low degree of polymerization) give the most rapid wetting in the Draves Test.

The differences in wetting time between the three series can be attributed to a solubility "block" effect similar to that already noted for cloud points and not to the ethylene oxide content, per se. Thus, the copolymers prepared with the 17.5 EO poly(ether) exhibit significantly

| Table 3. The relationship of ethylene oxide content to Draves wetting | | | |
|---|---|---|---|
| Degree of Polymerization | #EO in Chain | %EO in Copolymer | 0.1% Draves Wetting, sec |
| 10 | 7.5 | 43 | 125 |
| 10 | 7.5 | 55 | 44 |
| 10 | 17.5 | 64 | 712 |
| 20 | 7.5 | 43 | 272 |
| 20 | 7.5 | 55 | 157 |
| 20 | 17.5 | 64 | 1720 |

increased wet-out times in the Draves Test when compared with the copolymers prepared from the mixed EO/PO poly(ether)s or the 7.5 EO polymers. This fact is attributable to the increased solubility of the 17.5 EO based copolymers, which make them less available at the interface to promote wetting. The copolymers based on the EO/PO mixed poly(ether)s are less soluble, are more readily pushed out of solution and, hence, give better wet-out times (Table 3).

This "block" effect of poly(ether) structure is similar to that observed for surface tension reduction and cloud points. That is, the length of the ethylene oxide chain is significantly more important than ethylene oxide content in determining wetting. Higher ethylene oxide content, due to an increase in the number of side chains, does not increase the wetting times as does higher ethylene oxide content due to an increase in EO chain length (Table 4). This suggests, as in the case of cloud points, that the hydrogen bonding nature of the poly(ether) chain is an important factor in determining the wetting character of the silicone poly(alkylene oxide) copolymers.

Wetting is important to the cosmetics and personal care formulation in a number of applications. In the area of skin care, rapid wetting by lotions and creams is often desirable. Liquid coatings such as fingernail polish, mascara and perhaps lipsticks are easily and more uniformly applied if a wetting agent is incorporated. Penetration of shampoos, cream rinses and other hair treatments into thick hair is improved by

| Table 4. The relationship of ethylene oxide content to foam heights | | | |
|---|---|---|---|
| Degree of Polymerization | #EO in Chain | %EO in Copolymer | 1.0% Foam Heights Initial mm |
| 5 | 7.5 | 43 | 96 |
| 5 | 7.5 | 54 | 135 |
| 5 | 17.5 | 64 | 151 |
| 5 | 17.5 | 74 | 154 |
| 10 | 7.5 | 43 | 33 |
| 10 | 7.5 | 55 | 123 |
| 10 | 17.5 | 64 | 143 |
| 40 | 17.5 | 49 | 84 |
| 40 | 17.5 | 64 | 138 |
| 40 | 17.5 | 72 | 140 |

addition of these wetting agents. Rapid wetting and penetration of beards by shaving foams is observed by incorporation of the silicone copolymers in the formulation. Through out the cosmetic industry, one of the major uses for the subject copolymers is as wetting agents.

4. Solubility:

Degree of solubility is a factor of EO chain length with longer chains providing greater solubility.

5. Foam heights:

(a) Short silicone backbones generally provide greater foam heights;
(b) As the backbone increases from 3 to 20 silicones, foam heights steadily decrease;
(c) The decline of foam height is slower for backbones over 20 silicones until the polymer backbone increases significantly in length, at which point the foam height begins to increase again;
(d) Mixed EO/PO poly(ether)s generally yield poorer foaming characteristics due to the incompatibility of the PO moiety with the aqueous phase;
(e) Higher %EO content provide greater foam heights.

As stated above, the subject silicone polymers exhibit clearly defined cloud points, above which temperature they tend to act as antifoams. Because these copolymers are as likely to be used at temperatures below the cloud points, their foaming characteristics also are important to the cosmetic formulator.

The relationship of foaming characteristics to structural variations is exceedingly complex, but the data suggests that the degree of polymerization, percentage of ethylene oxide and poly(ether) structure play major roles. Along a homologous series, foam heights tend to decrease rapidly as the degree of polymerization increases to about 20. As the length of the silicone backbone increases further, the decline in foam heights is slowed and may eventually reverse at significantly long chain lengths.

A possible explanation for this observance is based on somewhat opposing ideas. The first, noted under the Surface Tension Section, suggests that the smaller molecules pack more uniformly at the liquid/air interface. The efficient packing results in lower aqueous surface tensions for the smaller molecules. At the same time, such packing would allow for more stable film formation and greater foam heights. At the other end of the scale, the higher molecular weight copolymers should be able to stabilize foam better due to "bulk/size" effects. Drainage of the films would be reduced simply due to the sheer bulk of the copolymers. The result would be greater film stability and foam heights. A combination of these two opposing trends results in the observed foam height profiles - initial rapid decrease followed by a leveling off or gradual increase in foam height.

The "bulk/size" effect may be used to explain the relatively flat response of foam height towards degree of polymerization in the mixed EO/PO poly(ether) systems. Because of the high molecular weight of these poly(ether)s, the molecular weight of the copolymers increases rapidly along the series. As a result, decreased drainage of the films due to copolymer bulk comes into play at a lower degree of polymerization than all EO based copolymers. This contribution combined with the decrease in foam stabilization with increasing degree of polymerization results in an observed flat response.

It should also be noted that these mixed EO/PO based copolymers produce inferior quality aqueous foams when compared with the all EO based systems. The former give light, course foams while the latter give dense, fine celled foams. It appears that these copolymers are only weakly capable of sustaining such aqueous foams. It is suggested that the cosmetic formulator, concerned with the generation of dense high quality foams, investigate the use of all EO based copolymers initially.

One unusual, and potentially useful property was displayed by the mixed system, however. At low degrees of polymerization, the generated foams rapidly collapsed within five minutes. Such collapse was not observed with any of the all EO based copolymers. This suggests these materials have some inherent defoaming ability even at temperatures below the cloud point. That this defoaming behavior disappears, as the degree of polymerization is increased, is consistent with the idea that increases in copolymer bulk contribute to foam quality and stability by reducing drainage.

Why these defoaming and poor foam characteristics should occur at all probably is associated with the additional presence of the hydrophobic propylene oxide moieties at the air/water interface. This tends to interfere with the ideal orientation of the silicone at the surface, resulting in weakly stabilized foams of poor quality throughout the series. At lower degrees of polymerization, the packing of these copolymers at the surface is insufficient to stabilize the weak foams generated. Likewise, copolymer bulk is insufficient to stabilize even these poor quality foams by retarding drainage. As a result, defoaming character is observed at the lower degrees of polymerization and the foam collapses. The higher molecular weight materials, on the other hand, are sufficiently bulky to retard drainage and preserve even these poor quality foams. This suggests that copolymers with low degree of polymerization prepared with high molecular weight mixed EO/PO poly(ether)s might be used effectively as mild defoamers for certain formulations; defoamers which are totally compatible in that system. The formulator would not have to worry about separation due to insoluble components generally associated with traditional antifoams or defoamers. At the same time, he would maintain benefits generally associated with the silicone poly(alkylene oxide) copolymers such as low surface tension, rapid wetting, a non-oily silky feel and water solubility. For applications where foam either is not necessary or is undesirable, this approach may provide significant benefits.

The quality of foam generated by the all EO based copolymers is superior to that generated by the mixed EO/PO materials. Regardless of foam height, which varies widely, the foams are dense and fine in cell structure. Because there is no disruption of the surface film by propylene oxide hydrophobes, the all EO materials are better able to sustain high quality foam structures.

However, ethylene oxide content does exert a direct influence on the ultimate foam heights of all EO copolymers (see Table 4). As a simple rule, the higher the ethylene oxide content, at constant degree of polymerization, the greater the foam height attained. This is the same as saying that the greater the water solubility of the copolymers, the greater the foam height. Note, this is fundamentally different than the cloud point relationship discussed earlier. Water solubility correlates directly with the amount of water held at the air/liquid interface. Greater water solubility implies a greater ability to maintain water at the interface. The copolymers act to pull more water into the film structure resulting in stronger, more uniform cells which can support physically a greater film height. Thus, in all cases, an increase in ethylene oxide content results in a greater foam height.

It has been demonstrated that all EO based silicone copolymers containing greater than 60% ethylene oxide are at least good profoamers. Further, those copolymers satisfying the above requirements, and which have a low degree of polymerization, are superior foaming agents which can compete directly with traditional organic profoamers in applications requiring dense, high quality foams. Use of these silicone poly(alkylene oxide) copolymers in such foaming applications is also likely to impart improved wetting and foam penetration, improved lubricity, and improved feel of the skin or hair surface-properties generally associated with silicones. Thus, these types of copolymers are suggested for use in aerosol shave creams and shampoo formulations.

6. pH Stability:

(a) Poly(ether) poly(siloxane)s containing numerous short EO side chains provide the greatest pH stability, (range 3-11);
(b) pH stability is a factor of concentration of copolymer in solution and wettability;
(c) Si-O-C bonded copolymers and short chain Si-O bonded copolymer (tri-siloxanes) are subject to hydrolytic attack and are, therefore, stable only in the range of 6.5-8.5.

The general physical properties of Dimethicone Copolymers are illustrated in Table 5.

The essential feature of the Dimethicone Copolymer surfactants compared to purely organic surfactants is the much higher surface activity provided by the poly(dimethylsiloxane) moiety. For this reason, the Dimethicone Copolymers are more effective surface tension depressants and provide superior wetting and lubricating properties than hydrocarbon based organic surfactants. To illustrate the surface activity of Dimethicone Copolymers, Table 6 gives a comparison of the surface tensions of 1% solutions of various types of surfactants in water.

The outstanding surface activity of Dimethicone copolymers is realized both in aqueous and organic solutions. Dependent on their overall poly(dimethylsiloxane)/poly(oxyalkylene) ratio and on the ethylene oxide/propylene oxide ratio in the poly(ether) chains, the solubility behavior of the Dimethicone Copolymers in water, alcohols and other organic solvents can be varied over a wide range. Starting with a certain poly(dimethylsiloxane) moiety, water soluble copolymers which are insoluble in mineral oil are obtained by linking it to poly(ether)s predominantly

Table 5. Physical properties of Dimethicone Copolyols

| | |
|---|---|
| Appearance | colorless to slightly colored, viscous liquids |
| Viscosity, 25°C, $mm^2/s$ | 20-2000 |
| Specific gravity, $g/cm^3$, 25°C | 1.02-1.07 |
| Flash point, DIN 53213 | >80°C |
| Cloud point, 1% in water | 0-100°C |
| Surface tension, dynes/cm | 21-25 |

| Surfactant (1% aq.) | Surface tension dynes/cm |
|---|---|
| Pure water | 72.8 |
| Anionic surfactants | 30-36 |
| Nonionic surfactants | 29-35 |
| Poly(siloxane)-Poly(ether) copolymers | 22-29 |

Table 6. Surface tension of various surfactants in 1% water solutions

composed of ethylene oxide. On the other hand, copolymers which contain poly(ether)s consisting mainly of propylene oxide are easily soluble in oils but less soluble in water.[11]

Table 7 demonstrates the solubility behavior of Dimethicone Copolymers, consisting of a linear poly(dimethylsiloxane) moiety with an average molecular weight of about 1800 and four pendent poly(ether) chains with different ethylene oxide/propylene oxide ratios in different solvents.

Provided a copolymer is sufficiently soluble, it can act as a foam builder in both aqueous and organic phases. If, on the other hand, the solubility is low or when, as a result of the inverse solubility behavior of the copolymers in water, their solubility is reduced by warming aqueous solutions, less foam is formed and it is even possible to obtain foam suppressing properties.

The chemical modification of poly(dimethylsiloxane)s with pendent poly(ether) groups does not change the basic function attributed to the poly(siloxane) moiety. However, due to the solubility imparted by the poly(ether) part, the Dimethicone Copolyols can perform their beneficial effects in a broader range and without the drawbacks of pure poly(dimethylsiloxane)s. The copolymers lubricate without oiliness and provide softness and gloss properties without being greasy and tacky.

Furthermore, the Dimethicone Copolyols containing ethylene oxide units in their poly(ether) part are able to exert intermolecular bonding forces with polar organic groups like amide or ester structures. These bonding forces can contribute to viscosity building.

Depending on the ethylene oxide content, they adhere, therefore, more

Table 7. Solubility of Dimethicone Copolyols as a function of the composition of the poly(ether) part

| Solvent | Poly(ether) in the copolymer (mole% EO/PO) 100 EO | 50/50 EO/PO | 20/80 EO/PO |
|---|---|---|---|
| Water | soluble | soluble | insoluble |
| Ethanol | soluble | soluble | soluble |
| Mineral oil | insoluble | insoluble | soluble |
| Lanolin | insoluble | insoluble | soluble |
| Stearyl alcohol | soluble | soluble | soluble (70°C) |

or less strongly in these organic substrates. However, as the bonding forces are less strong than with cationic surfactants, the Dimethicone Copolyols do not build up and are easily washed out by water.

The "Hydrophile-Lipophile Balance", or HLB, is often used as a selection criterion for predicting the effectiveness of a nonionic surfactant or emulsifier in a given application. Since the experimental determination of an HLB value for a particular surfactant is quite tedious and somewhat system-dependent, several procedures for estimating the HLB have been devised.[16]

The most commonly used method involves dividing the weight percent of ethylene oxide (EO) in the nonionic surfactant by five. This method is acceptable for simple hydrocarbon (alkyl, aryl, or aralkyl) alcohols that have been modified with EO. This method does not, however, consider the nature of the hydrophobic species or its functionality. Nor does it consider variations in the hydrophobic nature of the attached poly(ether) because of its molecular weight distribution, the presence of propylene oxide (PO) units, and the presence of a capping group in the poly(ether).

The unique structure of Dimethicone Copolymers includes most of the aforementioned exceptions. Consequently, HLB ranges for these products have been estimated from their aqueous solubility and cloud point, using the method described by Griffin and later by Schott.[16,17] These guidelines should serve as a good starting point for applications for which the HLB range of effective emulsifiers have already been established.

The Dimethicone Copolyols can be soluble in ethanol, stearyl alcohol and water; as a result, they have applications in nondurable skin care formulations where they provide the necessary silicone properties, then wash off. They may serve as lubricants in nondurable skin creams and lotions, as wetting agents in aerosol shaving lather, and as emollients in liquid conditioning soaps. And, also because of the solubility, many Dimethicone Copolyols are useful in shampoo formulations requiring clarity.

A use level of 0.1 to 1.0 weight percent Dimethicone Copolyol is sufficient to provide lubricating and wetting characteristics to skin lotions. From 1.0 to 5.0 percent improves (decreases) wetting time and adds lubricating and foam stability properties to shaving products. In hair styling mousses, levels ranging from 0.1 to 1.0 percent make products spread easily and reduce resin tack. Addition of silicone glycol can lower the surface tension of a water-based mousse to 28 dynes/cm, the critical surface tension needed to wet virgin hair.[18,19] These silicones have also found use in ethnic hair care products where they add nongreasy lubricating and softening qualities, act as humectants, and help reduce tackiness.

In skin preparations, the Dimethicone Copolyols enhance the water repellency and improve the holding of humidity because of the strong interaction between the ethylene oxide units and water. This is done without an adverse influence on the natural skin transpiration.

Like pure poly(dimethylsiloxane)s, the Dimethicone Copolyols are tasteless, oxidation-resistant and have little odor. They have been found to be nonirritating, nonsensitizing, and nontoxic. These properties and their unique behavior as surfactants make the Dimethicone Copolyols a very versatile and valuable group of personal care chemicals. The Dimethicone Copolyols are recognized as safe for use in cosmetics and, in fact, studies show that their inclusion in anionic surfactant based products can reduce both eye and skin irritation.[1,20]

## DIALKYL AND POLY(ALKYL SILICONE) COPOLYMERS

Poly(siloxane)s have also been modified by the addition of alkyl side chains. These are also considered to be surfactants, as they can reduce the surface tensions of pure organic systems such as mineral oil. These modifications are referred to as silicone waxes. The poly(dimethylsiloxane)s modified with long-chain alkyl groups are at ambient temperature liquid or solid compounds depending on the alkyl side chain length of the modifying group R. Silicone waxes may be linear or comb-like, the alkyl groups is attached to the poly(dimethylsiloxane) by way of Si-O-C or Si-C bonds (Figure 8).[21,22] These silicone waxes are compatible with mineral oils and other types of hydrocarbons. The starting point for these organo-modified silicones are Si-Cl- or Si-H- functional poly(siloxane)s which are converted by reaction with fatty alcohols or propoxylated alkyl phenols or by the addition of alpha-olefins.

These products are compatible in oils and waxes and form excellent additives to improve the glide properties of personal care formulations and other industrial applications.[23] Silicone waxes are used in creams and lotions as well as in water free decorative products such as lipsticks, pressed powders, eyeshadows and mascaras. In lipsticks, $C_{22}$ dialkoxy poly(siloxane)s have been demonstrated to improve the spreadability, gloss and feel on the lips. The $C_{22}$ modification is used due to its melting point of about 35°C, which is the temperature of skin. This $C_{22}$ wax also can be used to thicken low molecular weight poly(siloxane)s and gel the higher molecular weight ones. Smaller $C_{16}$ and $C_{18}$ modified poly(siloxane)s show a unique compatibility with cyclic $D_4$ and $D_5$ poly(siloxane)s which then allows them to have greater compatibilities in personal care products. The silicone waxes also find use as hair glossing and conditioning agents in emulsified hair conditioners.

Comb-like silicone waxes with alkyl groups higher than $C_{18}$, when n is low and m is greater than 40, are efficient viscosity builders for mineral oils and mineral spirits (Figure 9). As shown in Figure 9, the viscosity of a thickened oil decreases with increasing shear rate. At temperatures higher than 45-50°C, the viscosity building effect is lost. At higher levels of silicone wax, mineral oils can be gelled.

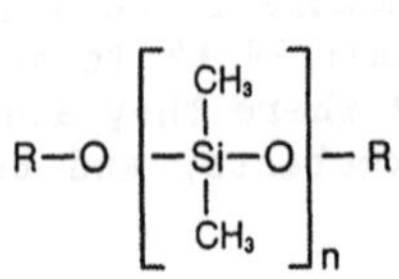

| R | n | Melting Point (°C, ± 5) |
|---|---|---|
| $C_{12}H_{25}$ | <10 | -30 |
| $C_{18}H_{37}$ | <10 | +25 |
| $C_{22}H_{45}$ | <10 | +35 |

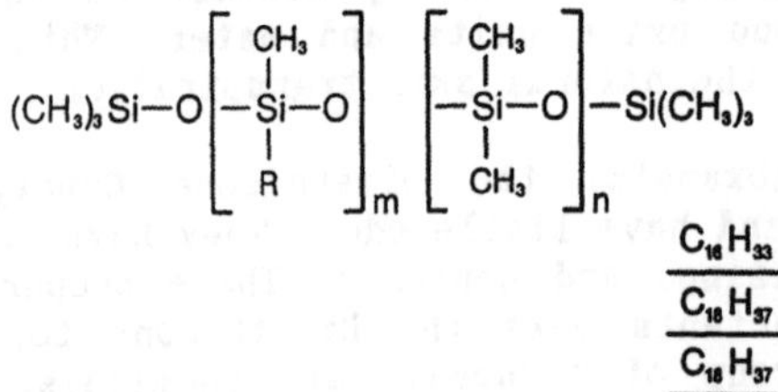

| R | m | n | Melting Point (°C, ± 5) |
|---|---|---|---|
| $C_{16}H_{33}$ | <5 | <5 | +10 |
| $C_{18}H_{37}$ | <5 | <5 | +20 |
| $C_{18}H_{37}$ | >40 | <5 | +40 |
| $C_{26}H_{53}$ | >40 | <5 | +60 |

Figure 8. Linear and comb-like silicone waxes.

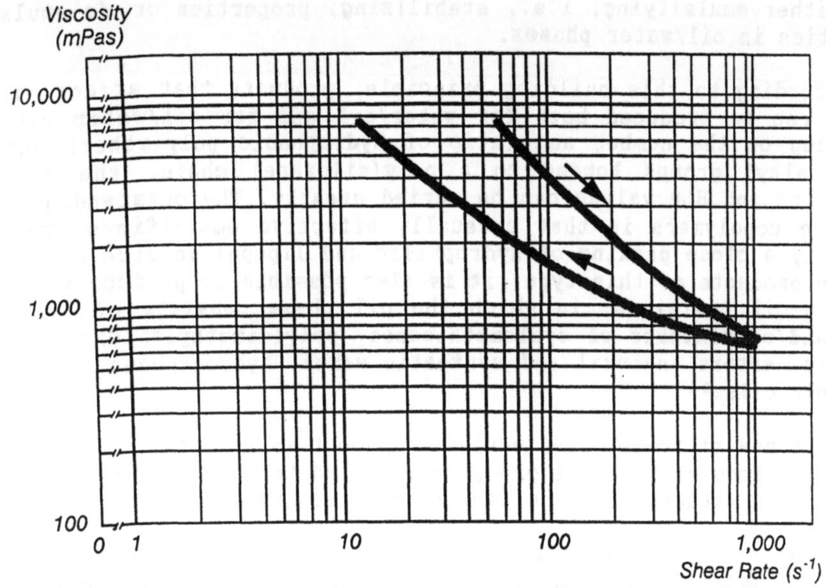


Figure 9. Viscosity of paraffinic oil (150 mPas), thickened with 5% silicone wax (melting point 60°C).

The melting point of silicone waxes is mainly governed by the alkyl chain length R, bonded to the silicone. Silicone waxes where R is $C_{16}$ or lower are liquids at ambient temperature, and those with R = $C_{18}$ or higher are semisolids or solids.

The silicone waxes demonstrate compatibility with many esters and oils commonly used in the personal care market, as well as with many natural oils and fatty triglycerides. Their use is expanding rapidly not only in cosmetic products, but also in pharmaceutical products due to their exceptional toxicity profiles.

## POLY[(ALKYL) POLY(ETHER) POLY(SILOXANE)] COPOLYMERS

Another obvious development that exists is the attaching of both poly(ether) and poly(alkyl) or alpha-olefins to one poly(siloxane) chain. The copolymers formed (Figure 10) can be designed in such a way that they

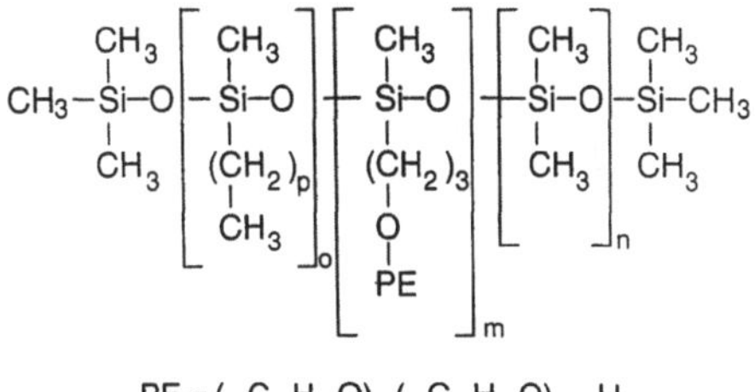

$PE = (-C_2H_4O)_x(-C_3H_6O)_y-H$

*CTFA: Cetyl Dimethicone Copolyol*

Figure 10. Poly(siloxane)-, poly(alkyl)-, poly(ether)-copolymers.

have either emulsifying, i.e., stabilizing, properties or de-emulsifying properties in oil/water phases.

According to this building principle, products that affect emulsification can be produced both for water/oil and for oil/water emulsions. Depending on the number and ratio of hydrophobic poly(ether) and lipophilic alkyl groups bonded to a poly(siloxane) chain, the solubility properties and HLB values can be varied greatly. The outstanding feature of these copolymers is that unusually effective emulsifiers can be designed by a close packing of hydrophilic and lipophilic side groups. With certain products of this type, it is also possible to produce very stable water-in-oil emulsions, in which the oil phase consists of cyclomethicones and/or mixtures of cyclomethicones, poly(dimethylsiloxane)s, mineral oils, esters, natural and synthetic waxes, hydrocarbons, and natural vegetable oils.[24]

These new silicone based emulsifiers containing both poly(ether) side chains and long chain alkyl groups are called ternary copolymers.[2] Figure 11 shows the structural principle of such ternary copolymer.

At the oil/water interface, the poly(ether) groups are dissolved in the water phase, whereas the alkyl groups are dissolved in the oil phase. In between are the poly(dimethylsiloxy) units, which are at certain chain length neither oil-nor water soluble and, thus, force the molecule to be ideally fixed at the water/oil interface. To some extent ternary copolymers (as shown in Figure 11) form, in combination with water and oil, a 3-phase system: water phase, silicone phase, oil phase.

In addition, the oligomeric character of the emulsifier results in the copolymer not being as easily desorbed out of the interface as monomeric compounds are. Emulsifiers based on poly(ether siloxane)s with long chain alkyl groups enable the preparation of very stable creams and lotions. Stabilizing hydrocarbon waxes do not have to be used, thus permitting production of emulsions in a cold/cold process.

These emulsions based on this silicone polymer tend to form a film on the skin. During the application and rub out of the emulsion, there is some rapid evaporation of the water as well as some absorption of the water through the skin. When properly formulated for stability and aesthetics, water-in-oil emulsions can offer their benefits to a number of personal care products. In protective creams, films formed by water-in-oil emulsions can help guard against water soluble irritants and help prevent chapping and dehydration of the skin. Emulsions designed for cleansing creams can be formulated with mineral oil or a combination of mineral oil and cyclomethicone. Emulsions of this type break on the skin and the oil phase easily emulsifies dirt and sebum. When formulated as

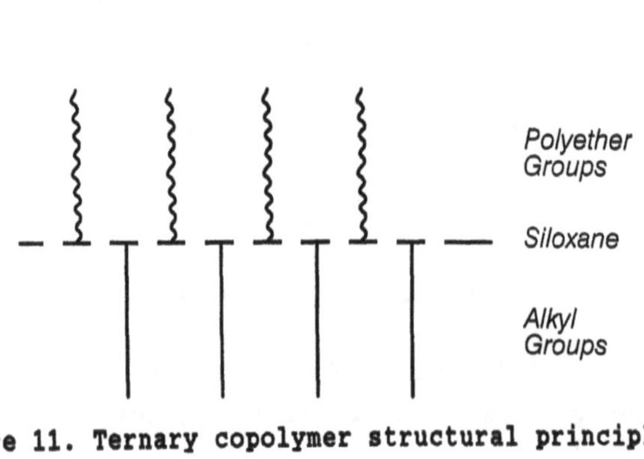


Figure 11. Ternary copolymer structural principle.

moisturizers, water-in-oil emulsions provide both water and oils for dry skin. They can be formulated to form occlusive films that retard trans-epidermal water loss. Also, the lubricating properties of the external phase make water-in-oil systems suitable for baby products, particularly lotions that prevent irritation and are resistant to urine.[25]

In sunscreens, the water-resistant films of water-in-oil systems give better protection than oil-in-water systems. And, because, they are not hygroscopic, they do not readily take up and retain moisture. They form a continuous film with better resistance to wash-off.

Silicone emulsions allow formulators to prepare stable water-in-oil formulations that provide maximum moisturizing and protective benefits as well as excellent aesthetics. Emulsions made with the copolyol have a lighter, less greasy and nonwaxy feel compared to conventional water-in-oil systems. And, when formulated properly, the formulations do not separate or lose viscosity. Emulsions formulated with poly[(alkyl) poly-(ether) poly(siloxane)] copolyols are very white and glossy, without a waxy appearance. These emulsions can be cost effective because of their high water content. In emulsions based on mineral oil or petrolatum, addition of the silicone as a primary emulsifier at levels as low as 2% or less by weight, helps form stable emulsions with water content as high as 82%. Formulations of this type require as little as 2% coemulsifier. A small amount of sodium chloride, magnesium sulfate, sodium citrate or other electrolyte in the aqueous phase also is required for emulsion stability.[26]

Occlusive formulations, usually based on petrolatum, are used in treating chronic conditions where the stratum corneum may be dry and thickened. Occlusion hydrates the stratum corneum to benefit dry skin and enhance drug permeation. Many drugs are soluble in petrolatum or mineral oil, and they can also act as permeation enhancers.[27] Occlusion therapy is typically used with anti-inflammatory steroids to enhance clinical response. Water-in-oil emulsions, and particularly those based on these silicone copolyols as an emulsifier, can provide the necessary long-term occlusivity required for topical drug delivery. The improved aesthetics of these water-in-oil formulations can enhance patient compliance, and the stability of formulations based on the silicone emulsifier helps ensure longer shelf life for pharmaceuticals.[28]

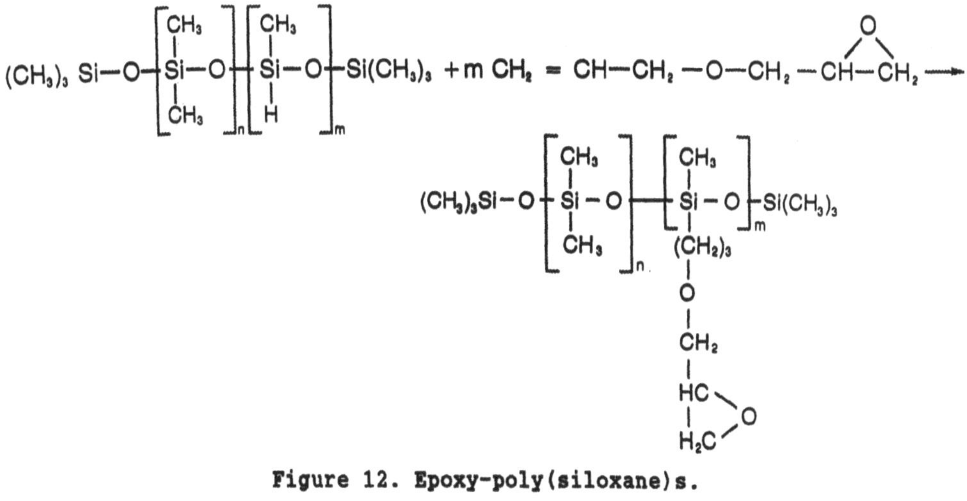


Figure 12. Epoxy-poly(siloxane)s.

The poly[(alkyl) poly(ether silicone)] polymers are being widely investigated for use in antiperspirants. The emulsions can be used to deliver the actives in a more usable state, as they are already solubilized.[29] The actives are further held to the skin via the occlusive layer which reduces friction and the resulting irritation. The tremendous stability of these W/O systems avoids many rheological problems.[30]

Use of oil-in-water and water-in-oil emulsions offers flexibility to formulators of personal care products. Important parameters for formulation of these emulsions are the proportions of the two phases, type of oil used, and amount and choice of emulsifying agents. Although products based on oil-in-water systems have typically been more popular with consumers because of their aesthetic qualities, water-in-oil emulsions can provide greater protection and moisturizing benefits. New silicone emulsions allow formulators to combine the best features of both emulsion types. These emulsions now allow formulators broader utility for topical creams and lotions formulated for delivery of drug products.

## IONIC ORGANO-POLY(SILOXANE) COPOLYMERS

The majority of silicone polymers in use in industry today are of the nonionic type. It is also possible to bond ionic side chains to poly(dimethylsiloxane) either additionally or exclusively. This opens up a completely new range of property combinations that is still untapped for the most part. These are relatively new development and are still in the developmental stage, however, new uses or applications are being developed each day.

One method of synthesis that can be used to produce ionic silicone polymers is via epoxy siloxanes by reaction with special nucleophiles (Figures 12,13).

The attractiveness of copolymers having ionic side groups lies certainly in their ability to form far more intense physical-chemical interactions with liquid or solid substrate phases compared to the nonionic products.

Cationic and amphoteric groups lead to an increased substantiveity on polar substrates, providing a very durable covering of surfaces. The products, thus, have a potential to not only form protective films, but also act as emulsifiers for finely dispersed polar particles in organic media. Additionally, poly(methylsiloxane)s modified with cationic or amphoteric groups combine a highly gliding ability with marked antistatic properties, which are particularly interesting for personal care and textile applications. Poly(ether) side chains can also be included in addition to the ionic groups to achieve a solubility behavior corresponding to the individual requirements.

The reaction of secondary amines with epoxy siloxanes results in tertiary amino siloxanes, which when reacted further with sodium monochloroacetate produce silicone betaines (Figure 14).[31,32] If the epoxide ring is reacted with tertiary amines in the presence of acids, e.g., acetic acid, the silicone quats are obtained (Figure 15).[33,34]

Silicone betaines and silicone quats are excellent additives in hair care preparations.[35] The silicone quats are more highly substantive than the silicone betains. These polymers show good compatibility with anionic surfactants and, therefore, are suitable as additives in shampoos. Silicone betaines and silicone quats are good antistatics. They improve the

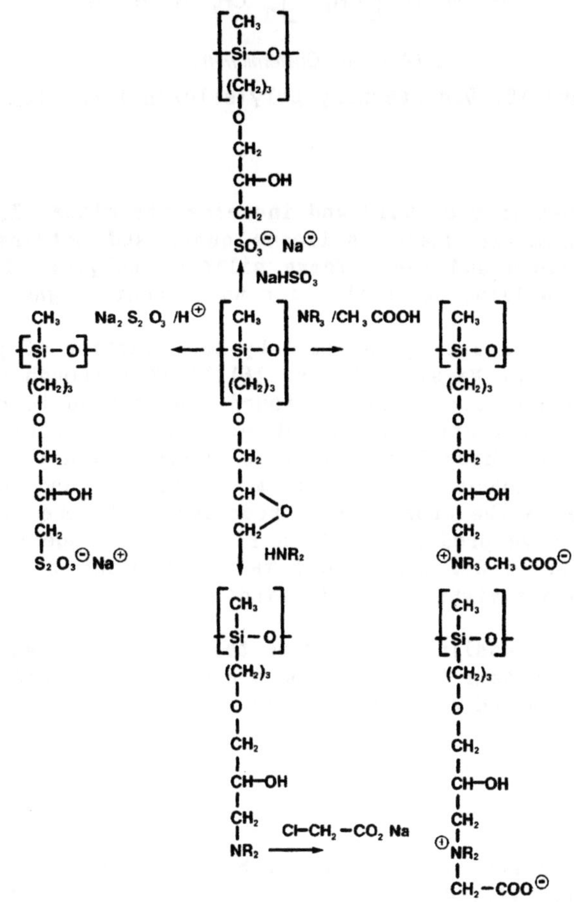

Figure 13. Synthesis of ionic organo-poly(siloxane)s.

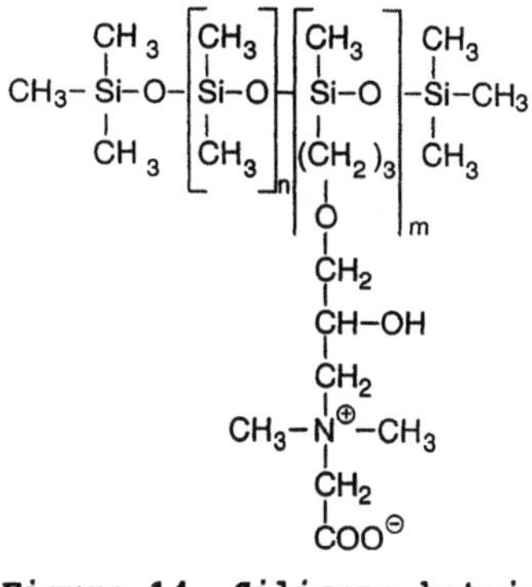

Figure 14. Silicone betaine.

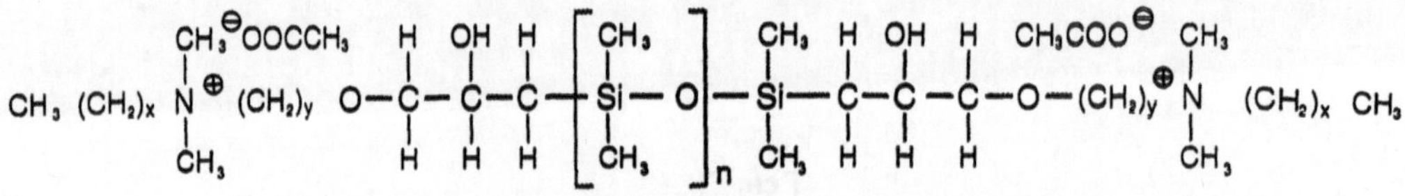

*CTFA Name: Quaternium-80*

**Figure 15. Diquaternary poly(siloxane) copolymer.**

wet and dry combability of hair and increase its gloss. In comparison to pure organic polymeric quats, silicone quats and betaines give hair a silky silicone gloss and feel. These silicone polymers do not have the disadvantage of building up on the hair as do their organic counterparts.

Anionic siloxanes are synthesized by the reaction of poly(siloxane)s with either $NaHSO_3$ or $Na_2S_2O_3$ (Figure 16).[36,37] Although the corresponding thiosulfates have not until now found application in cosmetic preparations, the thiosulfates are a subject of the latest application research. Silicone thiosulfates can be crosslinked via S-S bonds by reacting with thioglycolic acid, sodium sulphide or cystein on the hair (Figure 17). Due to the crosslinking reaction, silicone thiosulfates may be used in cold wave products as thin film forming agents. In permanent wave products, mixed S-S bonds to human hair might be possible, which would lead to wash resistant conditioning.

Also, the non-crosslinking silicone thiosulfate was shown in preliminary experiments on hair tresses to be a good anionic conditioning agent, especially for bleached and damaged hair.

## CONCLUSION

Recent developments in silicone technology have introduced silicone polymers with specialized physical properties that can enhance personal care products. Poly(dimethyl organo siloxane)s can be specifically modified towards a desired application by the choice of a suitable organic moiety. Hydrophobic derivatives find application in emulsions or nonaque-

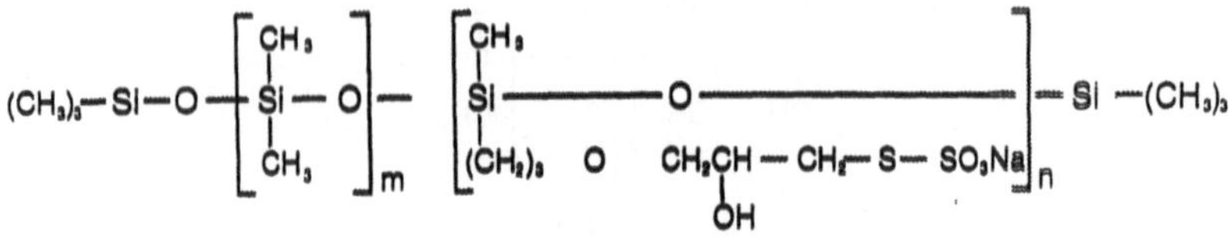

*CTFA Name: Sodium Poly PG-Propyl Dimethicone Thiosulfate*

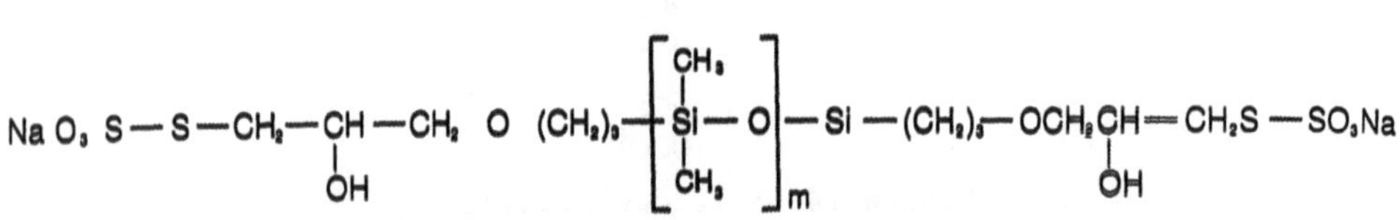

*CTFA Name: Disodium Poly PG-Propyl Dimethicone Thiosulfate*

**Figure 16. Poly(siloxane) poly(thiosulfate), sodium salt (bunte salt).**

$$n \equiv Si-R-S-SO_3^{\ominus} \xrightarrow{[Cat]} \frac{n}{2} \equiv Si-R-S-S-R-Si \equiv$$

**Figure 17. Crosslinking of silicone thiosulfates.**

ous preparations. Hydrophilic nonionic and ionic products, on the other hand, find uses in aqueous systems. In skin care products, improved silky feel is obtained by the addition of a suitable silicone compound. In hair treatments, improved wet and dry combability, improved silicone specific gloss, combined with a silky feel completes the desired properties. Silicone polymers in pharmaceutical and cosmetic bases can produce superior emulsions, contribute to barrier protection of the skin and maximize the delivery of other active substances to the skin.

Silicone polymers have low toxicity profiles and their inclusion in many personal care products has shown that the resulting formulation itself has a better toxicity profile. As the potential uses for silicone polymers expand, questions will certainly arise regarding their environmental safety. The environmental fate of silicone fluids has been evaluated by a variety of studies.[38-40] These studies show that silicone and silicone polymers can degrade in the environment.

There is a large number of chemical methods available to modify silicones, which even in this article cannot be totally covered. It is quite obvious that further product optimization will take place with the ever increasing requirement for silicone compounds in personal care products.

## REFERENCES


1. A. DiSapio and P. Fridd, Preprints 15th IFSCC Congress, 1, 89 (London, 1988).
2. D. Schaefer, Preprints 15th IFSCC Congress, 1, 103 (London, 1988).
3. M. S. Starch and C. R. Krosic, Cosmetic Technology, 20 (November, 1982).
4. Final report of the safety assessment for dimethicone copolyol, prepared by The Expert Panel of the Cosmetic Ingredient Review (April, 1981).
5. W. Noll, "*Chemistry and Technology of Silicones,*" Academic Press, NY, 1968.
6. I. Yilgor, et al., Polymer, 25(1), 800 (1984).
7. J. L. Hedrick, et al., Polymer Preprints, 27, 203 (1986).
8. I. Yilgor, et al., ACS Symposium Series No. 282, Chapter 14, 1985.
9. R. Meals, "*Encyclopedia of Chemical Technology,*" 18, 2nd edition, p. 221, 1969.
10. S. R. Wendell and A. J. DiSapio, Cosmetics and Toiletries, 98, 103 (1983).
11. W. Wolfes, Parfuemerie und Kosmetik, 68, 195 (1987).
12. DP 1595730 Th. Goldschmidt AG.
13. DP 3133869 Th. Goldschmidt AG.
14. DP 2533074 Th. Goldschmidt AG.
15. W. C. Davis and D. P. Jones, Polymer Preprints, 11, 447 (1970).
16. W. C. Griffin, Off. Dig. Fed. Paint and Varnish Production Clubs, 28, 446 (1956).
17. H. Schott, J. Pharm. Science, 58, 1443 (1969).
18. Y. K. Klamath, et al., J. Appl. Polymer Science, 29 (3), 1011 (1984).

19. M. S. Starch, Drug and Cosmetic Industry, 138, 38 (June, 1984).
20. R. L. Goldemberg, J. Soc. Cos. Chem., 30, 415 (1979).
21. USP 4417068 Th. Goldschmidt AG.
22. USP 4609752 Th. Goldschmidt AG.
23. USP 4275101 RCA Corp. USA.
24. USP 4698178 Th. Goldschmidt AG.
25. D. G. Krzysik and I. Van Reeth, Drug and Cosmetic Industry, 144, 28 (April, 1990).
26. P. Hameyer and C. Gould, Manufacturing Chemist, 20 (January, 1990).
27. W. R. Pfister and D. S. T. Hsieh, Pharmaceutical Technology, 14 (9), 132 (1990).
28. H. Tsutunei, et al., J. Soc. Cos. Chem., 30, 345 (1979).
29. D. T. Floyd, J. Soc. Cos. Chem., 29, 717 (1978).
30. D. T. Floyd, Cosmetics and Toiletries, 96 (10) 21 (1981).
31. USP 4609750 Th. Goldschmidt AG.
32. USP 4654161 Th. Goldschmidt AG.
33. USP 4833225 Th. Goldschmidt AG.
34. USP 4891166 Th. Goldschmidt AG.
35. H. Meyer, Cosmetics and Toiletries Manufacturers & Suppliers, Issue XV, 5 (January 1990).
36. USP 3392182 Th. Goldschmidt AG.
37. USP 4537595 Th. Goldschmidt AG.
38. C. L. Frye, Soap, Cosmetics, Chemical Specialties, 59, 32 (August 1983).
39. C. Anderson, et al., Chemosphere, 16 (10-12), 2567 (1987).
40. R. R. Buch and D. N. Ingrebrightson, J. Environ. Science, 13, 676 (1979).

# EFFECTS OF WATERPROOFNESS ON THE CLINICAL EFFICACY AND IRRITATION POTENTIAL OF SUNSCREEN PRODUCTS

Jonas C. T. Wang*, Jue-Chen Liu, Victoria Dole and Chung-ye Tseng

Johnson & Johnson Consumer Products, Inc.
Grandview Road
Skillman, NJ 08558

This study explored the effect of waterproofness, by using five commercially available polymers, on the clinical efficacy and irritation potential of sunscreen products. A marketed o/w SPF 30 sunscreen cream was selected as the model vehicle. To achieve the SPF 30 of the sunscreen formula, 7.5% octyl methoxycinnamate (Parsol MCX), 4.0% oxybenzone, and 3.0% octyl salicylate were the sunscreens chosen. Ganex V-220, Pemulen TR1, PVP K-90, Polymer X, and Carboset XPD1616 were the polymers chosen in the study. The study results suggested that the release of sunscreens from the waterproof formula is the rate-limiting step in absorption process. It also suggested that the waterproofness may provide an alternative way of reducing the irritation potential of the sunscreen products, without detriment to its clinical efficacy.

## INTRODUCTION

For many years, sunscreen products have been formulated to provide protection from UV radiation damage. To deliver an adequate protection to the consumers, the sunscreen products must be designed in such a way that, after application, the sunscreen agents should remain on the skin for a reasonable period of time under conditions of actual use, e. g., swimming, exercising, and/or perspiration. In addition to the initial static efficacy of the products, the resistance of the sunscreen actives to removal by wash-off is also an important feature of which the sun protection benefits can be warranted during actual use.

Conventionally, an anionic soap emulsifier system, in conjunction with a film former polymer, was used to yield waterproofness of the sunscreen products. R. S. Berger, et al.,[1] described the use of an acrylate film-forming polymer solubilized by forming the ammoniated salt in an anionic soap emulsion. When the sunscreen preparation dried on the skin, the ammonia evaporated, fixing the insoluble polymer to the skin. As a result of this "fugitive amine" transformation, the film was insoluble in water and resistant to rub-off. Ideally, the substantive film formed does not interfere with transepidermal water loss (TEWL) of normal sweat gland

*Cosmetic and Pharmaceutical Applications of Polymers*
Edited by C.G. Gebelein *et al.*, Plenum Press, New York, 1991

function. However, this type of formula has been reported to be irritating to the skin and/or stinging to the eyes and face, as a result of either the soap ingredients, the enhanced absorption of sunscreen by the soap system, or the combination of both.

For the last five years, a number of polymers materials have been utilized as alternative systems for waterproofness as well as skin protection of sunscreen and skin care products. This study was designed firstly to evaluate the effectiveness of several commercially available waterproof polymers used in the sunscreen products. The polymer systems selected in this study were tested for their sun protection efficacy by both *in vitro* and *in vivo* SPF determination.

This study was also designed to assess the safety of the waterproof sunscreen products with higher SPF (30 or higher) products in comparison with regular sunscreen products. Despite questions by the Food and Drug Administration as to the need for SPF numbers higher than 15, which the FDA considers a total sun block, sunscreen products with SPF numbers as high as 39 were still introduced into the market in 1988. The higher SPF sunscreen products were derived on the belief that extra-protection may be needed for patients with sensitive skin, the actual protection is less than the SPF number on the bottle, and that decreasing protection occurred due to washing off in water or perspiration. However, regulatory authorities, as well as dermatological and scientific communities, do not agree with these reasons because of the high irritation potential and the systemic safety concern of the use of the high SPF sunscreen products. Since those sunscreen actives have also been known to be irritating and even carcinogenic at higher doses.

In a previous study,[2] we observed that a measurable amount of the sunscreen active agent permeated through the skin membrane from the regular sunscreen products. It was believed that the irritation of the sunscreen could be associated with the percutaneous absorption of these actives.[3] Therefore, the second objective of this study was to assess and determine if the addition of the waterproof polymers to a higher SPF sunscreen formula could reduce (or enhance) the irritation of the sunscreen product. The safety evaluation of the selected formulas was performed by the primary dermal irritation testing, *in vitro* release and skin permeation rate.

## EXPERIMENTAL PROCEDURES

### 1. Chemicals

USP grade oxybenzone, purchased from American Cyanamide Co., Wayne, NJ, USP grade octyl salicylate from, Felton International Inc., Brooklyn, NY, and octyl methoxycinnamate (Parsol MCX) from Givaudan Corp., Clifton, NJ were used in the study. The waterproof polymers selected in this study were Carboset and Pemulen TR1 from B. F. Goodrich Chemical Co., Cleveland, OH, Ganex V-220 and PVP K90 from GAF Corp., Wayne, NJ. Polymer X is a proprietary phosphate emulsifier developed by J & J Consumer Products, Inc. DEA-Cetyl phosphate (Amphisol), purchased from Givaudan Corp., Clifton, NJ, was used as the primary emulsifier in the model formula. All other chemicals and reagents were used as purchased. All the solvents used in the chromatography analysis were HPLC grade.

## 2. Chromatography System and Conditions

A Vista 5500 High Performance Liquid Chromatograph (Varian Corp., Walnut Creek, CA), equipped with Varian's Polychrom 9060 variable wavelength UV detector and M-9090 Autosampler, were used in the study. The analytical column, a Supelcosil LC-18 (5 µ, 25 cm x 4.6 mm ID), reversed phase column, from Supelco Inc., Bellefonte, PA, was used. The chromatographic data were stored and analyzed by the Maxima 820 Data System, from Water/Millipore, Milford, MA.

A gradient system consisting of methanol and 0.5% acetic acid was used to separate the sunscreen actives. The initial mobile phase was methanol-acidic water (70:30), and followed by a linear gradient to 90% methanol in 12 min. The flow rate was maintained at 1.5 mL/min and the column temperature at 35°C. Oxybenzone was monitored at 280 nm, whereas octyl salicylate and Parsol-MCX were measured at 302 nm. The injection volume was 30 microliter.

## 3. Formulation Design

A marketed o/w sunscreen cream base was selected as the model formula in this study. The qualitative list of ingredients is shown in Table 1. The formula contains 7.5% Parsol MCX, 4.0% oxybenzone and 3.0% octyl salicylate as sunscreen actives, DEA-cetyl phosphate as the key emulsifier, propylene glycol as the solubilizer, and Carbopol 940 as the thickening agent, isopropyl isostearate, isopropyl myristate as emollients, and parabens and Kathon CG as preservatives. Polymers were the only variable in the formulation. The waterproof polymers and the used concentration are listed in Tables 2 and 3. All prepared formulation were chemically assayed for the sunscreen content by the HPLC method described in the above section.

Table 1. The o/w sunscreen formulation selected in this study

| Ingredient | Vendor | % by wt |
|---|---|---|
| Parsol MCX | Givaudan | 7.50 |
| Oxybenzone | Cyanamid | 4.00 |
| Octyl Salicylate | Felton Int. | 3.00 |
| Isopropyl isostearate | Emery | |
| Amphisol | Givaudan | |
| Glyceryl myristate | Gattefosse | |
| Isopropyl myristate | Inolex | |
| Propylene glycol | Union Carbide | |
| Novata B | Henkel | |
| Fragrance | | |
| Carbopol 940 | B. F. Goodrich | |
| Methyl paraben | Arsyoco | |
| Triethanolamine | Union Carbide | |
| Propyl paraben | Arsyoco | |
| Kathon CG | Rohm & Haas | |
| D. I. water | | |
| Waterproof polymer | | 0.30-4.00 |

| Formula no. | Waterproof Polymer | % by wt. |
|---|---|---|
| 1675-61 | Ganex V-220 (GAF) | 4.0 |
| 1675-62 | Pemulen TR1 (BFG) | 0.3 |
| 1675-64 | Pemulen TR1, Methocel | 0.4 |
| 1675-65-1 | Polymer X | 2.0 |
| 1675-66 | Carboset XPD 1616 | 1.9 |
| 1675-67 | PVP K90 (GAF) | 0.3 |

Table 2. The waterproof polymer and their concentration used in this study

## 4. *In Vitro* Release Studies

The release of sunscreen actives from the seven selected formulas were determined with the use of a 9-mm inside diameter modified Franz Cell Unit. A piece of PM-1 #33 microporous film with a 0.4 micron opening was used in the study. Normal saline (0.9% sodium chloride) solution maintained at 37°C ± 0.1°C was used as the receptor medium. A 400 micron thickness of each formula was spread in an even layer onto the membrane. One hundred fifty microliters of receptor fluid was sampled from the cell at 1/2, 1, 2, 3, 4, 6, 7, 8 hour after initation and analyzed by HPLC. The receptor fluid volume was replenished with the normal saline after each withdrawal.

## 5. *In Vitro* Hairless Mouse Skin Permeation Studies

Nine-mm inside diameter flat type modified Franz diffusion cells with open top caps were used for the hairless mouse skin permeation study. Normal saline solution maintained at 37°C ± 0.1°C was used as the receptor fluid. A piece of freshly excised hairless mouse skin was carefully mounted on the top of the cell. A known weight (20-25 mg) of the test formula was applied ontó the skin surface at the center of the cell covering the entire cell, then the cap was secured in place with a clamp. One hundred and fifty microliters of receptor fluid was drawn at the following intervals; 1, 2, 4, 5, 7, 22, and 24 hours. The withdrawn volume was replenished with fresh normal saline to maintain the initial volume. The sample solutions were immediately ready for HPLC analysis. The skin specimen was examined visually to ensure that no leakage occurred during the experiment.

Table 3. Chemical description of polymers

| Ingredient | Description |
|---|---|
| Ganex V 220 | Copolymer of PVP and Eicosene |
| Pemulen TR1 | Acrylates/C10-C30 Alkacrylate Copolymer |
| Methocel | Hydroxypropyl Methylcellulose |
| PVP K90 | Polymer of 1-Vinyl-2-pyrrolidone |
| Polymer X | N/A |

## 6. *In Vitro* and *In Vivo* SPF Testings

A new method of *in vitro* sunscreen SPF testing developed by Dr. Curt Cole, et al., was adopted for screening purpose in this study.[4] The proposed *in vivo* sunscreen product testing procedures published in the Federal Register in 1978 was adopted for *in vivo* SPF testing in this study.[5]

# RESULTS AND DISCUSSION

## 1. Formulation Evaluation

Six waterproof sunscreen formulas were prepared for the study. Accelerated conditions, i. e., 40°C, 50°C, 75% relative humidity/35°C, and three freeze/thaw cycles were employed for stability evaluation. Among the six formulas, the formula consisting of Polymer X was found to be physically unstable at 50°C in 8 weeks. The carboset formula was found to be cosmetically unacceptable due to its bad smell. Therefore only the formulas with Ganex, Pemulen and Methocel, and PVP were selected for *in vivo* waterproof SPF determination.

## 2. *In Vitro* Release Study

According to the previous study,[2] approximately 80% of the sunscreen actives released easily from the solution and non-waterproof cream base. In contrast to the results we previously observed, the study findings revealed that negligible amounts of octyl salicylate and octyl methoxycinnamate (Parsol MCX) were released from the waterproof sunscreen cream. Only oxybenzone, varied from 0% to 10%, released from the creams.

Figure 1 shows the release profiles of oxybenzone from the six creams over 8 hr. As shown in Figure 1, the Pemulen formula demonstrates the

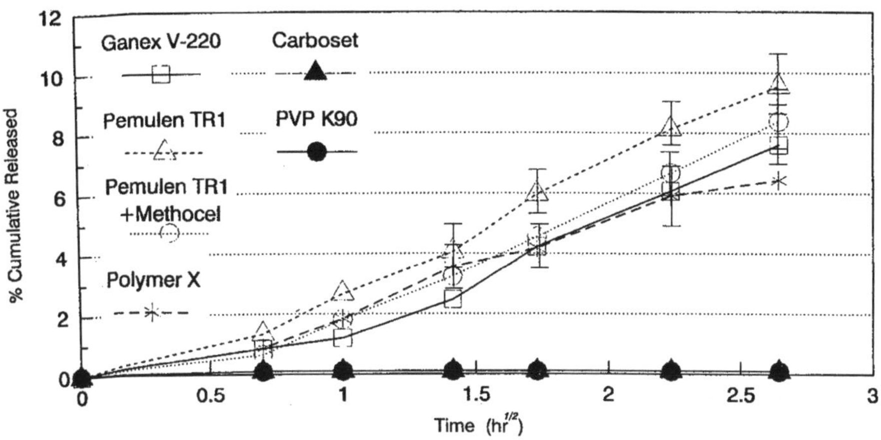


Figure 1. The percent (%) of oxybenzone released through 6.36 x $10^{-1}$ $cm^2$ of PM-1 #33 microporous film (pore size = 0.4 microns) at 37°C ± 0.1°C. Bars indicate S. E. of means (n = 3).

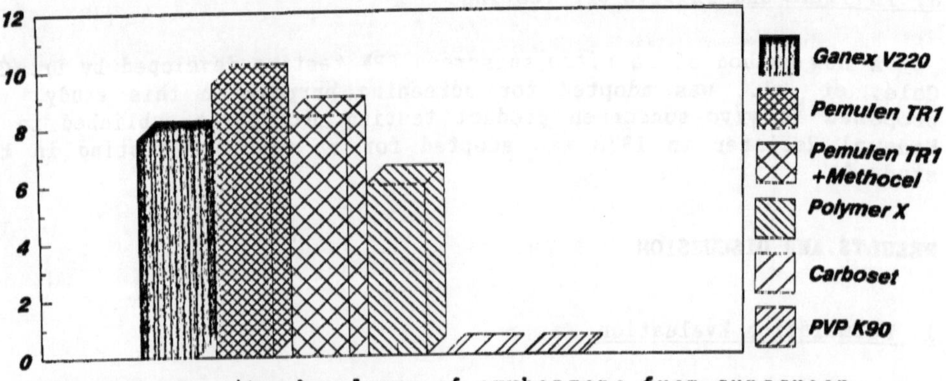


Figure 2. *In vitro* % release of oxybenzone from sunscreen formulations.

highest release rate, while the Carboset and PVP shown no release at all in 8 hr. It was speculated that a complexation reaction between the actives and PVP and Carboset could be responsible for the slow release behavior. Ganex, Pemulen and Methocel, and Polymer X cream show intermediate release profiles with 6-8% of oxybenzone released in 8 hr. The addition of methocel in the Pemulen formula seemed to slow down the release of oxybenzone from the cream base. The percentage of oxybenzone released from six formulas over 8 hr. is shown in Figure 2. The results suggests that the release of sunscreen actives from the waterproof cream could be the rate-limiting step in the absorption process, rather than the percutaneous penetration.

## 3. *In Vitro* Skin Permeation Studies

The hairless mouse skin membranes were chosen in this skin permeation study mainly because the results obtained would be consistent with the results from the *in vitro* SPF test method. The results of the *in vitro* hairless mouse skin permeation study is shown in Table 4-6. As can be seen, only a negligible amount of oxybenzone permeated through the skin from formulas containing Pemulen/methocel, Polymer X and PVP K90 was detected in receptor fluid in 24 hr. The slow permeation of the sunscreens through the skin membrane is consistent with the findings of the release study. The release of the sunscreen active from the waterproof creams is considered to the the rate-limiting step. The barrier function of the stratum corneum may have also contribute to the slow penetration of the sunscreen actives.

Table 4. Percentage distribution of oxybenzone in hairless mouse skin after 24 hr application

| Ingredient | % Permeated | % Retained in skin | % Recovered on skin |
|---|---|---|---|
| Ganex V 220 | 0.00 | 9.35 | 90.65 |
| Pemulen TR1 | 0.00 | 6.89 | 93.11 |
| Pemulen TR1 + Methocel | 0.00 | 8.80 | 91.20 |
| Polymer X | 1.26 | 10.22 | 88.52 |
| Carboset | 0.36 | 27.07 | 72.57 |
| PVP K90 | 0.95 | 29.78 | 69.27 |

| Table 5. Percentage distribution of octyl salicylate in hairless mouse skin after 24 hr application | | | |
|---|---|---|---|
| Ingredient | % Permeated | % Retained in skin | % Recovered on skin |
| Ganex V 220 | 0.00 | 3.72 | 96.28 |
| Pemulen TR1 | 0.00 | 2.78 | 97.22 |
| Pemulen TR1+Methocel | 0.00 | 2.55 | 97.45 |
| Polymer X | 0.00 | 4.62 | 95.38 |
| Carboset | 0.00 | 7.89 | 92.11 |
| PVP K90 | 0.00 | 11.58 | 88.42 |

The mass balances of sunscreen actives were shown in Tables 4-6. Figures 3-5 show the percent (%) of sunscreen retained in the hairless mouse skin after 24 hours application. The low recovery of oxybenzone in Carboset and PVP K90 formulas suggested that there might be some active protein complexation occurring in the permeation process. The extraction method used in this study might not be sufficient to extract the sunscreen completely from the skin membrane. The amount of oxybenzone, octyl salicylate, and octyl methoxycinnamate (Parsol MCX) remained in the skin varied from 2% to 30% of applied dose. In Figures 3-5, an unexpected result was observed, where the sunscreens concentration (approx. 30% of applied dose) in the skin from the PVP and Carboset creams formula was found to be much higher than that in other creams. The true mechanism which resulted in this unexpected behavior is yet unknown. It was speculated that PVP and Carboset might affect the hydrophobicity of skin membrane in favor of the partitioning of the sunscreen into the skin.

## 4. Primary Dermal Irritation Evaluation

Due to the physical instability, the Polymer X formula was excluded from the further study. Since higher sunscreen concentration was required to achieve the high SPF 30, the irritation score was expected to be high. However, only mild irritation with very low scores (0.2-0.8) was observed for all five formulas, as shown in Table 7. The addition of the waterproof polymers to the cream vehicle seemed to reduce the irritation potentials of sunscreen agents.

| Table 6. Percentage distribution of Parsol MCX in hairless mouse skin after 24 hr application | | | |
|---|---|---|---|
| Ingredient | % Permeated | % Retained in skin | % Recovered on skin |
| Ganex V 220 | 0.00 | 3.98 | 96.02 |
| Pemulen TR1 | 0.00 | 2.48 | 97.52 |
| Pemulen TR1+Methocel | 0.00 | 2.73 | 97.27 |
| Polymer X | 0.00 | 1.82 | 98.18 |
| Carboset | 0.00 | 3.26 | 96.74 |
| PVP K90 | 0.00 | 10.71 | 89.29 |

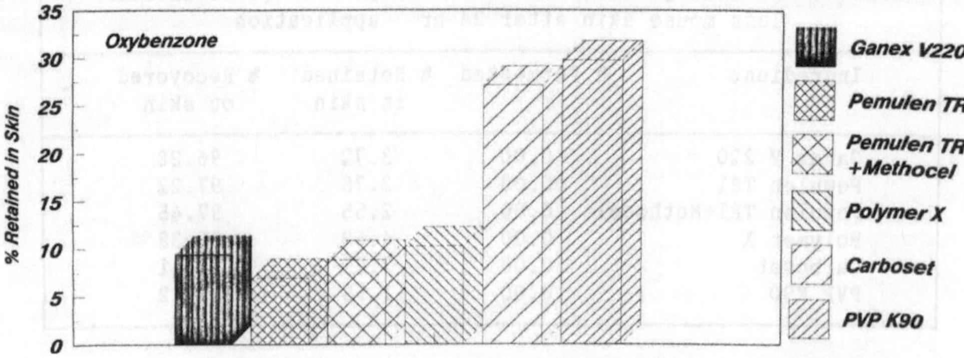


Figure 3. The percent (%) of the applied dose of oxybenzone retained in skin after 24 hr application. Bars indicate S. E. of means (n = 3).

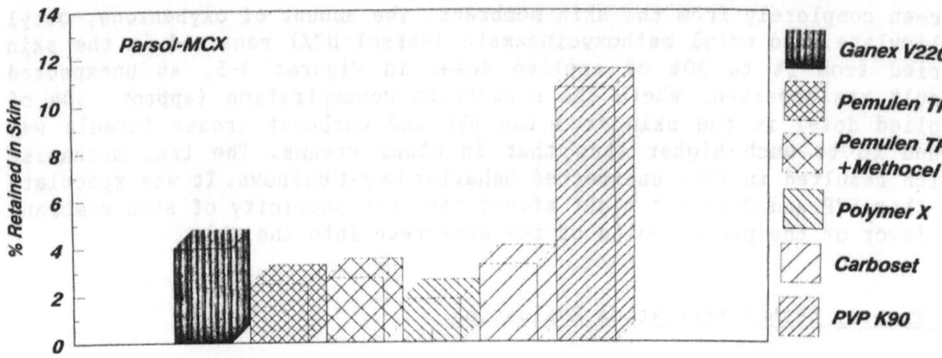


Figure 4. The percent (%) of the applied dose of octyl salicylate retained in skin after 24 hr application. Bars indicate S. E. of means (n = 3).

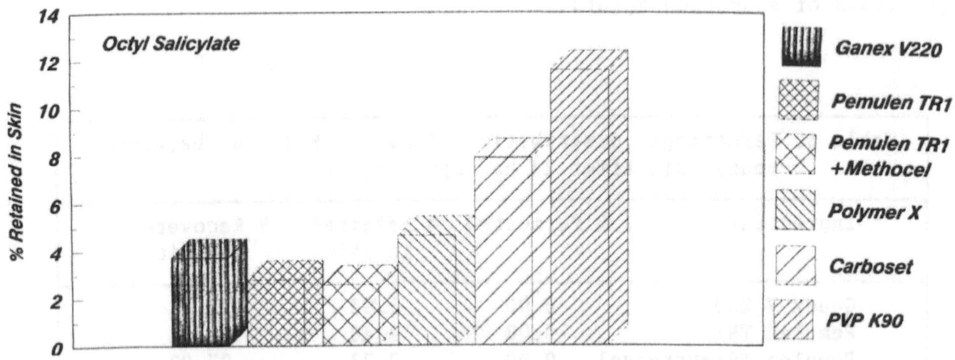


Figure 5. The percent (%) of the applied dose of Parsol-MCX retained in skin after 24 hr application. Bars indicate S. E. of means (n = 3).

Table 7. Primary dermal irritation (PDI) values of waterproof sunscreen formulas tested in this study

| Ingredient | PDI |
|---|---|
| Ganex V 220 | 0.8 |
| Pemulen TR1 | 0.2 |
| Pemulen TR1+Methocel | 0.7 |
| Carboset | 0.4 |
| PVP K90 | 0.6 |
| | 0-2.0 Mild |

## 5. *In Vitro* and *In Vivo* SPF Determination

A new method of *in vitro* sunscreen SPF testing developed by Dr. Curt Cole, et al., was adopted in this study.[4] According to Dr. Cole, the validated *in vitro* method has provided a good *in vitro-in vivo* SPF correlation. It is currently considered to be a potential alternative for *in vivo* SPF testing. The proposed *in vivo* sunscreen product testing procedures published in Federal Register in 1978 were adopted for *in vivo* SPF testing in this study.[5]

The results of the *in vitro* and *in vivo* SPF testings is shown in Table 8. Polymer X was excluded from the *in vivo* SPF testing due to its physical instability. The Carboset formula was also excluded from the further testing due to its unpleasant ammonium smell, resulting from neutralization, and its potential irritation to the skin. According to the supplier's information, Pemulen alone would not provide waterproofness. Therefore, only the Pemulen plus methocel formula was included for the further evaluation for *in vivo* SPF testing.

The results suggested that all the three formulas tested provided the photo-protection in both static and waterproof measurements as they have claimed. The *in vitro* and *in vivo* data also seemed to correlate well, although in most cases the *in vitro* values appeared to be higher than the *in vivo* values.

Table 8. SPF value of six waterproof sunscreen formulas tested in this study

| Ingredient | *in-vitro* static | *in-vitro* wp | *in-vivo* static | *in-vivo* wp |
|---|---|---|---|---|
| Ganex V 220 | 36.92 (4.35) | 36.92 (4.35) | 32.92 (5.69) | 31.70 (6.54) |
| Pemulen TR1 | 35.58 (8.35) | 34.27 (8.19) | | |
| Pemulen+ Methocel | 31.73 (6.12) | 31.73 (6.12) | 33.70 (5.92) | 29.10 (7.16) |
| Polymer X | 42.14 (4.24) | 42.14 (4.24) | | |
| Carboset | 30.20 (7.96) | 28.98 (7.29) | | |
| PVP K90 | 29.28 (7.17) | 29.28 (7.17) | 36.10 (6.70) | 28.80 (6.14) |

## CONCLUSIONS

In conclusion, the findings from both *in vitro* and *in vivo* SPF testings suggested that the waterproofness of the o/w sunscreen formulation could be achieved by the addition of Ganex V-220, Pemulen TR1 and Methocel, PVP K-90 waterproof polymers. The film-forming property of these polymers is considered to be the main mechanism providing waterproofness of the sunscreen products.

The study results of the *in vitro* release and skin permeation experiments also suggested that release of the sunscreens from the waterproof cream was the rate-limiting step in the absorption process, rather than the percutaneous penetration. The irritation potential of the waterproof sunscreen formulas might also be reduced resulting in a decrease in the release and skin permeation of the sunscreens due to the film-forming property of the waterproof polymers.

The study clearly suggested that the waterproofness could provide an alternative way of reducing the irritation potential of the high SPF (30) sunscreen products, without detriment to their clinical efficacy.

## ACKNOWLEDGMENTS

The authors gratefully acknowledge Dr. Curt Cole, Ms. Rory Van Fossen, Mr. John McCormack and Mr. Vinnie Walczak for providing the excellent services *in vitro* and *in vivo* SPF testing, and primary dermal irritation determination. Special thanks to Mrs. Melba Kiser for her help in the permeation study.

## REFERENCES

1. R. S. Berger, J. A. Mezick, and C. M. Papa, J. Soc. Cosmet. Chem., 29, 641-649 (1978). "Design and Evaluation of a water-resistant sunscreen preparation."
2. M. Corbo, unpublished results (1990).
3. K. Hideo and S. Masami, Cosmetics and Toiletries, 93, 39-48 (1978). "Prevention of adverse skin reaction to cosmetics via percutaneous absorption control."
4. C. A. Cole and R. L. Van Fossen, in: "*Sunscreens - Development, Evaluation, and Regulatory Aspects,*" N. J. Lowe and N. A. Shaath, Eds., Marcel Dekker Inc., New York and Basel, 1990, pp. 395-404. "*In Vitro* Models for UVB and UVA Photoprotection."
5. Federal Register, 43, (No. 166), 38259-38269 (1978). "Sunscreen product testing procedures for determination of the sun protection factor (SPF) value and related labeling claims."

# MELANIN, NATURE'S OWN SUNSCREEN POLYMER

Sergio Nacht

Research & Development
Advanced Polymer Systems, Inc.
3696 Haven Ave.
Redwood City, CA 94063

Melanin is a colored polymer widely distributed in species as diverse as mushroom, squid, sturgeon and chicken. In man, it is produced within the epidermis as a response to ultraviolet (UV) radiation injury. It protects the skin because it absorbs a wide range of UV and visible light. However, it has never been used as a sunscreen because (a) it was not available in commercial quantities and (b) it is difficult to formulate. Biosynthetic melanin, produced by genetic engineered organisms, was entrapped in a MICROSPONGE[R] polymeric system and formulated in a cream base containing also regular sunscreens to boost the UVB absorbance. When this formulation was tested in humans, melanin provided better UVA protection than any other sunscreen available and at a lower concentration. Furthermore, studies conducted with radioactive melanin showed no penetration through human skin, thereby supporting its safety.

## INTRODUCTION

In this day and age, it is nothing short of amazing that this year, one half of all new cancer cases in the U. S. A. will be of the skin type. This amounts to half a million new cases of skin cancer this year alone, out of which 27,000 will be of the frequently fatal type of malignant melanoma.[1] This situation has reached epidemic proportions even though it is well known that the cause of this problem is overexposure to ultraviolet (UV) radiation from the sun, probably compounded by a gradual and continuous depletion of the ozone layer;[2] with this relationship between UV radiation and reduction in the ozone in the atmosphere, skin cancer is still on the rise.

## REVIEW

Let us briefly review the relationship between the UV spectrum and its biological effects. The sun emits short-wave radiation which, since

*Cosmetic and Pharmaceutical Applications of Polymers*
Edited by C.G. Gebelein *et al.*, Plenum Press, New York, 1991

it is shorter than the violet color, is called ultraviolet. The UV range covers from about 200 to 400 nanometers in wavelength. It is invisible and can only be detected with appropriate instruments or by its biologic effects.

The shortest UV (below 290 nm), the so-called UVC, is the most biologically potent. It is very carcinogenic and it can even kill living organisms like bacteria; that is why it is used to sterilize surgical rooms, obviously when nobody is present. The UVB (between 290 and 320 nm) produces acute sunburn and is also carcinogenic. This is the type of radiation commonly filtered out by regular sunscreens.

The longer UV wavelength (above 320 nm) is called UVA and is responsible primarily for suntanning and, while it is known that it can cause premature aging of the skin, until not too long ago it was thought not to be carcinogenic. However, this situation has changed since recent animal studies demonstrated the carcinogenic potential of UVA.[3]

Figure 1 shows the different skin penetration properties of these types of radiation. So far, UVC does not reach the surface of the earth because it is absorbed by the ozone layer in the atmosphere. But UVB does reach us in significant amount and it can penetrate to the basal layer of the epidermis, the layer that divides the epidermis and the dermis. There, it can induce cancerous changes in the cells and it is primarily responsible for the so called "basal cell carcinoma."

UVA is more insidious because it does not produce the acute effects of UVB and it can penetrate deeper into the dermal layers of the skin. There it can break down collagen, it can affect the activity of the cells in the dermis and induces changes that we perceive as "wrinkles"; it has also been shown recently that it can affect specific cells that are related to our immunological system (the Langerhans cells) which are probably involved in the initiation of malignant melanoma.[4]

Thus, the biologic effects of the UV radiation can either be acute or chronic (Table 1) and the incidence of these long term effects keeps rising in spite of an increase in sunscreen usage of about 20% every year and of the gradual shift to the usage of products with higher sun protection factors (SPF numbers).

Not too long ago, there was a flurry of activity in the media because investigators at University of California-San Diego found that there is an amazing relationship between the usage of sunscreens and the increase in a certain type of skin cancer malignant melanoma.[5]

Is it possible that sunscreens cause cancer? It does not make sense and certainly we believe this is not the case. But regular sunscreen

Table 1. Biologic effect of ultraviolet radiation on human skin

| |
|---|
| A. Acute |
| Sunburn |
| Phototoxic Reactions |
| Photosensitive Reactions |
| B. Chronic |
| Photoaging |
| Skin Cancers |

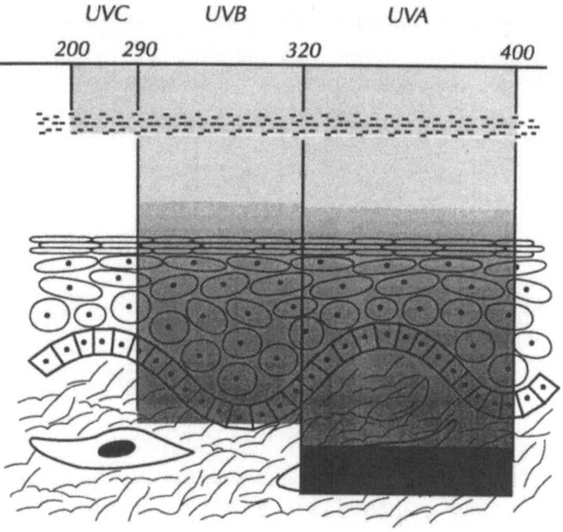


Figure 1. Skin penetration properties of ultraviolet radiation of various wavelengths.

products, like those used for many years containing PABA derivatives, cannot provide broad protection against UV radiation. Their absorption spectrum is sharply tuned to absorb UVB, to prevent sunburning; in reality, the SPF scale has been based on this particular property. But if we look at this scale in relationship to the amount of UVB that these products absorb, we recognize that we quickly reach a point of diminishing returns (Figure 2). As we moved into higher SPF values, the product absorbs incrementally less UVB. For example, to obtain an SPF of 4, it is necessary to absorb 75% of UVB radiation. For an SPF of 10, a 90% absorption is needed but to achieve an SPF of 20, we only need an extra 5%. Furthermore, a product that absorbs 98% of the UVB radiation, can result in an SPF of 50.

This seems pretty impressive, but in order to obtain these very high numbers the product needs to be loaded with more and more chemical sunscreens, which results in an increase in side effects like allergies and irritation. On the other hand, these high SPF products allow people to stay all day long in the sun without sunburning, but they give a false sense of protection because, in the meantime, they are bombarded by UVA, which is present in large amounts in solar radiation and which is not screened by regular sunscreens.

Therefore, there is a distinct possibility that the use of high SPF products that cannot protect at the same time against UVA may be more deleterious than beneficial, since they may induce an overexposure to UVA. This could help explain the constant rise in malignant melanoma.

At the present time, there is only one product that has been approved by the Food and Drug Administration for broad spectrum (UVB and UVA) protection. It contains a new chemical called Parsol 1789 [butyl methoxydibenzoylmethane, Givaudan Labs] which can provide protection throughout the UVA spectrum. However, there are still some questions about its long term safety since skin sensitization to this compound has been reported.

And yet, perhaps something can be learned by looking at what Mother

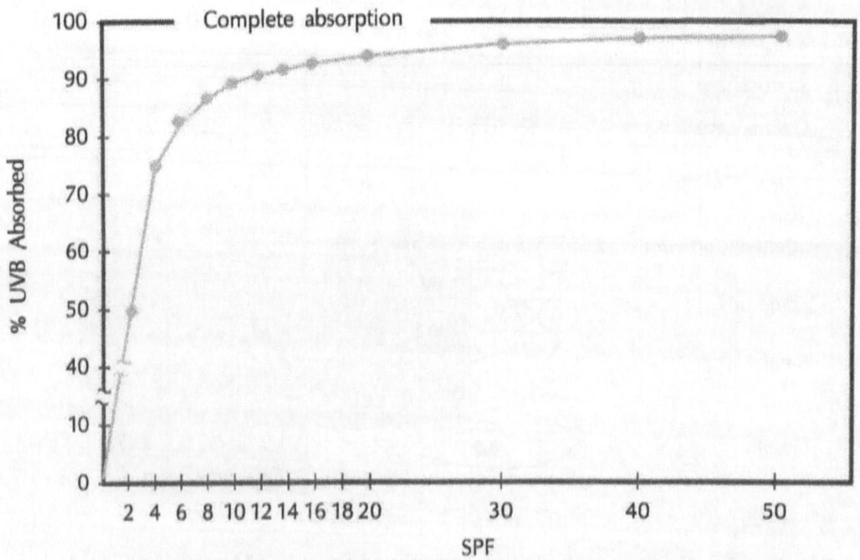


**Figure 2.** Relationship between the percent of UVB absorbed by a sunscreen formulation and its Sun Protection Factor or SPF value.

Nature teaches us (Table 2). Near the coast of Panama there are some islands called the San Blas Islands where native Cuna Indians live. A sub-population of this tribe are all albinos; namely, they cannot produce melanin, the natural pigment in our skin. All of these natives develop skin cancer by the age of 30.[6] In Australia it is a well known fact that more than 80% of its Anglo-Saxon population will develop skin cancer by the age of 50.[7]

In the U. S. A. it is said that about 50% of the general population over the age of 50 will develop skin cancer. And yet, in India where the exposure to the sun is much more intense, less than 1% of its population eventually develops skin cancer.[8] And the real extreme is the native black population in Africa, the area of the planet most heavily bombarded by UV radiation, where nobody develops skin cancer except in cases having etiologic factors such as trauma or chronic ulceration.[9]

This is telling us that the pigment that colors our skin can protect us against sun radiation and, if we look at the original racial distribution on our planet, we realize that the closer to the equator that man lives, the darker the color of his skin. Thus, the synthesis of melanin

**Table 2.** Skin cancer incidence among different populations

| Country | Population | Incidence |
|---|---|---|
| Panama | Albino Indians | 100% |
| Australia | Anglo-Saxons | >80% |
| U. S. A. | Mixed | 50-80% over 50 y.o. |
| India | Native | 0.3-1% over 50 y.o. |
| Africa | Black | Zero |

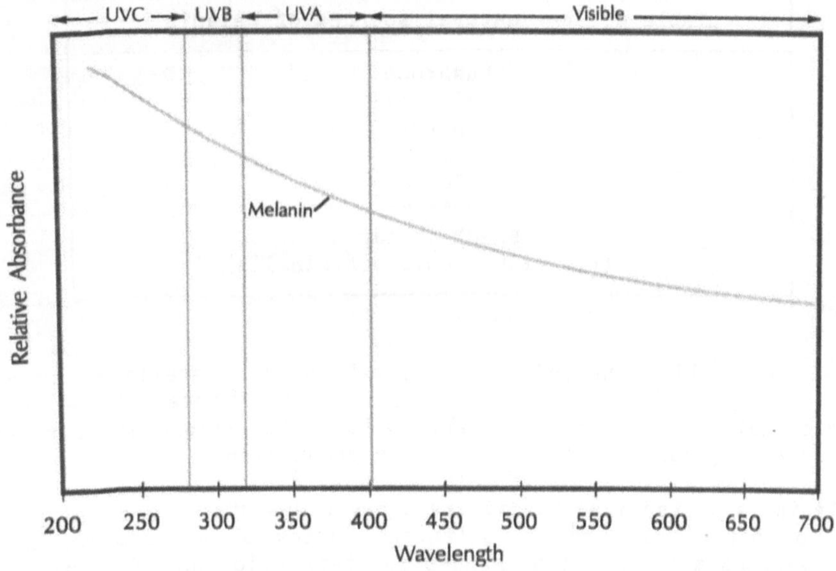


Figure 3. Light absorption spectrum of melanin in the ultra-violet and visible range.

in our skin can be considered a survival mechanism. This material is produced in reaction to the sun in specialized cells within our skin, called melanocytes, to protect against very broad ultra-violet radiation as depicted in Figure 3.

Melanin has multiple photo-protective actions (Table 3): it absorbs throughout the spectrum acting as a neutral density filter. It can also absorb energy in the UV and visible ranges and it can also scatter impinging radiation, primarily because it is present in small "packets" of solid melanin called melanosomes which can "bounce" UV rays and cause them to lose their intrinsic energy. Very importantly, melanin can also act as a trap for unstable electrons preventing the formation of damaging free radicals which are known to be one of the main mechanisms of cancer initiation.

In addition, melanin is safer than chemical sunscreens because it cannot penetrate into the skin since it is a polymer of large molecular weight and, therefore, it is neither irritating nor allergenic.

However, there was a significant technological challenge in trying to develop a sun-protection product with melanin to mimic Mother Nature's

| Table 3. Photoprotective actions of melanin |
| --- |
| Acts as a neutral density filter |
| Scatters impinging radiation |
| Absorbs energy in the UV and visible ranges, dissipating it as heat |
| Acts as a "trap" for unstable electrons, preventing the formation of unstable free radicals |

| Table 4. Some natural sources of melanin |
|---|
| Mushroom |
| Bovine eye |
| Birds |
| Hair |
| Squid |
| Octopus |
| Black caviar |
| Cuttle fish (sepia officinalis) |

own way. Two problems needed to be solved: melanin availability and to formulate it properly for optimal efficacy. In nature, there are many sources of melanin that go from mushroom to caviar (Table 4), but none of these are really suitable for large scale production.

Melanin is synthesized by different organisms starting with tyrosine, L-dopa, or other amino acids. Different ratios of various amino acids lead to various melanins with changes in their color and absorption properties. For instance, eumelanin is black while pheomelanin is red and they appear in different members of the animal kingdom. As of today, the complete detailed structural formula of melanin is still unknown. Figure 4 gives an idea of its complexity since it depicts only a partial structure of the whole molecule.

Melanin has also been chemically synthesized simply by oxidation of L-dopa.[10] However, this leads to a black material of variable and undetermined composition and, therefore, this route of synthesis is not suitable for drug application. For these reasons, we identified a genetic engineering firm, Biosource Genetics Corporation in Vacaville, California, which succeeded in bio-synthesizing melanin in industrial quantities.

There were still two more problems to solve. One, its cosmetic attributes since black melanin, although effective, would not be successful in a commercial market. But, since there are other colors of melanin, it seemed reasonable that suitable colors for melanin could be achieved to make it cosmetically acceptable.

Thus, a proprietary color modification process was developed to convert black melanin into various shades of brown melanin in a highly reproducible fashion. Advanced Polymer Systems has filed a patent application on this process that leads to MICROSPONGE® systems loaded with color modified melanin, with a more pleasant beige tone.

Is there a need for using microsponges as a support for melanin? In the skin, melanin is produced in the basal layer of the epidermis as an insoluble compound. It is made by the melanocytes in packets, the melanosomes, which are extruded into the skin surrounding the cells of the epidermis, the keratinocytes.

Figure 5 presents an electromicrograph of black skin where the melanosomes can be distinctively seen surrounding the cell. At a higher magnification, (Figure 6) these melanosomes look even more discreet and yet they are very effective as a UV protection.

Other people have tried in the past to produce effective sunscreening formulations using soluble melanins, but these formulations did not pro-

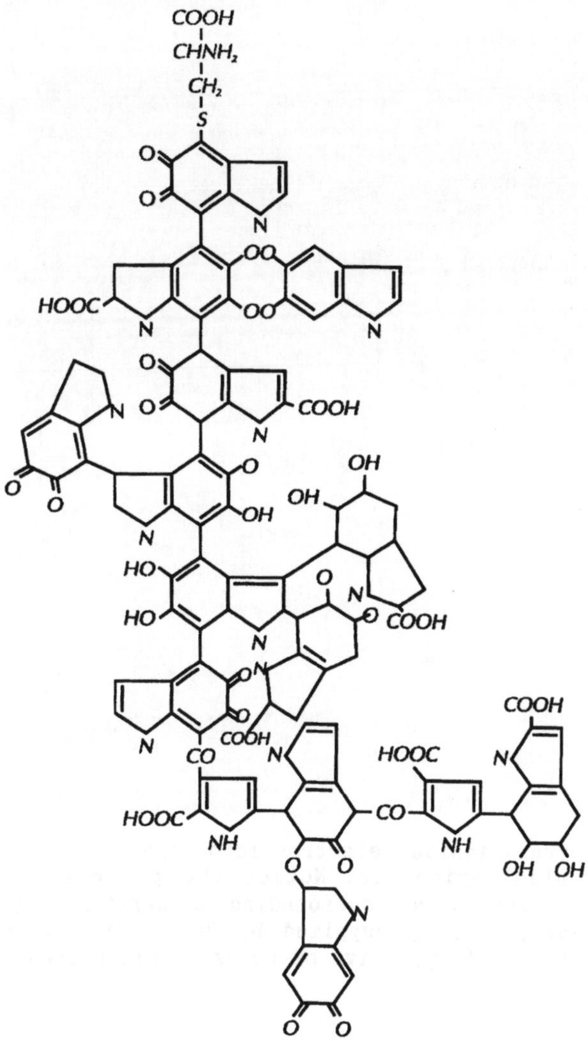


Figure 4. Partial structure of sepia melanin molecule. (After Nicolaus).

vide the expected efficacy, probably because this is not the way that melanin is supposed to work.

To mimic Mother Nature closer, we used a very appropriate technology called Microsponge Systems. Microsponges are small microspheres with large amounts of pores where we can entrap active ingredients. Figure 7 shows a scanning electromicroscope (SEM) view of a single microsponge, 25 microns in diameter.

By cutting one microsponge in half, we can see in Figure 8 that its inside contains large amounts of void spaces that can be filled with the appropriate active ingredient, in this case melanin. If a melanin-loaded microsphere is then sliced in one-micron thick sections and these sections are observed by transmission electromicroscopy, the structure indicates that the voids have been homogeneously filled with melanin (Figure 9); coincidentally, each of these melanin particles are of the same order of magnitude as a melanosome. Therefore, these particles can probably be

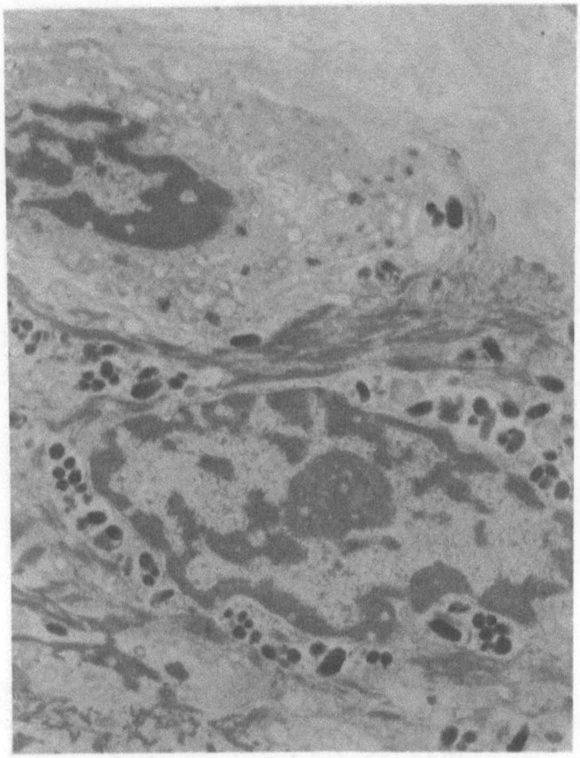

Figure 5. Transmission electronmicrograph (TEM) of human black epidermis. Notice the packets of melanin (melanosomes) surrounding a keratinocyte (photograph kindly supplied by Dr. R. M. Lavker, Dept. Dermatology, Univ. of Pennsylvania Medical School).

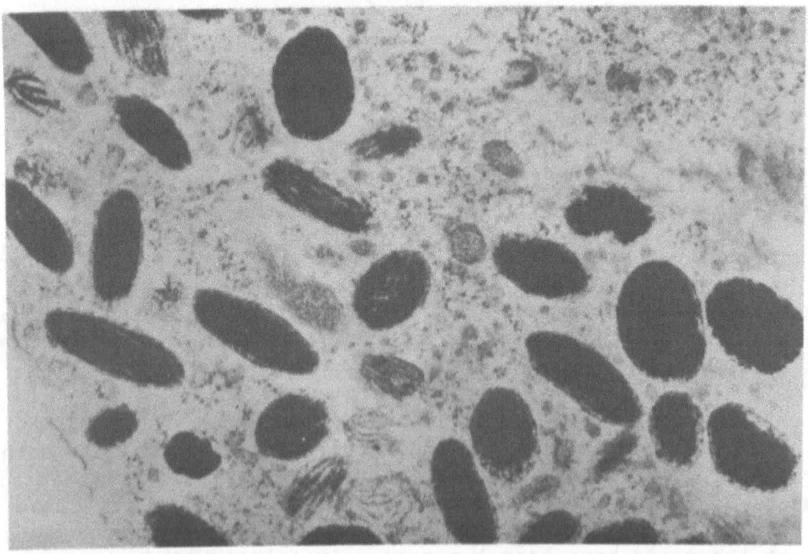

Figure 6. A higher magnification TEM of melanosomes from Figure 5.

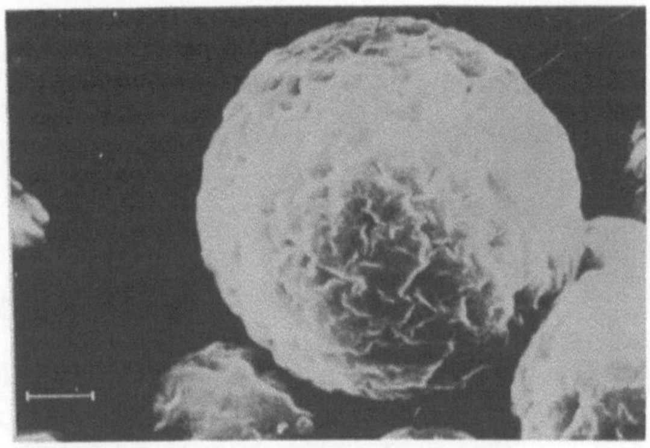

Figure 7. Scanning electromicrograph (SEM) of a polymeric MICROSPONGE[R] particle. Bar = 4µm = 16/100,000 inch).

considered as artificial melanosomes which can now be formulated into appropriate vehicles to be deposited onto the skin. The final formulation is a cream of a pleasant beige color that can be easily applied to the skin.

The Microsponge Systems can provide programmed release; in this case, we have selected a "zero" or no release to make this product even more effective as a sunscreen by retaining the melanin on top of the skin. Advanced Polymers Systems has already completed a series of animal and human studies that confirm the efficacy of this product.

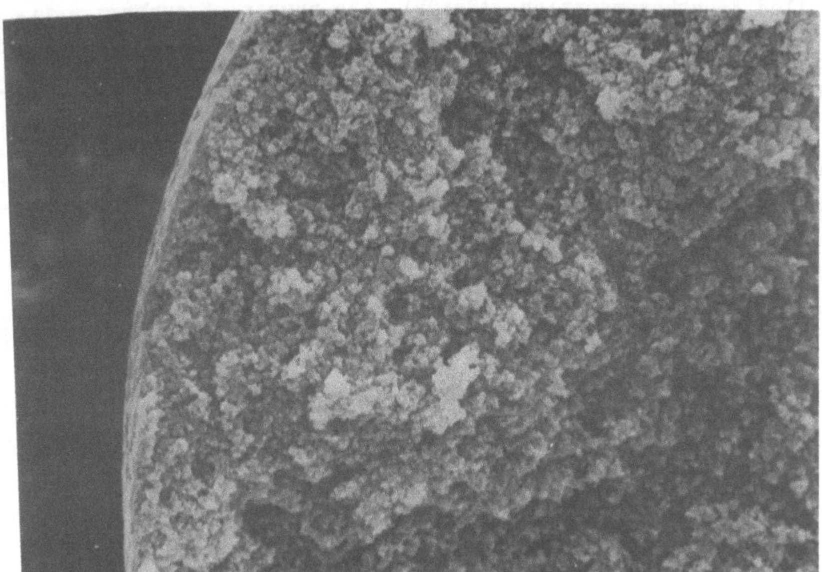

Figure 8. SEM of a fractured microsponge showing its internal structure. Notice the void spaces present in this microsponge.

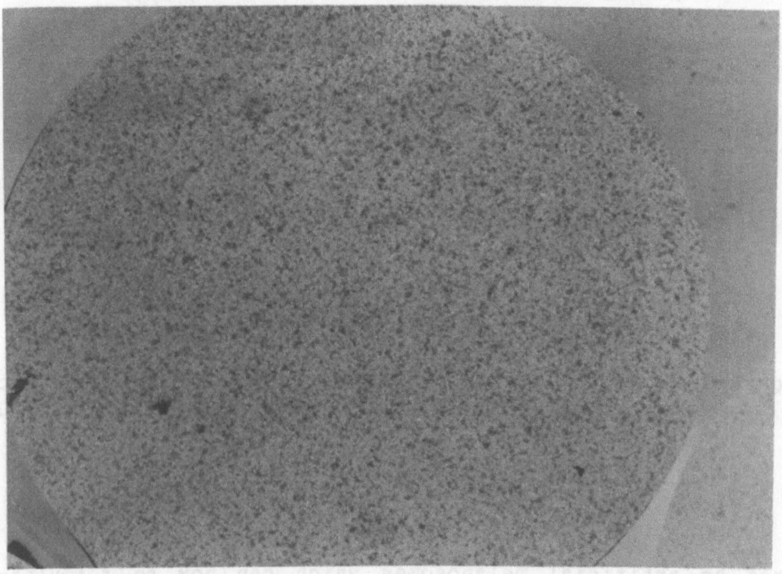

Figure 9. TEM of a 1-micron thick section of a microsponge filled with melanin. The black dots represent melanin aggregates (approximate diameter of the microsphere is 15 um).

Studies in hairless mice were done to evaluate the UVA protection that this formulation provides using a so-called sunburn cell assay in which the animals are treated topically with a phototoxic compound, a psoralen, to make them more sensitive to UVA. If the animals are not treated and just exposed to UVA, there are a certain number of sunburn cells that appear (Table 5). If they are previously treated with a compound called 8-methoxypsoralen, (8-MOP), the number of sunburn cells more than doubles. All the other groups of animals were also treated with 8-MOP but, in addition, different treatments were applied; in the first case, just the cream vehicle was applied: this resulted in a smaller number of sunburn cells representing about a 15% protection. When regular sunscreens were added to this vehicle the level of protection increased to about 30%. But, if on top of that, 1% melanin in Microsponges were added, the level of protection is greater than 100%; namely, it even

Table 5. Prozone[R] UV-A protection in hairless mice (Sunburn Cell Assay)

| Treatment | No. Sunburn Cells | %Protection* |
|---|---|---|
| Untreated | 142 | - |
| 8-MOP | 368 | 0 |
| 8-MOP + Vehicle | 333 | 15 |
| 8-MOP + Vehicle with Sunscreens | 297 | 31 |
| 8-MOP + Prozone[R] | 119 | 110 |

* % Protection = 100(8-MOP - Treatment)/(8-MOP - Untreated)

| Table 6. Prozone® UV-A protection in human | | | | |
|---|---|---|---|---|
| | Commercial Sunscreen* | Prozone® | Prozone s Melanin | Vehicle |
| Phototoxic Protection Factor (PPF) | 4.7 | 6.0 | 3.4 | 1.2 |
| Immediate Pigment Darkening (IPD) | 3.8 | 4.5 | 2.2 | 1.1 |

* With Parsol 1789

reduces the number of sunburn cells that UVA alone without a phototoxic agent can produce.

These assays were repeated in humans, under FDA guidance and Table 6 presents the results obtained.

These tests were conducted with two different methodologies; one called a "phototoxic protection factor," also involves the application of 8-MOP to the area of the skin to be exposed and a protection factor can be defined, very much like a sun protection factor or SPF value.

In the other technique, the Immediate Pigment Darkening, no treatment is applied other than the test products and the ability of the skin to generate a "quick melanin" made out of existing precursors within the skin layers is evaluated; this is an indication of UVA exposure. In both cases we see that the Prozone formulation was between 20 and 30% more effective than a commercial sunscreen with Parsol 1789, the most effective broad-spectrum sunscreen available until today. Moreover, the complete Prozone formulation was twice as effective as the same formulation without melanin, thus clearly supporting the benefit of the melanin added.

In addition, standard SPF determinations showed that the current formula has an SPF of 18 and is waterproof; namely, it allows more than 90 minutes immersion in water without significant loss of efficacy. It is worth mentioning that the vehicle for this product is also an excellent moisturizer. The clinical program is due for completion in the section half of 1991 and the New Drug Application should be filed before the end of that year.

In conclusion, melanin is Nature's own way to protect the body against UV radiation since it gives broader protection than any other sunscreen and it is safer because since, it is a macromolecule, it cannot penetrate into the skin, primarily when entrapped in a Microsponge System to make this even less likely.

This development is supported by a proprietary melanin color modification process that allows the formulation of a product with pleasant cosmetic attributes and yet does not reduce any efficacy properties.

## REFERENCES

1. Source: American Cancer Society, 1989.
2. R. T. Watson, M. A. Geller, R. S. Storlaski, R. H. Hampson (1986). "Present state of knowledge of the upper atmosphere: Processes that

control ozone and other climatically important trace gases," Washington, DC, National Aeronautics and Space Administration.
3. B. Staberg, H. C. Wulf, P. Klemp, T. Poulsen and H. Brodthagen, J. Invest. Dermatol., 81, 517-519 (1983). "The carcinogenic effect of UVA irradiation."
4. J. Alcalay, J. N. Craig and M. L. Kripe, Cancer Res., 49, 4591-4596 (1989). "Alterations in Langerhans cells and Thy-1+dendritic epidermal cells in murine epidermis during the evolution of ultraviolet radiation induced skin cancers."
5. The New York Times, August 9, 1990; San Francisco Chronicle, August 9, 1990; San Jose Mercury News, August 9, 1990.
6. A. W. McFadden, Arch. Dermatol., 84, 1013-1023 (1961). "Skin disease in the Cuna Indians."
7. H. O. Lancaster Med. J. Aust., 1, 1082-1087 (1956). "Some geographical aspects of mortality from melanoma in Europeans."
8. A. K. Khare, private survey in India supported by an APS grant, 1988.
9. A. G. Oettle, in: "*Skin cancer in Africa*," F. Urback, Ed., National Cancer Institute Monograph No. 10, 349-359 (1963).
10. S. Ito, E. Novellino, F. Chiocarra, F. Misuraca and G. Prota, Experimentia, 36, 822-823 (1980). Copolymerization of Dopa and cysteinyldopa in melanomagenesis *in vitro*."

# COSMETIC AND PHARMACEUTICAL USES OF CHITIN AND CHITOSAN

Shigehiro Hirano, Keiji Hirochi, Ken-ichiro Hayashi,
Toshiya Mikami and Hisashi Tachibana

Department of Agricultural Biochemistry and Biotechnology
Tottori University
Tottori 680, Japan

Chitin, chitosan and derivatives have been tested for use in pharmaceutical and cosmetic fields. Beads, powders, tablets, films and sponge sheets have been prepared from chitin and chitosan. Several water-soluble and enzymatically digestible derivatives of chitin and chitosan have been prepared, and the rate of their enzymic hydrolysis was controlled by the substitution degree and structure of *N*-acyl groups in these derivatives. These derivatives are usable as cosmetic ingredients because of their biodegradable, biocompatible, viscosity and moisture-holding properties. Some commercial products are hair care products containing chitosan, skin care products containing carboxymethylchitosan, and chitin suture absorbable in tissues.

## INTRODUCTION

Chitin [(1→4)-2-acetamido-2-deoxy-β-D-glucan] and chitosan (*N*-deacetylchitin) are naturally abundant biopolymers (polysaccharides), and they are commercially manufactured from shellfish wastes. Chitin and chitosan are biofunctional, biocompatible, have low toxicity, and almost no antigenicity.[1] Chitin is hydrolyzed by both lysozyme (EC 3.3.1.17) and chitinase (EC 3.3.1.14), and chitosan is hydrolyzed by chitosanase (EC 3.3.1.?). Lysozyme is widely distributed in animal tissues. An artificial skin-covering for burns was manufactured from chitin fabric. Chitosan salts with organic acids are soluble in water and aqueous methanol, and their aqueous solution has viscous, moisture-holding and emulsifying properties. Our intention has been directed to the use of these unique properties of chitin and chitosan for cosmetic and for drug delivery system.

Here we report the preparation of beads, fine powders, films, sponge sheets, and some novel water-soluble derivatives of chitin and chitosan, and their pharmaceutical and cosmetic uses.

*Cosmetic and Pharmaceutical Applications of Polymers*
Edited by C.G. Gebelein *et al.*, Plenum Press, New York, 1991

**EXPERIMENTAL**

## 1. Methods

IR spectra (KBr) were recorded with a Hitachi 215 grating spectrometer, specific rotation with a JASCO Dip-181 digital polarimeter, and $^{13}C$-NMR spectra with a JEOL FX-200 FT NMR spectrometer. Viscosity was measured with an Oswald viscometer in water at 25°C.

## 2. Hydrolysis by Lysozyme, Chitase, and Chitosanase

One-tenth of one mL of lysozyme (5 mg of lysozyme/mL in 0.05M citric acid - 0.1M $Na_2HPO_4$, pH 6.8) or 0.1 mL of chitinase (2 mg of chitinase/mL in the same buffer solution) was added to a solution of the *N*-acyl derivative (7.5-7.8 mg) of CM-chitin in 1.6 mL of the buffer solution. The mixture was incubated in a test tube at 40°C for 40 min. The enzymic reaction was stopped by addition of 0.2 mL of 1 N NaOH, and the increase of reducing-sugar value was analyzed by a modified Schales and Schales method and calculated as µmol of *N*-acetyl-D-glucosamine.[2]

One-tenth of one mL of chitosanase (0.2 mg/mL) was added to chitosan (10 mg) solution in 2% aqueous acetic acid (1.5 mL), which had been adjusted at pH 6.5 by addition of 1 N NaOH (0.45 mL). The mixture was incubated at 50°C for 30 min. The reaction was stopped by heating at 100°C for 15 min, and 0.05M $Na_2CO_3$ (1.25 mL) was added. The precipitate was removed by centrifugation at 1500 g for 15 min, and 0.5 mL of the supernatant solution was withdrawn. Its reducing sugar-value was analyzed as described above, and calculated as µmol of D-glucosamine.

## 3. Materials

Crab shell chitosan (d.s. = 0.10 for NAc) was supplied by Tsuchiyoshi Sangyo Co., Ltd., Shimane. *N*-Acetylchitosan (a regenerated chitin; d.s. = 1.0 for NAc) was prepared by chemical *N*-acetylation of chitosan.[3] Lysozyme (Grade III, Sigma) from hen egg-white (Grade III, Sigma), Chitinase (Sigma) from *Streptomyces griseus*, and chitosanase (Wako Chemicals) from *Bacillus pumilus* were commercial products.[4]

## 4. Chitosan Beads, Powders, and Tablets

Chitosan solution (3-12%) in aqueous acetic acid was injected, with stirring, into 10% aqueous NaOH solution through a syringe. The precipitate was collected and air-dried to give the beads (porous diameter 0.1-0.3 µ). Chitin and chitosan were mechanically powdered (>200 mesh), and were molded into tablets.

## 5. Chitosan and N-acylchitosan Films and Sponge Sheets

Chitosan solution (2-3%) in 2% aqueous acetic acid-methanol was spread as a thin layer over a glass plate. The layer was air-dried at room temperature to afford chitosan film. On the other hand, the chitosan solution in 2% aqueous acetic acid-methanol was mixed with carboxylic anhydride (5 moles/GlcN), and it was spread as a thin layer over a glass plate to afford a thin gel layer. The gel layer was air-dried at room temperature to afford *N*-acylchitosan film.[5]

A layer of chitosan solution in aqueous acetic acid or a layer of *N*-acylchitosan gel layer was freeze-dried to give the corresponding sponge sheet.

## 6. N-2-Acetoxybenzoyl Derivative of Chitosan

Chitosan was treated with 2-acetoxybenzoic anhydride (0.2, 0.4 and 0.6 moles/GlcN) in aqueous acetic acid-methanol at room temperature to afford the product in 60-90% yields (d.s. 0.97 for acetoxybenzoyl).[6]

$\nu_{max}$ (KBr): 1760 (C=O of OAc), 1650 and 1550 (C=O and NH of NAc), 760 $cm^{-1}$ (phenyl).

Anal. Calc. for $[C_6H_{10}NO_4(C_9H_7O_3)_{0.42}(H)_{0.97}(H)_{0.03}]_n$:

C, 55.56; H, 5.29; N, 4.40.
Found: C, 55.95; H, 5.76; N, 4.45.

The *N*-2-acetoxybenzoyl derivative was treated with acetic anhydride (3 moles/GlcN) in methanol at room temperature overnight to give partially *N*-acetyl derivatives.[3]

## 7. O-Carboxymethylchitosan Na Salt

Alkali *N*-acetylchitosan was prepared from powdery *N*-acetylchitosan (4.0 g) in 42% NaOH in the presence of 0.2% LiCl. The alkali *N*-acetylchitosan was dissolved in 14% NaOH and treated with chloracetic acid to afford *O*-carboxymethyl (CM) derivative of *N*-acetylchitosan in 3.8 g yield (77%).[7] $[\alpha]^{26}_D$ -12° (c 0.7, 0.05% NaOH). $^{13}$C-NMR data: δ = 180.0 (C=O of COOH), 177.5 (C-6), 57.0 (C-2), 25.5 ($CH_3$ of NAc). The *O*-CM-*N*-acetylchitosan (6.6 g) was dissolved in 10% NaOH (350 mL), and the mixture solution was heated at 80°C for 9 h. The reaction product was neutralized with aqueous acetic acid and was dialyzed in distilled water for 5 days. The product was precipitated by addition of three volumes of ethanol to give the product in 4.8 g yield (87%). $[\alpha]^{25}_D$ -4° (c 1.0, 0.5% NaOH).

Anal. Calc. for $[C_6H_{10}NO_4(C_2H_2O_2Na)_{0.61}(H)_{0.39}\cdot 0.10H_2O]_n$:

C, 40.97; H, 5.67; N, 6.63.
Found: C, 41.09; H, 5.97; N, 6.57.

## 8. N-2-Acetoxybenzoyl Derivative of O-Carboxymethylchitosan

CM-chitosan was treated with 2-acetoxybenzoic anhydride (5 moles/GlcN) in aqueous acetic acid/methanol at room temperature to afford the corresponding product in a gel form (70-95% yields).[3]

$\nu_{max}$ (KBr): 1730 (C=O of OAc), 1650 and 1560 (C=O and NH of NAc), 760 $cm^{-1}$ (phenyl).

Anal. Calc. for $[C_6H_9NO_4(C_2H_3O_2)_{0.71}(C_9H_7O_3)_{0.42}(H)_{0.58}\cdot 041H_2O]_n$:

C, 48.47; H, 5.58; N, 5.05.

Found: C, 48.39; H, 5.81; N, 5.07.

## 9. N-Acyl Derivatives of O-Carboxymethylchitosan

*O*-CM-chitosan was *N*-acylated with carboxylic anhydride (5 moles/GlcN) in aqueous methanol at room temperature to afford the products in 59-97% yields (d.s. = 0.25-0.76 for N-acyl) (Table 1).[3]

## 10. Di-*O*-methylchitin

Alkali *N*-acetylchitosan was methylated with dimethyl sulfate and NaOH,[8] the product was dialyzed against running water for one day and lyopilized to give the product in 70-90% yield. It was soluble in water and aqueous methanol, but insoluble in ethanol. $[\alpha]^{26}_D$-26° (c 0.5, water). $\nu_{max}$ (KBr): 2950 (C-H), 1660 and 1550 $cm^{-1}$ (C=O and NH of NAc). $^{13}$C-NMR data: δ = 175.6 (C=O of NAc), 102.4 (C1), 81.8 (C4), 74.7 (C5), 59.8 and 59.7 ($OCH_3$), 54.9 (C2), 23.2 ($CH_3$ of NAc).

Anal. Calc. for $[C_{10}H_{17}NO_5\cdot 0.69H_2O]_n$:

C, 49.30; H, 7.55; N, 5.75; $OCH_3$, 25.5.

Found: C, 49.13; H, 7.47; N, 5.66, $OCH_3$, 24.4

# RESULTS AND DISCUSSION

## 1. Chitin and Chitosan as Materials for Cosmetic and Pharmaceutical Uses

Chitin and chitosan have been processed into beads, gels, powders, fibers, sheets, tablets, and sponges, which are usable as basal materials for cosmetics, and drugs (Figure 1). Oral administration of chitosan or chitin with drugs enhanced absorption of the drugs from intestines into blood in animals.

## 2. Control of the Rate of Enzymic Hydrolysis of Chitin and Chitosan for Use in Drug Delivery System

Chitin and chitosan are biocompatible,[1] and the chemical modification of amino and carboxyl groups in the molecule give rise to novel molecular and biological functions. Chitosan is usable as a carrier of drugs to specific organs, because chitosan is digested in animal intestine.[9] The 2-acetoxybenzoyl derivative of chitosan, which is orally administered, moves to the intestine without any trouble in stomach, and the chitosan is digested and aspirin is released in the intestine.

A sheet of thin chitin film (32-50 um) was implanted in the back muscle of a rabbit, and it was completely digested within three months. We have attempted to control the digestive period of chitin and chitosan by chemical modification. Lysozyme hydrolyzed a partially *N*-acetylated derivative (d.s. = 0.8 for NAc) of chitosan about four times faster than

Table 1. Some *N*-acyl derivatives of *O*-carboxymethylchitosan

| Carboxymethyl chitosan | D.s. for *N*-acyl | $[\alpha]_D^{18}$ (0.05% NaOH) | Solubility[a] in 0.05% HCl | water | 0.05% NaOH |
|---|---|---|---|---|---|
| *N*-Acetyl | 0.75 | -12° (c 0.5) | S | S | S |
| *N*-Propionyl | 0.54 | - 9° (c 0.5) | S | S | S |
| *N*-Butyryl | 0.60 | -10° (c 0.7) | S | S | S |
| *N*-Pentanoyl | 0.50 | -18° (c 0.7) | S | S | S |
| *N*-Hexanoyl | 0.40 | -16° (c 0.7) | S | S | S |
| *N*-Octanoyl | 0.51 | - | S | S | S |
| *N*-Decanoyl | 0.41 | - | W | I | W |
| *N*-Lauroyl | 0.35 | - | W | I | W |
| *N*-Palmitoyl | 0.25 | - | I | I | I |
| *N*-Benzoyl | 0.35 | - | W | I | W |
| *N*-Succinyl | n.d. | - 8° (c 0.6) | S | I | S |
| *N*-Phthaloyl | n.d. | - 2° (c 0.7) | S | S | S |

(a) Key: S, soluble; I, insoluble; W, swell.

*N*-acetylchitosan (d.s. = 1.0 for NAc) (Figure 2). Chitosan hydrolyzed partially *N*-acetyl derivatives (d.s. = 0.4-0.8 for NAc) of chitosan about 1.5 times faster than *N*-acetylchitosan (d.s. = 1.0 for NAc) (Figure 3). Chitosanase also hydrolyzed partially *N*-acetyl derivatives (d.s. = 0.6 for NAc) 3-4 times faster than chitosan (d.s. = 0.9 and 0.20 for NAc) (Figure 4).

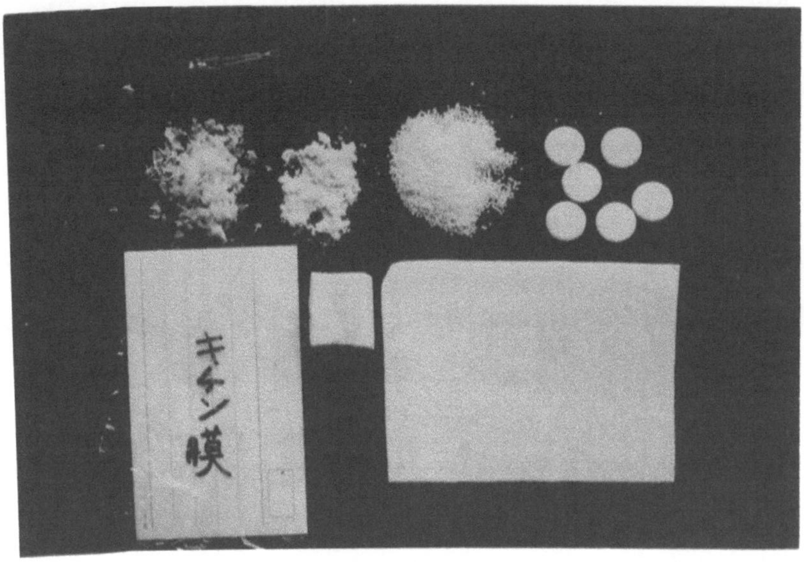


Figure 1. Chitosan flakes, chitosan powders, chitosan beads, chitosan tablets (from left to right on the upper row), and chitosan membrane, chitosan sponge, and chitosan sponge sheet (from left to right on the lower row).

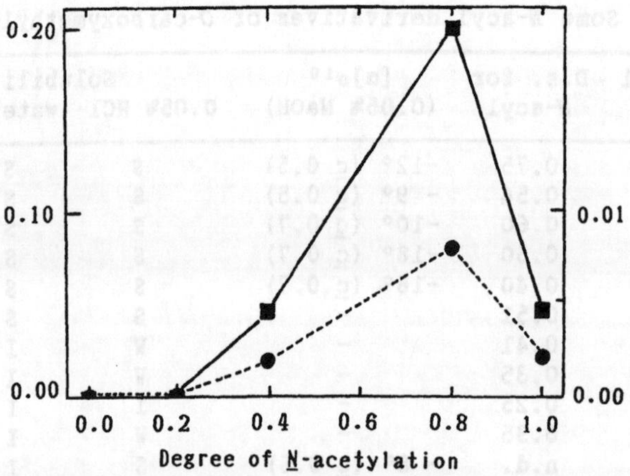


Figure 2. The activity of lysozyme from hen egg-white towards chitosan with various degrees of *N*-acetylation. Reducing sugar value (■) is shown in µmol of *N*-acetyl-D-glucosamine/mL; *N*-acetyl-D-glucosamine value (●) is shown in µmol.

The rate of enzymic hydrolysis was controlled by the structure of the *N*-acyl group in the *N*-acyl derivatives of CM-chitosan. The relative rates of enzymic hydrolysis with respect to *N*-acyl groups were in an order of *N*-acetyl > *N*-propionyl > *N*-butyrl > *N*-pentanoyl > *N*-hexanoyl (Figure 5). The relative order is the same as that of the water-insoluble *N*-acyl derivatives of chitosan.[10] Partially *N*-acetylated derivatives of CM-chitosan were digested by these enzymes at higher rate than the CM-derivatives of chitin and chitosan.

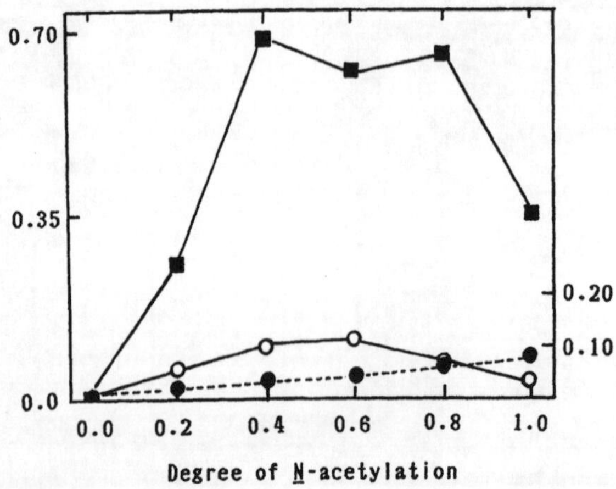


Figure 3. The activity of chitosan from *Streptomyces griseus* towards chitosan with various degrees of *N*-acetylation. Reducing sugar value (■) is shown in µmol of *N*-acetyl-D-glucosamine; *N*-acetyl-D-glucosamine (●) and D-glucosamine (o) values are shown in µmol/mL.

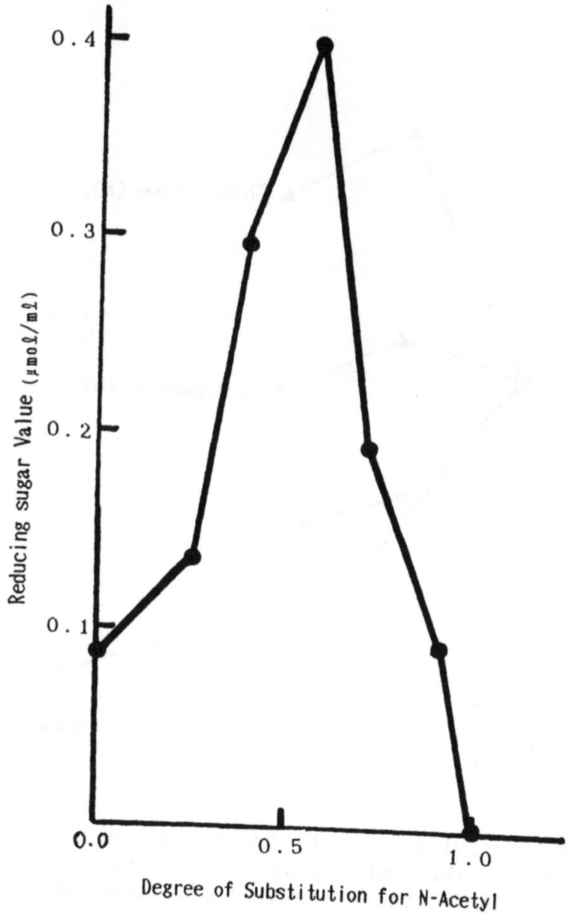


Figure 4. The activity of chitosanase towards chitosan with various degrees of *N*-acetylation. Reducing sugar value (●) is shown in μmol/mL of D-glucosamine.

## 3. Water-soluble Derivatives of Chitin and Chitosan for Novel Cosmetic Ingredients

*N*-Acetyl, *N*-propionyl, *N*-butyryl, *N*-pentanoyl and *N*-hexanoyl derivatives of *O*-carboxymethylchitosan were soluble in water, but *N*-octanoyl, *N*-decanoyl, *N*-lauroyl, *N*-palmitoyl and *N*-benzoyl derivatives were insoluble in water. Strong hydrogen bondings in chitin molecule decrease in solubility in water, but their destruction by chemical modification, e.g., *O*-methylation increases in solubility in water. In the literature, *O*-methylchitin was prepared by methylation of chitosan with dimethyl sulfate and NaOH, followed by *N*-acetylation and by repeated methylation.[11] We have found that alkali chitin in 14% NaOH solution was easily methylated with dimethyl sulfate and NaOH to afford di-*O*-methylchitin. Chitosan has a cationized amino group, CM-chitin has an anionized carboxyl group, and chitin has no ionized group. The *O*-methylchitin has neither cationized nor anionized group in the molecule (Table 2). It was soluble in water and in aqueous alcohol, and electrically neutral and mild to skin. Its aqueous solution was transparent, moisture-holding, viscous, and emulsifying. The viscosity of a 0.1% aqueous solution at 25°C was 1.27, 1.93 and 3.99 c.p. for di-*O*-methylchitin (MW = 15,000),

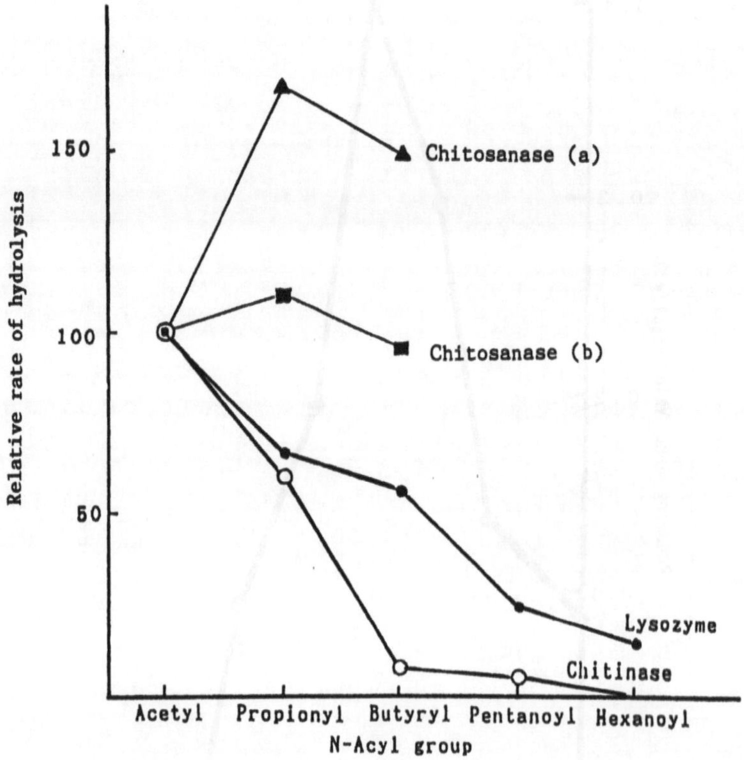


**Figure 5.** **The activities of chitosanase at pH 6.2 (a) and at pH 5.6 (b), and those of chitinase and lysozyme towards *N*-acyl derivatives of carboxymethylchitosan with various structures of the *N*-acyl group.**

**Table 2. Water-soluble derivatives of chitin and chitosan usable as cosmetic ingredients**

| Compounds | Groups Anionized | Groups Cationized |
|---|---|---|
| *O*-Carboxymethylchitin[a] (*N*-Acetyl-*O*-carboxymethylchitosan) | $COO^-$ | none |
| *O*-Carboxymethylchitosan | $COO^-$ | $NH_3^+$ |
| *N*-Propionyl-*O*-carboxylmethylchitosan | $COO^-$ | $NH_3^+$ |
| *N*-Butyroyl-*O*-carboxymethylchitosan | $COO^-$ | $NH_3^+$ |
| *N*-Pentanoyl-*O*-carboxymethylchitosan | $COO^-$ | $NH_3^+$ |
| *N*-Hexanoyl-*O*-carboxymethylchitosan | $COO^-$ | $NH_3^+$ |
| *O*-Methylchitosan[b] | none | $NH_3^+$ |
| *O*-Methylchitin (*N*-Acetyl-*O*-methylchitosan) | none | none |

(a) The commercial name is "chitin Liquid".
(b) Details of this compound and its *N*-acyl derivatives will be reported elsewhere.

Figure 6. A skin care product containing "Chitin Liquid" (center), and four hair care products containing chitosan (left and right ends).

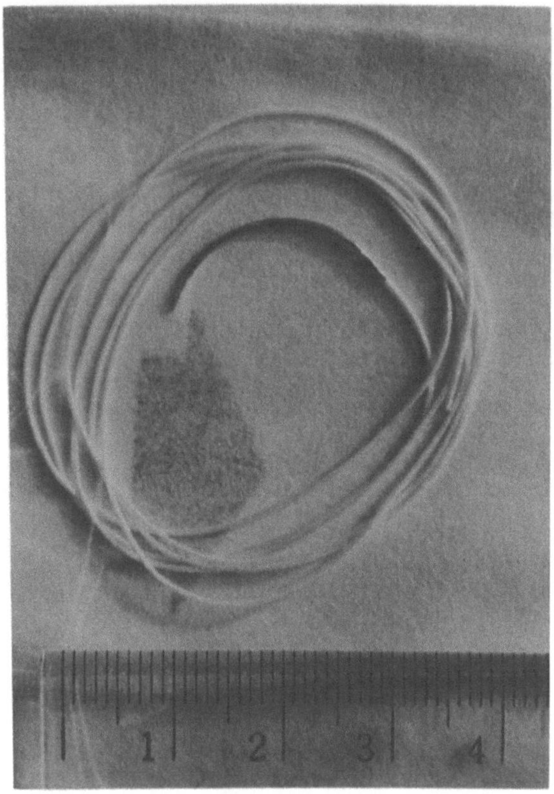


Figure 7. Chitin suture absorbable in tissues. The rate of absorption is controlled by the substitution degree and the structure of *N*-acyl groups. A product of Yunitika Co., Ltd., Uji, Kyoto, Japan.

hyaluronic acid Na salt and CM-chitin Na (MW = 250,000), respectively. The viscosity was controlled by the degree of polymerization of chitin. These properties strongly indicate that the di-*O*-methylchitin is usable as a novel cosmetic ingredient.

## 4. Commercial Products

Figure 6 shows commercial hair care products containing chitosan, and a commercial skin care product containing *O*-CM-chitin (its commercial name is "chitin liquid"). These products are now on the market. Chitin suture (Figure 7) is one of the applications of digestibility of chitin by lysozyme in tissues.

## REFERENCES


1. S. Hirano, H. Seino, Y. Akiyama, and I. Nonaka, in: "*Progress in Biomedical Polymers,*" C. G. Gebelein and R. L. Dunn, Eds., Plenum Press, New York, 1990, p. 283-290, "Chitosan: A Biocompatible Material for Oral and Intervenous Administrations."
2. T. Imoto and K. Yagishita, Agric. Biol. Chem., **35**, 1154 (1971).
3. S. Hirano, Y. Ohe, and H. Ono, Carbohydr. Res., **47**, 315 (1976).
4. Y. Tominaga and Y. Tsujisaka, Biochem. Biophys. Acta, **410**, 145 (1975).
5. S. Hirano, Agric. Biol. Chem., **42**, 1939 (1978).
6. S. Hirano and Y. Ono, Carbohydr. Polym., **4**, 15 (1982).
7. R. Trujillo, Carbohydr. Res., **7**, 483 (1968).
8. W. N. Haworth, E. L. Hirst, and H. A. Thomas, J. Chem. Soc., 1885 (1931).
9. S. Hirano, C. Itakura, H. Seino, Y. Akiyama, I. Nonaka, N. Kanbara, and T. Kawakami, J. Agr. Food Chem., **38**, 1214-1217 (1990).
10. S. Hirano and Y. Yagi, Carbohydr. Res., **83**, 103 (1980).
11. M. L. Wolfrom, J. R. Vercellotti and D. Horton, J. Org. Chem., **29**, 547 (1964).

# POLYSACCHARIDES: "NATURALS" FOR COSMETICS AND PHARMACEUTICALS

George L. Brode

Union Carbide Chemicals and Plastics Company Inc.
P. O. Box 670
Bound Brook, NJ 08805

Certain polysaccharides and polysaccharide derivatives are "natural" products for use in cosmetic and therapeutic applications. Polysaccharides are ubiquitous in nature, found extensively in both plants and animals. Largely of carbohydrate structure, this class of polymers is essentially derived from sugar-type molecules by cyclodehydration/addition reactions. Polymer properties and the resulting formulations containing them depend on several key structural parameters, including: (1) molecular weight and molecular weight distribution; (2) conformation and stereochemistry in the monomeric and oligomeric repeat units; (3) type of functional group(s); (4) derivatization - both the type and distribution; (5) charge density and distribution in ionic polymers; (6) hydrophile/lipophile ratios and distribution patterns, etc. These factors are discussed with a view to capitalizing on those features important for cosmetic and therapeutic applications. Emphasis is placed on cellulose derivatives, chitin/chitosan, and hyaluronic acid.

## OVERVIEW

Polysaccharides are found widely distributed in nature and are "naturals" for use in various cosmetic, pharmaceutical and food products.[1-6] This paper focuses on the first two application areas and is divided into three segments: (1) Polysaccharide Categories and Sources; (2) Key Molecular Parameters and Structure Property Relationships; (3) Applications in Cosmetics and Therapeutics.

## DISCUSSION

### 1. Categories and Sources

Polysaccharides can be arbitrarily divided into four categories: nonionic, anionic, cationic, and hydrophobe modified; there has been

| Table 1. Polysaccharide categories and sources (examples) | |
|---|---|
| NONIONIC | ANIONIC |
| Cellulosics (P-l)<br>Starch (P-l)<br>Guar (P-l)<br>Agarose (P-m)<br>Chitin (A-c)<br>Dextran (M) | Cellulosics (P-l)<br>Starch (P-l)<br>Carrageenan (P-m)<br>Alginic Acid (P-m)<br>Carboxymethyl Chitin (A-c)<br>Hyaluronic Acid (A-p;M)<br>Xanthan (M) |
| CATIONIC | HYDROPHOBE MODIFIED |
| Cellulosics (P-l)<br>Starch (P-l)<br>Guar (P-l)<br>Chitosan (A-c) | Cellulosics (P-l)<br>Emulsion (M) |
| P = Plant; A = Animal; M = Microbial<br>l = land; m = marine<br>c = crustacea; p = poultry, etc. | |

minimal work on amphoterics. The number of applications roughly parallel this order. Major sources of polysaccharides include plant, animal, and microbial, with plant sources being the largest. Table 1 gives a breakdown of examples in the various categories, together with their primary sources.

In the cases of cellulose, chitin and chitosan, derivatization of the starting polymer is required for solubilization in hydrophilic media, regardless of molecular weight. By comparison, polymers such as starch, guar, hyaluronic acid, etc., are soluble in the underivatized state, though in the former two cases, derivatization is often used to obtain specific properties. Finally, hydrophobe modified polysaccharides,[7,8] as a category, is included because of the unique properties that can be conveyed by incorporating such functional groupings; applications are just now emerging but can be expected to expand in the future.

What Table 1 does not portray is the enormous range of properties available with polysaccharides. Numerous molecular variations found in nature, or designed by synthetic chemists, provide this array of possible structures. Many papers in this book deal with various polysaccharides that are useful in cosmetics and pharmaceuticals, and they represent but a fraction of the research and development worldwide. In a superficial attempt to shed some light on the unlimited number of permutations and combinations, key molecular parameters are briefly discussed in the next section as an introduction to the properties and applications.

## 2. Key Molecular Parameters

There are three key parameters that govern polysaccharide properties:

(1) Repeat unit:
Structure and conformation.

(2) Substituents:
Type
Concentration
Pattern (DS, degree of substitution and MS, molar substitution)

(3) Molecular weight

## 2A. Structure and Conformation

The importance of structure and conformation of the repeat unit (or sequence) can be gleaned by considering the repeat unit in the most abundant polysaccharide, cellulose (III), (Figures 1 and 2). This repeat unit is essentially that derived from pyranose (II). Removal of a molecule of water from the 1,4-hydroxyl groups leads to cellulose. Depiction of pyranose in a planar form (II) fails to distinguish conformational features, more clearly seen from the "chair" representation (II'). In pyranose (and in the anhydroglucose unit of cellulose), all four hydroxyl groups at carbons 1, 2, 3 and 4, together with the methylol group at carbon 5, are in the equatorial position. Other polysaccharides are similarly related to carbohydrate entities which can display other axial and equatorial combinations. These are extremely important in determining properties and end-use applications. Figure 3 shows the second most abundant polysaccharide, chitin, also a highly crystalline, water insoluble, β-1,4-polymer.[9,10]

The anhydro polymeric structures ($-H_2O$) can result from other hydroxyl groups in various combinations, as, for example, alternating 1,3- and 1,4- in hyaluronic acid (Figure 4); 1,5- in dextran (Figure 5); and so forth.

While equatorial linkages are common, they are not unique. The β-1,4-linkage in cellulose, resulting in a linear structure (Figure 6), can be contrasted with the axial-equatorial linkage in starch (Figure 7). The latter yields a coiled structure with less crystallinity and greater solubility.

Such molecular asymmetry can also occur from substituents in the polysaccharide. Guar is an example. Approximately every other anhydroglucose unit is substituted on the 6-hydroxyl. Despite the β-1,4- conformation Guar (Figure 8) is highly water soluble.

(1)

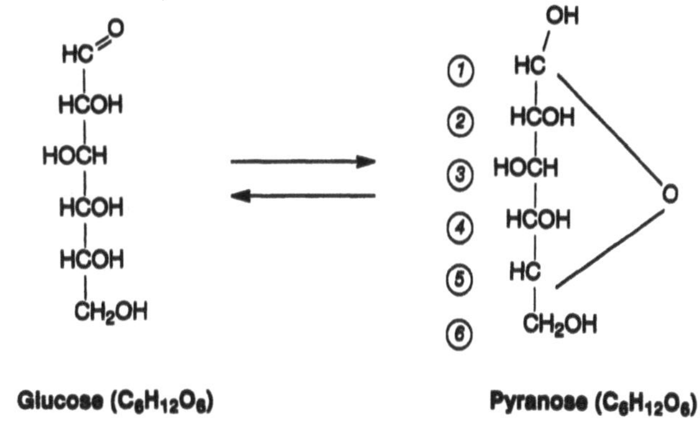


Figure 1. Glucose ($C_6H_{12}O_6$) to pyranose ($C_6H_{12}O_6$) equilibrium.

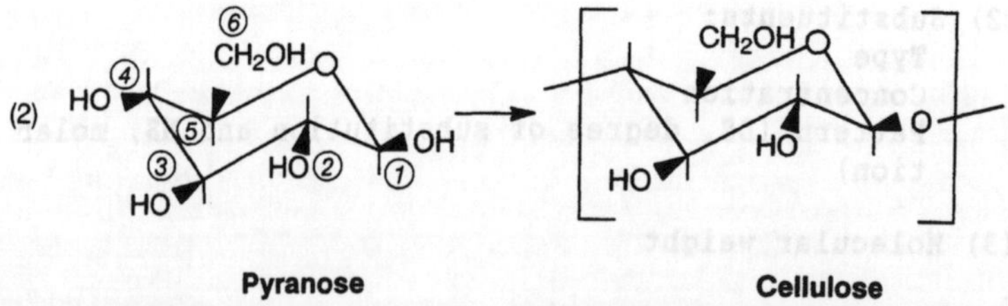


**Figure 2. Pyranose to cellulose transformation.**

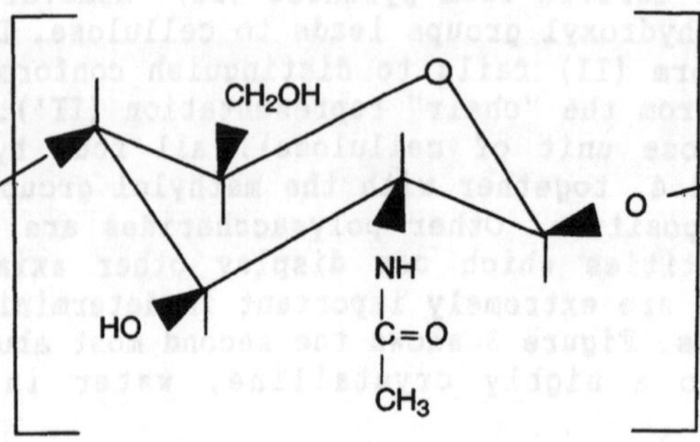


**Figure 3. Chitin (β-1,4).**

**Guar is unique in a second respect. It is one of the few natural polysaccharides with a cis-2,3-hydroxyl conformation (axial equatorial). Such a conformation provides close proximity of the hydroxyl groups which allows for complexation with multivalent cations such as $Ti^{+4}$. Strong, thermally and hydrolytically stable gels can be obtained as a result. Certain glycol modified polysaccharides have been recently described[11] which display similar gellation characteristics. In both cases these gels are quite different from those observed with, for example, carrageenan (ionic), agarose or xanthan (hydrogen bonding), or hydrophobe modified systems (domain interaction).**

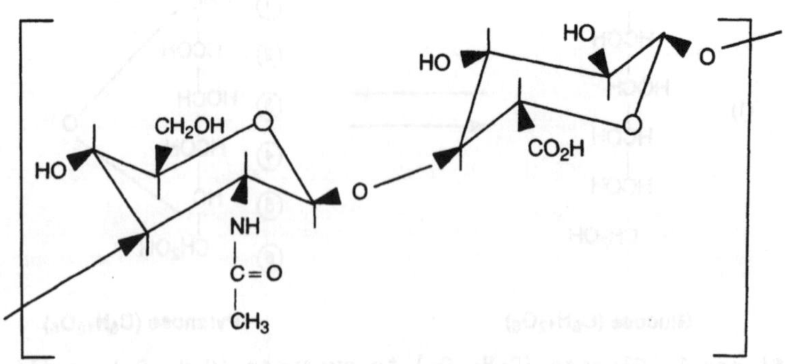


**Figure 4. Hyaluronic acid (β-1,3-; β-1,4-).**

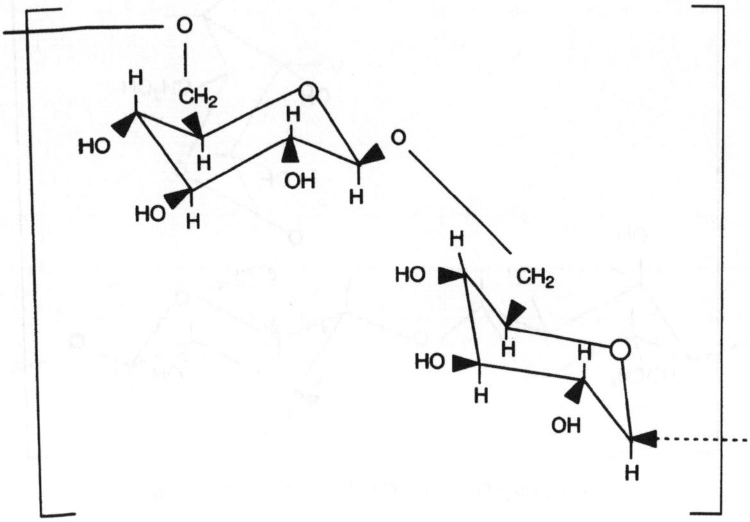


Figure 5. Dextran (β-1,6-).

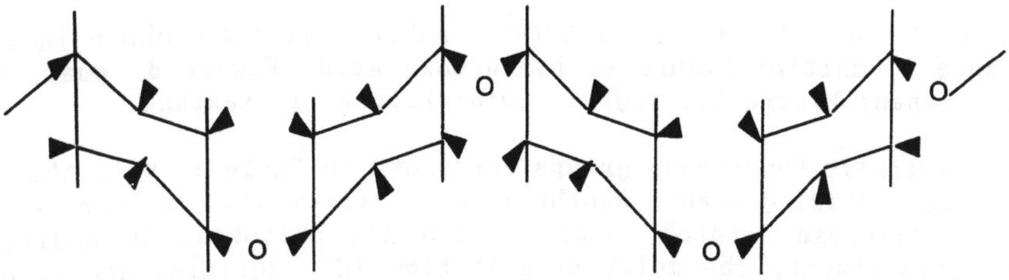


1,4 Equatorial-Equatorial: Linear, Crystalline, Insoluble

Figure 6. A β-1,4-conformation (cellulose).

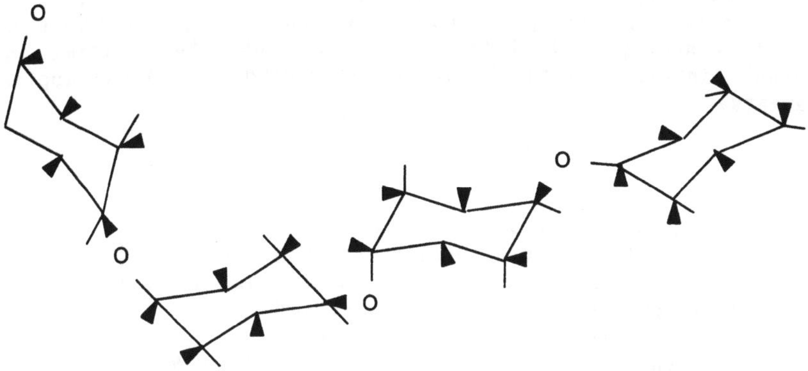


1,4 Axial-Equatorial: Coiled, Less Crystalline, Soluble

Figure 7. An α-1,4-conformation (starch).

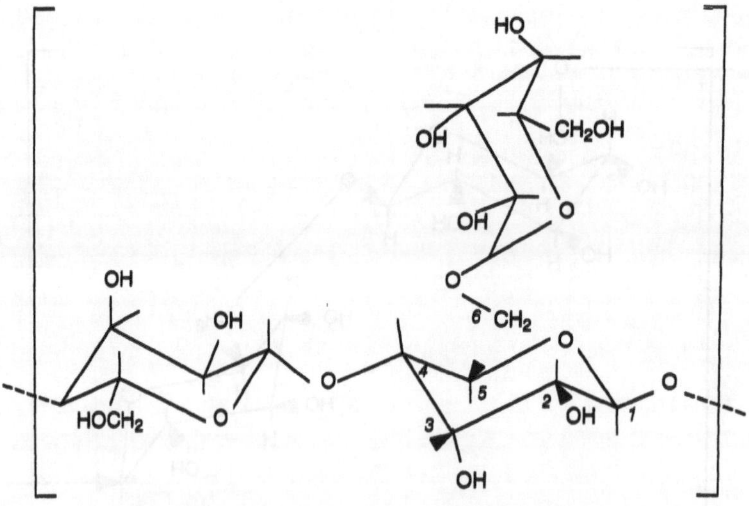


2,3 Hydroxyl Groups in *Cis* Conformation (axial, eq.)

**Figure 8. Guar (β-1,4-; 6-substituted).**

## 2B. Substituents

The most commonly occurring *natural* substituents are shown in Table 2 and Figure 3, chitin; Figure 4, hyaluronic acid; Figure 8, guar; Figure 9, carrageenan; Figure 10, alginic acid; Figure 11, xanthan.

Key *synthetic* functional groups are shown in Table 3, with the corresponding derivatizing agent. Synthetic derivatives are confined, essentially, to cellulose, starch, guar, chitin and chitosan. In addition to type of substituent, the molar substitution (MS, defining amount of incorporation) and the degree of substitution (DS, defining the distribution pattern) are critical parameters.

Carboxymethyl cellulose (Figure 12) and chitosan pyrrolidone carboxylate (Figure 13) are examples of, respectively, derivatized anionic and cationic polysaccharides.[12] As with other derivatized polysaccharides (for example, ethyl cellulose, hydroxypropyl cellulose, hydroxyethyl cellulose, hydroxypropyl guar, cationic starch, hydrophobe modified cationic hydroxyethyl cellulose, etc.), the substituent level and (in some cases) distribution can be varied according to end-use application requirements.

**Table 2. Key natural substituent groups**

| | |
|---|---|
| Primary Hydroxyl | $-CH_2OH$ |
| Secondary Hydroxyl | $-CHOH-$ |
| Carboxylate | $-CO_2^-$ ($M^+$ or $H^+$) |
| Sulfonate | $-OSO_3^-$ ($M^+$) |
| Acetamide | $-NHCOCH_3$ |
| Amine | $-NH_2$ |
| Saccharide | $-C_6H_{12}O_6$ |

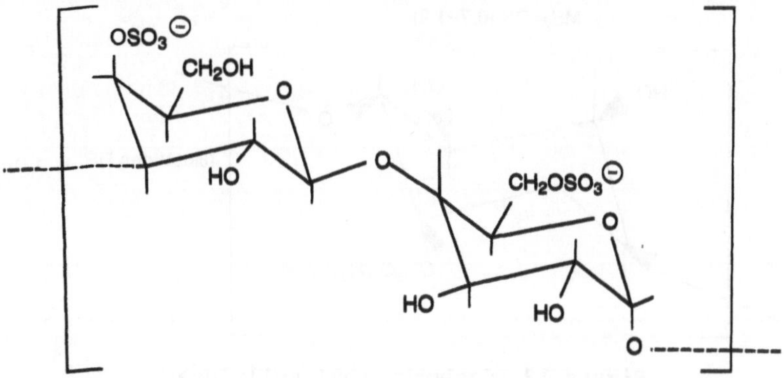


Figure 9. Lambda-carrageenan [β-1,3-; α-1,4-]

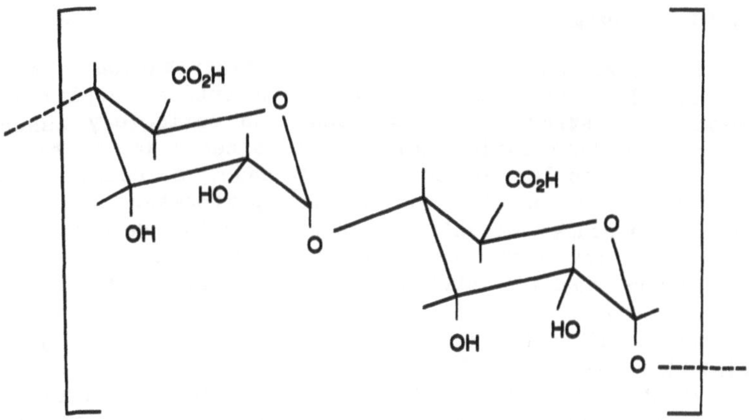


Figure 10. Alginic acid (α-1,4-; cis-2,3-hydroxyl groups).

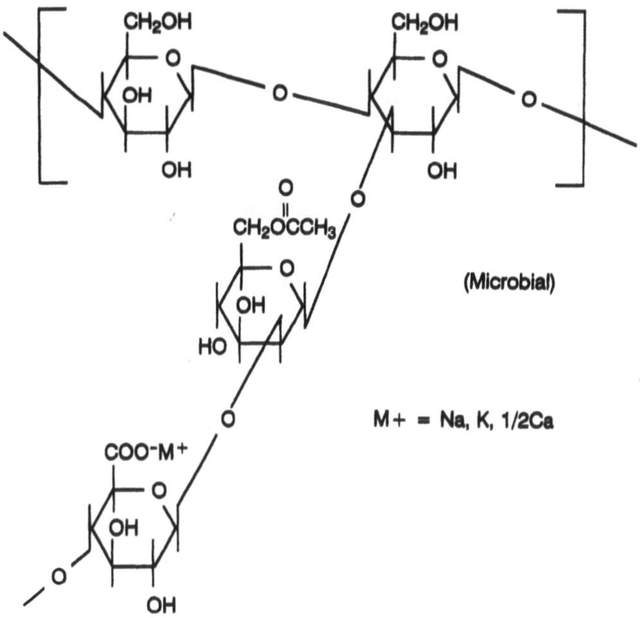


Figure 11. Xanthan.

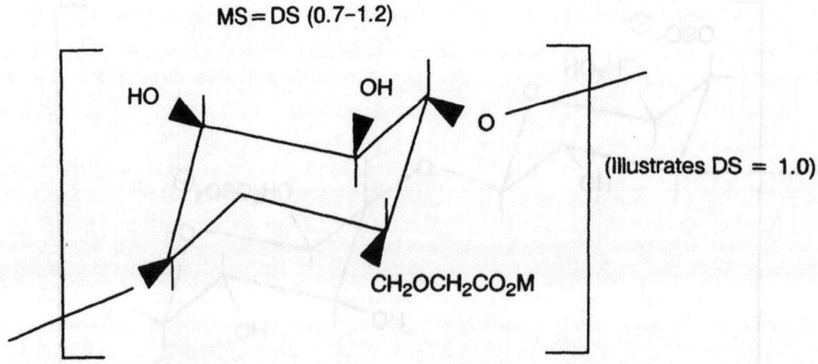


Figure 12. Carboxymethyl cellulose.

## 2C. Molecular Weight

Molecular weight and molecular weight distribution are important parameters, but for many cosmetic and pharmaceutical applications, the polysaccharide unit structure, substituent and toxicology issues are of greater importance. Formulation expertise is often a key to bridge available options in molecular weight. In any case, molecular weights are generally above the threshold required for performance, i.e., ~>20,000 weight average molecular weight. At such molecular weights, properties such as film formation mechanical properties, transmission coefficients, moisture retention, cohesive strength, etc., are adequate.

Exceptions to this generality include viscosity control, gel strength or (certain) drug delivery. In those cases, narrower molecular weight limitations (high or low) may be required. High pressure liquid chromatography is useful in developing precise molecular weight limits for nonionic polysaccharides. Ionic polymers still present a significant challenge in establishing absolute molecular weights and most certainly molecular weight distributions.

Table 3. Key synthetic functional groups

| Reactant | Substituent | Primary Substrate |
|---|---|---|
| $CH_2CH_2$ \ / O | $-(CH_2CH_2O)_nH$ | A, B |
| $CH_2$-CH-$CH_3$ \ / O | $-CH_2CH(OH)CH_3$ | A, B, C, (E) |
| $CH_3Cl$; $C_2H_5Cl$ | $-CH_3$; $-C_2H_5$ | A (E) |
| $ClCH_2CO_2M$ | $-CH_2CO_2^-M^+$ | A, B, (D) |
| $ClCH_2CH(OH)CH_2{}^+N(CH_3)_3Cl^-$ | $-CH_2CH(OH)CH_2{}^+N(CH_3)_3Cl^-$ | A, C |
| $ClCH_2CH_2NR_2\cdot HCl$ | $-CH_2CH_2NR_2$ | B |
| Chitin (IV) | $-NH_3{}^+X^-$ | D |
| $C_nH_{2n+1}Br$ | -R (Hydrophobe) | A |

A = cellulose; B = starch; C = guar; D = chitin; E = chitosan () indicates an emerging class of materials.

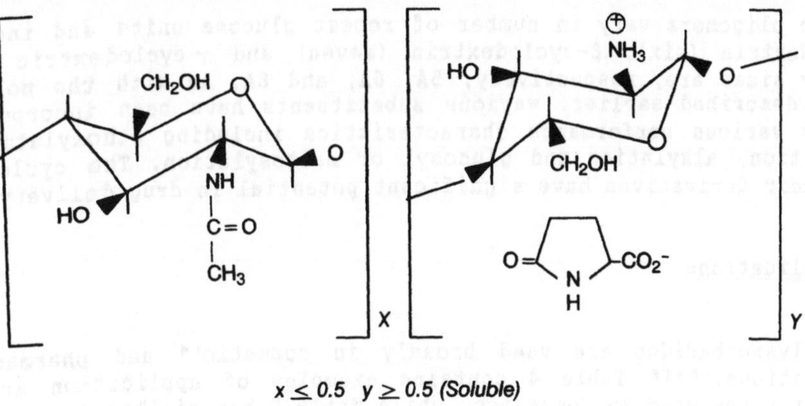


**Figure 13. Chitosan pyrrolidone carboxylate.**

In concluding this segment of the paper, a class of very low molecular weight "polysaccharides" not discussed earlier should be mentioned. The cyclodextrins are an emerging class of unique carbohydrates.[13] These

**Table 4. Polysaccharides in cosmetics**

| Application | Polysaccharide examples |
|---|---|
| Skin moisturizers/ Conditioners | Hyaluronic acid<br>Chitosan salts/derivatives<br>Cationic guar |
| Soap | Starch<br>Starch derivatives |
| Sunscreens | Hydroxyethyl/propyl cellulose<br>Xanthan |
| Cleansing creams/ Body lotions | Hydrophobe modified nonionic and cationic polysaccharides<br>Cationic guar |
| Gels/ Lotions | Cellulosics - cationic/nonionic<br>Gellan<br>Hydrophobe modified cellulosics |
| Toothpaste | Carrageenan<br>Sodium carboxymethylcellulose<br>Xanthan |
| Foam builders/ Mousses | Hydrophobe modified polysaccharides<br>Methyl cellulose |
| Shampoo | Hydroxyethyl cellulose<br>Cationic cellulosics<br>Xanthan |
| Nail lacquer | Nitrocellulose |

cyclic oligomers vary in number of repeat glucose units and include: α-cyclodextrin (six), β-cyclodextrin (seven) and γ-cyclodextrin (eight). Cavity sizes are, respectively, 5A, 6A, and 8A. As with the polysaccharides described earlier, various substituents have been incorporated to modify various performance characteristics including ethoxylation, propoxylation, alkylation and glucosyl or maltosylation. The cyclodextrins and their derivatives have significant potential in drug delivery.

## 3. Applications

Polysaccharides are used broadly in cosmetic[14] and pharmaceutical applications.[15-18] Table 4 contains examples of application areas and polymer types used in cosmetics, while Table 5 has similar information on pharmaceutical applications. New applications are continuously emerging and are expected to do so in the future as a result of structure property versatility inherent in this class of naturally based polymers.

## CONCLUSION

This brief overview of an important class of excipients, namely the

Table 5. Polysaccharides in pharmaceuticals

| Application | Polysaccharide Examples |
|---|---|
| Gastro-intestinal Delivery | Alginates<br>Chitosan |
| Sustained release capsules | Alginates<br>Hydroxypropyl methyl cellulose |
| Tablet binders (vitamins, etc.) | Ethyl cellulose<br>Methyl cellulose |
| Suspending agents | Carboxymethyl cellulose<br>Xanthan |
| Sutures | Chitin<br>Chitosan |
| Enteric coatings | Hydropropyl methyl cellulose<br>Hydropropyl methyl |
| cellulose phthalate | |
| Periodontal tissue repair | Chitosan derivatives |
| Ethical moisturizers\ Shampoo | Xanthan gum<br>Hyaluronic acid<br>Cationic cellulosics<br>Chitosan derivatives |
| Ophthamology | Hyaluronic acid |
| Protective lotions | Cationic cellulose/guar<br>Hydrophobe modified cellulosics |

polysaccharides, gives a fleeting perspective on the versatility of polymer structures and their broad utility. Available from several abundant natural sources, the almost unlimited options available with these materials, together with the inherent biocompatible feature of carefully manufactured and purified products, make the polysaccharides a "natural" choice for a wide range of cosmetic and pharmaceutical applications.

## REFERENCES


1. R. L. Whistler and J. N. BeMiller, Eds., "*Industrial Gums: Polysaccharides and Their Derivatives,*" 2nd Ed., Academic Press, New York, 1973.
2. R. L. Davidson, Ed., "*Handbook of Water Soluble Gums and Resins,*" McGraw-Hill, New York, 1980.
3. A. Hebeisch and J. T. Guthrie, "*The Chemistry and Technology of Cellulosic Copolymers,*" Springer-Verlag, Boca Raton, 1981.
4. MIT Sea Grant Program, "*Biotechnology of Marine Polysaccharides,*" Mass. Institute of Technology, 1984.
5. P. A. Sandford, et al., *Pure and Applied Chemistry,* 56(7), 879-892 (1984). "Microbial Polysaccharides: New Products and Their Commercial Applications."
6. O. B. Wurzburg, "*Modified Starches: Properties and Uses,*" CRC Press, Boca Raton, 1986.
7. L. Landoll, U.S. Patent 4,529,523 (to Hercules Inc.), July 16, 1985, "Hydrophobically Modified Polymers."
8. G. L. Brode, R. L. Kreeger, E. D. Goddard, F. M. Merritt and D. B. Braun, U.S. Patent 4,663,159 (to Union Carbide Chemicals and Plastics Company Inc.), May 5, 1987, "Hydrophobe Substituted, Water-Soluble Cationic Polysaccharides."
9. R. A. A. Muzzarelli, "*Chitin,*" Pergamon Press, London, 1978.
10. Chitin and Chitosan, 4th International Conference, August 1988, Trondheim, Norway.
11. G. L. Brode, J. P. Stanley and E. M. Partain, U.S. Patent 4,579,942 (to Union Carbide Chemicals and Plastics Company Inc.), April 1, 1986, "Polysaccharides, Methods for Preparing Such Polysaccharides and Fluids Utilizing Such Polysaccharides."
12. E. M. Partain and G. L. Brode, U.S. Patent 4,929,722 (to Union Carbide Chemical and Plastics Company Inc.), May 29, 1990, "Acid Decrystillization of Aminopolysaccharides and Derivatives Thereof."
13. D. Duchene and D. Wouessidjewe, *Pharmaceutical Technology,* 14, pp. 26, 28, 32, 34, June 1990. "Physiochemical Characteristics and Pharmaceutical Uses of Cyclodextrin Derivatives, Part I."
14. R. Y. Lochhead, "The History of Polymers in Hair Care (1940-Present)," in: *Encyclopedia of Polymers and Thickeners for Cosmetics: Cosmetics and Toiletries,* 103, 23-61, December, 1988.
15. J. M. McGinity, "*Aqueous Polymeric Coatings for Pharmaceutical Dosage Forms,*" Drugs and the Pharmaceutical Sciences Monographs, Marcel Dekker, Inc., New York, 1989.
16. A. G. Nigalaye, P. Adusumilli and S. Bolton, *Drug Development and Industrial Pharmacy,* 16(3), 449-467 (1990). "Investigation of Prolonged Drug Release From Matrix Formulations of Chitosan,"
17. D. S. Jackson, U.S. Patent 4,659,700 (to Johnson & Johnson Products, Inc.), April 21, 1987, "Chitosan-Glycerol-Water Gel."
18. G. L. Brode and E. M. Partain, U.S. Patent 4,946,870 (to Union Carbide Chemicals and Plastics Company Inc.), August 7, 1990, "Delivery Systems for Pharmaceutical or Therapeutic Actives."

# CATIONIC POLYSACCHARIDES FOR COSMETICS AND THERAPEUTICS

G. L. Brode, E. D. Goddard, W. C. Harris and G. A. Salensky

Union Carbide Chemicals and Plastics Company Inc.

Chitosan salts, covalent chitosan derivatives and quaternary ammonium salts of cellulose ethers are examples of cationic polysaccharides with significant potential in cosmetic and/or therapeutic applications. Film forming properties, moisture retention, viscosifying properties and substantivity are features that make this class of materials of special interest. The substantivity of cationic polymers to hair and skin has been demonstrated by several methods. More recently, substantivity from fully formulated products has been studied *in vitro* and *in vivo* using a combination of tape stripping and Electron Spectroscopy for Chemical Analysis (ESCA). The definitive assignment of substantivity from a fully formulated product is particularly noteworthy in view of the step change in the complexity of the systems. Details of these experimental results will be presented.

## INTRODUCTION

Skin and hair substrates vary among individuals, but essentially such surfaces carry a net negative charge and are anionic/hydrophobic in character. Carefully designed cationic, or, hydrophobic modified cationic polymers are especially useful for application to these substrates. Examples of cationic polysaccharides include:

Cationic cellulosics
Cationic guar
Amine modified starches
Hydrophobe modified cationic cellulosics
Chitosan and chitosan derivatives

This paper discusses certain aspects of the last two examples which display the following unique advantages: biocompatibility; substantivity; film formation; good oxygen/moisture transmission coefficients; moisture retention without tackiness; water and water/alcohol solubility; cosmetic elegance. More specifically, the paper focuses on substantivity of certain members of these two classes with the aim of differentiating characteristics that make for substantivity to skin or hair.

*Cosmetic and Pharmaceutical Applications of Polymers*
Edited by C.G. Gebelein *et al.*, Plenum Press, New York, 1991

**EXPERIMENTAL**

Analysis of samples involves either: (a) direct treatment of stratum corneum taken from the skin of neonatal rats or; (b) tape stripping of human subjects or; (c) anionic vinyl substrates, exposed to either neat or formulated polymer systems. In the former, the stratum corneum samples were washed three times with distilled water and dried *in vacuo* before ESCA analysis. In the latter experiments, polymer solutions or formulated polymer compositions were applied to the forearm sections of human subjects or simulated anionic vinyl substrates. Tape stripping and ESCA evaluation of unwashed and washed substrates together with starting substrate were then performed. In case (b) material transferred to the tape was thus examined.

Electron spectroscopic procedures such as those described in the literature were employed.[1,2] A surface Science Instrument SSX-100 Spectrometer, equipped with a low energy electron flood gun, was used to obtain high resolution spectra, including angular measurement for depth penetration (sample homogeneity).

**DISCUSSION**

**1. Hydrophobe Modified Cationic Polysaccharides**

Cationic polysaccharides have long been used in hair care, with skin care also receiving much focus recently. To provide for both the hydrophobic and ionic nature of these substrates, structure optimization has been under way to factor in both features. One class of polymers receiving attention has been hydrophobe modified cationic cellulosics.[3] In these systems the hydrophobe can be separate from the quaternary nitrogen or, part of the quaternary nitrogen functional group hypothetically depicted by Figure 1.

The hydrophobe group and concentration have remarkable effects on solution rheology, salt sensitivity, surfactant properties and substantivity. Regarding substantivity, the combination of both hydrophobic and cationic groups in the same molecule allows a sharp reduction in cationic content without compromising other key features desirable in cationic polysaccharides.

Table 1 demonstrates the effect on solution viscosity as a function

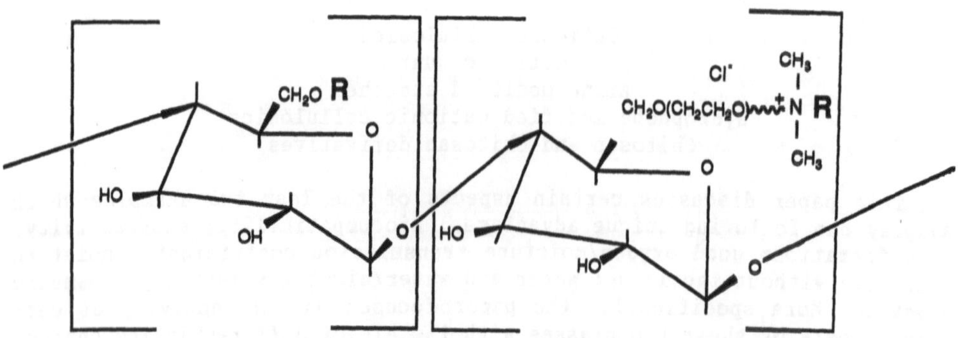


Figure 1. Hydrophobe modified cationic cellulose.

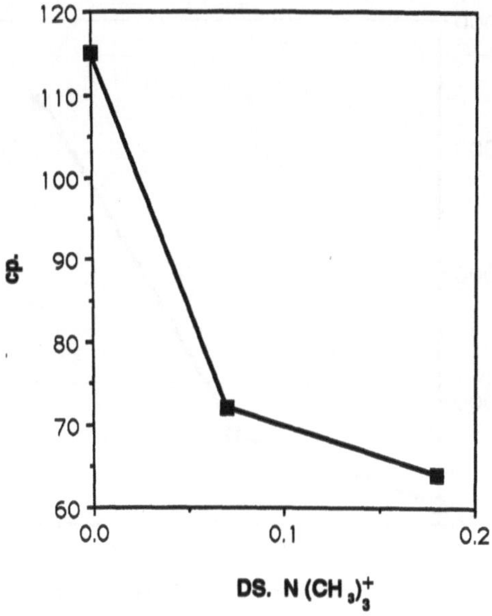


Figure 2. Viscosity vs. $-N^+(CH_3)_3$.

of several different substituent groups. Obviously, hydrophobes of >$C_8$ can have a profound effect on parameters such as solution viscosity. Depending on size and concentration of hydrophobe, water or water/alcohol solubility is greatly affected, as well as viscosity as seen by viscosity profiles of 2% solutions in water, whereas in Figure 2, with no hydrophobe present [$+N(CH_3)_3$ substitution only] a drop in viscosity occurs as the weight fraction of quaternary groups increase.

Figure 3, shows the opposite effect ($+N(CH_3)_2$ $C_{12}H_{25}$ substitution) due to incorporation of hydrophobe. Figure 4 shows the effects of these different hydrophobes, namely: (1) with $C_8$, a relatively weak hydrophobe, some compensation in viscosity reduction is seen with increasing hydrophobe concentration (compare $C_8$ of Figure 4 with $N(CH_3)_3$ in Figure 2): (2) with $C_{12}$ there is a significant increase in viscosity above 0.05 degree of substitution; (3) with a $C_{18}$ hydrophobe a very sharp increase in viscosity is seen at a substitution level of 0.01 moles of hydrophobe/mole of anhydroglucose unit.

Table 1. Effect of R structure/concentration on solution viscosity

| R | Degree of substitution | 1% soln. viscosity (cp) |
|---|---|---|
| (Hydroxylethyl cellulose) | - | 115 |
| $-CH_3$ | 0.1 | 72 |
| $-CH_2\phi$ | 0.1 | 100 |
| $-C_8H_{15}$ | 0.1 | 105 |
| $-C_{12}H_{25}$ | 0.1 | 634 |
| $-C_{18}H_{37}$ | 0.03 | 12,000 |

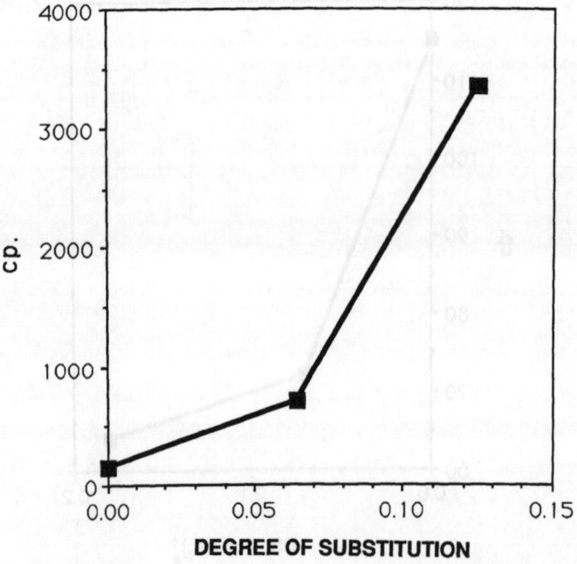


Figure 3. Viscosity vs. $-N^{+}(CH_3)_2(C_{12}H_{25})$ concentration.

Finally, in Figure 5 the effect of starting polysaccharide molecular weight at constant hydrophobe structure is seen, vis., hydrophobe is more effective as molecular weight of the substrate polymer increases, in accord with Staudinger, Mark-Houwink equations.

Figure 6 shows a very interesting salt sensitivity profile for three quaternary ammonium derivatives. The presence of a hydrophobe shows an increase in viscosity with added salt, unlike the more conventional inverse solution viscosity with non-hydrophobe containing ionic polymers.

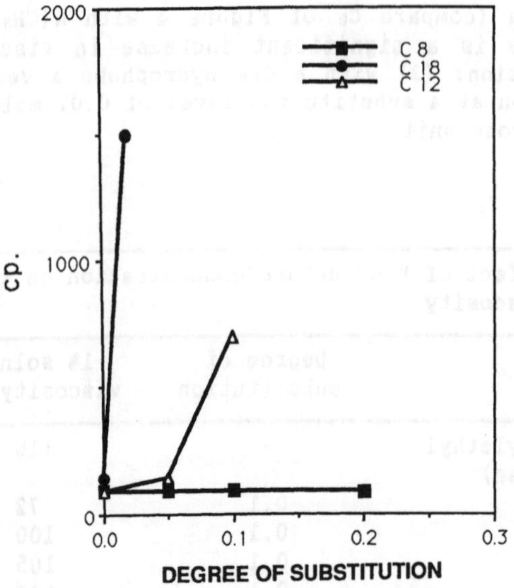


Figure 4. Viscosity vs. hydrophobe and hydrophobe concentration.

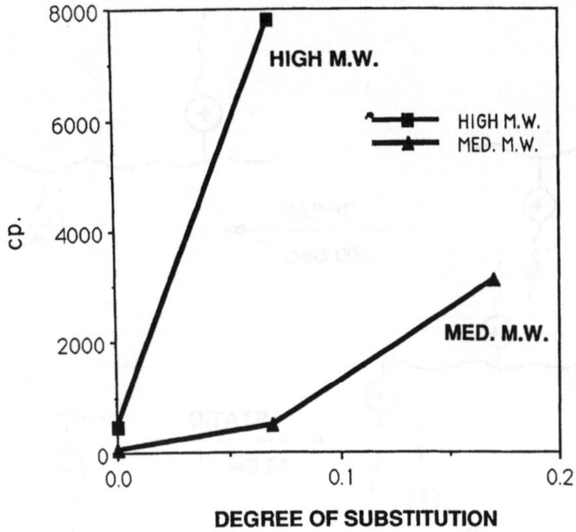


Figure 5. Viscosity vs. Hec M. W. [$-N^{+}(CH_3)_2C_{12}H_{25}$].

We believe one mechanism to explain this behavior results from enhanced hydrophobe association as the median polarity increases (hydrophobe domains "salting out" phenomenon). Such association would be relatively weak and would be expected to show reversible shear sensitivity. Measurements made on solutions of these polymers show high viscosity under "static" or low shear environments, i. e., the hydrophobe domains result in Newtonian, high viscosity solutions with mild pseudoplasticity. At higher shear rates, this association rapidly decreases (reversibly to lower viscosity; Figure 7).

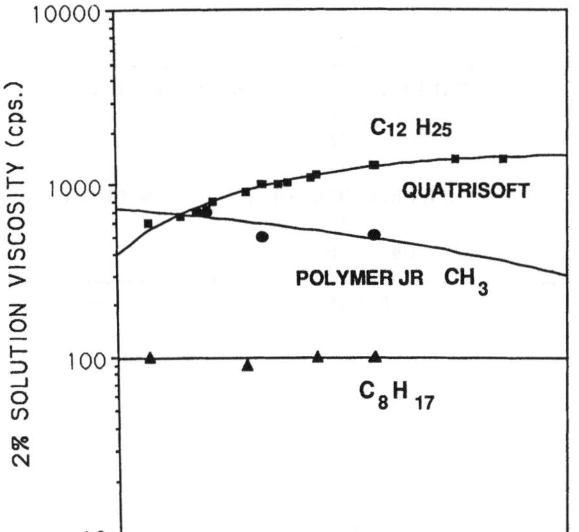


Figure 6. Effect of salt content on solution viscosities as function of R.

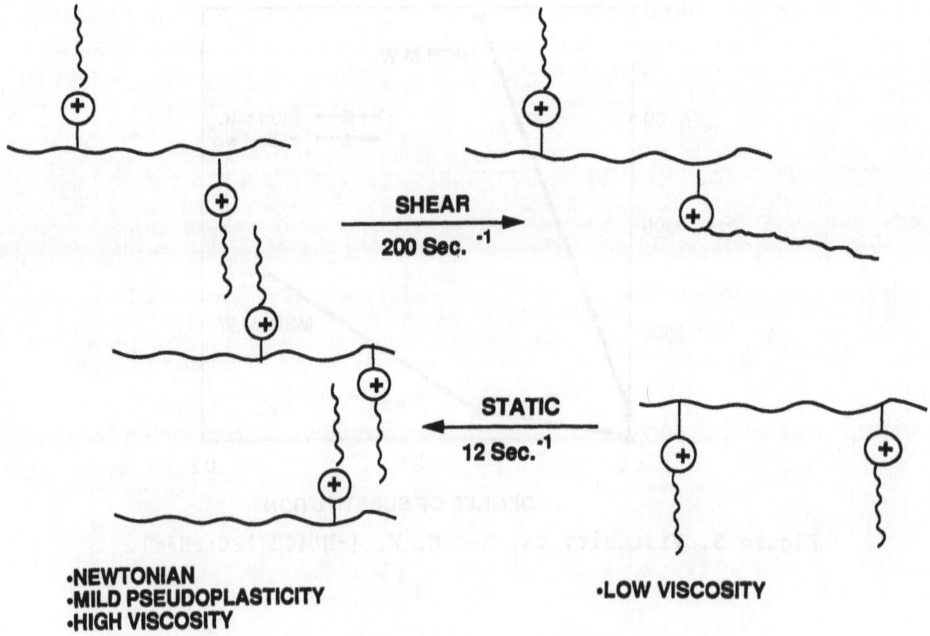


**Figure 7. Mechanism of hydrophobe activity.**

Such hydrophobe domain structure should and does lead to altered surfactant behavior. Note in Table 2 that the presence of the dodecyl group increases the surface activity, as seen by lower surface tension values. Also shown in Table 2 is the ability to stabilize a foam (formed by shaking) where again the performance is improved with the long chain hydrophobe. Foam stability in this case is undoubtedly assisted by reversibility of absorption: using a Langmuir film balance we showed that an

**Table 2. Surfactant properties**

| R | $CH_2CHCH_2N^+(CH_3)_2C_{12}H_{25}$ $Cl^-$ (OH on the central CH) | $CH_2CH_2N(CH_3)_2$ |
|---|---|---|
| D. S. | 0.036 | 0.03 |
| Time (hrs) | | Foam height (mL) |
| 0 | 24 | 18 |
| 1 | 15 | 3 |
| 3 | 10 | 0 |
| 24 | 4 | 0 |
| Time (mins) | Surface tension (dyes/cm) | |
| 1 | 59 | 67 |
| 30 | 57 | 66 |

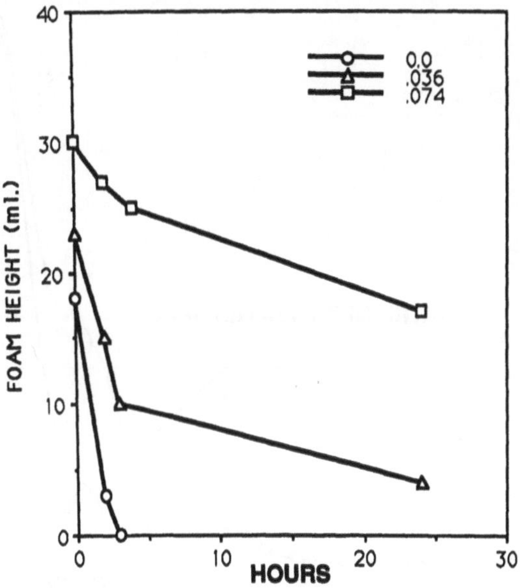


Figure 8. Foam height vs. concentration $[-N^{+}(CH_3)_2C_{12}H_{25}]$.

aged surface of a dilute (0.01%) solution of this polymer developed a surface pressure of 20 dyne/cm on compression.

Also of value in hydrophobe modified cationic polysaccharides is the substantivity of such cationic polysaccharides to hair and skin. Figure 9 shows Electron Spectroscopy for Chemical Analysis (ESCA) spectra for unwashed neonatal skin (a), and the skin after immersion in 0.1% aqueous solution of a $C_{12}$ modified cationic polysaccharide (Quatrisoft; Polyquaternium 24) followed by washing three times with distilled water (b). The increased intensity in the region of the spectra assignable to ether carbons is conclusive evidence for deposition of the polysaccharide. The emergence of a pronounced "C-O shoulder" on the main C peak is evident. Since the cationic polysaccharide contains much higher oxygen content than natural skin, polymer absorption would be accompanied by a decrease on C/O ratio. Table 3 contains atomic % data corroborating deposition and substantivity of this class of materials.

## 2. Chitosan Derivatives

A significant amount of work is under way in both academic and industrial laboratories on another class of cationic polysaccharides known as

Table 3. ESCA analysis of neonatal skin substrates

| | Atomic % | | | | |
|---|---|---|---|---|---|
| | C | O | N | C/O | C/N |
| Neonatal skin control | 81.0 | 13.0 | 2.3 | 6.2 | 35.2 |
| $C_{12}$ $H_{25}$ | 75.9 | 19.6 | 1.3 | 3.8 | 63.3 |

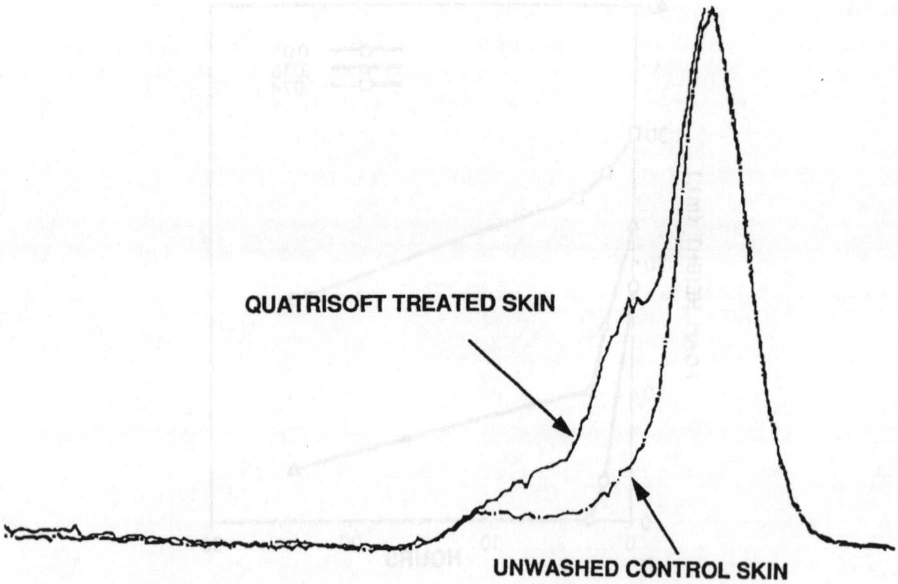

Figure 9. ESCA - carbon high resolution spectra.

chitosans (Figure 10). Structurally these materials are related to the cellulosics but are significantly different in several respects:

(1) the number two hydroxyl group is substituted by a primary amine group
(2) the cationicity is predominantly of protonated amine type; and
(3) the cationic charge density is higher.

Both chitosan salts[4] and covalent chitosan derivatives[4,5,6,7] are receiving attention. The results with chitosan salts, as regards substantivity to stratum corneum, are shown in Tables 4 though 6.

Table 4 shows the ESCA analyses of neonatal stratum corneum treated with a 0.1% solution of chitosan pyrrolidone carboxylate (C-PCA m = 0.8; n = 0.2), and subsequently washed three times with distilled water. Whereas surface changes are observed, in the expected direction, indicating the polymer to be substantive, that conclusion is somewhat equivocal since N+ was not detected on the treated outer skin surface; the treated inner surface did show both N+ and N, however. This inner surface is less hydrophobic and more protein-like in character.

Table 4. ESCA analysis of neonatal rat skin treated with chitosan pyrrolidone carboxylate (C-PCA)

| Sample | Atom% C | N | N | $N^+$ | C/N |
|---|---|---|---|---|---|
| Neonatal skin (outer)-I | 81.1 | 2.3 | 2.3 | N.D. | 35 |
| (I+C-PCA) | 79.9 | 3.9 | 3.9 | N.D. | 21 |
| Neonatal skin (inner)-II | 79.8 | 3.0 | 3.0 | N.D. | 27 |
| (II+C-PCA) | 79.3 | 2.8 | 2.0 | 0.8 | 25 |

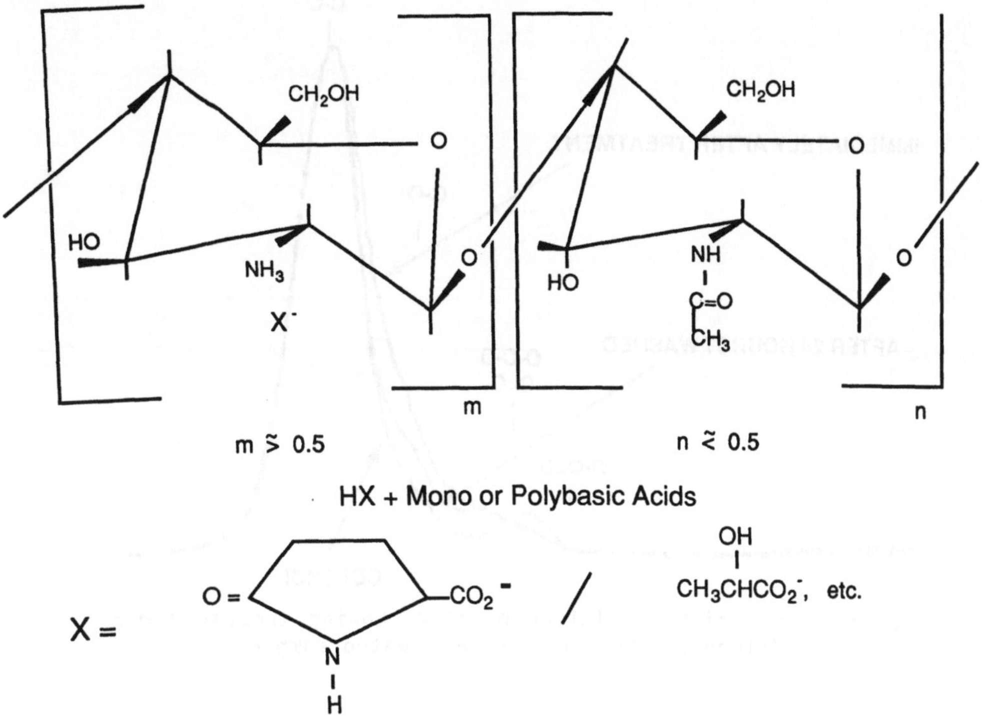


Figure 10. Chitosan derivatives.

After washing the neonatal skin in sodium dodecyl sulfate, (sebum removed), both surfaces showed high concentrations of N+ and N after chitosan salt treatment.

An *in vivo* study on human skin was also very revealing. In this case, human subjects were treated with chitosan lactate, tape-stripped before and after washing, and analyzed by ESCA. Table 5 and Figures 10 and 11 summarize the results. Clearly, chitosan lactate was retained 24 hours after application, during which subjects followed normal routines, including showering. Figure 12 is specifically interesting in that the Nitrogen spectrum overlays clearly shows the N+ pattern characteristic of the ammonium ion.

Table 5. ESCA analysis of tapestrips from human skin treated with chitosan salt solution (*in vivo*)

| | Atom % | | | | |
|---|---|---|---|---|---|
| Sample | C | O | N | C/O | C/N |
| Stripped human skin | 85.0 | 12.5 | 1.6 | 6.8 | 53.1 |
| I + chitosan salt (after application) | 80.4 | 15.0 | 3.7 | 5.4 | 21.7 |
| I + chitosan salt (after washing/24 hrs) | 81.8 | 13.7 | 3.0 | 6.0 | 27.2 |
| Tape surface | 83.7 | 16.3 | - | - | - |

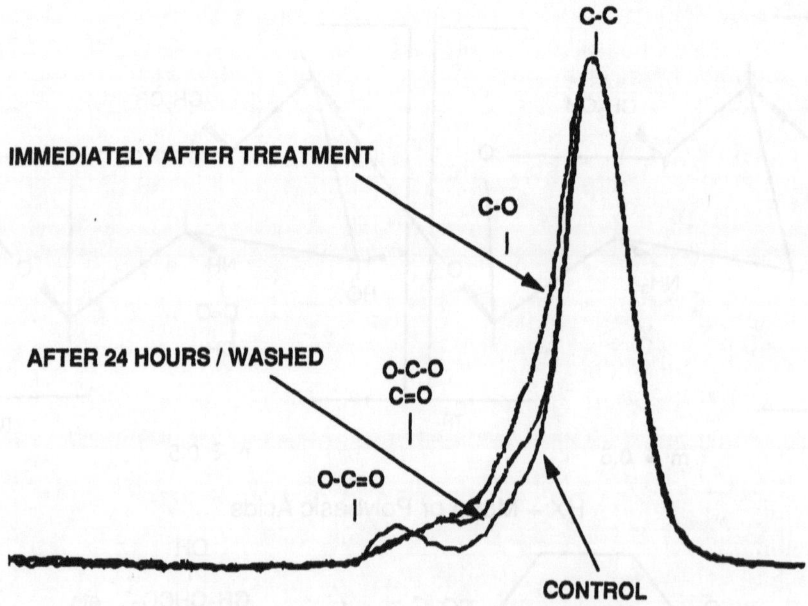


Figure 11. C1s high resolution overlay showing increased C-O (alcohol/ether) carbon on treated samples.

Finally, a study was made on formulated (silicone containing) chitosan salt compositions to determine if substantivity could be expected in more complex systems. In these experiments, an anionic vinyl substrate was used with direct ESCA analysis of the coated and washed (sodium dodecyl sulfate) specimens. The emollient cream contained a significant silicone content. As shown by the results in Table 6, such products can be expected to display a graduated composition, wherein the chitosan salt is concentrated at the anionic interface (skin) with the silicone portion predominantly at the outer surface.

## CONCLUSION

1. Cationic polymers are substantive to hair and depending on its initial condition may also be substantive to skin. This paper has concentrated on skin substantivity with hydrophobe modified cationic cellulosics and chitosan salts. A recent

Table 6. ESCA analysis of formulated chitosan emollient cream (anionic vinyl substrates)

| Sample | C | Cl | O | N | Si |
|---|---|---|---|---|---|
| Anionic vinyl | 74.3 | 17.5 | 8.2 | - | - |
| Vinyl treated with chitosan emollient (II) | 61.2 | 0.3 | 20.7 | - | 17.8 |
| II washed with SDS | 67.7 | 0.3 | 23.4 | 2.2 | 4.1 |
| II washed 2x with SDS | 69.4 | 10.6 | 16.0 | 1.1 | 2.2 |

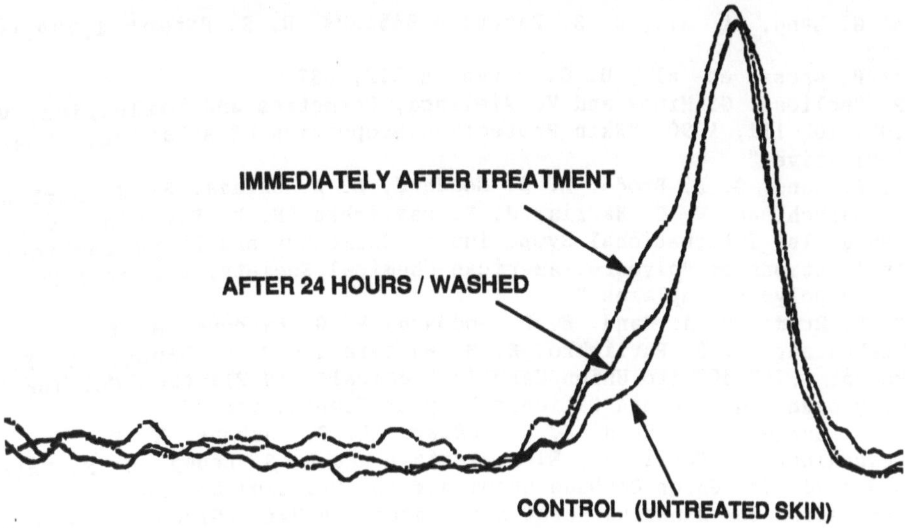


Figure 12. N1s high resolution spectra overlay showing increased $NR_4^+$ form of nitrogen on treated samples.

publication[8] has described the use of cationic guar as a skin protective agent.

2. The combination of cationicity and hydrophobic structure provides enhanced substantivity to oleophilic/anionic substrates such as skin, even at low cationic charge density and hydrophobe content.

3. Substantivity can be expected to be observed in formulated products, or even with anionic polymers when properly formulated.[9,10,11]

4. Electron Spectroscopy for Chemical Analysis is a powerful tool for determining hair and skin surface composition before and after treatment with various compositions.

## REFERENCES


1. M. K. Bahl, J. Soc. of Cos. Chemists, 36, 287-296 (1985).
2. E. D. Goddard and W. C. Harris, J. Soc. Cosmetics Chem., 36, 295-306 (1987).
3. G. L. Brode, R. L. Kreeger, E. D. Goddard, F. M. Merritt and D. B. Braun, U. S. Patent 4,663,159 (to Union Carbide Chemicals and Plastics Co. Inc.). "Hydrophobe Substituted, Water-Soluble Cationic Polysaccharides."
4. E. M. Partain and G. L. Brode, U. S. Patent 4,929,722 (to Union Carbide Chemicals and Plastics Co. Inc.). "Acid Decrystallization of Aminopolysaccharides and Derivatives Thereof."
5. R. A. A. Muzzarelli, "*Chitin*," Pergamon Press Publisher, 1978.
6. Chitin and Chitosan, 4th International Conference, August 1988, Trondheim, Norway.
7. a) G. Lang, G. Maresch, H. Wendel, E. Konrad, H. R. Lenz, J. Titze, U. S. Patent 4,835,266 (to Wella AG)."Hydroxypropylchitosan Useful for Hair or Skin Treatment."

b) G. Lang, et al., U. S. Patent 4,845,204, U. S. Patent 4,772,690, etc.
c) P. Gross, et al., U. S. Patent 4,202, 881.
8. P. Pugliese, G. Himes and W. Wielinga, Cosmetics and Toiletries, Vol. 105, 105-111, 1990. "Skin Protection Properties of a Cationic Guar Derivative."
9. P. A. Band, G. L. Brode, E. D. Goddard, E. A. Balazs, A. G. Barbone, E. Lashchiner, W. C. Harris, J. P. Pavlichko, E. M. Partain and P. S. Leung, 1st International Symposium on Cosmetics and Pharmaceutical Applications of Polymers, American Chemical Society, August 1990. "Interpolymer Complexes."
10. G. L. Brode, P. A. Band, E. D. Goddard, A. G. Barbone, A. E. Leshchiner, J. P. Pavlichko, E. M. Partain and P. S. Leung, U. S. Patent 4,767,463 (to Union Carbide Chemicals and Plastics Co. Inc.). "Glycosaminoglycan and Cationic Polymer Combinations."
11. G. L. Brode, P. A. Band, E. D. Goddard, A. G. Barbone, A. E. Leshchiner, J. Pavlichko, E. M. Partain and P. S. Leung, U. S. Patent 4,913,743 (to Union Carbide Chemicals and Plastics Co. Inc.). "Processes for Managing Keratinous Materials Using Glycosaminoglycans and Cationic Polymer Combinations."

## INTERPOLYMER COMPLEXES BETWEEN HYALURONAN AND CATIONIC CELLULOSE POLYMERS

P. A. Band,[1] G. L. Brode,[2] E. D. Goddard,[2] A. G. Barbone,[1] E. Leshchiner,[1] W. C. Harris,[2] J. P. Pavlichko,[3] E. M. Partain III [2] and P. S. Leung[2]

(1) Biomatrix Inc.
65 Railroad Ave.
Ridgefield, NJ 07657
(2) Union Carbide
(3) Amerchol

The past decade has marked a rapidly growing interest in topical applications of both hyaluronan and cationic cellulose polymers. Because of their polyelectrolyte character and opposite charge, they can form strong interpolymer complexes with novel properties and significantly enhanced utility. This paper will describe the physiochemical properties and applications of complexes formed between hylan (a modified hyaluronan) and cationic cellulose polymers. The effect of charge density and hydrophobe content of the cellulosics on water solubility and stability of the complexes will be outlined. For stable water soluble complexes, the rheological properties of their solutions will be described, and conditions for optimizing their viscoelastic properties will be defined. The utility of the interpolymer complexes between hylan and cationic cellulosics will be demonstrated in terms of their rheological synergy and their enhanced substantivity to keratinous substrates. The latter has been determined by ESCA analysis and radiotracer studies.

## RESULTS AND DISCUSSION

Water soluble polymers are widely used in topical formulations, and can physically modify the surface of keratinous tissues.[1] They can be designed to provide specific functionalities in hair and skin treatment products through derivatization and complexation. This paper will describe the development of a multifunctional polymer complex with distinctive topical benefits.

Hyaluronan (HA, hyaluronic acid) has become well established for its ability to form a hydrated viscoelastic matrix on the surface of the skin.[2,3] Polycations are likewise established for their substantivity to keratinous substrates.[4,5] By combining the viscoelasticity and hydration capacity of a derivitized hyaluronan preparation (hylan, U.S. Patent

4,713,448)[6] with the substantivity of a dodecyl substituted cationic cellulose polymer (U.S. Patent 4,663,159),[7] an interpolymer complex can be formed which optimizes the durability and effectiveness of these polymers to physically modify keratinous surfaces (U.S. Patent 4,767,463).[8]

## OVERVIEW OF HYLAN-POLYCATION INTERACTIONS

Because of its polyanionic character, hylan displays a strong interaction with charged cationic polymers. The strength of this interaction, and its physical manifestations, depend on many factors. Structural parameters, charge density, concentration ratios and preparation methods all contribute to the outcome of hylan complexation with particular polycations.

Different classes of cationic polymers were tested for their compatibility with hylan. The polycation hylan complexes can be categorized in terms of their apparent solubility in salt-free aqueous solution. As judged by solution turbidity, water soluble complexes can be formed with Polyquaternium 4 (diallyldimethyl ammonium chloride/ hydroxyethylcellulose copolymer), Polyquaternium 11 (dimethyl sulfate quaternized vinylpyrrolidone/dimethylaminoethyl methacrylate copolymer), and Polyquaternium 24 (N-dodecyl,N,N-dimethyl quaternary ammonium chloride substrated hydroxyethyl cellulose). Water insoluble complexes are formed with Polyquaternium 6 (poly dimethyl diallylammonium chloride), Polyquaternium 10 (N,N,N-trimethyl ammonium chloride substrated hydroxyethyl cellulose), and Polyquaternium 7 (acrylamide/ dimethyl diallylammonium chloride copolymer). In general, the polycation's charge density seems to be the most important factor determining the solubility of its complex with hylan.

Polyquaternium 24 (PQ24) is of particular note because of its pendent alkyl chains. When combined with a high molecular weight hylan preparation, it forms a stable interpolymer complex with interesting and important properties. The concentration ratio dependence for formation of a hylan-PQ 24 complex is shown in Figure 1. Its formation can be followed by turbidity measurements. At high PQ 24:hylan ratios, the complex causes little scattering of visible light. As the PQ 24:hylan ration is lowered, the aggregate size increases, as does the turbidity. Below a PQ 24:hylan ratio of 20, water miscibility decreases and an aqueous coacervate phase forms. This results in the observed sharp drop in turbidity measurements.

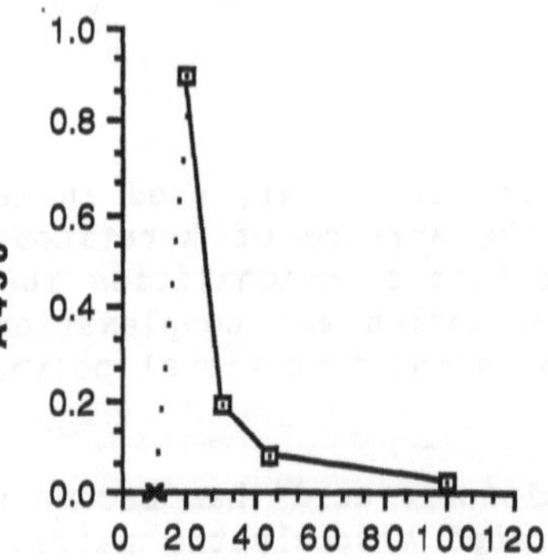


Figure 1. Turbidity of hylan-P24 compositions.

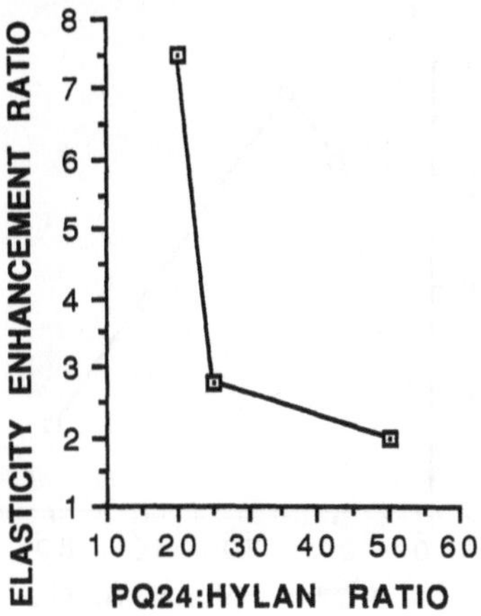


Figure 2A. Elasticity of hylan-PQ24 compositions.

## RHEOLOGICAL SYNERGY

The formation of a hylan-PQ 24 complex is also evident from rheological measurements. When hylan and PQ-24 solutions are combined, there is a dramatic enhancement in both viscosity and elasticity compared to that of the separate polymer solutions. This rheological synergy is evident in Table 1. Both the viscosity and elasticity of the hylan-PQ24 are enhanced by approximately an order of magnitude compared to that of the separate polymers. Figures 2A and 2B illustrate the dependence of these rheological properties on the PQ24:hylan concentration ratio. For both the viscosity and elasticity, the synergistic interaction between the polymers is optimized as the polymer concentration ratio approaches the range where coacervation occurs.

The ability of a hyaluronan preparation to provide this type of viscoelastic synergy with PQ24 is dependent on both its molecular weight and purity. The importance of these factors is evident from Table 2, which compares an impure hyaluronan preparation of moderate molecular weight (approximately 2 million) with a pure high molecular weight hyaluronan derivative (hylan, molecular weight approximately 5 million) and an hyaluronan prepared by bacterial fermentation (molecular weight is approxi-

Table 1. Rheological synergy of the HY-PQ24 complex

| | Shear Viscosity at 0.05 $s^{-1}$ (Pas) | Modulus of elasticity at 0.5 Hz (Pa) |
|---|---|---|
| 2% PQ24 | 1.27 | 0.152 |
| 0.1% HY | 2.02 | 0.565 |
| Additive: | 3.29 | .717 |
| Measured: | 33.1 | 6.91 |
| Rheological enhancement | 10 fold | 9.6 fold |

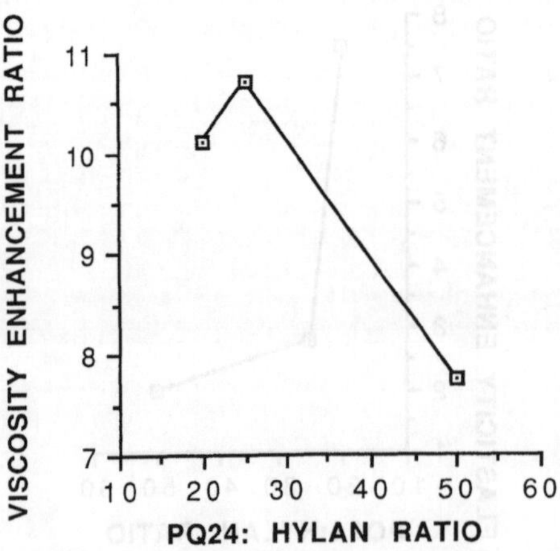


Figure 2B. Shear viscosity of hylan-PQ24 compositions.

mately 0.5 million). The highly significant differences in the viscoelasticity of these hyaluronan-PQ24 complexes is due to the effect of protein impurities on the complex, and the endogeneous viscoelastic properties of the hyaluronan preparation tested.

## DEPOSITION ONTO KERATINOUS SUBSTRATES

An important property of the hylan-PQ24 complex is that it significantly enhances the substantivity of hylan to skin and hair. This enhanced substantivity is dependent on the Polyquaternium 24 to hylan ratio, as shown in Figures 3A for hair and 3B for neonatal rat stratum corneum. The ratio providing for optimal deposition on these two substrates is very different. This may be due to compositional and electrostatic differences between skin and hair, as well as differences between the top and bottom surfaces of the stratum corneum membranes.

To evaluate the top surface of stratum corneum separately, and to determine how the substantivity varies when delivered from o/w emulsions, deposition of the tritiated hylan was determined by spreading a radiolabeled emulsion onto only the top surface of the stratum corneum membranes. Both dry (ambient humidity), and hydrated (100% humidity) mem-

Table 2. Comparison of HA preparations in complex with PQ-24 (PQ-24 concentration constant at 2.5%)

| | 0.1% hylan[1] | 0.1% Animal HA[2] | 0.1% Bacterial HA[3] |
|---|---|---|---|
| shear viscosity ($0.19\ s^{-1}$) | 77 Pas | 13 Pas | 2.5 Pas |
| elastic modulus (5Hz) | 60.3 Pa | 19.7 Pa | 8.5Pa |

(1) MW 5-7 million, protein content 0.001%
(2) MW 2-3 million, protein content 0.2%
(3) MW 500,000, protein content 0.002%

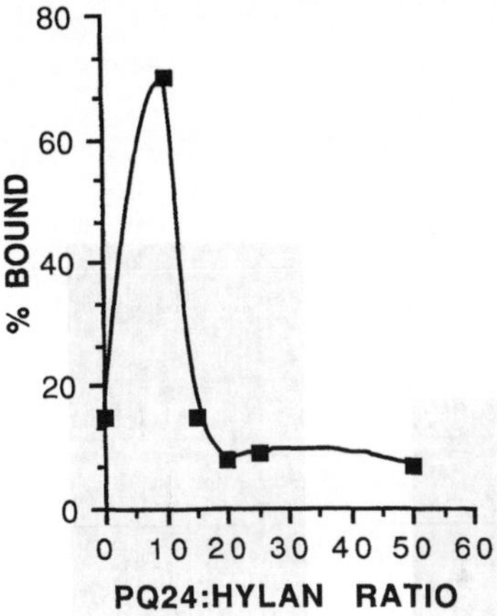


Figure 3A. Substantivity of hylan-PQ24 compositions to hair.

branes were tested for comparison, to evaluate the effect of stratum corneum swelling in these experiments. This data, presented in Figure 4, demonstrate significant deposition of the hylan from a non-ionic formulation. Of particular interest is the improvement of deposition when the stratum corneum is swollen by hydration, in which case nearly complete deposition is achieved (92% of the $^{3}H$-hylan applied), despite extensive post-application washing.

The substantivity of hylan-PQ24, to hair and skin was also demonstrated by measuring electrokinetic streaming potentials and by photoelectron spectroscopy (ESCA). Using the latter technique, the complex was found to provide a uniform surface coverage as can be seen from the data in Table 3; namely the untreated skin control, (I), being primarily proteinaceous displays no detectable quaternary nitrogen ($N^{+}$). A similar spectrum is seen in (II), i.e., skin treated with hylan and washed in distilled water; hylan, being anionic in nature, is not inherently substantive to skin and is removed by rinsing. By comparison, skin treated with PQ-24, a hydrophobe modified cationic polysaccharide containing quaternary nitrogen, show a very different ESCA profile, (III), after identical treatment with distilled water; note specifically the presence

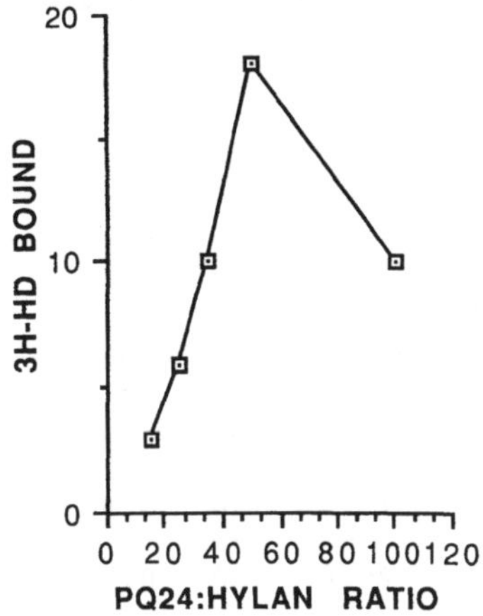


Figure 3B. Substantivity of hylan-PQ24 compositions to skin.

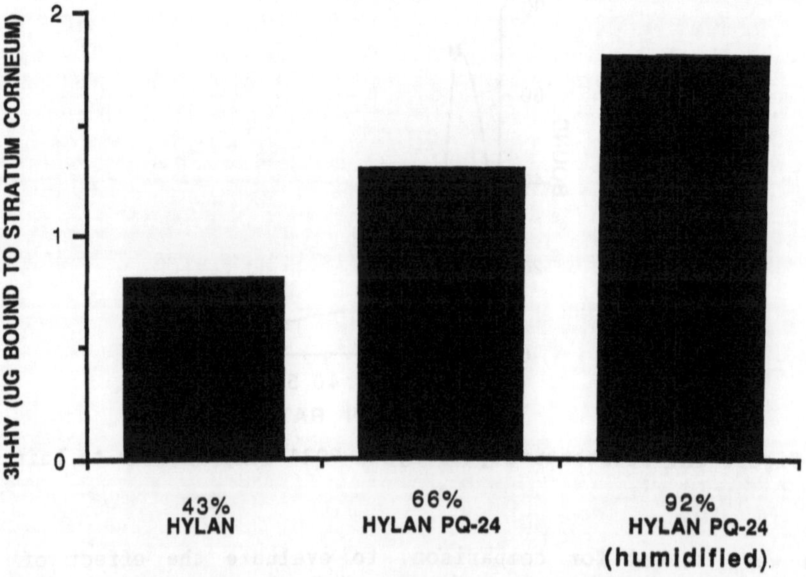


**Figure 4. Substantivity of HY-PQ24 when delivered from an o/w emulsion.**

of $N^+$ species. Finally, quaternary nitrogen is also seen when skin is treated with hylan PQ-24 complex and again treated with distilled water, (IV). In this case, $N^+$ is reduced, and the $N/N^+$ increased since hylan contains no quaternary nitrogen, though it does contain amide nitrogen. Significant differences are also observed with C and O weight percents. Both hylan and PQ-24 are polysaccharides with high O/C ratios and relatively low N content.

In subsequent experiments aimed at examining the composition of the substantive films at various depths, angular resolved electron spectroscopy showed a consistent $N/N^+$ ratio to a depth of 56 A with the hylan/PQ-24 complex. By contrast, sequential applications of PQ-24, followed by treatment with hylan, showed, as would be expected, increased glycosaminoglycan content at the surface (18 A), with higher $N^+$ at 56 A. In either case, PQ-24 is an effective anchor for the unique moisturizing properties of hylan.

Table 3. Skin substantivity measurements using electron spectroscopy

| Sample | C | O | N | C-O | $N^{\pm}$ | $N/N^{\pm}$ |
|---|---|---|---|---|---|---|
| (I) Untreated Skin | 83.8 | 11.4 | 3.1 | - | 0 | NA |
| (II) Hylan +I | 82.5 | 11.8 | 3.9 | 10.7 | 0 | NA |
| (III) PQ-24 +I | - | - | - | 19.8 | 0.4 | 2.0 |
| (IV) Hylan/PQ-24 +I | 80.8 | 18.0 | 0.5 | 21.9 | 0.2 | 2 |

## REFERENCES


1. R. Y. Lockhead, Cosmetics and Toiletries, 103, 23-61 (1988). "The History of Polymers in Hair Care."
2. E. A. Balazs and P. Band, Cosmetics and Toiletries, 99, 65-71 (1984). "Hyaluronic Acid: Its Structure and Use."
3. P. Band, Drug and Cosmetic Industry, 137, 54-55 (1985). "Effective Use of Hyaluronic Acid."
4. J. A. Faucher and E. D. Goddard, J. Soc. Cosmetic Chem., 27, 542-553 (1976). "Sorption of a Cationic Polymer by Stratum Corneum."
5. J. A. Faucher, E. D. Goddard and R. B. Hannan, Textile Research Journal, 47, 616-620 (1977). "Sorption and Desorption of Cationic Polymer by Human Hair: Effects of Salt Solution."
6. E. A. Balazs, A. Leshchiner and P. Band, U.S. Patent 4,713,448 (1987). "Chemically Modified Hyaluronic Acid Preparation and Method of Recovery Thereof from Animal Tissues."
7. G. L. Brode, R. L. Kreiger, E. D. Goddard, F. M. Merritt and D. B. Braun, U.S. Patent 4,663,159 (1987).
8. G. L. Brode, P. Band, E. D. Goddard, A. Barbone, A. Leshchiner, J. Pavlichko, Jr., E. Partain and P. Leung, U.S. Patent 4,767,463 (1990). "Glycosaminoglycans and Cationic Polymer Combinations."

# ALKYLATED POLYVINYLPYRROLIDONE POLYMERS IN COSMETICS AND PHARMACEUTICALS

Dr. S. L. Kopolow, Dr. R. Login and Dr. M. Tazi

GAF Chemicals Corporation
1361 Alps Road
Wayne, New Jersey 07470

Alkylated poly(vinyl pyrrolidones) are a line of polymers which impart perceivable benefits to the customer in skin care and cosmetic applications. These benefits are: water and wear resistance, pigment dispersion, rich non-greasy feel, and improved stick integrity. Due to the flexibility of this chemistry, new polymers have been investigated. These products are an improved waterproofing product and an ethanol soluble product. The improved waterproofing product, has all the features desirable in the current product, and a significantly improved retention of a UV absorber in a sunscreen formulation. An additional feature of this new product is that it can be supplied in bulk or flaked form, a significant benefit to the manufacturer of cosmetic products. The ethanol soluble product was investigated for use in hair and skin applications, in which unique delivery systems are desired.

## INTRODUCTION

Alkylated poly(vinyl pyrrolidone) polymers are a class of compounds which were developed in the mid-1960s. They have been used for over twenty years in cosmetic applications to disperse and protect active raw materials safely. Today, these polymers are used in sunscreens, as the waterproofing agent, and in lipsticks, as the pigment dispersant and/or wear-proofing agent.

This article will describe some new results from our laboratory, based on these polymers. A description of two new alkylated PVP products, being investigated for introduction to the cosmetic and pharmaceutical industry, will also be given.

First, however, a brief description of the alkylated PVP polymer chemistry. The structure is such that the film forming properties of the PVP and the hydrophobic character of a long chain alpha-olefin are combined to form unique products. The structure is represented in Figure 1, where the R group may be either a hydrogen or an alkyl chain.

This structure represents a unique class of products in which one may

*Cosmetic and Pharmaceutical Applications of Polymers*
Edited by C.G. Gebelein *et al.*, Plenum Press, New York, 1991

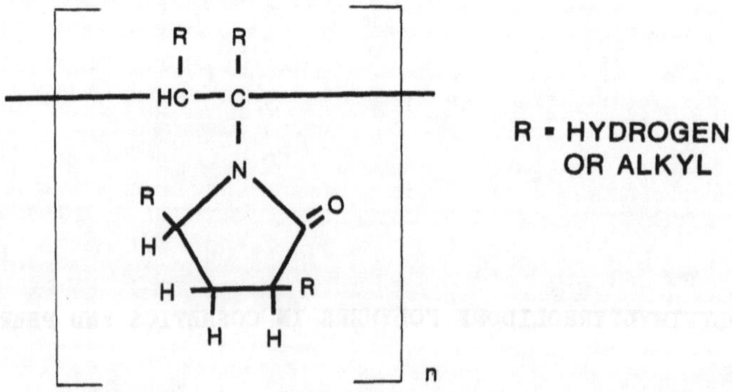


Figure 1. The structure of the alkylated poly(vinylpyrrolidines).

balance the hydrophobic/hydrophilic character of the polymer by adjusting the PVP/olefin molar composition and/or by using olefins of different chain lengths in the synthesis. As one can imagine, the number of possible products is extremely large, and products tailored to meet specific product attributes can be produced. This will be discussed further in a later part of this article.

The two most commonly used alkylated PVP polymers are the PVP/hexadecane and PVP/eicosene copolymers. Typical properties for these polymers are presented in Table 1. PVP/hexadecene copolymer is a pale yellow viscous liquid which has a HLB value of 10. PVP/eicosene copolymer is a waxy solid, in bulk form, which has a HLB value of 8. Both products are water insoluble and oil soluble and have been demonstrated to be safe in personal care applications.

These copolymers have been used to impart four primary benefits to skin care and cosmetic applications. These are:

Table 1. Alkylated poly(vinyl pyrrolidone) typical properties

| Copolymer | PVP/hexadecene | PVP/eicosene |
|---|---|---|
| Physical form | liquid | solid |
| Activity (%) | 100 | 100 |
| Appearance @25°C | pale yellow viscous liquid | off white waxy solid |
| HLB requirements | 10 | 8 |
| Solubilities: | | |
| Oil | soluble | soluble |
| Water | insoluble | insoluble |
| Ethanol | partially soluble | insoluble |
| Toxicity: | | |
| $LD_{50}$ | >64 g/kg | >17 g/kg |
| Skin irritation | slightly | none |
| Ocular irritation | minimal to none | none |

Water and wear resistance
Pigment dispersion
Rich feeling
Improved stick integrity

Additional benefits obtained when formulating with alkylated PVP products are: low toxicities, non-comedogenicitites, broad compatibilities, not sensitive to pH, not sensitive to salt, and stable to prolonged heating.

As previously mentioned, alkylated PVP copolymers are used mainly as waterproofing film formers in the cosmetic industry, especially in the sunscreen market. Water resistant and waterproof products can be developed without the use of high loading of heavy oils. To demonstrate this effect, the water resistance of a sunscreen lotion was determined by the use of the tape stripping method[1] using a basic formulation shown in Table 2.

The results shown in Figure 2 demonstrate that the retention of octyl dimethyl PABA on the skin can be significantly improved by the addition of PVP/eicosene copolymer.

One would predict from the alkylated PVP chemistry that a longer carbon chain would impart significantly better water resistance. This is observed upon comparison of PVP/hexadecene and PVP/eicosene copolymers, with the PVP/eicosene having a larger water resistance value. More water resistance translates to longer sunscreen efficacy, a consumer perceived benefit that has become more important in today's SPF race.

Both PVP/hexadecene and PVP/eicosene copolymers are compatible with nonionic, anionic and cationic materials over a wide pH range. Although they are not water soluble, they are easily emulsified. In some formulations, it has been found that they are self emulsifying. Both water in oil and oil in water emulsions are possible. The maximum benefits are achieved when using nonionic emulsifiers. Use of anionic emulsifiers are also effective, but can lessen the water resistance. This fact becomes an additional benefit, in that products containing alkylated PVP copolymers can be removed with soap and water.

The benefits of longer wearing, water resistant products is not limited to sunscreens. They are ideally suited for use in oil based eye and facial make-up products, (e. g., mascara, eyeliner, lipstick, etc.).

Table 2. Alkylated PVP sunscreen formulation

| Ingredient | weight % with copolymer | weight % without copolymer |
|---|---|---|
| PVP/eicosene Copolymer | 2.0 | 0.0 |
| Myristyl Myristate | 1.0 | 1.0 |
| Glyceryl Stearate SE | 3.5 | 3.5 |
| Octyl Dimethyl PABA | 4.0 | 4.0 |
| Dimethicone | 1.0 | 1.0 |
| Distilled Water | 85.0 | 87.0 |
| Carbomer 940 | 0.1 | 0.1 |
| Triethanolamine | 0.1 | 0.1 |
| Distilled Water | | |
| Preservative q.s. to | 100.0 | 100.0 |

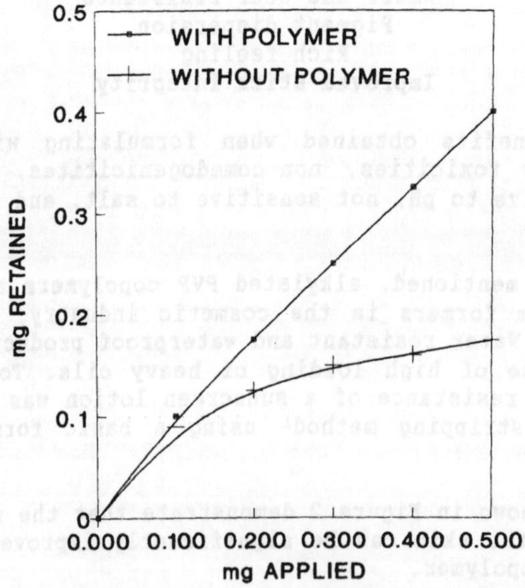


**Figure 2. Retention of octyl dimethyl PABA on the skin.**

**The use of alkylated PVP copolymers will improve the retention of skin care actives such as:**

**emollients and moisturizers in lotions**
**pharmaceutical additives in OTC preparations**
**pigments in cosmetics**

**The improved retention of pigments was demonstrated in our laboratory by preparing two foundation bases, one with PVP/eicosene and the other without. Figure 3 depicts a film of the two foundation bases on a glass plate, which was placed under running water for five minutes. The film with the alkylated PVP copolymer shows no removal, whereas the film without the alkylated PVP copolymer shows considerable degradation. This translates to improved retention of pigments by the alkylated PVP copolymers.**

**Alkylated PVP copolymers function as dispersants for many of the pigments used in colored cosmetics. Typical use levels of several pigments is given in Table 3. In aliphatic vehicles, typical use levels range from 2.5% to 9.0% based on the weight of pigment. This improves the efficiency and provides better coverage from make-up systems.**

**Table 3. Typical use levels of pigments by alkylated PVP copolymers**

| Pigment | Copolymer | Typical use |
|---|---|---|
| Titanium Dioxide | PVP/hexadecene<br>PVP/eicosene | 6.0% |
| Zinc Oxide | PVP/eicosene | 4.5% |
| Iron Oxide, yellow | PVP/eicosene | 9.0% |
| Iron Oxide, red | PVP/hexadecene<br>PVP/eicosene | 4.5% |

Figure 3. Films of two foundation bases.

In stick products, such as eye pencils, cover sticks, and lip care products, the stick integrity is improved by incorporation of alkylated PVP copolymers. A lipstick formulation is presented in Table 4, with and without alkylated PVP copolymers. Sticks that are firm, yet spread smoothly and easily, were obtained. PVP/hexadecene copolymer reduced the penetrometer reading, whereas PVP/eicosene increased the reading as compared to the product with no alkylated PVP copolymer. The selection of the best additive will depend on the performance requirements of the system under development. In eye and cheek pencils, this translates to easier sharpening while maintaining smooth application and excellent color blending.

As mentioned at the beginning of this article, the chemistry of alkylated PVP copolymers is extremely flexible. By modification of the PVP/olefin molar ratio and/or by use of various length olefins the properties of the polymer can be dramatically changed.

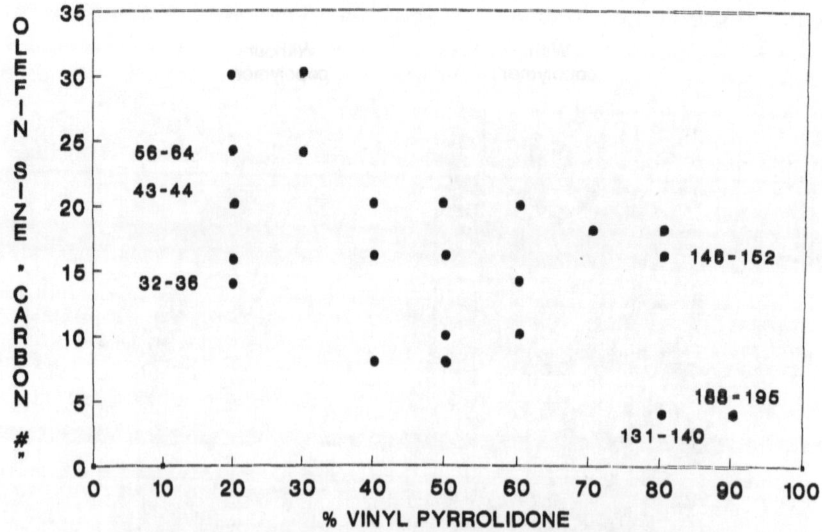


Figure 4. PVP/Olefin melting point grid.

These differences can be visualized if one generates PVP/Olefin grids based upon plots of Carbon number versus % Vinyl Pyrrolidone. Figure 4 is a PVP/Olefin grid of melting points. As the %VP increases the melting point increases and as the carbon # increases the melting point increases.

Another grid of interest is that of ethanol solubility, Figure 5. PVP/olefin compositions with a low level of VP are insoluble, whereas with high level of VP they are soluble. Increasing the number of carbon atoms decreases the solubility. Generally the solubility of PVP/olefin products can be shown in the solubility grid given in Figure 6.

In general the water solubility increases as the number of carbon atoms decreases and the %VP increases. Conversely, the solvent solubility increases with increased number of carbon atoms and decreased %VP.

**Table 4. Lipstick formulation with alkylated PVP copolymers**

| Ingredients | % weight 1 | 2 | 3 |
|---|---|---|---|
| PVP/hexadecene Copolymer | 0.0 | 2.3 | 0.0 |
| PVP/eicosene Copolymer | 0.0 | 0.0 | 2.3 |
| Candeilla Wax | 9.3 | 9.3 | 9.3 |
| Beeswax | 5.8 | 5.8 | 5.8 |
| Glyceryl Dilaurate | 3.5 | 3.5 | 3.5 |
| Poly(glyceryl-3-Di-isosterate) | 34.9 | 34.9 | 34.9 |
| Lanolin Oil | 23.3 | 23.3 | 23.3 |
| Isocetyl Alcohol | 11.6 | 11.6 | 11.6 |
| Tridecylneopentanoate | 8.1 | 5.8 | 5.8 |
| Carnauba | 3.5 | 3.5 | 3.5 |
| Penetrometer Reading | 105.0 | 92.0 | 107.0 |

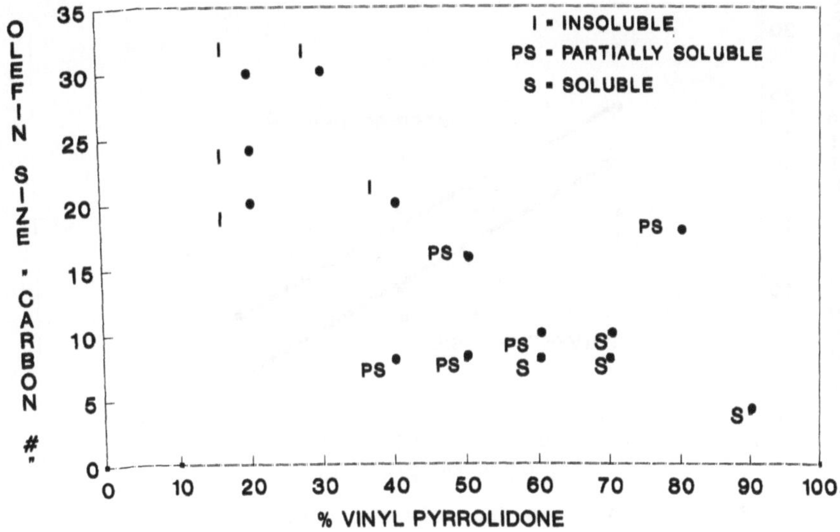


Figure 5. Ethanol solubility grid.

**Table 5. PVP/decene typical properties**

| CFTA designation | PVP/Decene Copolymer |
|---|---|
| Physical form | liquid |
| Solvent | ethanol |
| Activity (%) | 50 |
| Appearance | pale yellow solution |
| Viscosity (CPS) (Brookfield RVT #2 @ 50 RPM) | 100-200 |
| Molecular weight (GPC) | 4000 to 6000 |
| Solubilities: | |
| Ethanol | soluble |
| Water | insoluble |
| Mineral oil | insoluble |
| Castor oil | soluble |
| Lanolin oil | insoluble |
| Isopropyl myristate | insoluble |
| Isopropyl palmitate | insoluble |
| Cetyl octanoate | insoluble |
| Pentaerythritol pelagonate | insoluble |
| Octyl dodecanol | soluble |
| Ammonium lauryl sulfate (15% and 29%) | soluble |
| Sorbitol, 70% | insoluble |
| d-Limonene | soluble |
| Glycerine | soluble |
| Cocamide DEA | soluble |
| Acetone | soluble |

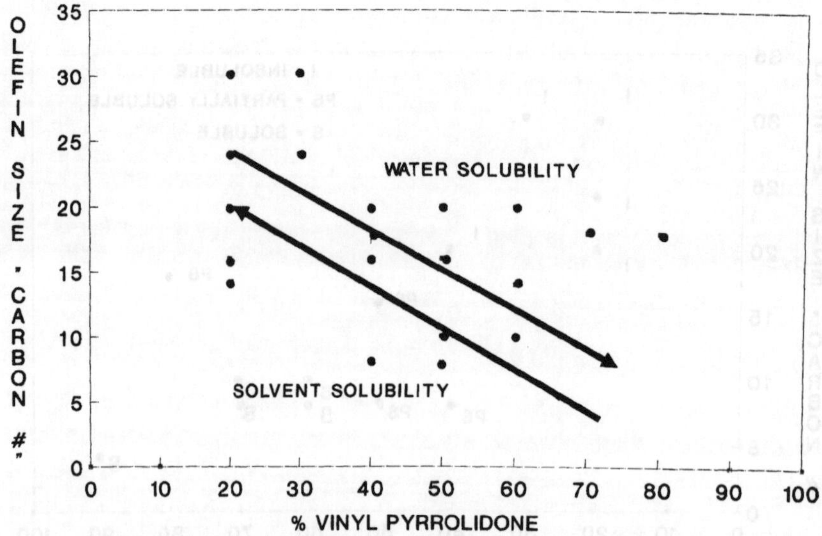

Figure 6. Aqueous and solvent solubility grid.

Two new polymers have been investigated for the cosmetic and pharmaceutical industry, which offer significant benefits over commercially available materials. PVP/decene copolymer was designed to be ethanol soluble for use in hair and/or skin care products. PVP/triacontene copolymer was designed to have improved waterproofing properties and to be available in flake form.

PVP/decene copolymer typical properties are presented in Table 5. PVP/decene will be supplied as a 50% solution in ethanol. It has limited solubility in cosmetic ingredients and is insoluble in water.

In hair care applications, especially in ethnic care products, PVP/decene copolymer was expected to exhibit the properties shown below. It may also repair split ends on ethnic hair when incorporated in a shampoo.

Shine/gloss enhancement
Mild conditioning
Water resistance
Body and hold

Toxicity studies on the PVP/decene copolymer indicate that it is safe for use in cosmetic and pharmaceutical uses. These toxicity results are presented in Table 6.

Table 6. PVP/decene toxicity results

| | |
|---|---|
| $LD_{50}$ | >5 g/kg |
| Skin irritation | mild |
| Ocular irritation | minimal |
| Ames assay | did not demonstrate mutagenic potential |
| Dermal $LD_{50}$ | >2.0 g/kg |
| Human skin clinicals | did not induce phototoxicity, photoallergy, or contact dermal sensitization |

| Table 7. PVP/triacontene copolymer typical properties | |
|---|---|
| CTFA designation | Tricontanyl PVP |
| Physical form | flake |
| Activity (%) | 100 |
| Appearance @ 25°C | off-white solid |
| HLB requirement | 6 |
| Solubilities: | |
| Water | insoluble |
| Ethanol | insoluble |
| Mineral oil | soluble |
| Propylene glycol | insoluble |

Due to its low melting point (32 to 36°C), PVP/eicosene copolymer is currently sold as a bulk product, which makes handling somewhat difficult, since it must be melted or broken up before it can be used. The PVP/triacontene copolymer was designed to have a high melting point (60 to 65°C), which allows the product to be produced in a stable flake form. This, therefore, means it is easier to use. PVP/triacontene copolymer typical properties are presented in Table 7. The low HLB requirement, 6, is a measure of its hydrophobicity. Toxicity results for PVP/Triacontene are presented in Table 8. This data indicates that it is safe for use in cosmetic and pharmaceutical applications.

Due to the structure similarity of PVP/decene and PVP/triacontene to PVP/hexadecene and PVP/eicosene one might assume that the aforementioned products are also non-comedogenic. Since PVP/triacontene is produced with a longer olefin it is most likely that it would have improved water resistance. Preliminary data indicates that it may be 2 to 2.5 times more water resistant than PVP/eicosene. PVP/decene, with a shorter olefin chain, will be expected to have lower water resistance. However, one might also expect to get some of this same characteristic in other solvent/delivery systems.

| Table 8. PVP/triacontene toxicity results | |
|---|---|
| Oral $LD_{50}$ | >5 g/kg |
| Skin irritation | slightly |
| Ocular irritation | minimal |
| Ames assay | did not demonstrate mutagenic potential |
| Dermal $LD_{50}$ | >2.0 g/kg |
| Human skin clinicals | did not induce phototoxicity, photoallergy or contact dermal sensitization |

## REFERENCES

1. R. Oteri, S. Johnson, and S. Dastis, Cosmetics & Toiletries, **102**(3), 107-109 (1987). "A New Waterproofing Agent for Sunscreen Products."

Table 1. PVP/triacontene copolymer typical properties

| | |
|---|---|
| CTFA designation | Triacontanyl PVP |
| Physical form | flake |
| Activity (%) | 100 |
| Appearance @ 25°C | off-white solid |
| HLB requirement | 1 |
| Solubilities: | |
| Water | insoluble |
| Ethanol | insoluble |
| Mineral oil | soluble |
| Propylene glycol | insoluble |

Due to its low melting point (32 to 36°C) PVP/eicosene copolymer is currently sold as a bulk product, which makes handling somewhat difficult, since it must be melted or broken up before it can be used. The PVP/triacontene copolymer was designed to have a high melting point (60 to 65°C), which allows the product to be produced in a stable flake form. This, therefore, makes it easier to use. PVP/triacontene copolymer typical properties are presented in Table 1. The low HLB requirement, 1, is a measure of its hydrophobicity. Toxicity results for PVP/triacontene are presented in Table 2. This data indicates that it is safe for use in cosmetic and pharmaceutical applications.

Due to the structure similarity of PVP/eicosene and PVP/triacontene to [illegible] one might assume that the [illegible] products are also non-comedogenic. Since PVP/triacontene is [illegible] [illegible] it would have [illegible] water [illegible]. Preliminary data indicate that it may be 2 to 2.5 times more water resistant than PVP/eicosene [illegible] will be expected to have [illegible] however, one [illegible] in other [illegible] delivery systems.

Table 2. PVP/triacontene toxicity data

| | |
|---|---|
| Oral LD50 | > 5 g/kg |
| Skin irritation | slightly |
| Eye irritation | minimal |
| [illegible] | [illegible] |
| Dermal LD50 | > 2.0 g/kg |
| Human skin [illegible] | did not induce [illegible] [illegible] |
| | [illegible] |

REFERENCE

1. [illegible] Johnson, and [illegible], Cosmetics & Toiletries 102(3), 119-122 (1987). "A New Waterproofing Agent, PVP/Eicosene Products."

# HYLAN AND HYLAN DERIVATIVES IN DRUG DELIVERY

Nancy E. Larsen, Edward A Leshchiner, Edward G. Parent and Endre A. Balazs

Biomatrix, Inc.
Ridgefield, New Jersey

Hylan and hylan derivatives have been developed as topical and implantable (or injectable) vehicles for the controlled and localized delivery of biologically active molecules. Hylan, a sodium hyaluronate derivative, is a hydrophilic, polyanionic, viscoelastic polymer which exhibits extraordinary biocompatibility and unique versatility as the basis for various release systems. Hylan vehicles include the soluble hylan polymer, hylan gels (insoluble, injectable, chemically crosslinked), hylan microparticles (insoluble, injectable, 10-40 µm particle size, chemically crosslinked), hylan membranes (insoluble, implantable, chemically crosslinked) and hylan gel composites (crosslinked hylan gels containing ingredients other than the drug to be delivered). Hylan gels were evaluated *in vitro* for their ability to sustain prolonged delivery of the antibiotic gentamicin. Using a column method with continuous flow through the hylan gel/gentamicin samples, it was observed that 20 hours was required to achieve 90% release, while in the control, >80% of the drug was eluted during the first hour. A microbiological assay supported these data; hylan gel/gentamicin was able to sustain bactericidal activity for over 50 days, while the control lost bactericidal activity after 3 days. The kinetics of release of a particular molecule can be influenced by the the chemical and physical characteristics of the hylan system.

## INTRODUCTION

Hylan and hylan derivatives have been developed as topical and implantable (or injectable) vehicles for the controlled and localized delivery of biologically active molecules. Hylan, a sodium hyaluronate derivative, is a hydrophilic, polyanionic, viscoelastic polymer which exhibits extraordinary biocompatibility and unique versatility as the basis for various release systems. Hylan vehicles include the soluble hylan polymer, hylan gels (insoluble, injectable, chemically crosslinked), hylan microparticles (insoluble, injectable, chemically crosslinked), and hylan gel composites (crosslinked hylan gels containing ingredients other than the drug to be delivered). Hylan and hylan deriva-

*Cosmetic and Pharmaceutical Applications of Polymers*
Edited by C.G. Gebelein *et al.*, Plenum Press, New York, 1991

tives have been evaluated in *in vitro* and *in vivo* models for their ability to optimize delivery of a variety of pharmacologically active molecules including the antibiotic gentamicin, the enzyme thrombin, and the vasodilator serotonin (5-hydroxytryptamine). The results from selected studies are described in this paper.

## MATERIALS AND METHODS

### 1. Preparation of Sample and *in vitro* Assay of Gentamicin Release

A column method was used to assess release of [$^{125}$I]-gentamicin (New England Nuclear, specific activity 0.95-1.3 mCi/µg) from hylan gel. The gentamicin solution used in testing consisted of a mixture of [$^{125}$I]-gentamicin and unlabeled gentamicin; the concentration of unlabeled gentamicin was 1-2 mg/mL; the concentration of [$^{125}$I-gentamicin was 1-2 µg/mL). Gentamicin solution was added to lyophilized hylan gel and rehydration allowed to take place over a period of 12-18 hours. An aliquot of the [$^{125}$I]-gentamicin/gentamicin solution (without hylan gel) was applied to a PD10 column (Sephadex G25, flow rate 10 mL/hour) and the elution profile was determined by measuring total radioactivity in each fraction by gamma spectroscopy. Hylan gel/[$^{125}$I]-gentamicin samples were applied to an identical column and elution of gentamicin measured as described above.

### 2. Preparation of Sample and Microbiological Assay of Gentamicin Release

A microbiological assay was used to evaluate the duration of bactericidal activity of gentamicin released from hylan gel and control delivery systems. Duration of bactericidal activity of gentamicin releasing systems was measured by monitoring the inhibition of growth of *Staphylococcus epidermidis* in the presence of test samples. One mL samples of a gentamicin/hylan gel mixture and gentamicin solution (control) were placed into dialysis tubing (loading dose = 2.5 mg/mL) and were incubated at 32°C in 2 mL Tryptic Soy broth (TSB) containing 2-4 x $10^4$ micro organisms; samples were transferred to fresh tubes containing new cultures of *S. epidermidis* and fresh TSB at 1-3 day intervals until growth was detected (by turbidity) in the test samples. Controls were tubes containing bacteria and TSB only, bacteria, TSB, and added gentamicin, and TSB alone.

### 3. Preparation of HHCT+T (Embolization Agent)

HHCT+T was prepared as described in U. S. Patent #4,795,741. Briefly, 1 g of sodium hylan was mixed with 20 mL of water and allowed to swell overnight and then 2.8 mL of 2 N NaOH was added. Then, 10 g of powdered tantalum (particle size 1.3 microns, Kennametal, Latrobe, PA) was mixed with 2 mL of water and was stirred into the hylan solution, followed by addition of 0.2 g of vinyl sulfone (Aldrich Chemical Company, Milwaukee, WI) in 2 mL of water. The mixture was stirred until gelation occurred (approximately 8 minutes). The gel was allowed to swell in sterile 0.15 M NaCl overnight, then washed four times, four hours per wash. Hylan content was determined on a hydrolyzed aliquot of gel by assay of hexuronic concentration after tantalum had been removed by centrifugation.[1] Acceptable concentrations of hylan and tantalum were 0.3-0.5% and 5-7% respectively. For each gram of hylan-tantalum gel produced, 0.06 g microcrystalline cellulose and 0.1 mL of 20% hexamethonium chloride (HMC) was mixed in. The final mixture was autoclaved (121°C, 20 minutes) and tested for sterility,[2] pyrogenicity,[3] and cytotoxicity (*in vitro*) (Agar Overlay Test, 3T3 fibroblasts).[4] Thrombin (bovine, KABI-Vitrum, Helena Laboratories, Beaumont, TX) was added at the time of testing (*in vitro* or *in vivo*).

## 4. *In Vitro* Evaluation of HHCT+T

### 4A. Clot Formation

Clotting activity of HHCT+T mixtures was tested *in vitro* by mixing 0.4 mL HHCT+T (2.5 USP units of thrombin) with 0.4 mL of fresh, human, whole, anticoagulated (citrated or heparinzed) blood. For citrated blood, 1 volume of 3.8% sodium citrate was mixed with 9 volumes of blood; alternatively, 5 mL of whole blood was added to a vial containing 300 USP units of sodium heparin (Sigma Chemical Company, St. Louis, MO). The following parameters were measured: time for clot formation, quality of the clot formed (firm, loose), compatibility with tantalum.

### 4B. Release of Thrombin and HMC

To measure thrombin release, a freshly formed gel/clot (0.8 mL volume) was placed in 0.15 M NaCl (10 mL volume) and aliquots removed at 0.5, 1, 2, 4 Hours. A fibrinogen clotting assay and a colorimetric assay were used to measure released thrombin. In the colorimetric assay, 1 mL of BAEE (N-benzoyl arginine ethyl ester; 1 mM final concentration in 10 mM Tris HCl 0.14 M NaCl, pH 8.75) was incubated with 50 µl of test sample or standard thrombin solution for 15 minutes at 37°C. The absorbance at 255 µm was measured, a standard curve constructed and the units of thrombin activity/test sample determined. In the fibrinogen clotting assay, 0.2 mL of fibrinogen solution (2.5 mg/mL) was added to 0.1 mL of test sample or thrombin standards (0.08-0.6 u/mL) and the time to clot measured. To measure HMC release, the assays were performed as described above except that aliquots (1 mL) were assayed for HMC content according to Mitchell and Clark.[5]

## 5. *In vivo* Evaluation of HHCT+T in Rabbits

### 5A. Subcutaneous and Intramuscular Injection

New Zealand white rabbits (female, specific pathogen-free, SPF) were anesthetized with ketamine/xylazine and 0.5 mL of HHCT (thrombin omitted) was injected subcutaneously into the back of each animal; 0.5 mL was also injected into the thigh muscle of the same animals. Three animals were used in this preliminary study; two rabbits were sacrificed at one week post injection, one rabbit was sacrificed four weeks post injection. Photographs of excised tissue were made, then each tissue was placed into 10% neutral buffered formalin. Fixed tissue was embedded in paraffin, sectioned, stained with hemotoxylin and eosin or toluidine blue and submitted to a veterinary pathologist for evaluation.

### 5B. Embolization-Auricular Artery, Rabbit

New Zealand white female SPF rabbits (14 total) were anesthetized with ketamine and xylazine and designated rabbits were heparinized. Pre-injection blood was removed from the auricular artery of the ear opposite the one assigned to be injected with HHCT+T. The area of injection was cleaned with 70% ethanol, and using a 25 g needle, approximately 0.05-0.1 g (0.05-0.1 mL) of HHCT+T was injected into the artery, in the direction of blood flow (toward the tip of the ear). An embolus of 2-6 cm was formed in the artery. Blood was removed at 0.5, 1.0, and 24 hours for analysis. At sacrifice (5 mL pentobarbital into marginal ear vein) each rabbit was weighed, the ear removed, photographed and placed in 10% neutral buffered formalin for histologic evaluation.

All surgical and other experimental procedures were carried out subject to the approval of the Institutional Animal Care and Use Committee and in accordance with the NIH guidelines for the Care and Use of Laboratory Animals.

### 6. Preparation of HY/PVA and *in vitro* evaluation of [$^3$H]-5 HT Release (Serotonin)

Hylan/PVA,[(polyvinyl acetal), Cooper Medical, was prepared according to procedures described in Patent #4,605,691. Briefly, PVA disks were mixed together with hylan fluid and vinyl sulfone in an aqueous alkaline solution, and polymerization allowed to occur. Once polymerization was completed the HY/PVA disks were washed extensively with sterile distilled water and stored in sterile 0.15 M NaCl at 4°C or lyophilized.

PVA and PVA/HY disks were loaded with [$^3$H]-5-hydroxytryptamine by incubating control and coated disks in a solution of 40 nM [$^3$H]-5HT for several hours at 4°C, followed by three quick (10 second) rinses in water, and immersion in 50 mL of 0.15 M NaCl. Then, 50 µl aliquots were removed at regular intervals, mixed with 5 mL Hydrofluor liquid scintillation fluid, (National Diagnostics) and the radioactivity was measured by beta spectroscopy. At the end of the study, the radioactivity remaining in the disks was determined and used to calculate total loading dose and percent release.

## RESULTS

### 1. *In Vitro* release of Gentamicin from Hylan Gel

Using the column method, the elution profile of gentamicin release from hylan gel indicates that approximately 20 hours was required to achieve 90% release of gentamicin from hylan gel, while in the control (no hylan gel), more than 80% of the gentamicin was eluted from the column during the first hour (Figure 1). When 0.3 M sodium chloride was used as the column eluent, there was rapid elution of the gentamicin from the hylan gel.

In the microbiological assay, hylan gel/gentamicin samples were observed to sustain bactericidal activity for over 50 days (the assay was stopped at 50 days, with no indication of bacterial growth in test samples), while in the control (no hylan gel) all bactericidal activity was lost after the first transfer on day three (see Methods section) (Figure 2). Positive and negative controls were run in parallel with test samples in order to demonstrate that the bacterial cultures used in these experi-

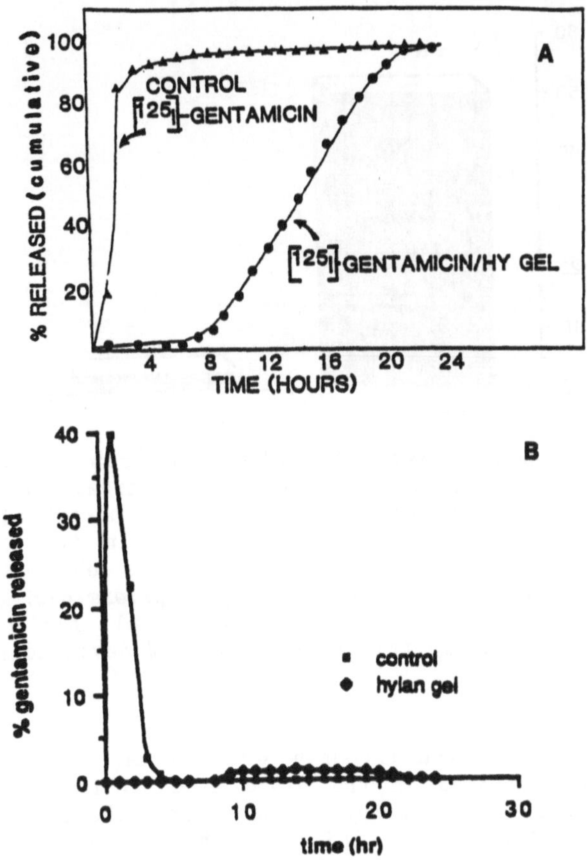


Figure 1. *In Vitro* release of $^{125}$I-Gentamicin from hylan gel, column method. Release of $^{125}$I gentamicin from hylan gel vehicle was measured using a column method as described in the Methods Section. (A) The data is presented as percent total gentamicin released (cumulative) versus time (hours) for the hylan gel gentamicin vehicle (0.1% gentamicin) and the control gentamicin solution (0.1% gentamicin). Release was measured at 23°C, the eluting solvent was 0.15 M NaCl. (B) The data is presented as percent gentamicin released versus time (hours); same conditions as A.

ments exhibited normal growth and normal susceptibility to the antibiotic gentamicin.

## 2. *In Vitro* Evaluation of HHCT+T (Hylan Gel Embolization Material

HHCT+T was found to be sterile, nonpyrogenic, noncytotoxic, and able to induce rapid clot formation without exclusion or loss of gel or blood components. In the *in vitro* release assays, thrombin levels were not detectable using the fibrinogen clotting assay; the colorimetric assay indicated that after 4 hours approximately 40% of the thrombin had been released (Table 1).

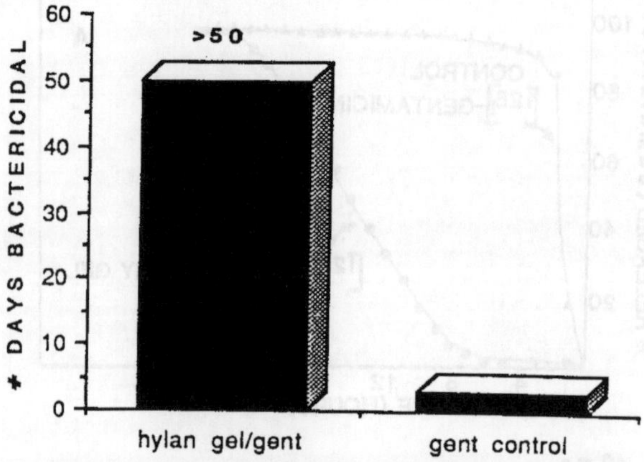


Figure 2. *In vitro* release of gentamicin from hylan gel; microbiological assay of bactericidal activity. Hylan gel/gentamicin vehicles and bacterial cultures were prepared as described in the Methods section. Bactericidal activity was equated with inhibition of microbial (*Staph. epidermidis*) growth (lack of turbidity in growth medium). All samples were tested in triplicate, standard deviations were <5%.

A portion of the measured thrombin activity may be a result of the presence of other esterases in the blood used to form the clot and, hence, the actual amount of thrombin released from the gel may be lower. *In vitro* assays of HMC release indicated that after 10, 60, and 180 minutes, saline aliquots contained 0.058, 0.068, and 0.o73 mg/mL HMC, representing 7, 8, and 9% respectively, of total HMC present in the gel (Table 1).

Table 1. *In vitro* thrombin and HMC release

| THROMBIN RELEASE[a] | | |
|---|---|---|
| Time, (min.) | Thrombin activity (u/mL) | % Total |
| 30 | 0.047 ± 0.014 | 29 |
| 60 | 0.048 ± 0.008 | 30 |
| 120 | 0.061 ± 0.005 | 38 |
| 240 | 0.067 ± 0.005 | 42 |
| **HMC RELEASE[b]** | | |
| Time, (min.) | HMC (mg/mL) | % Total |
| 10 | 0.0585 | 7 |
| 60 | 0.0682 | 8 |
| 180 | 0.0734 | 9 |

(a) Colorimetric assay using BAEE; no detectable thrombin measured by fibrinogen clotting assay.
(b) Method of Mitchell and Clarke.[5]

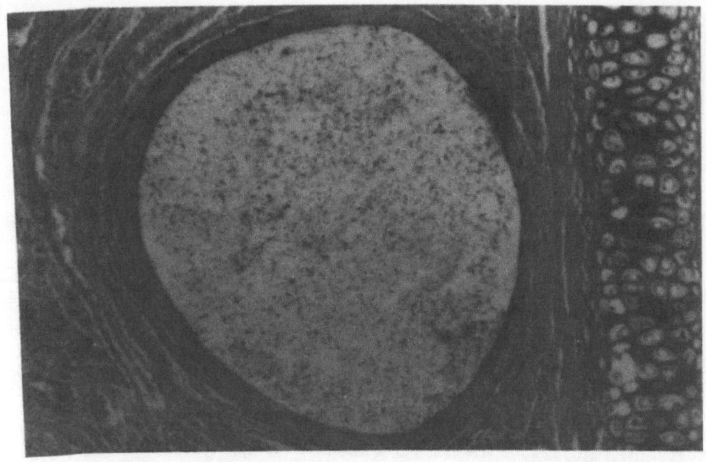

Figure 3. Photomicrograph of cross section through the auricular artery containing injected HHCT+T, 8 days after injection. The artery is occluded mainly by injected material, very few inflammatory cells are present. (Verhoeff-Van Gieson stain, 90 x magnification).

## 3. *In Vivo* Evaluation of HHCT+T

Subcutaneous injection of HHCT produced a minimal to moderate inflammatory response. Intramuscular injection of HHCT was associated with a mild to marked macrophage and giant cell reaction. Intra-arterial injection of HHCT+T resulted in occlusion of the blood vessel and was accompanied by a minimal or mild mixed inflammatory response (Figures 3-5).

Occlusion initially occurred as a result of the injected gel (HHCT+T) (Figure 3), while at longer time points the arteries were occluded mainly by a variety of inflammatory cells (including macrophages, giant cells, heterophils and lymphocytes) and fibrosis, although the gel was still present and most likely contributed to the maintenance of occlusion (Figures 4 and 5).

At six months the emboli were equal to or greater than 75% of their original gross size, an indication of the stability of the embolus. Also, when rabbits were subjected to whole body X-ray, no tantalum was observed except at the site of injection. Blood samples were analyzed for HMC and thrombin; there was no rise in levels of either component at 1 hour and 24 hours following injection (Table 2) suggesting that either levels released were below detectable limits, or that the release of these components was truly local and did not result in significant systemic exposure. Local release of thrombin and HMC was evidenced by local fibrin formation which occurred at the site of injection despite systemic heparinization.

In all *in vivo* embolization procedures performed, the radio-opaque component tantalum remained a part of the embolus and apparently did not interfere with clot formation.

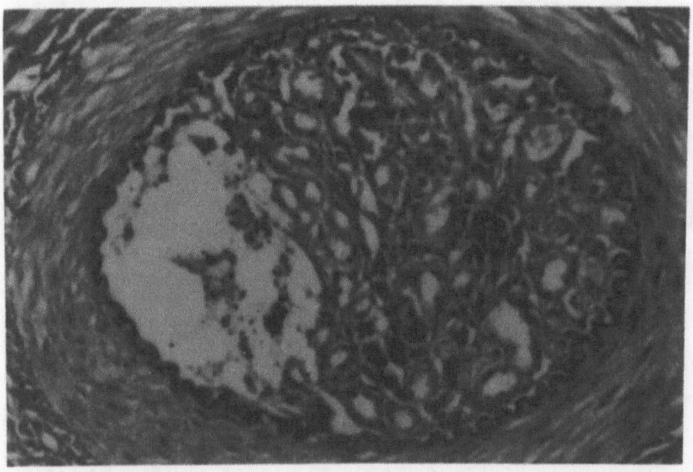

Figure 4. Photomicrograph of cross section of auricular artery containing injected HHCT+T, 7 weeks after injection. The artery is occluded by injected material, fibrotic tissue and inflammatory cells. The clear area in the arterial lumen was originally occupied by hylan gel, which shrunk during processing. (90 x original magnification, Verhoeff-Van Gieson stain.)

## 4. *In Vitro* Release of [$^3$H] 5HT From HY/PVA Pellets

[$^3$H]-5HT release kinetics from hylan coated PVA disks are shown in Figure s %5-HT release versus time (minutes). Release from the control disk exhibits an initial pulse release of 50% of the drug during the first 30 minutes followed by a slower phase of release. By three days >90% of the drug has been delivered to the surrounding saline from the control pellet. Hylan coated pellets, on the other hand, exhibit nearly zero order kinetics (constant rate of release) throughout the delivery period and require more than seven days to achieve >90% release.

Table 2. Blood analyses - thrombin and HMC release

| | | Esterase activity[a] (NIH u/mL) | HMC concentration (µg/mL serum) |
|---|---|---|---|
| Pre-injection | | 4.88 ± 0.27 | 5.20 ± 2.3 |
| Post-injection | 1 hour | 4.66 ± 0.29 | 5.70 ± 1.9 |
| | 24 hours | 5.01 ± 0.12 | 0.54 ± 0.2 |

(a) Colorimetric assay using BAEE; no detectable thrombin measured by fibrinogen clotting assay.
(b) Method of Mitchell and Clarke.[5]

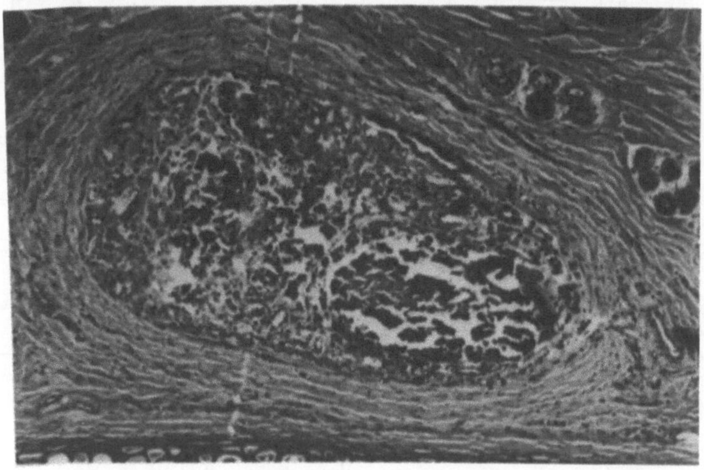

Figure 5. Photomicrograph of cross section of auricular artery containing injected HHCT+T, 24 weeks after injection. Blockage of the artery occurs as a result of a chronic inflammatory reaction, including fibrosis. (Verhoeff-Van Gieson stain, 90 x original magnification.)

## DISCUSSION

The suitability of hylan polymers for use in drug delivery systems was evaluated with *in vitro* and *in vivo* test systems. Results from these studies indicated that hylan matrices may provide the desired benefits of sustained and/or controlled release of various types of pharmacologically active molecules while furnishing an enhanced biological environment for the delivered molecules. The versatility of the hylan biopolymer allows chemical modification of the soluble polymer without destruction of its biological compatibility. Hylan and hylan derivatives may be developed for use as the delivery vehicle itself, for example, as a topical or injectable matrix, or may be developed in combination with other materials, for example, as coatings or copolymers for topical implantation, or injectable gels or other solids.

Several examples of hylan delivery systems were described in this report. Hylan gels were loaded with the broad spectrum antibiotic gentamicin. Gentamicin is an effective anti-bacterial drug, however, known hepatic toxicity may limit its use. *In vitro* results demonstrated the dramatic effect of hylan gel on the release kinetics of gentamicin. In the presence of hylan gel, the length of time required for 50% release was increased by a factor of 18 over the control. Release of gentamicin appears to be controlled by an ion exchange process (between cationic gentamicin and ionic hylan gel) and diffusion.

In another example, hylan gel formed the matrix for holding various components of an embolizing material. Hylan gel provided the necessary physical viscoelasticity and biological properties (biocompatibility) while combined with tantalum (a radio-opaque material necessary for X-ray visualization), microcrystalline cellulose (a binder material necessary for stability of the injected material), and the two drugs intended for local delivery, thrombin and hexamethonium chloride (HMC). Thrombin is an enzyme responsible for inducing rapid, local clot information by catalyz-

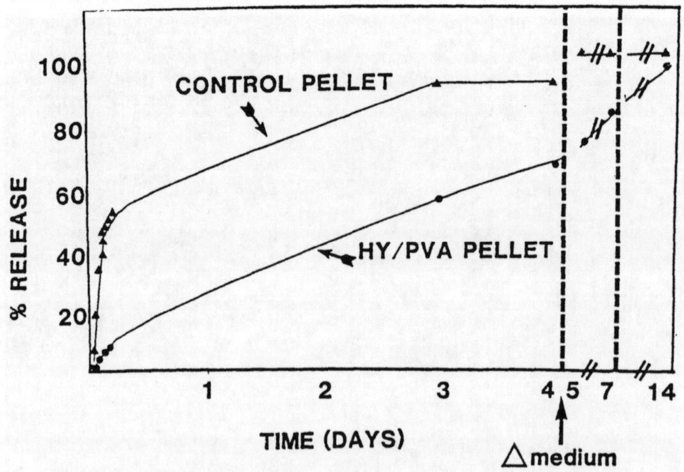


Figure 6. *In vitro* release of ($^3$H)-5HT from HY/PVA. Release kinetics of ($^3$H)-5HT were measured for hylan coated PVA disks (X) and for control untreated PVA disks (X) as described in the Methods section. Loading concentrations of 88 and 63 pg/disk were obtained for coated and uncoated disks, respectively. Assays were performed at 23°C, the reservoir fluid was 0.15 M NaCl. Four days after the start of the assay, reservoir fluid (50 mL 0.15 M NaCl) was replaced and the assay continued for an additional ten days. Each point is the mean of three values (three replicates); standard deviations were <10% of the mean value.

ing the polymerization of fibrinogen. HMC is a cationic molecule containing quaternary ammonium groups needed for the local neutralization of heparin. The results from *in vitro* and *in vivo* studies demonstrated that this embolizing material is sterile, non-pyrogenic, non-cytotoxic, and rapidly forms a stable, solid clot. Release of thrombin and HMC from the injectable material appeared to be restricted, with activity limited to the site of injection, with no significant systemic exposure. Local release of thrombin and HMC was evidenced by fibrin formation, which occurs at the site of injection despite systemic heparinization. Hylan gels for embolization may be preloaded with additional active ingredients for the local controlled release of substances whose systemic effects may be toxic or may interfere with their primary effect on target tissues.

In the third example, hylan was used to coat sponge-like poly(vinyl acetal) disks. The crosslinked hylan gel coating increased the loading dose of 5-HT (serotonin) and greatly reduced the rate at which 5-HT was released from the disks. 5-HT is a small cationic molecule, release from the hylan gel matrix is probably controlled by an ion exchange process and diffusion, and is similar in this way to the gentamicin-hylan gel system.

Controlled release of biologically active molecules provides a number of potential benefits which affect patient health (may reduce the total dose of drug administered, thereby avoiding or reducing toxic and other undesirable side effects), cost (may reduce total amount of drug required for therapy; may reduce the number of treatments or doses needed), and effectiveness (may avoid or reduce biodegradation of the active molecules; improve patient compliance and comfort). It is reasonable to conclude that hylan and hylan derivatives (gels, membranes, coatings, composites) will provide unique and effective drug delivery systems for a variety of pharmacologically active molecules. Often, the role of the hylan vehicle is two-fold; the physical and biological properties of hylan may be useful as tools in tissue augmentation, in anti-adhesion, and, as described in this report, in embolization.

The combination with drug molecules may serve to enhance the physical role of the hylan vehicle or may represent an additional function of the hylan material. For example, hylan gel embolization material (HHCT+T) may be used to block blood vessels which feed a tumor and also to deliver cytotoxic drugs to that tumor.

Additional *in vivo* studies are currently under way. The results from these studies will demonstrate more fully the potential uses for hylan and hylan derivatives in drug delivery.

## REFERENCES


1. E. A. Balazs, K. O. Berntsen, J. Karossa and D. A. Swann, Analytical Biochem., **12**, 547-558 (1965). "An Automated Method for the Determination of Hexuronic Acids."
2. "*United States Pharmacopia, XXI,*" United States Pharmacopieal Convention, Inc., 12601 Twinbrook Parkway, Rockville, MD 20852, 1985, Test 71.
3. "*United States Pharmacopia, XXI,*" United States Pharmacopieal Convention, Inc., 12601 Twinbrook Parkway, Rockville, MD 20852, 1985, Test 85.
4. W. L. Guess, S. A. Rosenbluth, B. Schmidt and J. Autian, J. Pharm. Sci., **54**, 1545-1547 (1965). "Agar Diffusion Method for Toxicity Screening of Plastics on Cultured Cell Monolayers."
5. R. Mitchell and B. B. Clark, Proc. Soc. Exptl. Med., 81, 105-109 (1952). "Determination of Quaternary Ammonium Compounds Including Acetylcholine, Tetraethyl Ammonium and Hexamethonium."

# POLYMERIC CONTROLLED DRUG DELIVERY SYSTEMS: RATIONAL DESIGN AND DEVELOPMENT

Avinash G. Thombre

Pfizer Central Research
Groton, CT 06340

Polymeric controlled drug delivery systems have emerged as important pharmaceutical dosage forms because it was recognized that drug delivery affords many medical and commercial advantages. Also, advances in polymer science and technology have enabled the economic manufacture and reproducible performance characteristics of drug devices. This paper provides an overview of oral controlled-release dosage forms: their rationale, principles governing their design and development, technologies available for drug delivery and their selection, and physiochemical principles governing the kinetics of drug release. Polymers studied for drug delivery applications and their selection, approaches to achieve constant release rates, and commercially important controlled drug delivery devices are discussed.

## INTRODUCTION

In the past few decades, polymeric controlled drug delivery systems have emerged as important pharmaceutical dosage forms. This success can be attributed to the following major contributing factors: (1) it was recognized that controlled drug delivery could afford many potential medical and commercial advantages, (2) novel concepts and designs for drug delivery devices were developed through a concentrated effort by interdisciplinary research teams, and (3) advances made in the fields of polymer science and engineering were successfully applied to drug delivery.

This paper offers a framework for the rational design and development of polymer-based controlled drug delivery systems. The focus is on the development of an oral dosage form for a specific drug as opposed to a broadly applicable generic technology. A variety of drug delivery systems have been studied. Although some of these systems are successful commercial products, many have been studied only in the laboratory.

*Cosmetic and Pharmaceutical Applications of Polymers*
Edited by C.G. Gebelein *et al*., Plenum Press, New York, 1991

## CONVENTIONAL VERSUS CONTROLLED RELEASE

A number of features distinguish controlled release systems from conventional pharmaceutical dosage forms (Table 1). Controlled release formulations attempt to alter drug absorption by modifying the release rate. This leads to reduced fluctuations in the plasma drug concentration, increased apparent half-life, and sustained drug effects. Assuming certain absorption and metabolism characteristics, a constant release rate may result in constant blood levels and maximize the duration for which drug levels are in the therapeutic range. Thus, much of the research in this area has been influenced by the desire to achieve a constant release rate or zero-order delivery.

The field of controlled release is heavily patented. Patents exist for most controlled release technologies, devices based on these techno-

| Table 1. Unique features of controlled release dosage forms |
|---|
| Altered pharmacokinetics because of modified release (input) rates:<br>Lower peak-to-trough blood level fluctuations<br>Longer apparent half-life<br><br>Heavily patented field: technology, devices, and manufacturing processes are frequently patented making legal issues very important:<br>Opportunity - Exclusivity and patent-life extension<br>Risk - Potential for infringement on issued patents<br><br>Research and development philosophy relative to controlled release is dependent on the company size:<br>Very small companies - generally pursue narrow single technology or specific application goals.<br>Medium size companies - generally aggressive and innovative pursuing several broadly applicable delivery technologies and patent strategies.<br>Large companies - have generally focused on specific products and relied on in-house development, contracts, licensing, collaborations and buy-outs.<br>Multidisciplinary research and development teams are important.<br><br>Emphasis on device performance and reliability<br>Medical rationale for controlled release<br>*in vitro* tests vs *in vivo* performance<br>Effect of food/pH of Gi tract on performance<br>Device reliability and reproducibility<br>Safety of novel dosage forms<br>Bioavailabilty and bioequivalence<br>Process scale-up and "bio-batch"<br><br>Manufacture relies on advances in polymer science and technology. |

logies, specific controlled release formulations of commercially important drugs, as well as the manufacturing processes. Patents afford the advantages of exclusivity and patent-life extension. They also increase the risk of potential infringement of issued patents. Therefore, the legal issues of patentability and infringement are very important in this field.

Companies involved in controlled release have differed widely in their philosophy towards research and their approaches to the development of delivery devices. Hundreds of small research companies have been typically driven by venture capital and have a single narrow goal of developing a specific patented drug delivery technology or application. The success of these companies has been dependent on their ability to demonstrate the utility of their product and find one or more interested customers. Intermediate-sized companies, e.g., Key and Alza, have typically pursued carefully planned and aggressive patent strategies with ideas generated by interdisciplinary research and development teams. The large well-established pharmaceutical houses have generally focused on specific commercially important drug entities and relied on contractual development, licensing, collaborations, and outright buy-outs for obtaining the technology.

From the regulatory viewpoint, the safety and reliability of drug delivery systems have been emphasized either because nontraditional excipients and manufacturing processes have been used or the formulations encounter changing physiological conditions (pH, water content, etc.) as they traverse the gastrointestinal tract, which may result in complex *in vitro-in vivo* correlations. Guidelines for evaluating the safety and performance of controlled release dosage forms are available.[1] Because the drug release characteristics may depend on processing variables, formulations made at a larger commercial scale may have to be shown to be bioequivalent to those made at a smaller pilot scale and used in clinical trials.

## 1. Design and Development

An outline for the design and development of a controlled release formulation is given in Table 2. The process begins with establishing the rationale and specifying the dose, delivery duration, and release kinetics. This is followed by performulation and formulation development. The technology and rate-controlling polymeric excipients are selected and prototype devices are tested *in vitro* and in animal models before progressing into humans. The formulation is then refined and the manufacturing process scaled-up and transferred to commercial production.

## 2. Rationale

The rationale for controlled drug delivery is based on medical and marketing needs. From the medical viewpoint (Table 3), drug delivery may allow a reduced dosing frequency leading to improved compliance. A uniformly efficacious response and a lower incidence of undesirable side-effects should result from reducing the fluctuations in plasma drug concentrations. Also, controlled release dosage forms, which release only a small portion of their dose in the stomach, and enteric systems, which release the drug in the small intestine, reduce the exposure of the gastric mucosa to the drug which may lead to reduced gastric irritation and improved stability of acid-labile drugs. It may be theoretically possible

**Table 2. Rational design and development of controlled drug delivery systems**

| |
|---|
| Establish rationale for release-rate modification |
| Set product specifications based on pharmacokinetics/pharmacodynamics:<br>Single/multiple dose<br>Effect of food<br>Half-life<br>Therapeutic index<br>Position-dependent absorption |
| Preformulation:<br>pH-solubility profile<br>Stability and excipient compatibility<br>Excipient characterization |
| Select technology and materials |
| Develop prototype formulations |
| Test *in vitro* and *in vivo* in suitable animal models |
| Determine performance in healthy individuals (Phase I) |
| Optimize design/refine formulation |
| Scale-up manufacturing process |
| Determine bioequivalency or clinical safety and efficacy (Phase II/III) |
| Transfer to production |

to reduce the total dose with controlled release formulations because the drug is more effectively utilized, which may also lead to a more favorable side-effect profile.

However, a catastrophic failure (dose-dumping) of a formulation containing two or four times the conventional dose could lead to serious side-effects, particularly if the drug has a narrow therapeutic index. If blood level peaks are pharmacologically important, pulsatile drug delivery or increasing release rates with time rather than zero-order release kinetics, may be required. Also factors such as inefficient absorption in the colon or increased first-pass metabolism can lower the bioavailabilty. If the therapy is highly individualized and the dose must be precisely titrated, controlled release formulations, particularly membrane-coated tablets, may be limited by the difficulty of dividing the dose.

From the commercial viewpoint (Table 4), drug delivery can afford increased patent-life, greater convenience, and improved compliance. It also may expand the market because physicians may be more comfortable prescribing a product that offers therapy with lower side-effects and also because the controlled release dosage form may allow the treatment of previously untreated disorders. For example, an immediate-release

Table 3. Controlled drug delivery systems: medical analysis

| PROS | CONS |
|---|---|
| Reduce dosing frequency | Potential for dose-dumping |
| Reduce blood-level fluctuations and decrease side-effects | Peaks may be pharmacologically important (need for pulsatile delivery) |
| Reduce total dose | BA may be poor due to various factors |
| Obtain uniform response | Down-regulation, or decrease in the number of receptors resulting from overstimulation |
| Reduce gastric irritation | Potential for delayed onset of activity |
| Increase compliance | Precise titration to individual needs may be difficult |
| Expand indications for the drug | |

formulation of nifedipine (Procardia) is used to treat angina, whereas constant-release nifedipine (Pfizer's Procardia-XL) is used to treat angina as well as hypertension. Drug delivery devices may provide cost-effective treatment because the dose may be lower. On the other hand, the use of novel processes and equipment may raise the production costs.

Table 4. Controlled drug delivery systems: marketing analysis

| PROS | CONS |
|---|---|
| Cost-effective treatment may be possible if dose can be reduced | Technology may be expensive |
| Increased convenience and patient compliance | Individual titration may be difficult |
| Increased physician comfort level | Potential for dose-dumping, long apparent half-life, prior experience with the conventional dosage forms do not favor controlled release dosage forms |
| Expanded market-treatment of previously untreated disorders | |
| Extended patent life and exclusivity | |

## 3. Specifications

The dose, delivery duration, and release kinetics for most controlled release formulations are based on known or assumed target blood-levels. Computer simulations can be used to predict the blood levels with several dose and release rate options. In certain cases, an immediate-release loading dose followed by a slow-release maintenance dose may result in an ideal blood-level profile. Data on drug absorption as a function of position in the gastrointestinal tract is valuable because it may influence the delivery duration and bioavailabilty. Before the *in vitro* drug release profile, which results in the desired target blood-level profile, is specified, the relationship between *in vitro* and *in vivo* release rates must be known. In some cases, the pharmacodynamics of the drug may be well-known and the specifications may be based on pharmadynamics rather than pharmacokinetics.

## 4. Technology

A number of drug delivery technologies are available or can be developed (Table 5). They can be classified by processing methods, e.g., coating, microencapsulation, and embedding, or by the drug release mechanism, e.g., diffusion devices, osmotic systems, and erodible matrices. Their selection depends on whether the product specifications can be met. Also, manufacturing issues including prior experience with the technology, in-house processing expertise, availability of commercial-scale equipment and reproducibility of performance must also be considered.

## 5. Polymers

The polymers that have been studied for drug delivery applications can be classified into the following four categories (Table 6): (1) inert, nonbioerodible hydrophobic polymers, e.g., ethylene vinylacetate, poly(dimethyl siloxane), poly(ether urethane), poly(vinyl chloride), and poly(ethylene); (2) crosslinked hydrogels which swell but do not dissolve in water, e.g., poly(hydroxyethyl methacrylate), poly(vinyl alcohol), poly(ethylene oxide), and crosslinked poly(vinyl pyrrolidone); (3) bioerodible polymers, e.g., poly(lactic acid), poly(glycolic acid), poly(caprolactone), poly(hydroxybutyrate), poly(amino acids), labile poly(esters), poly(orthoesters), and poly(anhydrides); (4) water soluble polymers, e.g., hydroxypropyl methylcellose, hydropropylcellulose, hydroxyethylcellulose, carboxymethylcellulose, and poly(vinyl pyrrolidone). Some polymers have a pH-dependent aqueous solubility, e.g., cellulose acetate phtalate.

In the case of diffusion-based reservoir devices, the polymer selection is based primarily on the permeability of the drug through the polymer. In contrast, if the drug diffuses through aqueous channels formed in a porous matrix device, the diffusivity of the drug in water may be more important than the barrier properties of the polymer. The polymer must either be inert or degrade into nontoxic products. Polymers used in implants must be biocompatible. Finally, the polymer should be inexpensive and possess good mechanical properties that allow easy fabrication and manufacture. A polymer approved for food and drug use is preferred as a dosage form component because this eliminates lengthy and expensive toxicity testing required to demonstrate the safety of the polymer.

| Table 5. Technologies for controlled drug delivery |
|---|
| **CHEMICAL**<br>Salt form selection<br>Pro-drugs<br>Carrier systems |
| **PHYSICAL**<br>Pan or fluid-bed coating<br>Microencapsulation by coacervation or spray-drying<br>Embedding by compression, extrusion, or injection-molding |
| **DELIVERY MECHANISMS**<br>Diffusion-based reservoir and matrix devices<br>Osmotic pressure-based OROS and GITS<br>Ion-exchange-based Pennkinetic system<br>Chemical erosion-based matrix devices<br>Complex (more than one) mechanisms |
| **TECHNOLOGY SELECTION**<br>SPECIFICATIONS<br>Dose<br>Duration<br>Release kinetics<br>DRUG PROPERTIES<br>Molecular weight<br>Solubility<br>pKa<br>Partition coefficient<br>Stability<br>LEGAL<br>Patentability<br>Infringement<br>MANUFACTURE<br>Prior experience<br>Equipment availabilty<br>Cost<br>Reliability |

## 6 Manufacturing Processes

Processes for manufacturing drug delivery systems range from the conventional, e.g., dry or wet granulation followed by compression, to those employed for processing polymers, e.g., extrusion and injection molding. Processing with equipment which is available in-house in development and production sizes is preferred from the economical stand-point and to utilize the expertise available in experienced personnel. Thus, versatile equipment and processes are highly desirable, e.g., fluid-bed processing can be used for coating, as well as for granulation, and drying.

| Table 6. Polymers for controlled drug delivery |
|---|
| **POLYMERS** |
| **NONBIOERODIBLE**<br>ethylene vinyl acetate, silicones<br>poly(ethylene), poly(vinyl chloride)<br><br>**BIOERODIBLE**<br>poly(lactic/glycolic acid)<br>poly(caprolactone), poly(hydroxybutyrate)<br>poly(ortho-ester)s, poly(anhydrides)<br><br>**HYDROGELS**<br>poly(hydroxyethyl methacrylate), poly(vinyl pyrollidone)<br>poly(vinyl alcohol), poly(ethylene oxide)<br><br>**SOLUBLE**<br>Most cellulose derivatives |
| **POLYMER SELECTION** |
| Barrier properties<br>Stability<br>Toxicity<br>Biocompatibility<br>Ease of fabrication<br>Regulatory status<br>Cost |

## 7 *In Vitro* and *In Vivo* Tests

The candidate formulations should be subjected to stability and release rate tests. The release rate test should be able to distinguish changes in the formulation or the process, and assure that the product is uniform, reproducible, and not prone to "dose-dumping". The release rate test may sometimes predict the *in vivo* performance and bioavailabilty, particularly if the release is independent of the external pH and hydrodynamics. A suitable animal model may be used to screen formulations for performance and safety before testing on humans. Phase I studies are ideally conducted with more than one dose strength and with more than one release profile to help establish an *in vitro-in vivo* relationship.

## 8 Optimization and Scale-Up

The optimization, scale-up and validation of novel manufacturing processes for drug delivery systems is frequently a very time-consuming step and requires significant resources. The drug release characteristics may be sensitive to the scale of manufacture or the equipment used. Therefore, these issues may be important in the demonstration of bioequivalence.

Examples of commercial novel drug delivery systems include the Pennkinetic system which is based on ion-exchange (e.g., Biphetamine, Pennwalt), the elementary osmotic pump (e.g., Accutrim, Ciba) and the GITS system (Procardia-XL, Pfizer) which are based on osmotic pressure, diffusion-based reservoir and matrix systems and those based on polymer erosion. The drug release profile from matrix devices based on the diffusion-dissolution mechanism is nonlinear and depends on the square-root of time. In an effort to obtain zero-order release, several modifications of the basic designs have been proposed, studied, published, and patented.[2] However, in most cases, the complexity of the design and difficulties anticipated in their large-scale manufacture have prevented the commercial development of most of these systems.

The future technical challenges for the field of polymeric drug delivery (Table 7) include: (1) the understanding and control of absorption and metabolism factors which may ultimately lead to the oral delivery of proteins and peptides; (2) the design of intelligent delivery systems with biofeedback, i.e., systems which regulate drug release based on a biological marker, e.g., the release of insulin depending on the blood-sugar levels; (3) tailored release rates including increasing release rates with time and pulsatile release; and (4) targeted and site-specific delivery.

These challenges must be met with innovative research by dedicated scientists and sound management by professionals. Factors that have previously contributed to the growth of this field, e.g., utilizing multidisciplinary project teams and the applying advances made in related fields, will continue to play an important role. Also important will be the early recognition of drug candidates that may benefit from controlled release technology, the systematic development of delivery devices and formulations in a rational, efficient, and cost-effective manner. This can be achieved by improving information and project management systems, and by investing resources to expand the scope of drug delivery.

Table 7. Future challenges for polymeric drug delivery

| SCIENTIFIC |
|---|
| Oral delivery of proteins and peptides |
| Biofeedback and intelligent delivery systems |
| Increasing release rates with time |
| Pulsatile delivery |
| Colonic and site-specific drug delivery |
| Targeted drug delivery |
| Chronotherapeutics and pharmacodynamics-based delivery |
| Penetration and absorption enhancers |
| **MANAGEMENT** |
| Early recognition of appropriate drug candidates |
| Expansion of the scope of drug delivery |
| Execution of rational design and development |
| Reduction in development costs and time |
| Improved information management systems |
| Products with greater reliability and safety profile |

## CONCLUSION

Starting as a simple curiosity and demonstrating elegant applications of physical and chemical principles, polymer controlled drug delivery systems are now accepted as commercially and medically important pharmaceutical dosage forms. In the future, developments in the delivery technology must keep pace with the discovery of new drug candidates such as proteins and peptides, which may need specialized formulations to control the rate and site of drug release. The rational design and development of delivery systems requires the ability to deal with a large number of complex and interrelated variables, an understanding of physiological and biological barriers, and an appreciation of engineering fundamentals. In addition to the technical component, medical, marketing, regulatory, legal, and economic factors play an important role which makes the successful development of drug delivery systems a highly specialized and challenging task. They also serve to make the design and development of drug delivery systems a rewarding activity.

## REFERENCES


1. J. P. Skelly, G. L. Amidon, W. H. Barr, L. Z. Benet, J. E. Carter, J. R. Robinson, V. P. Shah and A. Yacobi, Int. J. Pharm., 63, 83-93 (1990). "Reporting on the workshop on: *in vitro* and *in vivo* testing and correlation for oral controlled/modified release dosage forms."
2. J. R. Cardinal, et al., 2nd AAPS National Meeting, Boston, MA (June 1987). "Geometric approaches to constant rate controlled release drug delivery systems."


## ADDITIONAL READING


1. J. R. Robinson, Ed., "*Sustained and Controlled Release Drug Delivery Systems,*" Marcel Dekker, Inc., New York, 1978.
2. A. G. Thombre and J. R. Cardinal, in" "*Encyclopedia of Pharmaceutical Technology,*" James Swarbrick and James C. Boylan, Eds., Vol. 2, Marcel Dekker, Inc., 1990, pp. 61-88. "Biopolymers for Controlled Drug Delivery."

# MICROSPONGE[R] DELIVERY SYSTEMS (MDS): A TOPICAL DELIVERY SYSTEM WITH MULTIPLE MECHANISM FOR TRIGGERING THE RELEASE OF ACTIVES

Robert Eury, Rajesh Patel, Karen Longe, Tai Cheng and Sergio Nacht

Advanced Polymer Systems, Inc.
Redwood City, CA 94063

Recently the controlled release of topically active agents has been a subject of increasing interest. The use of an MDS (Microsponge[R] Delivery System) for the controlled administration of an active agent provides benefits that range from more efficient use of the active agent to a less irritating formulation. The average MDS is an active agent loaded in a macroporous bead with an average particle size of 25 µm. The use of an MDS confers to the formulation a sustained or controlled release and the capacity to either initiate or modify the release rate when exposed to some change in the external environment. The external stimuli can be invoked by the user in either an active or passive manner, i.e., rubbing, temperature, pH, etc. Thus the development of triggered-release MDS formulations results in "intelligent" responsive systems.

## INTRODUCTION

Before transdermal delivery methods were developed, the majority of active agents applied to the skin were those whose site of action was local. These were usually formulated as a solution or suspension of the active agent in a cream or ointment with no sustained or controlled release attributes. The advent of transdermal release devices has centered a great deal of attention on the skin, but primarily in how to overcome its barrier properties selectively. The application of controlled release methodologies to the administration of topically active agents has therefore been mostly overlooked. There are two points of consideration for these new topical delivery systems. First is that control of the release rate becomes quite important with active agents which are irritating at too high a local concentration. The second is that with many topical agents, e.g., antibiotics, it is desirable for the active agent to remain at the site of action (epidermis/dermis) and not to be absorbed into the systemic compartment. A Microsponge[R] Delivery System (MDS) is one of the few topical delivery systems that, in many cases, can satisfy the two previous points.

The ultimate goal is a release device in which the patient or environment influences the response of the device in a beneficial manner. The controlled release systems that have addressed this aspect have predominately been desired for systemic administration. Controlled release systems have been described where activating stimuli such as a change in temperature,[1,2] the presence or polarity of the external medium,[3,4] and even the presence of a specific molecule have been used to trigger or modulate the release of active agents.[5,6]

In order to be the most efficacious, an MDS should not release a significant amount of the trapped active until the system has reached the vicinity of the predetermined target. This requires that some external stimuli would result in the onset of release of the active agent. By building on the concepts outlined in the above systems, a more intelligent MDS can be developed which incorporates into itself one or more release switches. The changes in the external environment which have been investigated for use in the switching-on of the release from an MDS are rubbing, temperature, pH, presence of fluid, and polarity of external medium.

## SYNTHESIS OF MICROSPONGE[R] MATERIALS

The general synthesis of Microsponge[R] materials is a free-radical suspension polymerization.[7] The bases of the suspension polymerization process whereby macroporous materials are obtained are quite well known for styrenics,[8-11] (meth)acrylates,[12-13] and even hydrophilic hydrogel materials.[14] Even though macroporous materials have been studied for many applications including chromatography, ion-exchange, and reagent supports, only recently has the use of these materials for controlled release been pursued.

In general, the synthesis of a macroporous material involves the suspension polymerization of a dispersed phase consisting of monomers, crosslinking agent, initiator, and a cosolvent which functions as a porogen. The requirements for the most efficient porogens are that they must be miscible with the monomers but should be a precipitating solvent for the polymer to be formed. The components of the dispersed phase should not appreciably partition into the continuous phase. The particle size obtained is a function of the suspending agent used, its concentration, and the amount of shear induced during droplet formation. In many instances it is desirable that the suspending agent not bind to the surface of the beads. The properties of the resulting materials to be used in MDS can be varied over a broad range. The volume percentage particle size[15] can be varied from about 3 to 300 µm diameter with the one standard deviation unit of distribution as low as 22 to 25%. The surface area[16] can be varied from 20 to 500 $m^2/g$ and the pore volume[17] can range from 0.1 to 3.4 $cm^3/g$. The matrix polarity can be varied from fairly non-polar, i.e., styrenics, to very polar, i.e., polyacrylamides.

Scanning electron microscopy is quite enlightening as to the morphology of the resulting beads. The pore openings can be seen and the beads have good sphericity (Figure 1). The internal structure of the beads resembles a bag of marbles (Figure 2). The porosity is attributed to the continuous interstitial spaces between the "marbles".[18]

## LOADING OF ACTIVES

The actives can be incorporated into the beads by two methods. The

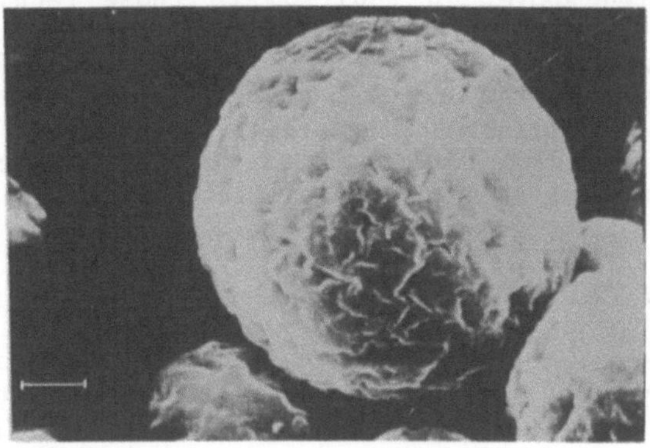

Figure 1. Scanning electron micrograph of Microsponge[R] materials, magnification x 990. Bar is 4 µm.

Figure 2. Scanning electron micrograph of freeze-fracture section, x 15,840.

first method is a one step process where the active agent functions as porogen. For this to be successful, the active must fulfill the porogen requirements listed above. In addition to those requirements, it must also not hinder the polymerization and not become inactivated by the polymerization process; i.e., it must be stable to radicals. If the active is not amenable to the one-step process, then a two-step process can be used. This is accomplished by impregnating preformed, clean, dry beads with the active according to techniques such as contact absorption. If necessary, solvents can be used to facilitate the absorption process.

## COMPARISON OF MDS TO OTHER TECHNOLOGIES

The MDS is quite different from other entrapment technologies like microencapsulation (Figure 3) and liposome (Figure 4). Microencapsulation has the advantage that a high payload can be attained. The disadvantages are that the coating materials and processes involved are not suitable for all actives. Microcapsules do not usually control the release rate of actives by themselves. Once the wall ruptures, all of the active agent contained within is released. Liposomes have the disadvantages of a low payload, the manufacturing process can be complex, chemical stability is limited, and microbiological stability requires the use of preservatives. Liposomes are also relatively expensive.

Microsponge® materials, in contrast to the two previous technologies, have the following attributes. They are stable over a broad range of pH and temperature, and are compatible with most vehicles and ingredients. Because the average pore diameter is less than 0.25 µm, they are self-sterilizing and do not require any preservatives. The payload can be as high as 50 to 60% by weight, e.g., mineral oil, and still be in the form of a superfine, free-flowing powder. An MDS can be cost effective even for the cosmetic mass market use where a material's cost is of utmost concern.

| | MICRO ENCAPSULATION | APS ENTRAPMENT SYSTEM |
|---|---|---|
| Concept |  Egg | 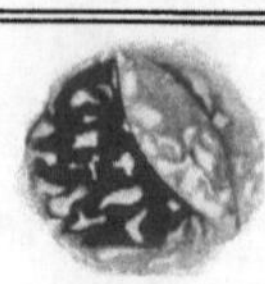 Sponge |
| Shell | Complete | Pores |
| Release Mechanism | Rupture of Shell | Pressure<br>Volatility<br>Temperature<br>Solvent<br>Partition Coefficient |
| Amount Released | 100% | Programmable<br>▪ Particle Size<br>▪ Pore Size & Volume<br>▪ Resiliency |

Figure 3. Comparison of microcapsule and Microsponge® characteristics.

| | Liposomes | APS Microsponge* |
|---|---|---|
| Concept |  |  |
| Chemical Composition | Phospholipids/Cholesterol | Inert, non-toxic synthetic polymers |
| Chemical Stability | Potential problems depending on pH, vehicle composition and nature of entrapped ingredient | Totally stable over broad range of pH; compatible with all useful vehicles and ingredients. |
| Microbiological Stability | Requires strong chemical preservatives | No preservative is required. Self-sterilizing |
| Release Mechanism | Diffusion or degradation of shell material | Pressure<br>Diffusion<br>Volatilization<br>Temperature<br>Solubility |
| Programmable Parameters | Chemical composition<br>Particle size | Chemical composition<br>Particle size<br>Resiliency (Elasticity)<br>Pore Volume<br>Pore Diameter |
| Entrapping Capacity | 30% or less | Up to 50 - 60% |
| Cost | Relatively expensive | Cost effective for Mass Market uses |
| Quality Control | Complex: requires tightly-controlled, ultra-pure raw materials | Simple: utilizes easily available raw materials of adequate purity |

Figure 4. Comparison of liposome and MDS technologies.

## ACTIVE AGENT RELEASE-RATE TRIGGERS

### 1. Rubbing

After the MDS has been applied to a given area, it is possible for the amount of active released to be modulated by the consumer. If the active is a non-volatile liquid, or is dissolved in such, then rubbing can trigger the release of additional agent. The renewable release of a sunscreen (octyl *p*-N,N-dimethylaminobenzoate, or Padimate O) due to rubbing has been demonstrated.[19] In this case, the amount of sunscreen on excised human skin was ascertained with respect to time and rubbing. The amount of Padimate O could be nearly restored to initial levels as compared to the "free" active formulation in which the amount of sunscreen on the skin surface gradually decreased (Figure 5).

Microencapsulation also offers the potential for rubbing triggered release. A comparison of skin softening (emolliency) as measured by a gas-bearing electrodynamometer[20] showed that both MDS and microcapules containing mineral oil were capable of increasing skin softness when rubbed five times after application. The MDS gave a greater amount of softening than the microencapsulation system for the entire time period. When the subjects rubbed the application areas three hours post-application, the MDS and microencapsulation systems both gave an increase in emolliency. The MDS gave a larger increase than the microencapsulation,

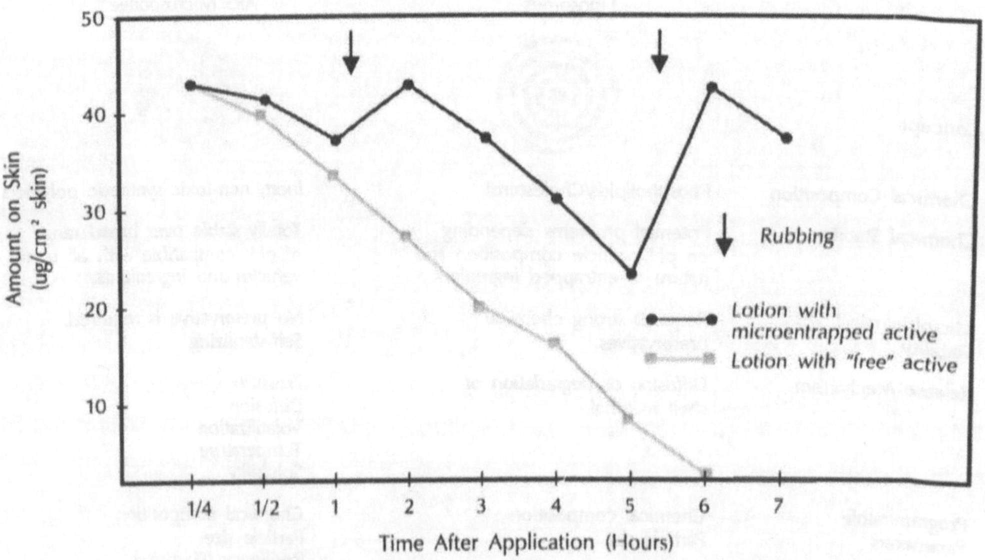


**Figure 5. Amount of $^{14}$C Padimate O recovered from the superficial skin layers after application of lotions containing sunscreen either dispersed (squares) or entrapped in an MDS (circles).**

**even though the MDS subject's skin were softer before the rubbing. The duration of emolliency was also greatly extended for the MDS as evidenced by the MDS treated skin being softened by 36% at the end of 6 hours whereas the microcapsule treated skin had returned to the baseline value (Figure 6).**

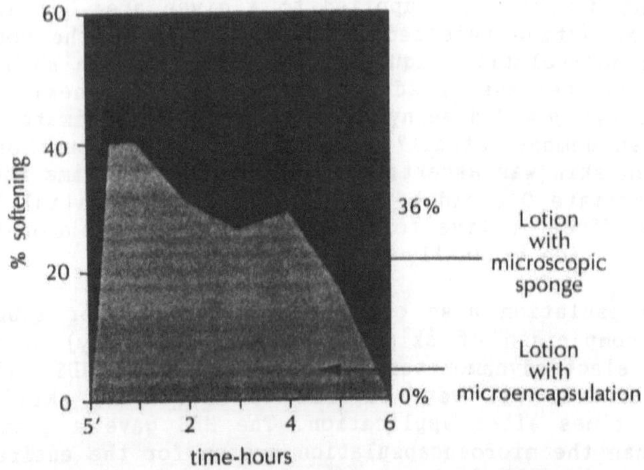


**Figure 6. Comparison of MDS and microencapsulation technologies on human skin emolliency as measured by gas bearing electrodynamometer.**

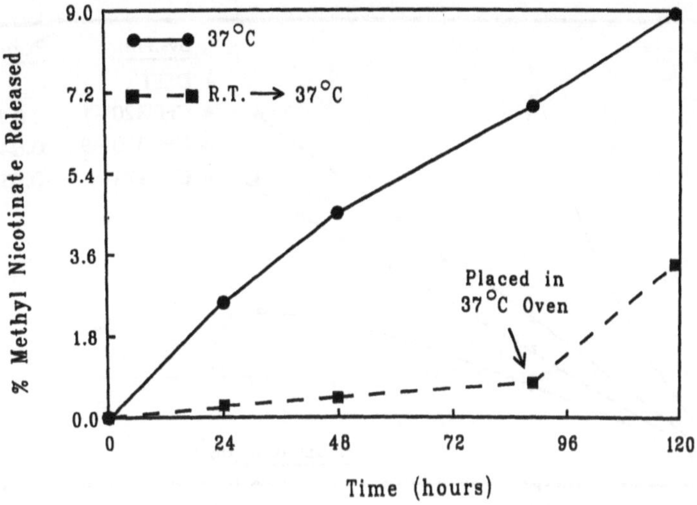


Figure 7. Effect of temperature on the triggering of release of methyl nicotinate from MDS.

## 2. Temperature

The release rate of an active agent from an MDS can be varied greatly by the temperature of the external environment. An MDS was formulated which contained 30% w/w of methyl nicotinate and gravimetric measurements of the release profiles at 37°C and room temperature were conducted. The material was divided into two portions, one which was placed into an oven at 37°C and one which was left at room temperature. After 3½ days, the room temperature samples were placed into the oven along with the other samples. The study showed that the release rate of methyl nicotinate at room temperature was much slower (*ca.* 0.3%/day) than that of the same MDS at 37°C (*ca.* 1.8%/day) (Figure 7). The release rate of the MDS held at room temperature could be increased to the rate of the system held at 37°C material when also placed in the oven. The difference between the on and off states for this temperature responsive MDS, although not as great as desired, is indicative of the feasibility of this approach.

The pore diameter of the Microsponge[R] employed can also modify the release rate of an active volatile. An isothermal release profile study of N,N-diethyltoluamide (DEET) as a function of pore diameter was performed (Figure 8). All of the samples released DEET at a slower rate than the "free" DEET. The release rate decreased from approximately 4.3%/hour for the "free" active to o.7%/hour for the MDS with a pore diameter of 0.018 µm. Thus, the use of an MDS was able to slow the rate of volatilization by as much as five fold.

## 3. pH

An MDS in which the active agent is not released until triggered by a change in the pH of the external medium has been developed.[21] This system involves an enteric coating-type of material which imparts pH responsiveness to the MDS. The dye used for these studies was D&C Red #28. This was chosen as the model active because it is very water-soluble and any lack of release rate control would be evident under these conditions. A sample

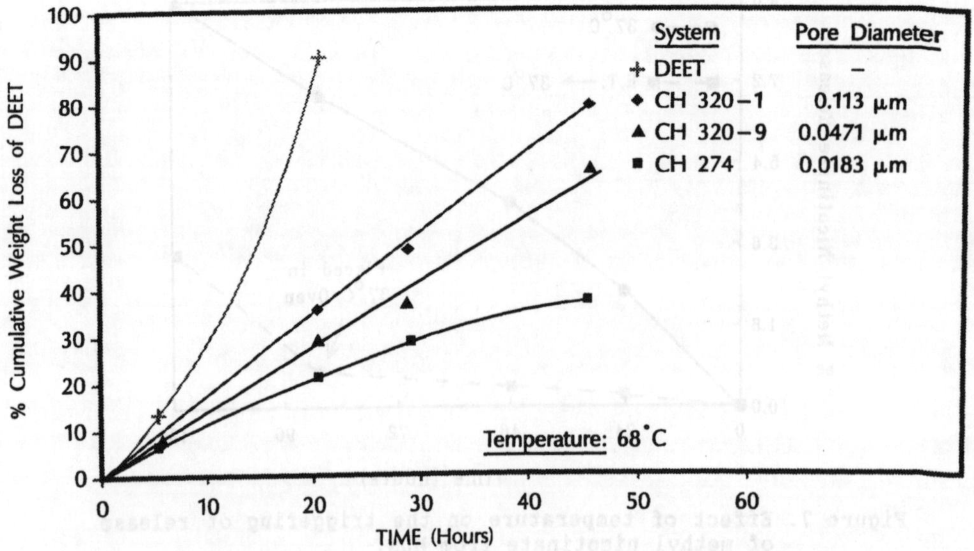


Figure 8. Effect of pore diameter on the release profile of DEET.

of the pH responsive MDS was placed into a modified USP spindle dissolution apparatus in which the release medium was a pH 3 buffer solution (Figure 9). No detectable amount of dye was released after 2 hours immersion. The spindles were then removed and immersed into a pH 8 buffer. This buffer induced the release of 80% of the dye over a 3 hour period.

The pH responsive MDS can also modulate release rates as a function of the pH of the external medium (Figure 10). Release studies were performed in a manner similar to that described above except that a constant pH was used throughout. At a lower pH, about 10 to 20% of the dye was released over a period of 3 to 6 hours. As the pH of the incubation

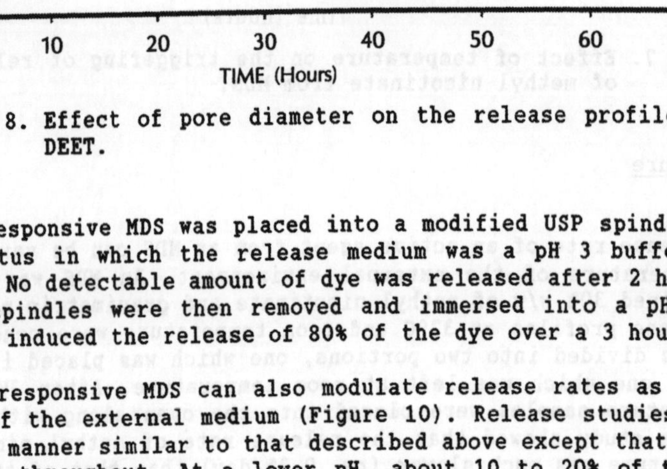


Figure 9. Release of D&C Red #28 from pH responsive MDS triggered by a change in the pH of the incubation medium.

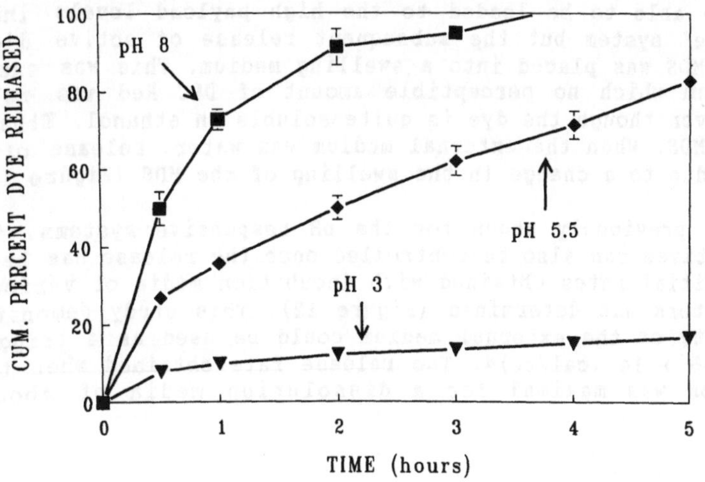


Figure 10. Release of D&C Red #28 from pH responsive MDS as a function of the pH of the dissolution medium.

medium increased, the release of dye also increased until at pH 8, about 80 to 100% of the dye was released over a period of 1 to 3 hours. Enteric materials usually do not exhibit a sharp dissolution pH, but rather they swell and eventually dissolve over a broader range due to the gradual ionization of side groups. As the fraction of groups in the ionized form is increased, the polymer swells due to ionic repulsions until enough groups have been ionized to impart water-solubility. The different amounts of MDS swelling at various pH accounts for the modulation of release with respect to pH.

## 4. Polarity of External Medium

An MDS in which the active agent is not released until triggered by a change in polarity of the external medium has also been developed.[22] Release profiles were again determined as described above. These mate-

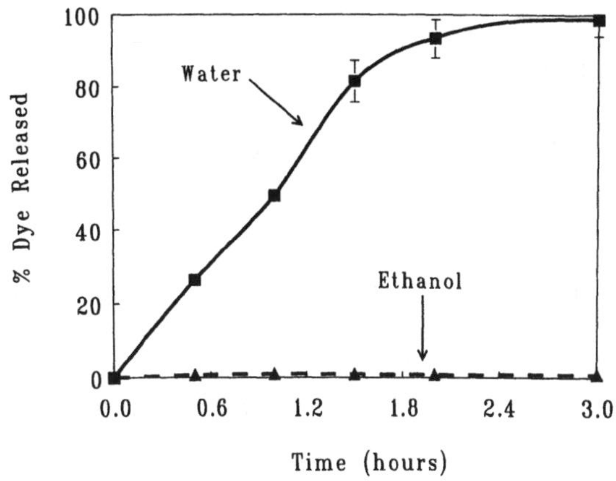


Figure 11. Effect of dissolution media on the release profile of D&C Red #28.

rials were able to be loaded to the high payload levels inherent to a Microsponge[R] system but the subsequent release of active did not occur until the MDS was placed into a swelling medium. This was exemplified by an MDS from which no perceptible amount of D&C Red was released into ethanol, even though the dye is quite soluble in ethanol. Ethanol did not swell the MDS. When the external medium was water, release of active was triggered due to a change in the swelling of the MDS (Figure 11).

As was previously shown for the pH responsive systems, the release rate of actives can also be controlled once the release has been triggered. The initial rates obtained with incubation media of various solubility parameters was determined (Figure 12). This study demonstrated that the polarity of the external medium could be used as a trigger from an MDS where $\delta > 14$ $(cal/cc)^{1/2}$. The release rate obtained when the MDS was switched-on was maximal for a dissolution media of about $\delta = 16$ $(cal/cc)^{1/2}$.

**CONCLUSIONS**

Microsponge[R] Delivery Systems are a practical and effective means of delivering topical agents. The MDS have many advantages over both microencapsulation and liposome technologies. The system has several parameters, e.g., surface area, pore volume, and particle size, which can be customized for a particular need. The MDS can also be tailored so that release of an active can be triggered by the external environment. One of the most effective triggers is rubbing, with other possible triggers being a change in pH, medium polarity, or temperature. "Intelligent" systems which modulate the release rate of the entrapped material after being switched-on are also possible. While the MDS technology is fairly new, several products utilizing these systems have already been marketed and more will be introduced soon, thus justifying their position as a very useful and unique technology in the repertoire of the controlled release topical formulator.

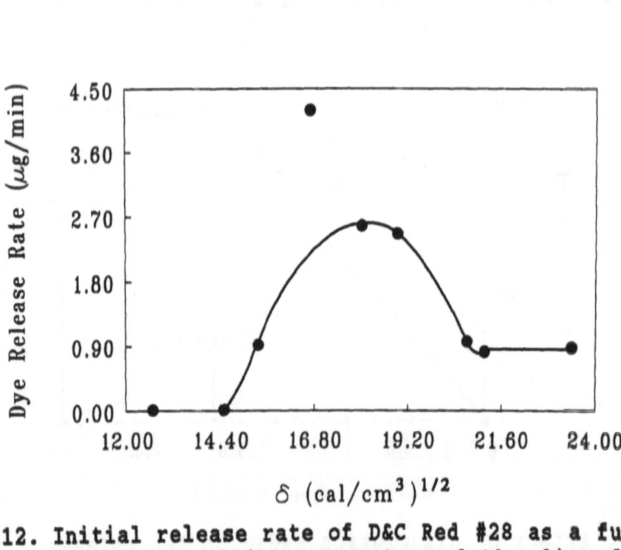


Figure 12. Initial release rate of D&C Red #28 as a function of the solubility parameter of the dissolution media.

## REFERENCES


1. Y. H. Bae, T. Okana, and S. W. Kim, Makromol. Chem., Rapid Commun., 9, 185 (1988).
2. Y. H. Bae, T. Okano, and S. W. Kim, J. Control. Rel., 9, 271 (1989).
3. R. S. Harland, A. Gazzaniga, M. E. Sangalli, P. Colombo, and N. A. Peppas, Pharm. Res., 58 (8), 488 (1988).
4. C. Kim and P. I. Lee, Proc. Intern. Symp. Control. Rel. Bioactive Mater., 17, 118 (1990).
5. J. Kost, T. A. Horbett, B. D. Ratner, and M. Singh, J. Biomed. Mater. Res., 19, 1117 (1985).
6. S. Y. Jeong, S. W. Kim, M. J. D. Eenink, and J. Feijen, J. Control. Rel., 1, 57 (1984).
7. R. Won, U.S. Patent 4,690,825 (1987).
8. J. C. Moore, J. Polym. Sci., A2, 835 (1964).
9. R. Kunin, E. Meitzner, and N. Bortnick, J. Am. Chem. Soc., 80, 305 (1962).
10. D. C. Sherrington, in: "*Macromolecular Syntheses,*" Vol. 8, E. M. Pearce, Ed., Wiley, New York, 1982, pp. 69-73.
11. G. J. Howard and C. A. Midgley, J. Appl. Polym. Sci., 26, 3845 (1981).
12. D. Horak, Z. Pelzbauer, M. Bleha, M. Ilavsky, F. Svec, and J. Kalal, J. Polym. Sci., 26, 411 (1981).
13. A Akelah, M. Hassanein, M. El-Sakran, E. R. Kenawy, and F. Abdel-Galel, Acta Polym., 40 (2), 129 (1989).
14. O. Mauz, K. Sauber, S. Noetzel, U.S. Patent 4,542,069 (1985).
15. Particle Sizer 3600 E type, Malvern Instruments, Malvern, England.
16. Flowsorb 2300, Micromeritics, Norecross, Georgia.
17. Poresizer 9310, Micromeritics, Norecross, Georgia.
18. D. Horak, Z. Pelzbauer, M. Bleha, M. Ilavsky, F. Svec, and J. Kalal, J. Appl. Polym. Sci., 26, 411 (1981).
19. D. Yeung, S. Nacht, D. Bucks, and H. I. Maibach, J. Am. Acad. Dermatol., 9, 920 (1983).
20. M. S. Christensen, C. W. Hargens, III., S. Nacht, and E. H. Gans, J. Invest. Derm., 69, 282 (1983).
21. U.S. Patent Appl. No. 5/41,111 (June 20, 1990).
22. U.S. Patent Appl. No. 7/416,567 (October 3, 1989).

# BIOELASTIC MATERIALS AS CHEMOMECHANICALLY TRANSDUCING ("SMART") MATRICES FOR DRUG DELIVERY

Dan W. Urry, Ph.D.

Bioelastics Research, Ltd.
1075 South Thirteenth Street
Birmingham, Alabama 35205

For biodegradable drug delivery systems, control of hydration can be a key to rate of degradation. When beginning with drug-doped condensed matrices, controlling the rate of hydration (swelling) can then become the key to drug release. One approach is to use responsive polymers coupled to chemical clocks with selectable half-lives. Polymers that exhibit chemically modulable inverse temperature transitions are an ideal material for such drug delivery matrices and elastometric polypeptides form such matrices. The chemical clocks can be peptide sequences containing Asn or Gln residues (or other suitably unstable chemical groupings) which, depending on the hydrophobicity of adjacent residues, can exhibit half-lives from days (or less) to decades. The half-lives of the [$CONH_2 \rightarrow COO^-$] reaction can be further shifted by hydrophobic tertiary structure. It is the rate of carboxylate anion or other the polar species formation which would then control the extent of swelling and rate of degradation.

Whether drug release is by controlled swelling with enhanced rate of degradation, or by contraction with expelling of contents, the devices can be designed with site specificity based on a unique chemical aspect of the target site much as proteins themselves are brought to fold or unfold, to assemble or disassemble and to function at specific sites with highly cooperative chemical response profiles. To the extent that protein structure is modulated and function is achieved by the mechanism considered here for drug delivery, that is, chemical modulation of the temperature of inverse temperature transitions, then the approach outlined here provides a step toward desired biological specificity with the capacity to release the drug being coupled to the unique chemistry of the diseased state. Such a bioresponsive drug delivery system would constitute a particular integration of the diagnostic-therapeutic pair in which a protein-based polymer is simultaneously the sensor of the need for, and the deliverer of, the drug and wherein the spent device can be degraded with release of simple amino acids and the chemical sensor.

*Cosmetic and Pharmaceutical Applications of Polymers*
Edited by C.G. Gebelein *et al.*, Plenum Press, New York, 1991

## INTRODUCTION

This manuscript conceptually addresses the potential, as drug delivery matrices, of a relatively new class of biomaterials within the general category of what has recently become termed protein-based polymers.* The specific protein-based polymers of interest here are described as elastometric polypeptide biomaterials or bioelastic materials which may be designed to exhibit efficient chemomechanical transduction.

## PROTEIN-BASED POLYMERS

A protein-based polymer is a chain molecule comprised of repeating peptide sequences where the repeating unit may be as small as a few residues or as large as hundreds of residues. The repeating unit could be as small as a repeating dipeptide, as in the poly(L,D)dipeptide of gramicidin A which forms ion selective transmembrane channels, or it could be a repeating domain of hundreds of residues as a model for the repeating domains of transmembrane pumps and channels and of multi-subunit enzymes. Protein-based polymers may be designed as simplified models for developing an understanding of the driving forces involved in protein function. And once the driving forces of protein function are adequately understood, protein-based polymers may be designed to perform many of the highly specific and cooperative functions of complex natural proteins. Protein-based polymers can then be developed as drug delivery systems not simply as inert carriers of drugs but rather as dynamic target specific deliverers of pharmaceuticals. Proteins are chemomechanical transducers in that a chemical signal (a chemical concentration change resulting for example in ligand binding) induces a conformational change (a mechanical response) resulting in function. This is most readily visualized in chemically-driven contraction and relaxation; and indeed protein-based polymers have been designed to exhibit chemically-driven contraction and relaxation, i.e. chemomechanical transduction.[1-5]

## BIOELASTIC MATERIALS AS TRANSDUCING MATRICES

For purposes of the following, the repeating peptide sequence of primary interest is $Val^1$-$Pro^2$-$Gly^3$-$Val^4$-$Gly^5$ which, as a protein-based polymer, is written poly(VPGVG). When $\gamma$-irradiation crosslinked by 20 Mrads, this protein-based polymer forms an elastomeric matrix designated as $X^{20}$-poly(VPGVG).[1,2] This matrix exhibits mechanothermal coupling by means of an inverse temperature transition, that is, at low temperature the matrix is swollen in water and on raising the temperature the matrix contracts, decreasing volume by an order of magnitude and length by a factor of greater than 2. This is thermomechanical transduction.[3] The matrix can be modified such that, when it is more hydrophobic (apolar), the transition occurs at a lower temperature and, when it is more hydrophilic (polar), the transition occurs at a higher temperature.

The change in hydrophobicity or hydrophilicity can be achieved in many ways. The addition of a functional group, that can be in either a more or less polar state, can be used to change the hydrophobicity and,

---

*The first symposium on protein-based polymers occurred as part of the April 23-29, 1990 National American Chemical Society meeting and was organized by D. Kaplan, M. Marron, and D. Tirrell who developed the term.

thereby, to change the temperature of the transition. A convenient functional group is the carboxyl group, which may either be in the less polar (more hydrophobic) [COOH] state, or the more polar (less hydrophobic) [COO-] state. Such a functional moiety is available in the glutamic acid residue, abbreviated Glu or E. In the elastomeric matrix, X$^{20}$-poly-[4(VPGVG)(VPGEG)], there are four Glu residues per 100 residues of polypentapeptide. At low pH, about 2, essentially all of the side chains of the glutamic acid residues are protonated giving only [COOH] moieties. In this case, in phosphate buffered saline [(0.15 N NaCl and 0.01M phosphate)], the transition is centered near 25°C and the matrix contracts on raising the temperature from 20° to 30°C as shown in Figure 1.[4] At pH 4.5, when there are about two carboxylate anions [COO-] per 100 residues, the contraction occurs on raising the temperature from 45° to 60°C (see Figure 1).

The capacity to change the temperature at which thermomechanical transduction occurs by changing the pH means that it is possible to remain at an intermediate temperature, say 37°C, and change the pH to achieve contraction and relaxation. At low proton concentration (high pH), the matrix is swollen. On increasing the proton concentration, i.e., on lowering the pH and protonating the carboxylate anions, the matrix contracts to less than one-half the swollen length and can lift weights in doing so that are a thousand times the matrix dry weight. The process is reversible such that the cycle can be repeated many times. This is chemomechanical transduction or mechanochemical coupling.[4]

## 1. Cooperativity in the Contraction/Relaxation Transition

A most interesting feature about this pH driven contraction/relaxation is shown in Figure 2.[5] When the elastomeric matrix is under load, the pH dependence of the contraction and of the degree of ionization α of the carboxyl moiety is very steep. In this case, 75% of the ionization, i.e., some 75% of the transition, is achieved in a ΔpH of 0.5, whereas, a

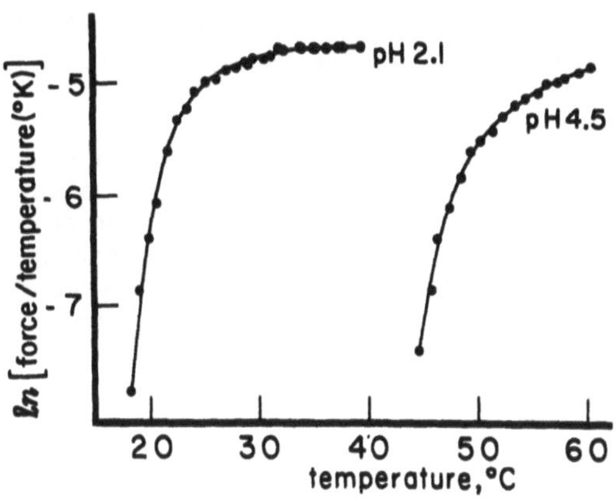


Figure 1. Thermomechanical transduction exhibited by X$^{20}$-poly[4(VPGVG),(VPGEG)]. At pH 2.1 contraction begins on raising the temperature from just below 20°C and is completed just above 30°C. At pH 4.5 contraction does not begin until above 40°C and is completed just above 60°C. (Reproduced with permission from reference 4).

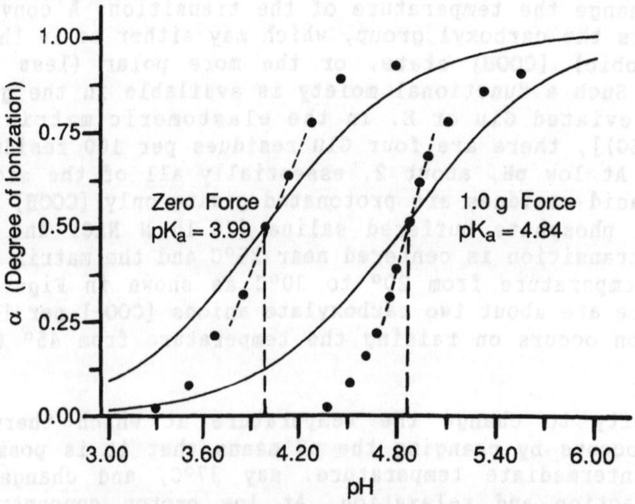


**Figure 2. When $X^{20}$-poly[4(VPGVG),(VPGEG)] is titrated without force being applied to the elastomer, an acid base titration curve is obtained with a $pK_a$ of 3.99. After the elastomer has been stretched by the application of a force, the $pK_a$ is raised, in this study, to 4.84 and the steepness of the curve is increased. See text for discussion. (Reproduced with permission from reference 5).**

$\Delta$pH of 2.5 is required to achieve the same change in the degree of ionization for a standard titration curve (see the solid lines of Figure 2). The steepness of the curve is a function of the hydrophobicity of the matrix, and it is the exposure of hydrophobic groups to the aqueous medium on stretching (applying a force) that causes the increase in steepness and the shift in $pK_a$ of Figure 2. Thus, the hydrophobicity of the matrix is one means of controlling the $pK_a$,[6] and of controlling the steepness of the transition.[5] The steepened transition is described as a positive cooperativity.

Cooperativity is a property of multisubunit enzymes and other multisubunit proteins which can be quantified from the steepness of the transition by means of a Monod-type analysis.[7,8] Using this analysis of protein functionality, the change in slope on stretching with exposure of more hydrophobic groups gives a change in Gibb's free energy per mole of ligand. With the ligand being the [$H^+$] moiety, analysis of the data in Figure 2 gives a value of about 2RT (S.Q. Peng and D. W. Urry, unpublished results). This is the same energy as determined from the shift in $pK_a$ on applying the force, i.e., $\Delta G/\Delta n = \Delta\mu = -2.3RT\Delta pK_a \approx 2RT$. Thus cooperativity, one of the more exquisite properties of protein function, is apparent in this protein-based polymer designed to exhibit chemomechanical transduction, i.e., chemically-induced contraction/relaxation.

The process of contraction or of relaxation (swelling) can be the controlling element for drug release. Accordingly, a drug delivery device could be designed to be responsive to a small change in pH and to a pH, though distinctly different, that is not greatly displaced from normal physiological pH.

## 2. Efficiency in the Conversion of Chemical Energy into Mechanical Work

An engine is "a machine designed for the conversion of energy into useful mechanical motion".[9] The contractile bioelastic matrices for drug delivery are engines of a type that convert chemical energy ($\Delta\mu\cdot\Delta n$) into the useful mechanical motion of drug delivery. Here $\Delta\mu$ is the change in chemical potential which may be written as $-2.3RT\Delta pH$ in Gibb's free energy/mole, when the chemical species is the proton and $\Delta n$ is the change in number of moles of protonated species in the protein-based polymer, both for the amount of work, w, achieved. The efficiency, $\eta$, of such an engine can be given by $w/\Delta\mu\cdot\Delta n$), where the work can be measured as a force, f, working through a distance, $\Delta l$, as for example in the lifting of a weight.[10]

Now it is of interest to compare the efficiency of the bioelastic matrix, $X^{20}$-poly[4(VPGVG),(VPGEG)], with that of crosslinked poly(methacrylic acid), $[(-CH_3CCOOH-CH_2-)_n]$.[10-12] The latter utilizes a charge-charge repulsion mechanism to extend the chain, wherein the chains become fully extended at 50 to 60% [$COO^-$]. On lowering the charge-charge repulsion by protonation, the extended chains collapse and achieve a contraction to about one-half length on lowering [$COO^-$] to 0 to 10%. Thus, to achieve the contraction of shortening to about one-half the length with the capacity to lift weights of about a thousand times the dry weight of the matrix requires the $\Delta n$ due to changing some 40[$COO^-$] moieties to [COOH] per 200 backbone atoms. To achieve this, the $\Delta pH$ needs to be about 2.5[13] giving 3.5kcal/mole. This is summarized in Table 1, along with the comparable values for $X^{20}$-[4(VPGVG),(VPGEG)], which uses a mechanism referred to as apolar-polar repulsive hydration free energies. This protein-based engine is capable of doing a similar amount of work in the contraction cycle, i.e., to contract to about half its length, and to pick up weights that are one thousand times its dry weight. But in this

| Table 1. Comparison of Chemomechanical Transduction Efficiencies |
|---|
| $\eta = w/(\Delta\mu\cdot\Delta n)$<br><br>Where $w$ = work = $f\Delta l$; $\Delta\mu$ = change in chemical potential; $\Delta n$ = change in moles of polymer species; $\Delta\mu\cdot\Delta n$ = the chemical energy. |
| $\eta_{c,c}$ = <u>efficiency of Charge-Charge repulsion mechanism</u> (as exemplified by polymethacrylic acid)<br><br>w: $\Delta f \approx 1/2$; $f \approx 1000$ x dry weight<br>$\Delta n$: 4($COO^- \rightarrow COOH$) per 300 backbone atoms<br>$\Delta\mu$: ($\Delta\alpha$ of $\approx 0.6 \rightarrow \Delta pH \approx 2.5$) $\equiv$ 3.5 kcal/mole<br><br>$\eta_{a,p}$ = <u>efficiency of Apolar-Polar repulsion free energy mechanism</u> (as exemplified by $X^{20}$-poly[4(VPGVG),(VPGEG)])<br><br>w: $\Delta f \approx 1/2$; $f \approx 1000$ x dry weight<br>$\Delta n$: 4($COO^- \rightarrow COOH$) per 300 backbone atoms<br>$\Delta\mu$: ($\Delta\alpha$ of $\approx 0.8 \rightarrow \Delta pH \approx 0.5$) $\equiv$ 0.7 kcal/mole |
| **RESULT:**<br><br>$\eta_{a,p}/\eta_{c,c} > 10$ |

case, only 4 [COO⁻] moieties are converted to [COOH] per 300 backbone atoms and the ΔpH required to do so can be as small as 0.5[5] such that the chemical energy, $\Delta\mu\cdot\Delta n$, required is more than an order of magnitude smaller. The contractile bioelastic matrices can achieve ten times the amount of mechanical work for a given change in chemical concentration. These materials would therefore, be much more sensitive (10x) in responding to the altered chemical composition of a diseased state and in delivering a drug.

## BIOELASTIC MATERIALS AND THE CONCEPT OF THE DIAGNOSTIC-THERAPEUTIC PAIR

For some time there has been an objective of integrating into a single drug delivery device both the diagnostic and therapeutic steps. The device would be a bioresponsive drug delivery system which, for example in cases of Type I diabetes, would have the capability of measuring the blood glucose level and in response would provide release of the appropriate dose of insulin.[14]

As noted above, bioelastic materials can be designed to contract or to relax in response to a chemical signal and, thereby, could be utilized to achieve drug delivery in response to a changed chemical environment.[15,16] Chemically driven contraction and relaxation has been demonstrated by a change in pH. Theoretically, chemomechanical transduction could be achieved by any functional moiety attached to the contractile protein-based polymer that could occur in both a less polar (more hydrophobic) state and a more polar (less hydrophobic) state. This would include the imidazole moiety of the histidine residue, in which case the effect of a ΔpH would be just the reverse of the glutamic acid residue, that is , for the histidine residue, relaxation would occur on lowering the pH and contraction would occur on raising the pH to physiological levels and presumably the $pK_a$ can be shifted by shifting the hydrophobicity of the protein-based polymer. This is depicted in Figure 3. The possible functional moieties could also include chemical groupings that could be oxidized and reduced as in many of the prosthetic groups which occur in enzymes. In this regard, it has been proposed that the enzymes that achieve function on changing the redox state of prosthetic groups (e.g., of the hemes of cytochromes, of the nicotinamide adenine and flavine adenine dinucleotides, etc.) do so by the change in polarity of the prosthetic group causing a change in the folded state of the protein wherein changing the folded state achieves function. The change in redox state shifts the temperature of an inverse temperature transition to a temperature above or below physiological temperature to result in a differently folded state as demonstrated in Figure 1, using pH for $X^{20}$-poly-[4(VPGVG),(VPGEG)]. With the appropriate redox potential for a redox couple in the contractile matrix, the chemomechanical transduction could be coupled to the oxidative state of an enzyme that is sensitive to the chemical to be modulated. For example, for glucose, there is an oxygenase and a dehydrogenase that use the dinucleotide prosthetic groups noted above.

The potential of the bioelastic materials is that the elastomeric matrix itself both could be the sensor of the altered chemical situation of the diseased state and, as an integral part of the sensing process, could be deliverer of the therapeutic dose of the drug required to correct or improve the diseased condition. Furthermore, the properly designed bioelastic matrix has the potential of doing so with the sensitivity, specificity and cooperativity that one has come to appreciate of enzymes and other proteins. In addition the protein-based polymer as a drug delivery device would degrade, once completing its task, with release of

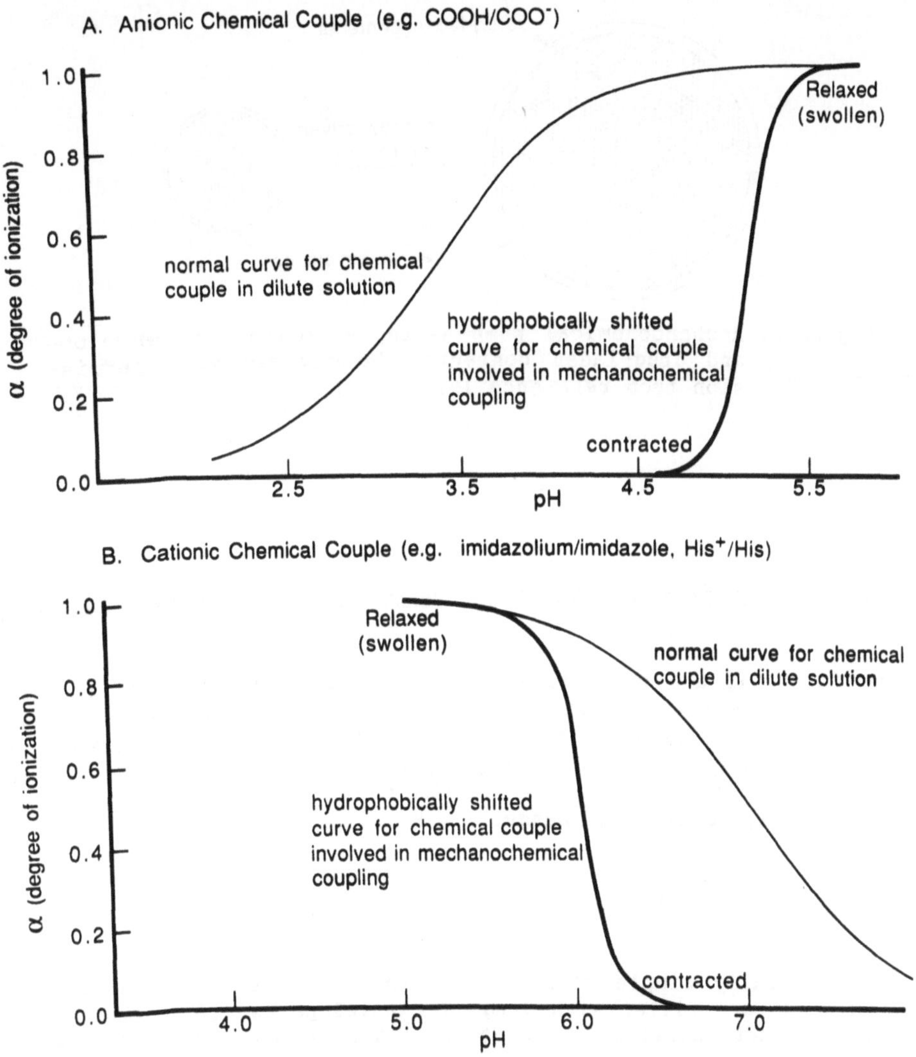


**Figure 3.** Schematic representations in $pK_a$ and increases in cooperativity on increasing hydrophobicity for [A], an anionic chemical couple and [B], a cationic chemical couple. (Reproduced with permission from reference 15).

amino acids and of what most commonly could be a physiological chemical sensing moiety when the sensing element is not an amino acid with a functional side chain. The downside of using protein-based polymers in this way is that they could elicit an antigenic or an immunogenic (foreign body) type of response. Such a possibility does not seem to have deterred the consideration of proteinuceous nanoparticles, i.e., gelatin, serum albumin, etc.,[17] which do not exhibit the transducing properties central to the bioelastic materials, and the primary bioelastic material, X[20]-poly(VPGVG), studied thus far appears to be quite innocuous and benign particularly in comparison to Dacron[R] and Dexon[R].[18]

In a series of ten tests required for devices and materials in contact with tissue, tissue fluid, and blood, the bioelastic material, poly-(VPGVG) and its $\gamma$-irradiation crosslinked matrix, has been found by an independent testing laboratory to be biocompatible.

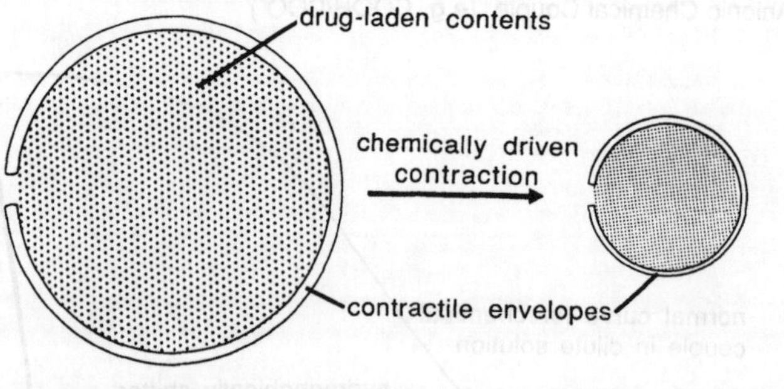


Figure 4. Mechanochemical pump having a contractile envelope and drug-laden contents. (Reproduced with permission from reference 15).

## EXAMPLES OF BIOELASTIC CONSTRUCTS AS "SMART" MATRICES FRO DRUG DELIVERY

The following are examples of three distinct bioelastic constructs which utilize the chemomechanical transducing (mechanochemical coupling) properties of the bioelastic materials. As they have been described previously,[15,16] they will only be briefly treated here. These may be compared to the innovative uses of polymeric materials as drug delivery systems recently reviewed by Langer.[19,20]

### 1. Mechanochemical Pumps

A conceptually simple construct is to have a contractile envelope surrounding a core filled with drug. The envelope can be chemically induced to contract expelling its contents. This is depicted in Figure 4.

### 2. Transducing Nanoparticles

By appropriate use of concentration, temperature and time, it is possible to form aggregates of the bioelastic materials with radii of controllable sizes ranging from 50 to 500 nm[21,22] (see Figure 5), which could be fixed by crosslinking at the desired size. This allows use of the differing sizes for the permeability barriers to particles of endothelium and basement membrane vasculature that occur normally and in diseased states. Thus a contracted nanoparticle could be sized to pass through the vasculature in order to approach the site of the diseased tissue better.[23] The contracted nanoparticle would be drug-laden and with the proper chemical sensing element could be induced to swell and release the drug in the chemical environment of the diseased state (see Figure 6).

### 3. Transducing Monoliths with Preprogrammed Rates

An interesting example of using chemomechanically transducing bio-

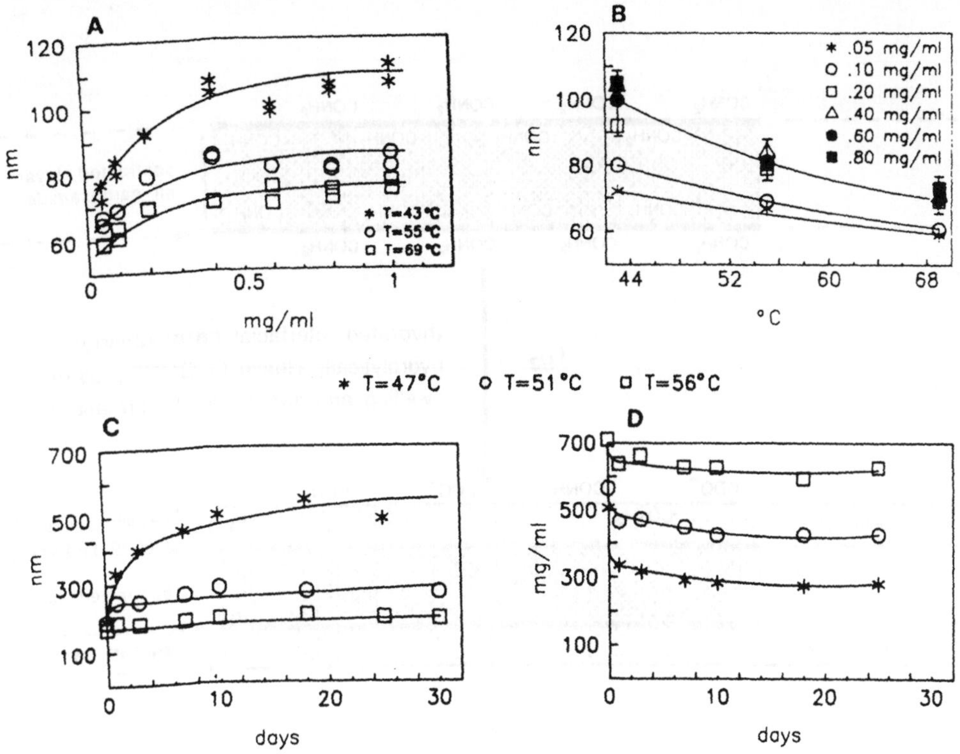


Figure 5. Concentration, temperature and time dependence for nanoparticles formed from poly(VPGVG). Recognizing that the ordinates are given as radii, particle diameters from 100 to 1000 nm are observed. (Reproduced from reference 21 with permission).

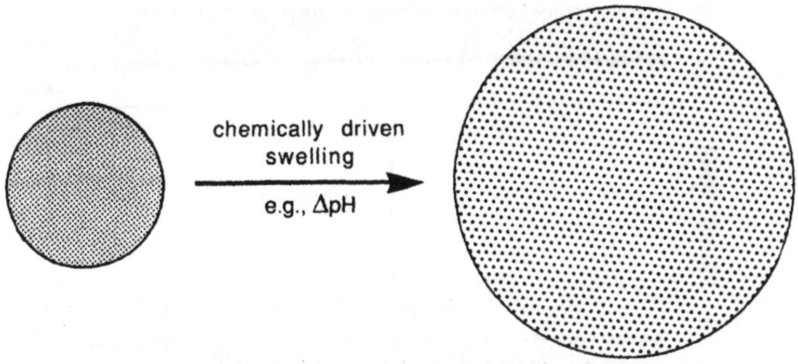


Diagnostic–Therapeutic Pair

1. Device uses altered (diseased) chemical environment to bring about drug release
2. Spent swollen device degrades releasing
   a. Peptides and amino acids – e.g., (Val, Pro, Gly)
   b. Sensor – e.g., (His, Glu, Prosthetic Group)

Figure 6. Protein-based nanospheres for drug delivery. The example is given wherein a contracted nanoparticle containing trapped drug is chemically induced to swell releasing its load of drug. (Adapted with permission from reference 15).

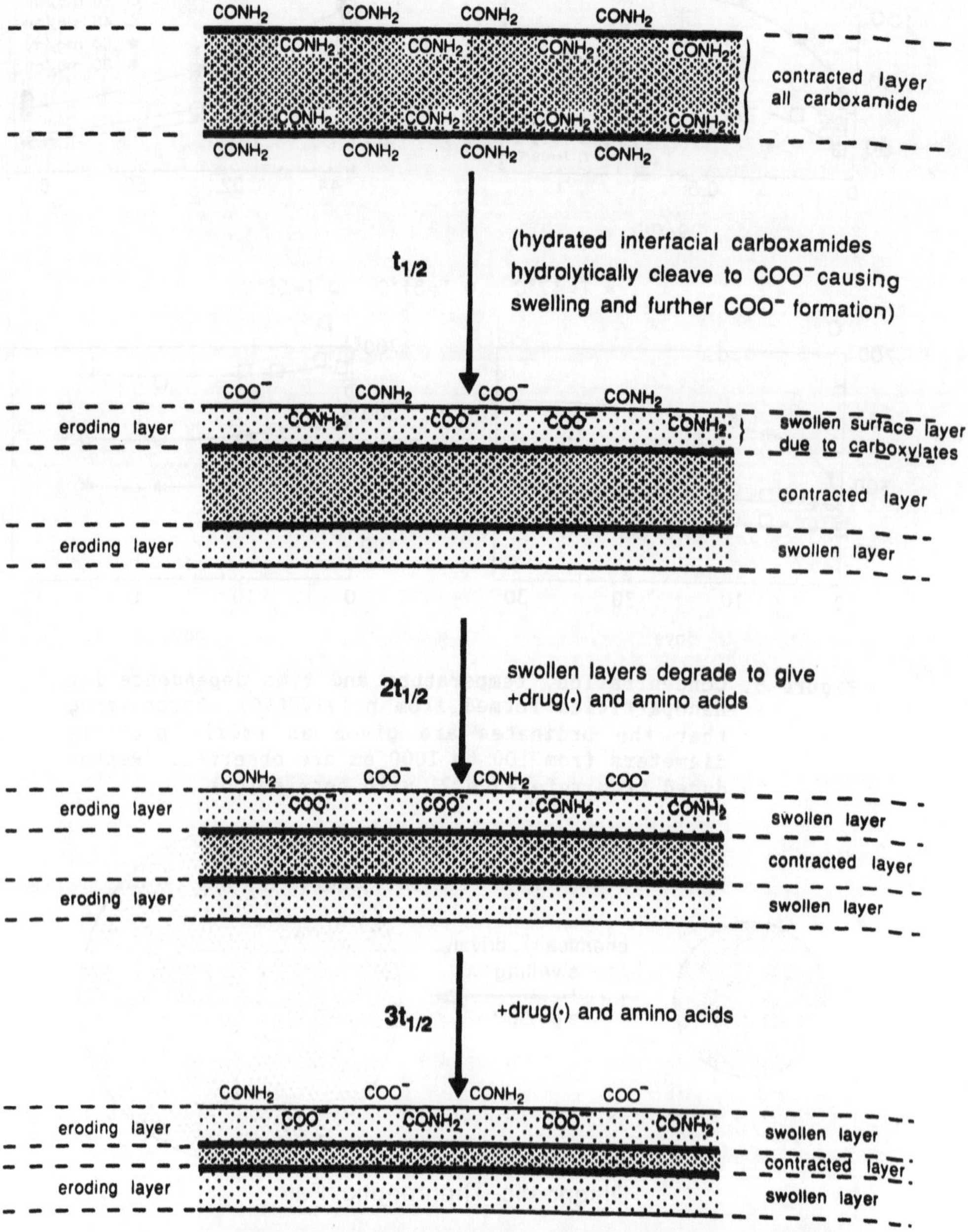


**Figure 7.** A monolith slab uniformly doped with trapped drug molecules, which contains as an example carboxamide [$CONH_2$] chemical clocks which, when at the surface, are spontaneously hydrolytically cleaved with predeterminable half-lifes, $t_{1/2}$, to give carboxylate anions [$COO^-$] which, in turn, cause the surface layer to swell releasing the trapped drug molecules. (Reproduced with permission from reference 15).

elastic matrices can be demonstrated by using a chemically labile moiety which, when it breaks down with a characteristic half-life, causes the matrix to swell and release its contents. For example, the amino acid residues, asparagine and glutamine, contain the carboxamide moiety [$CONH_2$] which spontaneously hydrolyzes in phosphate buffer at physiological pH to form the carboxylate anion [$COO^-$]. The formation of the carboxylate anion, as in the pH titration of Figure 2 to produce the [$COO^-$] moiety, raises the temperature of the inverse temperature transition to above physiological temperatures with the result of the local matrix-swelling and allowing release of drug. This is depicted in Figure 7. In addition, there could be any other chemically labile protective groups such as those commonly used in peptide chemistry. Also, of course, unstable linkages could be placed in the polypeptide backbone. An ester would be a structurally compatible unstable backbone linkage, but indeed, any chemically unstable spacer could be introduced into the chain of the protein-based polymer provided that it did not destroy the transducing properties and, of course, the drug could be covalently attached to the polymer chain by means of a linkage with the desired rate of cleavage.

## ACKNOWLEDGMENT

This work was supported in part by a Naval Research and Development Command ONR contract, #N00014-89-C-0282.

## REFERENCES

1. D. W. Urry, J. Protein Chem., 7, 1-34, 1988. "Entropic Elastic Processes in Protein Mechanisms. I. Elastic Structure Due to an Inverse Temperature Transition and Elasticity Due to Internal Chain Dynamics."
2. D. W. Urry, J. Protein Chem., 7, 81-114, 1988. "Entropic Processes in Protein Mechanism. II. Simple (Passive) and Coupled (Active) Development of Elastic Forces."
3. D. W. Urry, American Association for the Advancement of Science Publications, Protein Folding Symposium, 1989. "An Hydration Mediated Free Energy Driving Force for Protein Folding and Assembly."
4. D. W. Urry, B. Haynes, H. Zhang, R. D. Harris and K. U. Prasad, Proc. Natl. Acad. Sci. USA, 85, 3407-3411, 1988. "Mechanochemical Coupling in Syntheric Polypeptides by Modulation of an Inverse Temperature Transition."
5. D. W. Urry, Shao Qing Peng, L. Hayes, J. Jaggard and R. D. Harris, Biopolymers, 30, 215-218 (1990). "A New Mechanism of Mechanochemical Coupling: Stretch-Induced Increase in Carbolyl $pK_a$ as a Diagnostic."
6. D. W. Urry, D. K. Chang, H. Zhang and K. U. Prasad, Biochem. Biophys. Res. Commun., 153, 832-839 (1988). "pK Shift of Functional Group in Mechanochemical Coupling Due to Hydrophobic Effect."
7. J. Monod, "*Selected Papers in Molecular Biology by Jacques Monod*," York, 1978, pp. 593-623, 675-697.
8. J. Monod, J. Wyman, and J. P. Changeux, J. Mol. Biol., 12, 88-118, 1965.
9. "*McGraw-Hill Concise Encyclopedia of Science and Technology*," Second Edition, McGraw-Hill Publishing Co., NY, 1989, p. 711.
10. A. Katchalsky, S. Lifson, I. Michaeli, & M. Zwick, in: "*Size & Shape of Contractile Polymers: Conversion of Chemical & Mechanical Energy*," A. Wasserman, Ed., Pergamon Press, New York, 1960, p. 1-40.

11. W. Kuhn, B. Hargitay, A. Katchalsy, & H. Eisenberg, Nature, **165**, 514-516 (1950).
12. I. A. Steinberg, A. Oplatka, & A. Katchalsky, Nature, **210**, 568-571 (1966).
13. A Kachalsky, J. Polymer Science, 7, 393-412 (1951). "Solutions of Polyelectrolytes and Mechanochemical Systems."
14. R. E. Christoffersen, Federal Drug Administration/State of Maryland, "*International Biotechnology Conference*,", Bethesda, MD, March 27-28, 1990. "Biotechnology and Applications in Biomedicine: State of the Art and Future Developments."
15. D. W. Urry, Proc. Polymeric Mater. Sci. Eng., 63, 329-336, 1990. "Preprogrammed Drug Delivery Systems Using Chemical Triggers for Drug Release by Mechanochemical Coupling."
16. D. W. Urry, Federal Drug Administration/State of Maryland, International Biotechnology Conference, 1990 (in press). "Elastic-Protein-Based Materials as Biomedical Devices: Players in Biomaterials Revolution."
17. R. C. Oppenheim, in: "*Polymeric Nanoparticles and Microspheres*," P. Guiot and P. Couvreur, Eds., CRC Press, Boca Raton, Florida, Chapter 1, pp. 1-26, 1986. "Nanoparticle Drug Delivery Systems Based on Gelatin and Albumin."
18. J. E. Lemons, M. A. Igbal, and D. W. Urry, "Biocompatibility Comparisons of Dacron$^R$, Dexon$^R$, and Radiation crosslinked Poly(Val$^1$-Pro$^2$-Gly$^3$-Val$^4$-Gly$^5$)", Submitted for publication.
19. R. Langer, Pharmaceutical Technology, 13, 23-30, 1990. "Biomaterials in Controlled Drug Delivery: New Perspectives from Biotechological Advances."
20. R. Langer, Chemistry in Britain, **26**, 232-238, 1990. "Novel Drug Delivery Systems."
21. F. Sciortino, M. U. Palma, D. W. Urry and K. U. Prasad, Biochem. Biophys. Res. Commun., 157, 1061-1066, 1988. "Nucleation and Accretion of Bioelastomeric Fibers at Biological Temperatures and Low Concentrations."
22. F. Sciortino, D. W. Urry, M. U. Palma and K. U. Prasad, Biopolymers, 29, 1401-1407, 1990. "Self-Assembly of a Bioelastomeric Structure: Solution Dynamics and the Spinodal and Coacervation Lines."
23. S. S. Davis and L. Illum, in: "*Site Specific Drug Delivery*," E. Tomlinson and S. S. Davis, Eds., John Wiley & Sons. Ltd., 1986, Chapter 6, pp. 93-110. "Colloidal Delivery Systems-Opportunities and Challenges."

# NEW HYDROGELS FOR ORAL DRUG DELIVERY

You Han Bae and Sung Wan Kim

Department of Pharmaceuticals
and Center for Controlled Chemical Delivery
University of Utah
Salt Lake City, Utah 84112

One of the attractive applications of zero-order release devices could be using them as oral drug carriers. New hydrogels composed of two chemically independent networks [poly-(urethane) network and hydrophilic/hydrophobic balanced network], synthesized by an interpenetrating polymer networks (IPN) technique, showed versatile performance in release rate and drug loading with wide range of aqueous drug solubility. These results are obtained by simple modifications of polymer properties, such as crosslinking density of elastic network and hydrophilicity of the second network. The plausible explanation for the results obtained with highly soluble drugs is based on a resulted force balance between elastic contraction and osmotic swelling forces, while the hydrophilicity of the system directly effects the release rate of slightly soluble drugs. This new polymeric system could be a possible candidate for custom-tailored oral drug carrier.

## INTRODUCTION

Due to high patient compliance, oral administration continues to be the preferred route of pharmaceutical administration. Since numerous polymeric drug delivery devices have been developed, the application of this technology to oral drug delivery has become a subject of interest in pharmaceutical science.

Factors to be considered in the design of a polymeric oral drug carrier include aqueous drug solubility and stability, dose size, retention time, the drug's pharmacokinetics, adsorption site, safety, and cost.[1] Controlled release from simple polymeric monolithic devices would be beneficial for use in formulation of drugs for oral delivery. Time-independent (zero-order) release for a certain period would be preferred, but since zero-order release from a Fickian diffusion-controlled monolithic device would not be expected, several alternative approaches for zero-order kinetics have been attempted: introduction of diffusional barrier on the surface of monolithic devices,[2] relaxation-controlled release,[3-5] non-uniform drug distribution in initially dry and glassy matrices,[6] specific geometry,[7] and osmotically driven controlled delivery.[8-9] Most of these approaches, however, would be useful only in limited cases.

A heterogeneous interpenetrating polymer network (HIPN) system has been developed to provide versatility in drug loading and release performance. This system consists of two chemically independent crosslinked networks in which the proportions and properties of both networks can be varied independently. The first network consists of a hydrophobic poly-(urethane) network and the second network consists of a hydrophilic/hydrophobic balanced vinyl network.

The preliminary results of the release of incorporated drugs, the effect of changing matrix parameters, and the incorporation of drugs of varying solubility are reported in this paper. The results show that this polymeric formulation can be manipulated to provide a desired release rate for a given drug.

## EXPERIMENTAL

### 1 Polymer Synthesis

N,N-dimethylacrylamide (DMAAm) and styrene were distilled at 68°C/10 mmHg and at 86°C/35 mmHg under nitrogen environment. Ethylene glycol dimethacrylate (EGDMA) was used as received and 2,2'-azobisisobutyronitrile (AIBN) was recrystalized from methanol. Poly(tetramethylene ether glycol) (PTMEG, MW = 1020 and 3000) was dried at 80°C/1 mmHg for 2 hours before use. These chemicals were purchased from Polysciences, Inc. (Warrington, PA). A tri-NCO compound, 1,1,1-tris[N-(4'-methyl-3'-isocynatophenyl) cabamoyl-oxymethyl]-propane (Colonate L), was obtained in an ethyl acetate solution having 75 wt% solid and 13.2 wt% of NCO content from Japan Polyurethane Industry Ltd (Japan). Dibutyltin dilaurate (DBTL) from Aldrich Chemical Co. (St. Louis, MO) was used as a catalyst.

The HIPNs were synthesized by simultaneous crosslinking/polymerization in the composition shown in Table 1, where the sample code indicates the feed composition in weight percent and PTMEG molecular weight. The PTMEG was crosslinked using the trifunctional crosslinker, tri-NCO, with DBTL added as a catalyst. The vinyl phase was independently polymerized with AIBN as an initiator and EGDMA as a crosslinker.

The molten, dried PTMEG was allowed to cool in dry nitrogen atmosphere to 25-30°C at which point the premixed and degassed components of the vinyl polymer was added. Tri-NCO compound and DBTL were then incorporated. The reaction mixture was then degassed with an alternating vacuum/nitrogen purge procedure and filled into polyester lined glass molds with rubber gaskets. The polymers were cured at 70°C for 24 hours in a forced air draft oven. The polymerization and crosslinking reactions for both networks of the IPNs were conducted at 80°C for 24 hours. The membranes (0.62 - 0.72 mm thickness) were removed from the glass molds and punched into 10 mm diameter disks. The disks were Soxhlet extracted with ethanol for 48 hours. The ethanol swollen disks were deswollen in water, air dried for two hours, and finally vacuum oven dried overnight.

| Table 1. Feed composition for heterogeneous interpenetrating polymer networks | | | | | |
|---|---|---|---|---|---|
| | Sample | | | | |
| COMPOSITION | 100(1K)/ 0-0 | 50(1K)/ 40-10 | 50(1K)/ 30-20 | 60(1K)/ 40-0 | 60(3K)/ 40-0 |
| PTMEG (MW = 1020), g | 3.3 | 3.3 | 3.3 | 3.3 | |
| PTMEG (MW = 3000), g | | | | | 3.3 |
| Tri-NCO solution, g | 1.68 | 1.68 | 1.68 | 1.68 | 0.57 |
| DBTL, mg | 2.85 | 2.85 | 2.85 | 2.85 | 1.0 |
| DMAAm, g | | 3.6 | 2.7 | 4 | 4 |
| St, g | | 0.9 | 1.8 | | |
| EGDMA, mg | | 49 | 49 | 39 | 39 |
| AIBN, mg | | 12.9 | 12.9 | 10.3 | 10.3 |

## 2. Drug Loading and Release Experiments

The model drugs, pseudoephedrine HCl, sodium salicylate, and theophylline were loaded into HIPN disks by solvent sorption. Dry HIPN disks were loaded by swelling in loading solution of 20 w/v% drug in ethanol/water (80/20 v/v) mixture at room temperature for pseudoephedrine·HCl and sodium salicylate, and in an approximately 10 w/v% solution in ethanol/water (80/20 v/v) mixture at 65°C for theophylline. The high temperature for theophylline was used to increase its solubility in the loading solution. The polymer disks were allowed to swell for 2 to 4 days in the loading solution, blotted to remove excess surface associated solution, briefly rinsed in the solvent mixtures (the same composition for loading but without the drug), and then air dried for 2 days. The air dried disks were further dried in a draft oven at about 40°C until no further weight loss was observed.

Drug release into phosphate buffer saline (PBS, pH 7.4) was determined as a function of time at 37°C. Drug loaded polymer disks were placed in 10 mL of PBS, pre-equilibrated at 37°C, in 20 mL glass scintillation vials and agitated in a shaker bath at 80 strokes/min. The HIPN disks were removed from the vials and immersed in fresh, pre-equilibrated PBS for continued release. The drug concentration in the release media were quantitated by UV spectroscopy (Perkin Elmer Lambda, 4B UV spectrophotometer). At the end of the release experiment, the HIPN disks were removed from the media and sonicated in 10mL of ethanol. The ethanol extraction solution was replaced with a known volume of ethanol and repeated until no drug could be detected in the extract. The total drug in the extracts were quantitated by UV spectroscopy.

## RESULTS AND DISCUSSION

Many different types of hydrophilic/hydrophobic random copolymers were utilized in swelling-controlled release studies, resulting in n = 0.5 to n = 0.9 (where the n value is calculated from a curve fit to the empirical equation of $M_t/M_\infty = kt^n$).[4] However, n values near 1 were infrequently obtained and, in most cases, n was near 0.5. An advantage of using this kind of copolymer might be that these polymers have high swelling (sometimes with maximum swelling) in water/ethanol mixtures, useful for controlled drug loading with a non-toxic solvent.

Sodium salicylate loaded (approximately 16 wt% loading) poly(urethane) network (100(1K)/0-0) demonstrated an abnormal bimodal release pattern where an accelerated release rate was observed (Figure 1). By visual observation, the start of the accelerated release was close to the time that all the dispersed drug would be dissolved inside the matrix, by that time only 18% of the drug had been released. Maximum release rate was closely related to the time at which maximum water content in the device was observed. The equilibrium water content of the poly(urethane) without drug was less than 2%. This observation indicates that overswelling during drug release may be caused by dissolved drug, but build-up osmotic pressure inside the matrix is restricted by the elastic contraction force exerted by the poly(urethane) network.

In an attempt to utilize effects of the above two phenomena (first-order and bimodal release), interpenetrating polymer networks composed of a poly(urethane) (PU) network and a hydrophilic/hydrophobic (DMAAm/St) balanced vinyl copolymer network, were synthesized by varying the two network's weight ratios, the vinyl phase composition, and crosslink density of the PU phase.

When pseudoephedrine HCl was used as a model solute for highly water soluble drugs, drug loading under the same loading conditions varied considerably with changes in vinyl phase composition. This apparently results from different levels of swelling in the drug solution. Pseudoephedrine HCl release from IPN matrices with varying vinyl phase composition is shown in Figure 2. With 40% DMAAm in the feed composition,

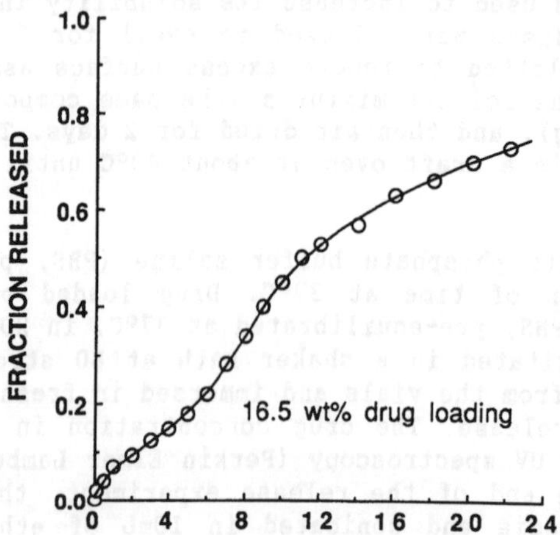


Figure 1. Fractional release of sodium salicylate from a 100(1K)/0-0 matrix.

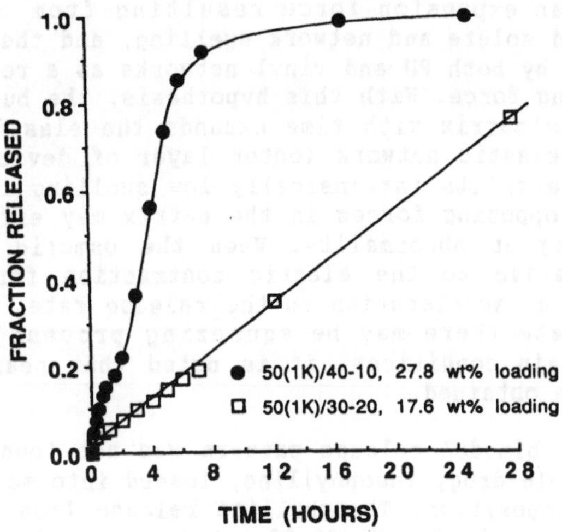


Figure 2. Effect of vinyl phase feed composition on drug loading and fractional release of pseudoephedrine HCl from HIPN matrices.

pseudoephedrine HCl loading was approximately 28 wt%, and the release pattern was bimodal. With 30% DMAAm, however, the drug release came closer to zero-order release kinetic. Another important observation is that changing the crosslink density of the PU network, by using different molecular weight PTMEG, while keeping the composition of the vinyl phase constant (DMAAm 40%), results in different release kinetics, as shown in Figure 3. A lower crosslink density showed first-order release kinetics, while a higher crosslink density caused a dramatic change in release kinetics, from first-order to a bimodal release pattern.

A hypothetical explanation for the observed results is based on the force balance inside the matrix during drug release. One can consider two

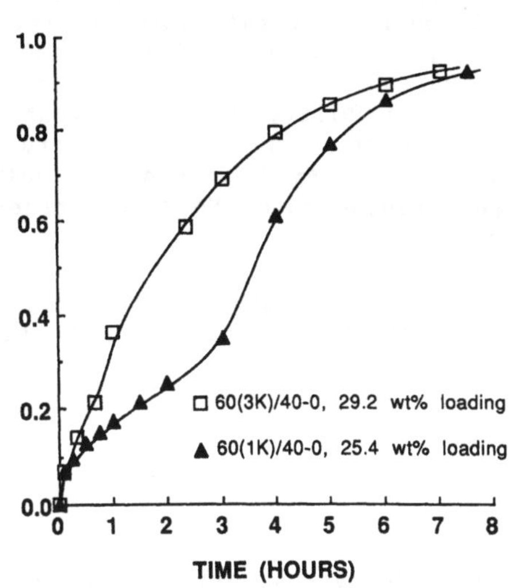


Figure 3. Crosslinking density effects on the release of pseudoephedrine HCl from HIPN matrices.

opposing forces: an expansion force resulting from osmotic pressure caused by dissolved solute and network swelling, and the elastic contraction force exerted by both PU and vinyl networks as a result of the overall osmotic swelling force. With this hypothesis, the build-up of osmotic pressure inside the matrix with time expands the elastic network, while the drug depleted elastic network (outer layer of device) may have tendency to shrink due to its intrinsically low swelling in water. The net result of the two opposing forces in the matrix may effect the drug release rate, causing an abnormality. When the osmotic force exceeds a certain level relative to the elastic contraction force, the network expands, resulting an acceleration in the release rate. After reaching a maximum release rate there may be squeezing process by contraction forces. Under certain conditions, it is noted that near zero-order release kinetics were obtained.

In Figure 4, a bimodal release pattern was not found in the release of a slightly soluble drug, theophylline, loaded into matrices of differing vinyl phase composition. Theophylline release from 50 (1K)/40-10 is much faster than from the 50(1K)/30-20 matrix, even though the amount of drug loading was similar. Here, the hydrophilicity of vinyl phase directly affected the release rate of theophylline, indicating that if there is not considerable osmotic pressure due to the loaded drug, the drug release is more closely associated with the HIPN properties. Near zero-order release was also obtained for theophylline.

A detailed investigation of the mechanism for the release of drug is in progress.

## CONCLUSION

Interpenetrating polymer networks composed of poly(urethane) network and HPL/HPB balanced vinyl network demonstrate a versatility in release of drugs with various aqueous solubilities and loadings.

The apparent release mechanism of highly water soluble drug is due to the overall balance in the monolithic device of osmotic expansion force caused by the dissolved solutes and the hydrophilicity of the vinyl network and the elastic contraction force generated by both networks. The major factors controlling release rate were, therefore, considered to be the crosslinking density of PTMEG network resulting in different contraction forces and the hydrophilic/hydrophobic monomer ratio of vinyl network adjusted for drug solubility and the amount loaded, resulting in balanced osmotic and contractive force. A more detailed mechanistic study is under investigation. With less soluble drugs, polymer properties themselves may affect the release rate. Near zero-order release can be obtained in both cases.

This system should be a candidate as a polymeric carrier suitable for most of oral drugs. The HIPN's properties can be tailored for each drug to achieve desirable release kinetics.

## ACKNOWLEDGMENTS

This work was supported by and collaborated with TherTech Inc. (Salt Lake City, Utah).

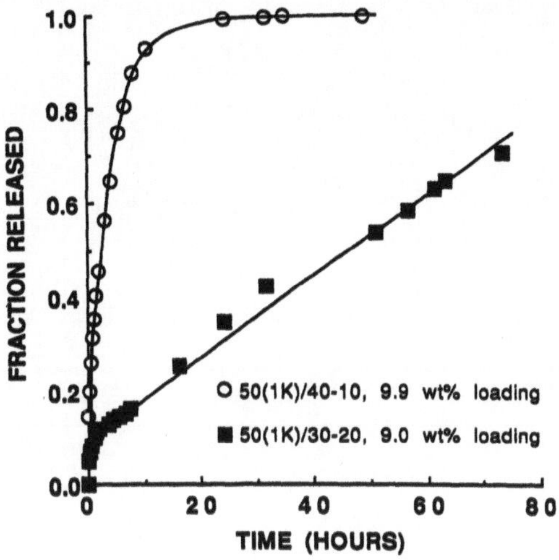


Figure 4. Theophylline release from HIPN matrices varying vinyl phase feed composition.

## REFERENCES


1. K. Park, R. W. Wood & J. R. Robinson, in: "*Medical Applications of Controlled Release,*" Vol. 1, R. S. Langer & D. L. Wise, Eds., CRC Press, Boca Raton, FL, 1984, p. 159.
2. E. S. Lee, S. W. Kim, S. H. Kim, J. R. Cardinal & H. A. Jacobs, J. Memb. Sci., 7, 293 (1980).
3. H. B. Hopfenberg, AIChE Symp. Ser., 77, 37 (1981).
4. N. A. Peppas & R. W. Korsmeyer, in: "*Hydrogels in Medicine and Pharmacy, Vol. III. Properties and Applications,*" N. A. Peppas, Ed., CRC Press, Boca Raton, FL, 1987, p. 109.
5. N. B. Graham & M. E. McNeil, Biomaterials, 5, 27 (1984).
6. P. I. Lee, Polymer, 25, 973 (1984).
7. D. S. T. Hsieh, W. D. Rhine & R. Langer, J. Pharm. Sci., 72, 17 (1983).
8. R. Gale, S. K. Chandrasekaran, D. Swanson & J. Wright, J. Memb. Sci., 7, 310 (1980).
9. G. M. Zentner, G. S. Rork & K. J. Himmelstein, J. Control. Rel., 1, 269 (1985).

# RELEASE OF PHENYLPROPANOLAMINE FROM DYNAMICALLY SWELLING POLY-(HYDROXYETHYL METHACRYLATE-CO-METHACRYLIC ACID) HYDROGELS

Jim H. Kou,[1] David Fleisher [2] and Gordon L. Amidon[2]

1. Syntex Research, 3401 Hillview Ave.
   Palo Alto, CA 94304
2. College of Pharmacy
   University of Michigan
   Ann Arbor, MI 48109

Poly(hydroxyethyl methacrylate-co-methacrylic acid) hydrogels swell extensively at high pHs due to the ionization of the carboxyl groups. The swelling leads to an increase of the free volume of the system and, thus, the release rate of the entrapped drug. Potential applications of the pH sensitive systems include delayed and controlled oral delivery, altered gastrointestinal transit following gastric emptying and targeted delivery to the small intestine. Phenylpropanolamine was used as a model drug to study the diffusional release characteristics of the swelling polymeric network. It was found that the release was a strong function of swelling and the apparent diffusivity of phenylpropanolamine followed an exponential relationship with the fractional water content of the matrix. These findings led to the development of a mathematical model which, based on experimentally measured swelling kinetics and dimensional changes, was capable of predicting release from a matrix undergoing pH induced swelling.

## INTRODUCTION

Hydrogels are hydrophilic polymers which can swell, but not dissolve, in an aqueous environment. If ionizable monomers are incorporated into the gel structure, the resulting polymer will have a pH dependent swelling property and the final equilibrium swollen structure will depend on the environmental pH. Potential applications of these pH dependent swelling system includes size discriminating solvent removal, usage as a chemicomechanical transducer, and controlled release of pharmaceuticals.[1-3]

Crosslinked poly(hydroxyethyl methacrylate) hydrogels were extensively studied because of their superior properties for use as biopolymers. In this work, we copolymerized hydroxyethyl methacrylate (HEMA) and methacrylic acid (MAA) to introduce the pH sensitivity. Tetraethylene glycol dimethacrylate (TEGDMA) was also incorporated to give the resulting polymers a three dimensional structure and mechanical strength. These weak acid polyelectrolyte hydrogels are potentially useful in oral delivery of drugs that are acid labile or have severe gastrointestinal side effects. In order to assess their controlled release potential, the matrix swelling and its effect on the diffusional drug release were examined. Phenylpropanolamine (PPA) was used as a model drug for the diffusion study because of its stability in the pH range of 1-7. In addition, the interaction between PPA molecules and anionic gel structure may serve as an additional mechanism for a prolonged release.

## EXPERIMENTAL

### 1. Hydrogel Synthesis

The hydrogels were prepared from monomeric materials via a radical mechanism. Hydroxyethyl methacrylate and methacrylic acid (Polysciences, Inc., Warrington, PA) were vacuum distilled before use while the crosslinker tetraethyleneglycol dimethacrylate (Polysciences, Inc., Warrington, PA) was used as received. The HEMA monomer was extracted with hexane and followed by vacuum distillation (250 μHg; 55°C). Cuprous chloride was used as the polymerization inhibitor in the distilling flask. The methacrylic acid was relatively pure and was only distilled under vacuum (1.5 mmHg; 45°C). The hydrogel slabs were synthesized via solution polymerization. The monomer solution contained 40 w/w% water initially. Various monomer and crosslinker concentrations were used to prepare gels of different compositions. The methacrylic acid concentration ranges from 1.50 w/w% and TEGDMA from 0.5 to 3 w/w% in the monomer mixtures. The monomer solution was initiated with a redox couple of ammonium persulfate and sodium metabisulfite. The reacting mixture was injected into a mold made of two glass plates spaced 1.8 mm apart by a silicone rubber gasket. The reaction was allowed to run for two hours at 60°C. The hydrogel cylinders were fabricated in the polyethylene tubing (3 mm I. D.) via bulk polymerization. The methacrylic acid compositions used in the monomeric mixtures was 20 w/w%, and the TEGDMA concentration was 0.2 w/w%. The monomeric solution was initiated by 1% 2,2'-azobisisobutyronitrile (AIBN) and then injected into a straight piece of polyethylene tubing. The two ends were sealed and the entire enclosure was maintained in a 60°C water bath for 12 hours. The resulting hydrogel slabs or cylinders were removed and washed in portions of fresh distilled water for at least one week before use to purge the residual reactants. Structural identification was not performed with these gels since the polymerization procedure for the methacrylate monomers was well established and documented.[4-6] For the rest of the discussion, whenever methacrylic acid composition is described, it is intended to mean the composition of the monomeric solution from which the polymer is made.

### 2. Equilibrium Swelling and Drug Release

Slab gels of different compositions were cut into squares of 1.8 x 1.8 $cm^2$ and equilibrated in 0.1 M HCl solutions until the weight was stabilized. These gels were then transferred to a pH 7 phosphate buffer. Sample weight was followed until a swelling equilibrium was reached. The ionic strength for all buffers used was maintained at 0.1 M.

In the release experiments, phenylpropanolamine was used as obtained without further purification. Drug loading was carried out by equilibrating the gel sample in 2% phenylpropanolamine solution buffered at a specific pH. The ionic strength for both loading and extracting media was kept at 0.2 M. In the diffusion coefficient measurement, gel discs of 3 cm diameter were used. The loaded hydrogel sample was clamped between a circular Plexiglas ring and base, permitting release over one surface. The entire composite was then anchored in the extracting medium by a Plexiglas rod. The PPA concentration in the bulk fluid was sampled and assayed until the completion of the release. The diffusion coefficient was then calculated from the experiments data according to a mathematical analysis by Lee.[7] Gel cylinders were used in the release experiments for model testing where the polymer samples were loaded in a 5% PPA buffered solution. The loaded sample was then held in a stainless steel wire basket submerged in a well stirred container. The gel dimensions, weight and PPA concentration in the bulk were followed as a function of time. PPA samples were assayed by an ion-pairing high performance liquid chromatographic method; the gel dimensions were monitored by a photographic technique described elsewhere.[11]

## RESULTS AND DISCUSSION

### 1. Swelling Equilibrium

Results of equilibrium swelling study are summarized in Figure 1. Here matrix hydration, H, is defined as the fractional water content, as shown in the equation below.

$$H = \frac{\text{gel swollen - gel dry weight}}{\text{gel swollen weight}}$$

In the unionized gel with the lowest methacrylic acid content, the matrix hydration is 0.38, which is in good agreement with the literature value for homogeneous poly(HEMA) gels.[4] A typical poly(HEMA-co-MAA) gel with 50% MAA and 0.5% TEGDMA can swell from 28% water content at pH 1 to 88% at pH 7. It is of interest to note that in Figure 1 the equilibrium state of gels of different methacrylic acid content is independent of the crosslinker concentration at pH 1. However, the crosslinking effect is readily revealed in the ionized gels at pH 7. This finding apparently suggests that the covalent crosslinkages are not the determining factor for the unionized swollen structure. Secondary structures were proposed to account for the equilibrium swollen structure in poly(HEMA) gels. Refojo attributed these structures to be the hydrophobic interaction between the α-methyl groups on the polymer backbone.[5] Ratner and Miller proposed an alternative explanation with the hydrogen bonding being the responsible interaction.[6] These interactions form physical crosslinkages between the polymer chains and shield the effect of the covalent crosslinks. Either of these interactions is likely to exist in the unionized poly(HEMA-co-MAA) gel matrices since both HEMA and MAA contain the necessary functionalities for these interactions to occur. When the secondary structures are disrupted upon the ionization of carboxyl groups at pH 7, the effect of the covalent crosslinks is readily revealed, as evidenced by the separation of the top two curves in Figure 1.

It is conceivable that the water content in the swollen polymeric matrix will have a significant effect on the solute diffusion. To quantitate the effect of water content on the solute diffusion rate in the matrix, the diffusion coefficient of PPA was measured at various levels of matrix hydration. The various hydration levels in these gels were effected by adjusting the environmental pH. A desorption method based on

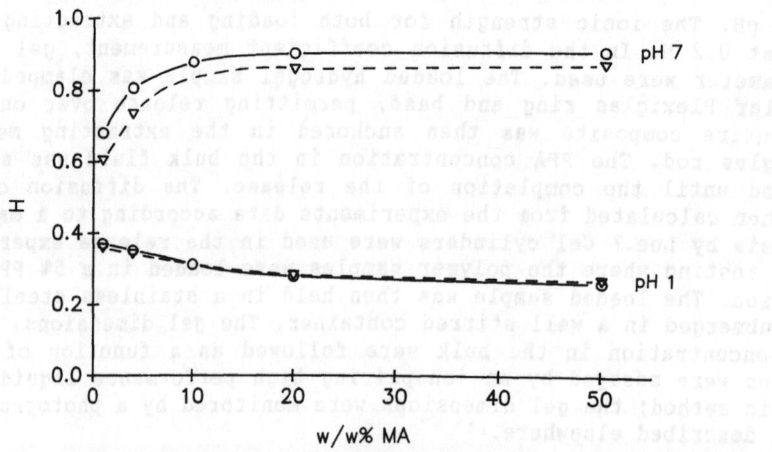


Figure 1. Effect of polymer composition on matrix hydration of poly(HEMA-co-MAA) gels at different pHs (o and x represent 0.5 and 3 w/w% TEGDMA respectively).

the analysis of Lee was used for the diffusion coefficient determination.[7] This method also renders the partition coefficient of PPA between the polymer gel and extracting fluid to be determined simultaneously. Table 1 summarizes the results obtained. The PPA diffusion coefficient has a very marked dependence on the matrix hydration. A 52-fold increase in diffusivity is noted when H increases from 0.35 at pH 1 to 0.88 at pH 7. Yasuda proposed a model based on free volume theory to describe the relationship between solute diffusivity in gel matrix and the matrix hydration.[8] It takes the form of Equation 1, where $D_{2,o}$ is the diffusion coefficient of the solute in pure solvent medium, and $k_f$ is a constant characteristic of the solute and the solvent molecules.

$$\ln D_2 = \ln D_{2,o} - k_f\left(\frac{1}{H} - 1\right) \qquad \text{(Equation 1)}$$

This model indicates that the plot of $\ln D_2$ vs $(1/H - 1)$ will be linear with the slope $k_f$ and intercept $\ln D_{2,o}$. Figure 2 is a semi-log plot showing the excellent agreement with the model. Over the range of hydra-

Table 1. Diffusion and partition coefficients of phenylpropanolamine in swollen poly(HEMA-co-MAA) gels[a] at various pH/hydration

| pH | H[b] | Dx10[8] (cm[2]/s)[b] | K[bc] | n[d] |
|---|---|---|---|---|
| 1 | 0.352 (0.0014) | 2.50 (0.453) | 1.41 (0.365) | 3 |
| 3 | 0.377 (0.0002) | 3.58 (0.351) | 1.59 (0.0403) | 4 |
| 5 | 0.639 (0.021) | 44.6 (3.83) | 4.42 (0.670) | 6 |
| 7 | 0.880 (0.004) | 139.0 (25.6) | 4.64 (0.402) | 5 |

(a) Polymer composition: 50 w/w% MAA and 0.2 w/w% TEGDMA.
(b) Numbers in () are standard deviations.
(c) Partition coefficients.
(d) Number of experiments.

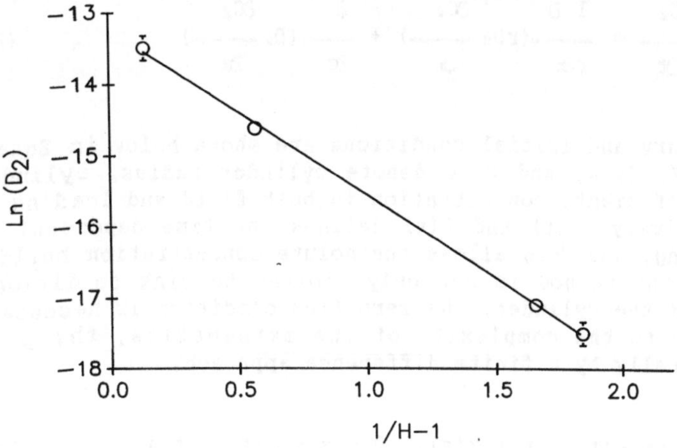


Figure 2. Dependence of apparent PPA diffusion coefficients in poly(HEMA-co-MAA) gels on matrix hydration: A semi-logarithmic plot showing the agreement with Yasuda's theory.

tion studied, the apparent linearity is excellent. The regressed parameter values are: $k_f$ = 1.98 and $D_{2,0}$ = 1.82 x $10^{-6}$cm²/s. For comparison, one can calculate the aqueous diffusion coefficient of PPA by the Hayduk and Laudie method.[12] The molal volume necessary for the calculation was the arithmetic mean of the Schroeder and Le Bas estimates. It was also assumed that the cationic PPA molecule was hydrated by six water molecules. The result gives a PPA aqueous diffusion coefficient of 5.51 x $10^{-6}$cm²/s Therefore, the experimental value is approximately three-fold smaller than the Hayduk and Laudie estimate, and this difference suggests that there exists an interaction between the PPA molecules and the poly(HEMA-co-MAA) network. It is also noted in Table 1 that the PPA partition coefficient increases with the matrix hydration. This is in support of the existence of specific interactions between cationic PPA molecules and the anionic hydrogel matrices. Such being the case, the diffusion coefficients measured are apparent in nature, since they include the interaction component. The presence of this interaction, however, does not perturb the apparent linearity of Yasuda's free volume relationship.

## 2. Modeling of PPA Release from the Swelling Hydrogel Matrix

The previous results show clearly that the solute diffusion in a swollen gel network depends on the available free volume in the solvent phase. One can further affirm this hypothesis by examining the solute diffusion in a dynamically swelling gel system. According to the Yasuda's theory, by knowing the time profile of the matrix hydration of a swelling gel, one should be able to predict the diffusional release profile of the loaded drug. Therefore, the following modeling and experimental work were designed to verify the applicability of the Yasuda's theory. A cylindrical geometry was selected for the study because it was relatively easy to fabricate the polymer cylinders, as compared to the suspension method for making spheres. The mathematical problem of drug release from a cylindrical matrix can be stated as Equation 2, where $C_2$ is the drug concentration in gel and $D_2$ is the penetrant concentration dependent PPA diffusion coefficient as expressed in Equation 1.[10]

$$\frac{\partial C_2}{\partial t} = \frac{1}{r}\frac{\partial}{\partial r}\left(rD_2\frac{\partial C_2}{\partial r}\right) + \frac{\partial}{\partial z}\left(D_2\frac{\partial C_2}{\partial z}\right) \qquad \text{(Equation 2)}$$

The boundary and initial conditions are shown below in Equations 3-7, where R, Z, K, $C_{2,b}$, and $C_{2,0}$ denote cylinder radius, cylinder length, partition coefficient, concentration in bulk fluid and loading concentration, respectively. R(t) and Z(t) defines the time dependent boundaries during swelling, and $C_{2,b}$ allows the solute concentration build-up in the bulk so that the method is not only limited to sink conditions. At the center axis of the cylinder, the zero flux condition is necessary for the symmetry. Due to the complexity of the mathematics, the problem was solved numerically by a finite difference approach.

$$C = KC_{2,b}(t), \quad r = R(t), \quad 0 \le z \le Z(t), \quad t > 0 \qquad \text{(Equation 3)}$$

$$\frac{\partial C_2}{\partial r} = 0, \quad r = 0, \quad 0 \le z \le Z(t), \ t > 0 \qquad \text{(Equation 4)}$$

$$C = KC_{2,b}(t), \quad z = Z(t), \quad 0 \le r \le R(t), \quad t > 0 \qquad \text{(Equation 5)}$$

$$C = KC_{2,b}(t), \quad z = 0, \quad 0 \le r \le R(t), \quad t > 0 \qquad \text{(Equation 6)}$$

$$C = C_{2,0}, \quad 0 \le r \le R(t), \quad 0 \le z \le Z(t) \quad t = 0 \qquad \text{(Equation 7)}$$

From the concentration profile obtained, the fraction released, $M_2(\tau)/M_{2,oo}$, can be calculated by Equation 8, where $\tau$ is the dimensionless time, defined by $\tau = D_{2,0}t/R(0)^2$.

$$\frac{M_2(\tau)}{M_{2,\infty}} = 1 - 2\left[\frac{R(\tau)}{R(0)}\right]^2 \frac{Z(\tau)}{Z(0)} \int_0^1 \int_0^1 \frac{C_2}{C_{2,0}} \frac{r}{R(\tau)} \frac{dr}{R(\tau)} \frac{dz}{Z(\tau)} \quad (8)$$

(Equation 8)

The model predicts the drug release rate is based on the independently determined parameters and swelling data, including quantities such as $k_f$, $D_{2,0}$, K, $V_g$, $V_b$, $R(\tau)/R(0)$, $Z(\tau)/Z(0)$ and $H(\tau)$ where $V_g$ and $V_b$ are the gel and bulk fluid volumes, respectively. The $k_f$, $D_{2,0}$ and K values were reported in the previous section. $V_g$ and $V_b$ values are 0.21 $cm^3$ and 150 $cm^3$ respectively. The time profiles of $R(\tau)$, $Z(\tau)$ and $H(\tau)$ were determined experimentally. $R(\tau)$ and $Z(\tau)$ were dedimensionalized, by their respective initial values, to obtain the $R^*(\tau)$ and $Z^*(\tau)$ functions.

Figure 3 illustrates the experimental data and predicted release curve for a typical poly(HEMA-co-MAA) gel consisting of 20% methacrylic acid and 0.2% crosslinker, TEGDMA. Also plotted in the graph are the experimental swelling data, i. e., H, $R^*$ and $Z^*$. The result shows that the model makes an excellent prediction of PPA release based on the swelling data. It is to be noted that the release curve is predicted strictly by swelling data and not by curve fitting. Figure 4 compares the release from the swelling and non-swelling matrices. The top and bottom curves are the non-swelling release at pH 7 and 1, respectively, and the middle is the same release curve as presented in Figure 3. It is apparent that these model-predicted curves do describe the release adequately in both swelling and non-swelling cases. The success of this model indicates

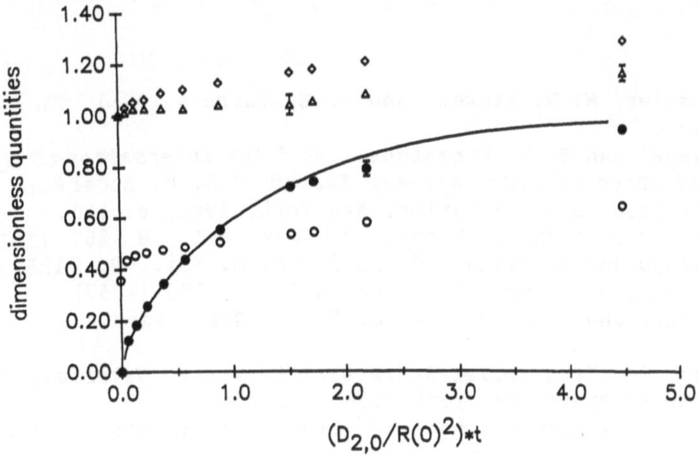


Figure 3. Dimensionless time profiles of experimental $M/M_\infty$ (●), H (o), R* (△) and Z* (◇) of PPA release from swelling poly(HEMA-co-MAA) gels and the model predicted release curve. Ppolymer composition is 20 w/w% MAA and 0.2 w/w% TEGDMA; PPA was loaded at pH 1 and released at pH 7.

that the diffusional release is indeed controlled by the available free volume in the swollen gel system.

**ACKNOWLEDGMENT**

The authors acknowledge the financial support from Smith Kline & Beckman Corporation.

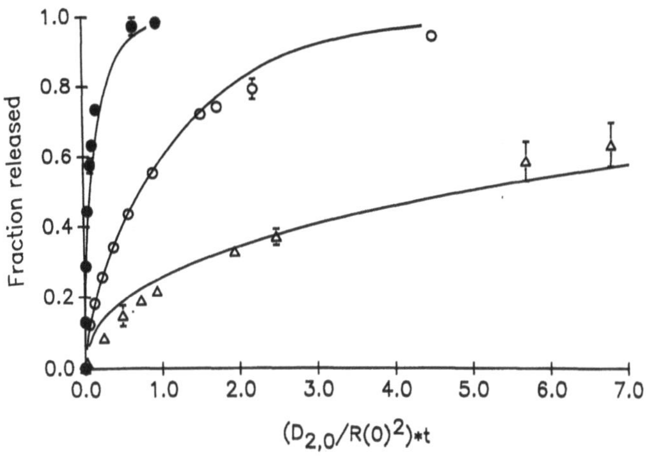


Figure 4. Fractional release of PPA from: non-swelling matrix at pH 1 (△; loaded at pH 1), swelling matrix at pH 7 (o; loaded at pH 1), and non-swelling matrix at pH 7 (●; loaded at pH 7); the curves are predicted from the model.

## REFERENCES


1. E. L. Cussler, M. R. Stokar, and J. E. Varberg, A. I. Ch. E. J., **30**, 578 (1984).
2. R. A. Siegel and B. A. Firestone, in: "*4th International Symposium on Recent Advances in Drug Delivery Systems*," J. M. Anderson, S. W. Kim and K. Knutson, Eds., Elsevier, New York, 199o, p. 181.
3. L. Brannon-Peppas and N. Peppas, J. Cont. Rel., **8**, 267 (1989).
4. M. F. Refojo and H. Yasuda, J. Appl. Polym. Sci., **9**, 2425 (1965).
5. M. F. Refojo, J. Polym. Sci.: Part A-1, **5**, 3103 (1967).
6. B. D. Ratner and I. F. Miller, J. Polym. Sci.: Part A-1, **10**, 2425 (1972).
7. P. I. Lee, in: "*Controlled Release of Bioactive Materials*," R. Baker, Ed., Academic Press, New York, 1980, p. 135.
8. H. Yasuda, C. E. Lamaze and L. D. Ikenberry, Makromol. Chemie, **118**, 18 (1968).
9. J. H. Kou, G. L. Amidon and P. I. Lee, Pharm. Res., **5**, 539 (1988).
10. J. H. Kou, D. Fleisher and G. L. Amidon, J. Cont. Rel., **12**, 241 (1990).
11. J. H. Kou, Ph.D. Thesis, University of Michigan, Ann Arbor, MI, (1987).
12. W. Hayduk and H. Laudie, A. I. Ch. E. J., **20**, 611 (1974).

# POLY(VINYL ALCOHOL) BEADS WITH CORE-SHELL STRUCTURE FOR DRUG DELIVERY

Cherng-Ju Kim[1] and Ping I. Lee[1,2]

(1) Faculty of Pharmacy
(2) Dept. of Chemical Engineering & Applied Chemistry
University of Toronto
Toronto, Ontario, Canada M5S 2S2

A new method of preparing large spherical poly(vinyl alcohol) (PVA) beads with a double-layer structure has been developed. The process involves a stepwise saponification of suspension polymerized PVAc beads and subsequent stepwise crosslinking of the PVA core and shell with glutaraldehyde. The resulting PVA beads have a thin, highly crosslinked outer shell and a lightly crosslinked inner core of different degrees of crosslinking. The kinetics of swelling and drug release from these beads in water were studied at 37°C using acetaminophen and proxyphylline as model drugs. The results show that the outer shell functions as a rate-controlling membrane upon increasing its crosslinking ratio, X, to about 0.5. This is reflected from the observed diffusional time-lags and the constant-rate regions during swelling and drug release. At lower shell crosslinking ratios ($X < 0.35$), the diffusional time-lags appear to be absent and the diffusional profiles are apparently first-order (Fickian) in nature, suggesting that the diffusion coefficients of both the drug and water have not been significantly affected by the small degrees of crosslinking in the shell. Based on the observed time-lags, the diffusion coefficient of water through the outer PVA shell with a high crosslinking ratio ($X = 0.5$) is estimated to be at least six times higher than that of acetaminophen and proxyphylline.

## INTRODUCTION

Poly(vinyl alcohol) (PVA) hydrogels have been studied extensively as carriers for controlled-release drug delivery in the form of either a single-unit compressed tablet or multiparticulate dose containing granules or particles.[1-3] To prolong the drug release from such a hydrophilic polymer network, PVA is often crosslinked to reduce the interstitial space available for drug diffusion. It has been well recognized that an increase in PVA crosslinking density results in a decrease in both the volume swelling and the rate of drug release. This is attributed to the decrease in diffusion coefficients of both the drug and solvent by the reduction of molecular weight between crosslinks and the associated macromolecular mesh size.[4]

As one increases the rate of water swelling of a dehydrated PVA hydrogel, the contribution of the slow macromolecular relaxation process at the swelling front may become important in determining the overall drug release kinetics. Despite the growing interest in utilizing glassy PVA hydrogels as swelling-controlled drug delivery systems, the kinetics of swelling of crosslinked PVA hydrogels during drug release has not been studied in detail, primarily due to limitations on available sample geometry. The frequently employed planar geometry and granular particles are limited in their usefulness in obtaining reliable swelling data because of the edge effect, the anisotropic swelling behavior and technical difficulties in the direct observation of swelling boundaries. On the other hand, spherical geometry eliminates these drawbacks and, therefore, permits accurate measurement of the transient dimensional changes during drug release. This aspect has been demonstrated by Lee using glassy poly(hydroxyethyl methacrylate) beads.[5]

The purpose of this study is to prepare spherical PVA beads from suspension polymerized poly(vinyl acetate) (PVAc) and to characterize the effect of crosslinking and core-shell structure on the kinetics of swelling and drug release. These PVA beads are sufficiently large (around 1 mm diameter) that they are potentially suitable for oral controlled-release drug delivery applications.

## EXPERIMENTAL

### 1. Synthesis of PVA beads

Double-layered PVA beads consisting of a highly crosslinked outer shell and a lightly crosslinked inner core were prepared from suspension polymerized PVAc by a process involving stepwise saponification with subsequent stepwise crosslinking. This is a modification of the process reported by Hirayama et al.,[6] for preparing micron size PVA particles as column packing materials in gel permeation chromatography. Specifically, the suspension polymerization of distilled vinyl acetate was carried out in an aqueous medium at 60 to 65°C for 5-6 hrs using PVA (fully hydrolyzed grade, MW 106,000-110,000, Aldrich Chemical) as suspending agent and AIBN as initiator. Intermediate gel beads with a thin crosslinked PVA shell and an uncrosslinked PVA core (designated as PVA I) were prepared by first partially saponifying the PVAc beads in an aqueous solution of $NaOH/Na_2SO_4$/methanol (2:1:1 by weight). The reaction time was controlled such that the thickness of the saponified surface layer is about 10% of the original radius. These beads were then crosslinked in an acidic methanolic solution of glutaraldehyde for 2 hrs at 60°C. Subsequent saponification in an alkaline methanolic solution converts the PVAc core into PVA (noncrosslinked).

The PVA I beads so obtained were then crosslinked further in an acidic methanolic solution of glutaraldehyde to produce PVA beads having highly crosslinked outer shells and lightly crosslinked inner cores (designated as PVA II). A range of crosslinking density was obtained by varying the crosslinking ratio X, defined as the number of moles of glutaraldehyde per mole of repeating vinyl alcohol units in PVA.

## 2. Swelling and drug release experiments

Acetaminophen and proxyphylline were selected as model drugs for swelling and drug release experiments because they cover a large range of solubilities (approx. 1.4% for acetaminophen and 50% for proxyphylline). Different acetaminophen and proxyphylline loading levels in PVA I and PVA II were achieved by equilibrating the PVA beads in aqueous drug solutions at 37°C for four days. After filtering, rinsing and drying, the drug-loaded beads were then used for swelling and drug release studies in distilled/deionized water. The swelling experiments were performed in a water-jacketed cuvette maintained at 37°C with a circulating water bath. The swelling front movement was observed with a WILD M420 microscope equipped with camera attachments. The *in vitro* drug release at 37°C under perfect sink diffusion condition was monitored continuously in a stirred cuvette on a Hewlett-Packard 8452A diode-array UV-Vis spectrometer equipped with a water-jacketed cuvette holder and a built-in magnetic stirrer, at 280 nm for acetaminophen and 260 nm for proxyphylline.

## RESULTS AND DISCUSSION

PVAc beads prepared by the present suspension polymerization procedures were generally smooth and spherical with a diameter range of 0.2-1.5 mm. Ordinarily, PVA is obtained directly from PVAc by the alkali or acid hydrolysis in the solution state. However, in order to preserve the spherical shape, it was difficult to use this direct process due to the lack of structural rigidity in the resulting PVA and the dissolution of converted PVA into the saponification medium. To overcome this problem, a stepwise saponification process, as described in the experimental section, was utilized. Initially, the PVAc beads were only saponified on the surface up to about 10% penetration of the original radius. At this stage, the rigidity of the beads was provided by the non-swelling PVAc core and the PVA surface layer was prevented from dissolving by the high concentration of $Na_2SO_4$ in the saponification medium. During the subsequent glutaraldehyde crosslinking, a reduction in the thickness of the PVA shell was observed; the extent of which depends on the concentration of the crosslinking agent and the reaction time. The crosslinked PVA shell, so formed, provides structural support for the PVA gel core in PVA I, which was formed by saponifying the remaining PVAc core.

PVA II beads, consisting of highly crosslinked outer shells and lightly crosslinked inner cores were prepared by further crosslinking of PVA I. The outer shell which was already crosslinked was further tightened by the additional crosslinking throughout the bead, making the crosslinking ratio of inner core to remain less than that of the outer shell by a constant factor. This factor is equivalent to the original shell crosslinking ratio in PVA I beads, which was estimated from swelling data to be $X = 0.3$. Such crosslinking process further reduces the shell thickness to an average of about 21 µm in PVA II at swelling equilibrium.

The dynamic swelling behavior of dehydrated PVA I and PVA II beads was studied by measuring the transient dimensional changes and solvent front penetrations. As shown in Figure 1, an increase in the PVA crosslinking ratio results in a lower water swelling capacity and a slower approach to the equilibrium swelling, both expressed in terms of changes in bead diameter. From results in Figure 2, the penetrating solvent front (the glassy/rubbery front) is observed to move slower with increasing crosslinking ratios. Some Fickian components are evident from the slight non-linearity in the initial portion of the time-dependence of penetra-

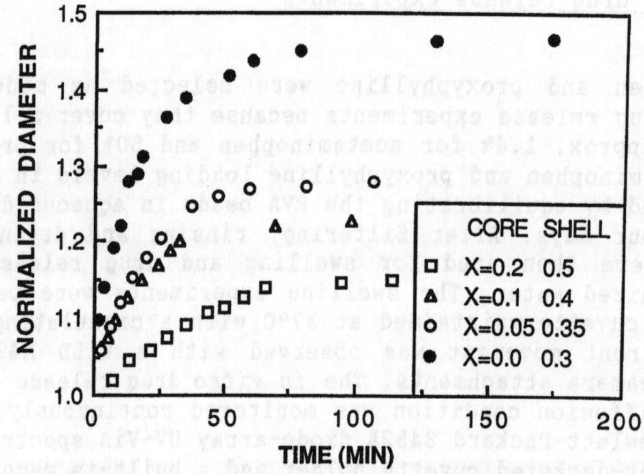


**Figure 1. Effect of crosslinking ratio on the transient dimensional changes in double-layered PVA beads.**

tion front position. These are followed by a linear (or constant-rate) penetration region and, eventually, an accelerated penetration near the core. The apparent acceleration of solvent front penetration near the core is a natural result of the spherical geometry[7] and not, as suggested by some authors,[8] a super-Case II transport behavior. Also observed in Figure 2 is the appearance of an initial time lag in the solvent front penetration at a higher shell crosslinking ratio of X = 0.5. The absence of observable time lags at lower crosslinking ratios suggests that the diffusion coefficient of water through the outer PVA shell was not significantly reduced by the small degree of crosslinking. However, the transport resistance of water through PVA shells with larger crosslinking ratio (X = 0.5) was significantly high to produce a measurable time lag and a slow approach toward swelling equilibrium.

Figures 3 and 4 show the effects of the crosslinking ratio on the release kinetics of acetaminophen and proxyphylline, respectively. Similar to the swelling front penetration results, the effects of crosslink-

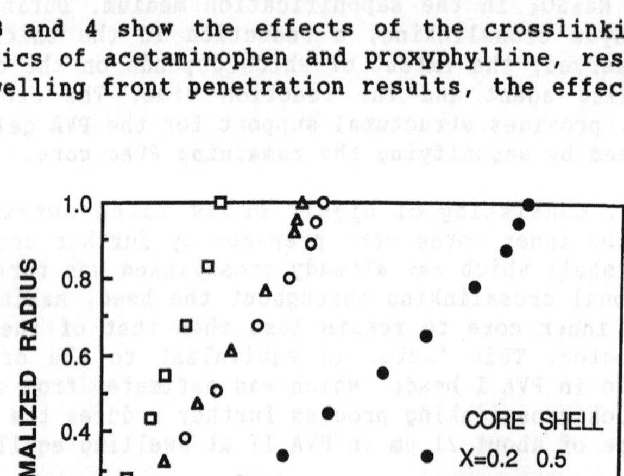


**Figure 2. Effect of crosslinking ratio on the transient swelling front position in double-layered PVA beads.**

ing ratio on the release of acetaminophen and proxyphylline were minimal at low shell crosslinking ratios (X < 0.35), where the drug release is generally completed within 1-2 hours. The corresponding release profiles exhibit a first order diffusion characteristic due to the spherical geometry. On the other hand, at a higher shell crosslinking ratio (X = 0.5), the cumulative release profiles for both drugs show an initial time lag of about 15 mins followed by a constant-rate region of 2-3 hrs before tailing off in a first order fashion. The overall duration for drug release in this case is extended to more than 10 hrs. It is apparent that at a higher crosslinking ratio of X = 0.5, the outer shell acts as a rate-controlling membrane which gives rise to a diffusional time lag characteristic of membrane-reservoir systems. As drug release proceeds, the diffusional distance and, therefore, the transport resistance through the core PVA matrix increases steadily and eventually exceeds the transport resistance through the surface layer. This gives rise to the observed first-order decline in the release rate during the latter part of the drug release. It has to be noted that the observed diffusional time-lag for PVA beads with a higher shell crosslinking ratio (X = 0.5) is longer (≈15 mins) for the release of acetaminophen and proxyphylline than that for the solvent front penetration (≈2 mins). Assuming the initial drug diffusion from such a composite system is analogous to that of a membrane reservoir system, these time-lag results suggest that the diffusion coefficient of water through the outer PVA shell with a crosslinking ratio of X = 0.5 is at least 6 times higher than that of acetaminophen and proxyphylline since the time lag is known to be inversely proportional to the diffusion coefficient D for a given membrane thickness.

## CONCLUSION

We have developed a new method of preparing large spherical PVA beads with a double-layer structure, which involves a stepwise saponification of suspension polymerized PVAc beads and stepwise crosslinking of the PVA core and shell to different degrees with glutaraldehyde. The kinetics of swelling and drug release were studied as a function of the PVA crosslinking ratio. The results demonstrate that the outer shell functions as a rate-controlling membrane upon increasing the shell crosslinking ratio to about X = 0.5. This is reflected from the observed diffusional time

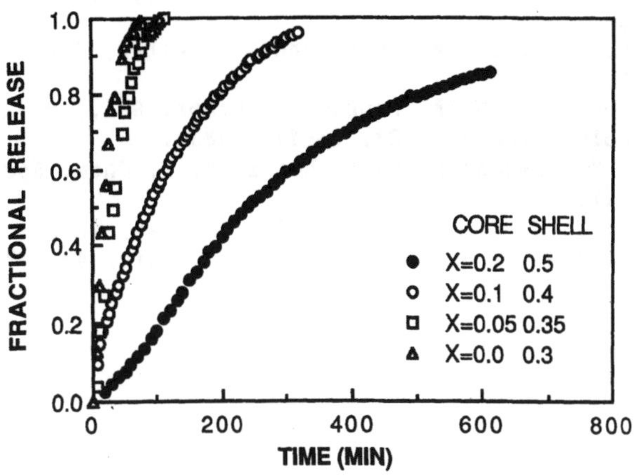


Figure 3. Effect of crosslinking ratio on the acetaminophen release from double-layered PVA beads at 5% drug loading.

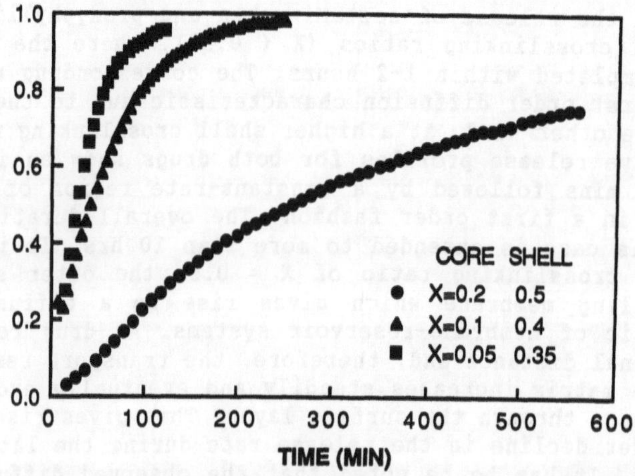


**Figure 4. Effect of crosslinking ratio on the proxyphylline release from double-layered PVA beads at 10% drug loading.**

lag and the constant-rate region during swelling and drug release. The double-layered PVA beads so formed are potentially useful for drug delivery applications and ideally suited as matrices for the elucidation of swelling and drug release mechanisms.

## ACKNOWLEDGMENTS

This work was supported by grants from the Medical Research Council of Canada (Grant Nos. DG-377, 378, and 379) and Ontario Center for Materials Research (Grant No. TCJ-372) to P. I. L.

## REFERENCES

1. P. Colombo, A. Gazzaniga, C. Caramella, U. Conte and A. LaManna, Acta Pharm. Technol., 33, 15-20 (1987).
2. B. Gander, R. Gurny, E. Doelker and N. A. Peppas, Pharm. Res., 6, 578-584 (1989).
3. B. Gander, V. Beltrami, R. Gurny and E. Doelker, Int. J. Pharm., 58, 63-72 (1990).
4. R. W. Koseymeyer and N. A. Peppas, J. Membr. Sci., 9, 211-227 (1981).
5. P. I. Lee, Polym. Commun., 24, 45-47 (1983).
6. C. Hirayama, K. Kawagachi and Y. Motozato, J. Chem. Soc. Jpn., 1974, 894-899 (1974).
7. P. I. Lee and C. J. Kim, J. Controlled Release, in press.
8. C. M. Klech and A. P. Simonelli, J. Membr. Sci., 43, 87-101 (1989).

# LIPOSOMES AS A TOPICAL DRUG DELIVERY SYSTEM EVALUATED BY *IN VITRO* DIFFUSION STUDIES

K. Egbaria and N. Weiner*

College of Pharmacy
University of Michigan
Ann Arbor, MI, 48109, USA

The topical delivery of several liposomal drug formulations was evaluated using *in vitro* diffusion experiments with hairless mouse and guinea pig skin in an effort to understand the mechanism by which liposomal entrapment enhances local activity of topically applied drugs. The uptake and distribution of hydrocortisone and fluocinilone acetonide into the skin strata from phospholipid liposomes was almost four times higher than from aqueous solution. The application of liposomes containing entrapped $^{14}C$-inulin and $^{3}H$-cholesterol, a lipid marker, revealed that the ratio of $^{14}C$ and $^{3}H$ remains unaltered in various layers of the stratum corneum and deeper strata of the skin. These results, coupled with the observation that there is no breakthrough of inulin into the receiver compartment, even after 24 hours, suggest that the bilayers of liposomes are able to carry entrapped solutes deep into the stratum corneum and even into skin strata of the living epidermis. The combined data strongly suggest molecular mixing of liposomal bilayers with those of the stratum corneum "bilayers". The extent of such mixing was greater with ceramide-based liposomes than with phospholipid liposomes since the former has a composition similar to that of stratum corneum "bilayers".

## INTRODUCTION

Liposomes are microscopic vesicles composed of one or more lipid bilayers arranged in a concentric fashion enclosing an equal number of aqueous compartments.[1] Various amphipathic molecules have been used to form the liposomes, and the method of preparation can be tailored to control their size and morphology. Drug molecules can either be encapsulated in the aqueous space or intercalated into the lipid bilayer; the exact location of a drug in the liposome will depend upon its physicochemical characteristics and the composition of the lipids.[2,3]

Liposomes have shown great potential as a drug delivery system. An assortment of molecules, including peptides and proteins, have been incorporated in liposomes, which can then be administered by different

* To whom correspondence should be directed.

*Cosmetic and Pharmaceutical Applications of Polymers*
Edited by C.G. Gebelein *et al.*, Plenum Press, New York, 1991

routes.[4-6] Due to their high degree of biocompatibility, liposomes were initially conceived of as delivery systems for intravenous delivery. It has since become apparent that liposomes can also be useful for delivery of drugs by other routes of administration.

Mezei, et al.,[7] reported that topical application of liposomal triamcinolone acetonide for 5 days resulted in a drug concentration in the epidermis and dermis 4 times higher than that obtained using a control ointment, while urinary excretion of the drug was diminished. Mezei, et al.,[8] compared also the deposition of topical applied gels of free and liposomally entrapped triamcinolone in rabbit skin. They found that application of the liposomal gel resulted in a concentration of triamcinolone acetonide approximately five times higher in the epidermis and three times higher in the dermis, than application of the free drug gel. The results of these studies suggested to them the inherent potential of liposomes as a selective drug delivery system for cutaneous application.

Vermorken, et al.,[9] tested the effect of topical application of dihydrotestosterone encapsulated in liposomes and as an acetone solution on the hamster flank organ. These authors reported that the systemic absorption of dihydrotestosterone from the liposomal system was negligible, whereas significant absorption was observed from the acetone solution. Since they were only concerned with systemic absorption, they reported no advantages of the liposome system over the acetone solution in achieving the desired biological effect.

Ganesan, et al.,[10] using liposomal formulations in *in vitro* diffusion experiments with hairless mouse skin found that neither intact liposomes nor the phospholipid of which they are comprised diffuse across the skin. They also reported that lipophilic drugs such as progesterone and hydrocortisone, which are expected to be intercalated within the bilayer structure of phospholipid multilamellar liposomes, passed through the skin with comparable facility to the free drug. The effect of application of liposomal progesterone in reducing the rate of hair growth in idiopathic hirsutism was reported by Rowe, et al.[11]

Patel, et al.,[12] suggested that liposomes can be used for the sustained release of drugs into the epidermis when applied topically. For example, when free and liposomally entrapped $^{3}$[H]-methotrexate was applied topically onto the skin of nude mice, percutaneous absorption of drug was greatly reduced by liposomal encapsulation. Furthermore, the retention of $^{3}$[H]-methotrexate in the skin was 2-3 fold higher from the liposomal form than the free form, again suggesting a localization of the liposomes in the epidermis, where a sustained release of methotrexate takes place.

The present study describes the evaluation of formulation effects on the deposition of various drugs into the strata of hairless mouse and guinea pig skin using *in vitro* diffusion cell studies, in an attempt to explain the mechanism of liposomal action on drug transfer into the skin, and ultimately, its improved therapeutic effect.

## EXPERIMENTAL

### 1. Materials

Cholesterol (CH), palmitic acid (PA), cholesteryl sulfate (CS), bovine brain ceramides (CM), hydrocortisone (HC), and HEPES free acid were obtained from Sigma (St. Louis, MO). Egg lecithin (PC) and phosphatidyl serine (PS) were obtained from Avanti Polar Lipids (Birmingham, AL). α-tocopherol (α-T) was obtained from Eastman Kodak (Rochester, NY). $^{14}$C-PA, $^{3}$H-CH, $^{3}$H-HC and L-3-phosphatidylcholine, 1,2-di{1-$^{14}$C}palmitoyl, {$^{14}$C-DPPC} were obtained from Amersham (UK). $^{14}$C-inulin was obtained from ICN Radiochemicals, inc. (Irvine, CA) and $^{14}$C-fluocinolone acetonide (FA) from Syntex (Palo Alto, CA). All other chemicals were of analytical grade.

## 2. Liposome Formulations

Multilamellar liposomes (MLV) containing PC:CH:PS at a mole ratio of 1:0.5:0.1 or CM:CH:PA:CS, referred to as "skin lipid", at a weight ratio of 4:2.5:2.5:1 were prepared using the conventional film method.[1] Briefly, the lipid mixtures were dissolved in a 2:1 (v/v) mixture of chloroform and methanol. Approximate amounts of drug were also dissolved in the solvent mixture. Trace amounts of $^{3}$H-HC, $^{14}$C-FA, $^{14}$C-inulin, $^{3}$H-CH and $^{14}$C-DPPC were incorporated in the phospholipid-based liposomes and trace amounts of $^{3}$H-HC, $^{14}$C-FA, $^{14}$C-inulin, $^{3}$H-CH and $^{14}$C-PA were added to the "skin lipid" liposomes. One percent of a α-T (based on total lipid amount) was added to each of these liposomal formulations as an antitoxidant. The lipids, drug and markers were then deposited as a thin film in a round-bottom flask by rotary evaporation under nitrogen. The flask containing the lipid film was stored in vacuum overnight to facilitate removal of any residual solvents. The films were hydrated by the addition of an isotonic 0.05 M HEPES buffer, pH 7.4, with mild agitation at a temperature above the phase transition temperature of the highest melting component in the mixture. After the film was homogeneously dispersed, the mixture was bath-sonicated for 5 minutes and the suspension was stored at 4°C overnight before use in the diffusion experiments. The final concentration of lipids was 50 mg/mL for phospholipid liposomes and 15 mg/mL for "skin lipid" liposomes. All of the liposomal preparations were examined with a Nikon Diaphot Light microscope to evaluate liposomal quality and integrity. additionally, the multilamellarity of the vesicles was confirmed by freeze fracture electron microscopy.

Large unilamellar vesicles (LUV) were prepared by a modification of the reverse-phase evaporation method of Szoka and Papahadjopoulos.[13] Appropriate amounts of the lipid mixtures, HC, and α-T, along with trace amounts of radiolabeled HC were dissolved in 10 mL of a chloroform-methanol mixture (2:1, v/v). Five mL of 0.05 M HEPES buffer (pH 7.4) and enough additional methanol (up to 1.5 mL) were added to yield a clear solution after brief sonication. The organic solvents and a small amount of water were then removed under nitrogen at a temperature above the phase-transition temperature of lipids by using a rotoevaporator. Solvent removal was continued until all foaming ceased. The resulting liposomal suspension was stored at 4°C overnight before using in the diffusion experiments.

Dehydration/Rehydration liposomes (DRV) were prepared by a modification of the method reported by Kirby and Gregoriadis.[14] Briefly, appropriate amounts of the various lipids, contained in a flask, were dissolved in chloroform/methanol (2:1; v/v). The solvents were removed using a rotoevaporator under vacuum at a temperature above the phase transition of the lipids. The resultant film was dried overnight in a desiccator to remove any residual solvent. An appropriate aliquot of HC stock solution in 0.05 M HEPES buffer (pH 7.4) containing trace amounts of radiolabeled

HC was then added and the mixture was hydrated at a temperature above the phase transition of the lipids for about an hour. Intermittent vortexing was required for complete hydration. The resultant dispersion was then dehydrated at 50°C under vacuum using the rotoevaporator. When the liposomal suspension became very viscous, and amount of water, equivalent to that removed, was reintroduced into the viscous suspension. The rehydrated liposomes were then allowed to equilibrate for about 45 minutes at a temperature above the phase transition temperature of the lipids. The dispersion was then stored at 4°C overnight before use in the diffusion experiments.

## 3. Diffusion Experiments

Full thickness hairless mouse and hairless guinea pig skin were excised from the fresh carcasses of animals sacrificed with carbon dioxide and used immediately after removing subcutaneous fat. The skin was mounted on a Franz diffusion cell with a nominal surface area of 2 cm² and a receiver compartment with a 7 mL capacity (Crown Glass, Somerville, NJ). The epidermal side of the skin was exposed to ambient conditions while the dermal side was bathed by a 0.05 M isotonic HEPES buffer, pH 7.4. The receiver solution was stirred continuously using a small teflon covered magnet. Care was exercised to remove any air bubbles between the underside of the skin and solution in the receiver compartment. The temperature of the receiver was maintained at 37°C. Following mounting of the section of skin, 200 µl of the test formulation was applied to the epidermis surface. A smaller amount of formulation was found to be sufficient to ensure uniform spreading across the entire exposed surface of the skin in the cell. A minimum of three cells were used for each formulation and duplicate experiments were carried out using sections of skin from different specimens for each formulation. All experiments were carried out with non-occluded donor compartments. At predetermined time periods, the experiments were stopped and the diffusion set-up was dismantled for assay of radiolabeled drug and lipids.

## 4. Assay of Radiolabeled Markers

Upon dismantling, the donor compartment of the cell was rinsed carefully 5 times with 0.5 mL HEPES buffer, pH 7.4. The skin was then removed and, it too was rinsed twice with 3 mL of the same buffer. The washing procedure was found to be sufficient to remove >99% of the formulation when determined at time zero. All washings were collected and assayed for radiolabel. Following the rinsing procedure, the skin patch was mounted on a board and a piece of adhesive tape (Scotch Magic Tape, 810, 3M Commercial Office Supply Division, St. Paul, MN), 1.9 cm wide and about 6 cm long, was used to strip the skin. The tape was of sufficient size to cover the area of skin that was in contact with the formulation. Nine strippings were carried out for each specimen and each strip was analyzed separately for radiolabeled drug and lipid. The remaining skin, as well as the receiver compartment solution, were also assayed for drug and lipid. Assay of the donor, skin rinses and receiver solutions were carried out after addition of about 15 mL of (Ecolite+, ICN Biomedical, Inc., Irvin, CA) to each system. The tape strippings and remaining skin were assayed as follows: each sample was placed in a combustion-cone and burnt in a tissue oxidizer (Model 306 Packard oxidizer, Packard Instrument Co., Downers Grove, IL). The separated radionuclides were then assayed using a scintillation counter.

**RESULTS**

The 24 hour *in vitro* uptake of HC from various formulations upon topical application to hairless mouse skin is shown in Figure 1. The *in vitro* cumulative uptake of HC into the various stripped layers of stratum corneum of hairless mouse skin from the phospholipid LUV, MLV, DRV and from aqueous solution at 24 hours is shown in Figure 2. A representative plot of the percent cumulative uptake of FA from liposomal formulation and from aqueous solution in the stripped layers of hairless mouse stratum corneum is shown in Figure 3. The percent cumulative uptake of inulin from a liposomal formulation and from solution in the stripped layers of hairless guinea pig stratum corneum is shown in Figure 4. The concomitant uptake of lipid marker and inulin for the liposomal MLV system is shown in Figure 5. Plots of the percent cumulative uptake in the stripped layers of hairless mouse stratum corneum of dual label negatively charged phospholipid liposomes and "skin lipid" liposomes after 24 hours of *in vitro* treatment are shown in Figures 6 and 7, respectively. In all cases, a mass balance of >95% was achieved after the donor compartment and the skin rinses were taken into account.

**DISCUSSION**

Although the use of liposomal drug formulations for topical application has been steadily increasing, few studies have been undertaken to explain the mechanism by which liposomal entrapment improves drug transfer into the skin. Most of the *in vitro* and *in vivo* studies have been concerned with the extent of percutaneous absorption rather than the extent of drug deposition in the deeper strata of skin. It should be pointed out that the treatment of many dermatological diseases by topical application is expected to be more efficient if a significant concentration of the drug is retained in the epidermis. The ability of the stratum corneum to act as a reservoir of drug transport through the skin was

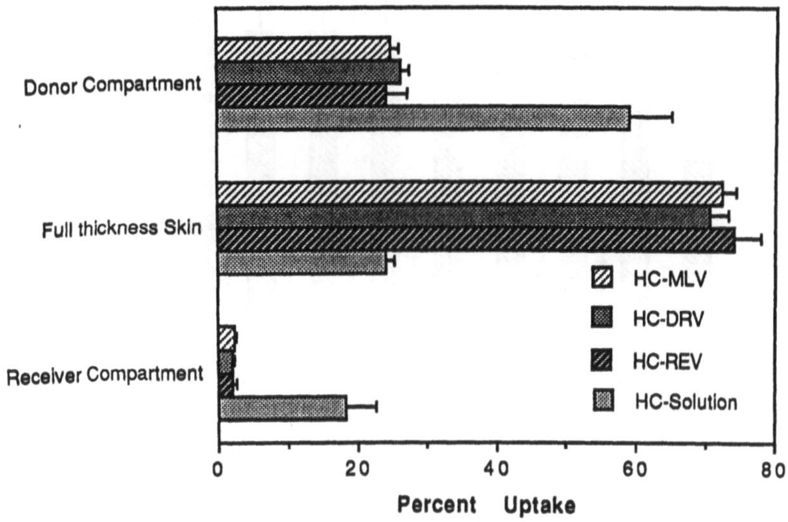


Figure 1. The 24 hour *in vitro* uptake of $^{3}$H-HC from phospholipid MLV, DRV and LUV and from an aqueous solution after topical application to hairless mouse skin.

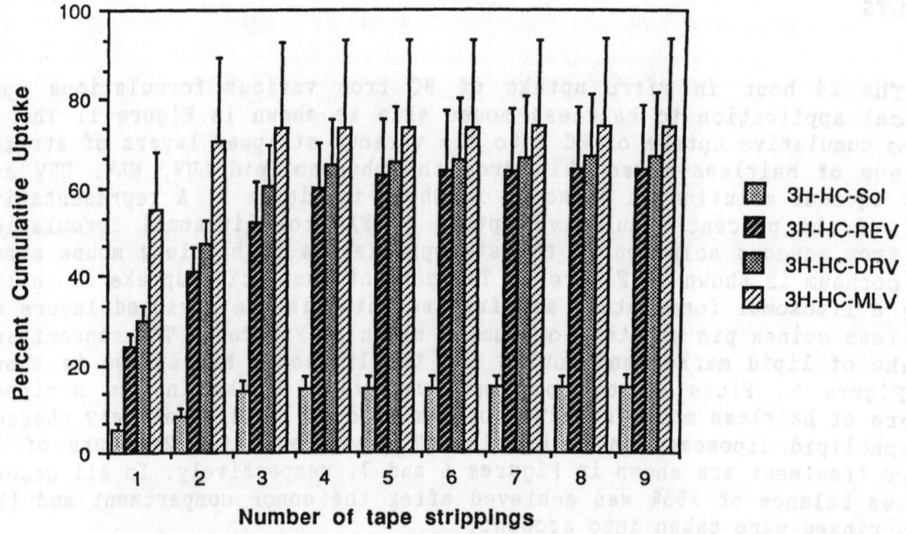


Figure 2. The 24 hour *in vitro* distribution of $^{3}$H-HC from phospholipid MLV, DRV and LUV and an aqueous solution in various strippings of hairless mouse stratum corneum.

amply demonstrated by Rougier, et al.,[15] who reported that the absorption of a variety of drugs through the skin was proportional to the amount of drug recovered in the stratum corneum following 30 minutes topical application.

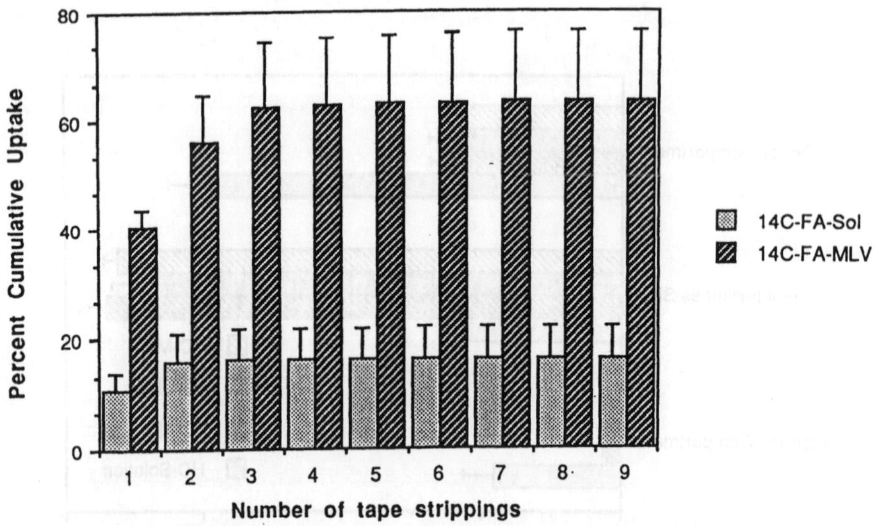


Figure 3. Comparison of the 24 hour *in vitro* uptake of $^{14}$C-FA from phospholipid MLV and from an aqueous solution in various strippings of hairless mouse stratum corneum.

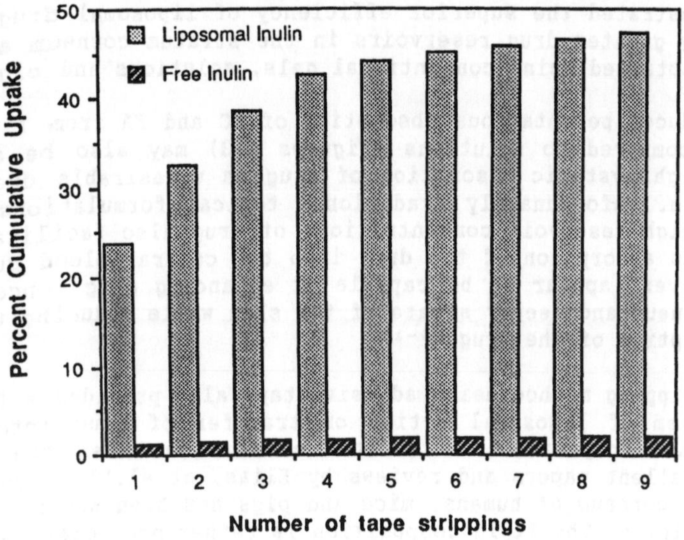


Figure 4. Comparison of the 24 hour *in vitro* uptake of $^{14}C$-inulin from phospholipid MLV and from an aqueous solution in various strippings of hairless mouse stratum corneum.

It is, therefore, important to examine the extent of drug accumulation in the various strata of the skin in addition to estimating percutaneous absorption profiles. The most common experimental procedure used to determine drug levels in the skin involves stripping the treated skin using adhesive tape. With appropriately radiolabeled drugs and lipids, it has been possible to ascertain the extent and distribution of drug and liposomal lipid in the various strata of the skin. Such studies have

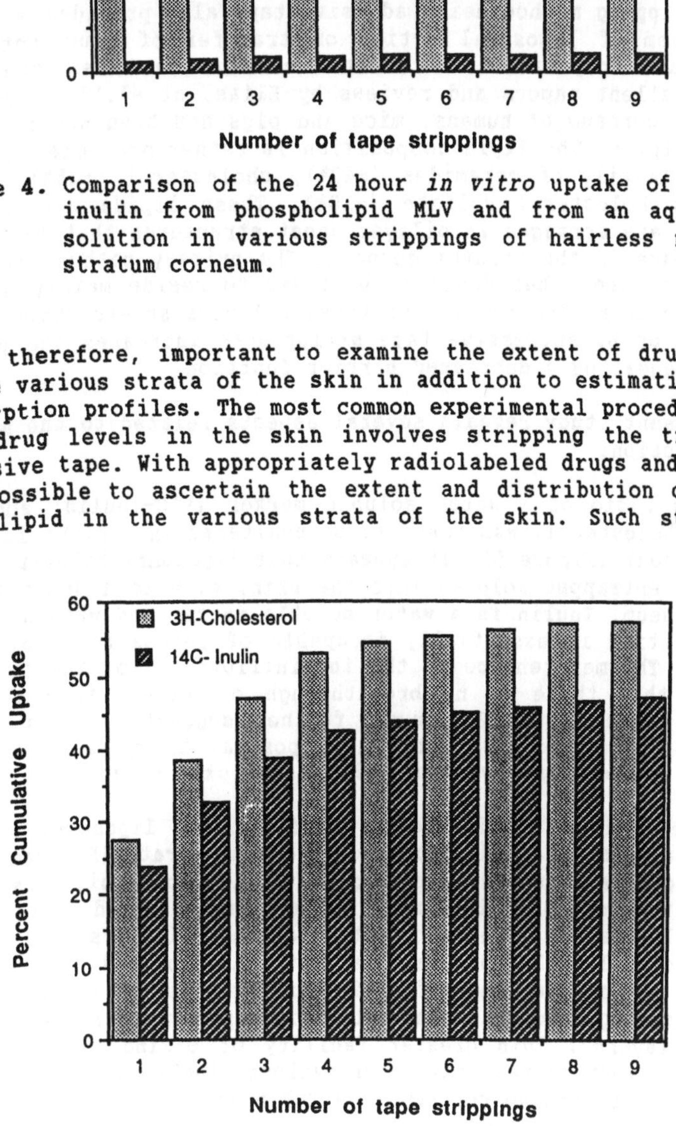


Figure 5. The 24 hour *in vitro* distribution of $^{3}H$-CH and $^{14}C$-inulin from phospholipid MLV in various strippings of hairless guinea pig stratum corneum.

amply demonstrated the superior efficiency of liposomal drug formulation in allowing greater drug reservoirs in the stratum corneum and epidermis than that obtained using conventional gels, solutions and ointments.[16,17]

The reduced percutaneous absorption of HC and FA from liposomal formulations compared to solutions (Figures 1-3) may also be an advantage wherever high systemic absorption of drug is undesirable due to harmful side-effects. Unfortunately traditional topical formulations capable of providing high reservoir concentrations of drug also facilitate enhanced percutaneous absorption of the drug into the central blood supply. Liposomes, however, appear to be capable of enhancing drug concentration in stratum corneum and deeper strata of the skin while reducing the percutaneous absorption of the drug.[16-18]

The stripping method using adhesive tape also provided a few clues to the mechanism of liposomal action on transfer of drug into skin. The structure and composition of stratum corneum has been the subject of several excellent papers and reviews by Elias, et al.[19-21] Very briefly, the stratum corneum of humans, mice and pigs has been shown to be devoid of phospholipids. The lipid composition is rather non-polar in nature and consists primarily of ceramides (≈40%), cholesterol (≈25%), fatty acids (≈25%) and cholesteryl sulfate (≈10%). These lipids are termed "skin lipids" and are arranged in bilayer sheet structures that fill the intercellular space in the stratum corneum. The primary pathway to the transport of water and other drugs is believed to reside mainly in these bilayers structures. The removal of these bilayers sheets either by solvent treatment[22] or by successive tape stripping[23] increases the permeability of water suggesting a decreased barrier function.

The present study reveals several aspects related to the mechanism of liposomal action:

(1) The ratio of a water soluble marker ($^{14}$C-inulin) and liposomal lipid ($^{3}$H-cholesterol) was found to be unaltered in various layers of the stratum corneum (Figure 5). It appears that liposomal bilayer are able to carry their entrapped solutes into the skin, even to a depth beneath the stratum corneum. Inulin is a water soluble marker and when applied as an aqueous solution is essentially incapable of permeating lipid membranes (Figure 4). The maintenance of the inulin:lipid ratio, together with the observance that there is no breakthrough of inulin into the receiver compartment, even after 24 hours, further suggests that a significant amount of water associated with the liposomal bilayers is carried into the stratum corneum and even into the deeper skin strata.

(2) The ratio of radiolabeled lipids of the liposomal preparation, was essentially maintained throughout the skin strata (Figures 6-7). This strongly suggests a molecular mixing of the liposomal bilayers with those of the stratum corneum "bilayers". It was also observed that such mixing of liposomal bilayers with the stratum corneum "bilayers" was more extensive with "skin lipid" liposomes than with phospholipid liposomes. It was argued that the greater ease of mixing of "skin lipid" liposomal bilayers is not surprising since its composition is similar to that of the stratum corneum "bilayers". This greater facility of mixing also explains the greater efficiency of penetration of cyclosporine[24] and interferon[17] into deeper strata of stratum corneum from "skin lipid" liposomes as compared to "phospholipid" liposomes.

• (3) The uptake and distribution of hydrocortisone and fluocinilone into the skin strata from phospholipid liposomes was almost four times higher than from aqueous solution (Figures 1-3).

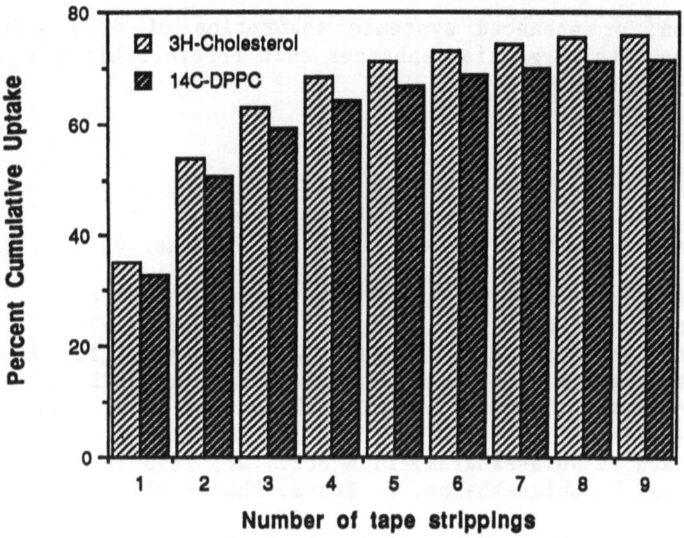


Figure 6. The 24 hour *in vitro* distribution of $^3$H-CH and $^{14}$C-DPPC from phospholipid MLV in various strippings of hairless mouse stratum corneum.

The combined results suggest that topically applied liposomal formulations, particularly those prepared from lipid mixtures of composition similar to the stratum corneum, may be an effective delivery system for the treatment of skin diseases. Since these liposomal formulations provide sustained, enhanced levels in strata associated with blood and lymph supplies, they have the capacity to meter a sufficient quantity of drug into the deeper tissue to treat skin symptomology. Such metering should also reduce the incidence of undesirable effects arising from systemic

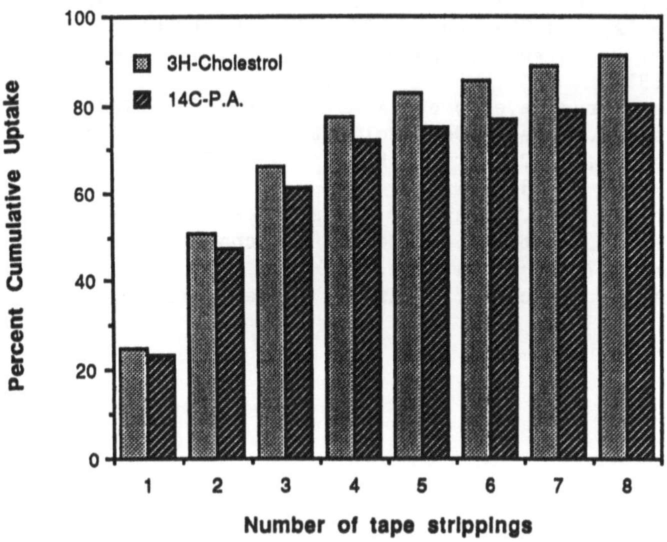


Figure 7. The 24 hour *in vitro* distribution of $^3$H-CH and $^{14}$C-PA from "skin lipid" MLV in various strippings of hairless mouse stratum corneum.

administration, or enhanced systemic absorption of drug after topical administration with permeation enhances that irreversibly disrupt stratum corneum bilayer structure.

## REFERENCES


1. A. D. Bangham, M. M. Standish, and J. C. Watkins, J. Mol. Biol., **13**, 138 (1965).
2. G. Gregoriadis, New Engl. J. Med., **295**, 704-710 (1976).
3. J. H. Fendler and A. Romero, Life Sci., **20**, 1109 (1977).
4. M. C. Finkelstein and N. G. Wessman, J. Lipid Res., **19**, 289 (1978).
5. M. B. Yatvin and P. I. Lelkes, Med. Phys., **9**, 149-175 (1982).
6. G. Gregoriadis, C. P. Swain, E. J. Willis, and A. S. Travill, Lancet, **1**, 1313 (1974).
7. M. Mezei and V. Gulasekharam, Life Sci., **26**, 1473-1477 (1980).
8. M. Mezei and V. Gulasekharam, J. Pharm. Pharmacol., **34**, 473-474 (1982).
9. A. J. Vermorken, M. W. Hukkelhoven, A. M. Vermeesch, C. M. Goos, and P. Wirtz, J. Pharm. Pharmacol., **36**, 334-336 (1983).
10. M. G. Ganesan, N. D. Weiner, G. L. Flynn, and N. F. H. Ho, Int. J. Pharm., **20**, 139-154 (1984).
11. T. C. Rowe, M. Mezei, and J. Hilchie, The Prostate, **5**, 346-347 (1984).
12. H. M. Patel, Dermatological ointment, U. K. Patent, GB2, 143 433A (1984).
13. F. Szoka and D. Papahadjopoulos, Biochem., **75**, 4194-4198 (1978).
14. C. J. Kirby, and G. Gregoriadis, in "*Liposome Technology*," G. Gregoriadis, Ed., CRC Press, Inc., Boca Ratan, FL, 1984, Vol. 1, pp. 19-28.
15. A. Rougier, D. Dupuis, C. Lotte, R. Roguet, and H. Schafer, J. Invest. Dermatol., **81**, 275-281 (1983).
16. N. Weiner, N. Williams, and G. Birch, Antimicrob. Agents Chemother., **33**, 1217-1221 (1989).
17. K. Egbaria, C. Ramachandran, D. Kittayanod, and N. Weiner, Antimicrob. Agents Chemother., **34**, 107-110 (1990).
18. W. Abraham, P. W. Wertz, and D. T. Downing, J. Invest. Dermatol., **90**, 259-262 (1988).
19. P. M. Elias, Drug Development Res., **13**, 97-105 (1988).
20. P. M. Elias, J. Invest. Dermatol., **80**, 044s-049s (1983).
21. P. M. Elias, E. R. Cooper, A. Kore, and B. E. Brown, J. Invest. Dermatol., **76**, 297-301 (1981).
22. S. Grayson and P. M. Elias, J. Invest. Dermatol., **78**, 128-135 (1982).
23. H. Schaefer, G. Stuttgen, A. Zesch, W. Schalla, and J. Gazith, Curr. Probl. Derm., **7**, 80-94 (1979).
24. K. Egbaria, C. Ramachandran, and N. Weiner, Skin Pharmacol., **3**, 21-28 (1990).

# POLYMERS AS STABILIZERS OF LIPID VESICLES

Tamotsu Kondo,[a,b] Yutaka Atsuta,[a] Atsushi Kato,[a] Koichiro Fukuda [a] and Hiroyuki Ohshima[a]

(a) Faculty of Pharmaceutical Sciences and
(b) Research Institute for Biosciences
Science University of Tokyo
Shinjuku-ku, Tokyo 162, Japan

Poly(acrylic acid) and carboxymethylchitin, respectively, were used to stabilize dipalmitoylphosphatidylcholine and egg yolk phosphatidylcholine vesicles in the micrometer range through complexation with the vesicle membranes. The stability of poly(acrylic acid)-complexed dipalmitoylphosphatidylcholine vesicles, as estimated by the release of a fluorescent marker into the surrounding medium, was found to depend on temperature as well as pH, ionic strength, and glucose concentration of the medium. These vesicles were quite stable to the action of various proteins, with no leak of the marker. On the other hand, carboxymethylchitin-complexed phosphatidylcholine vesicles were disintegrated to a certain extent through the interaction with serum albumin, while they were very resistant to serum globulins and fibrinogen. The vesicles were decomposed by the action of phospholipase C, an enzyme which can digest phosphatidylcholine, with remarkable release of the vesicle content. Phagocytosis of the vesicles by guinea-pig polymorphonuclear leucocytes was low, suggesting a high possibility that the vesicles when loaded with hemoglobin can be used as artificial red blood cells.

## INTRODUCTION

Phospholipid vesicles have been expected to be a suitable carrier for drugs and other bioactive substances. However, they have found few actual applications because of poor stability to various environmental conditions encountered in their use. In view of this, many attempts have been made to obtain vesicle membranes of enhanced stability by coating the vesicle surface with polymers.[1-6] This idea originated from the fact that bacteria and plant cells possess a cell wall consisting of polymeric materials, which strengthens the structure of cell membranes. In a similar sense, we have investigated polymer-complexed phosphatidylcholine vesicles with the aim of using them as carriers for water-soluble drugs and hemoglobin.[7-13] In the present article, the results obtained in our investigation on polymer-complexed phospholipid vesicles are summarized.

*Cosmetic and Pharmaceutical Applications of Polymers*
Edited by C.G. Gebelein *et al.*, Plenum Press, New York, 1991

## EXPERIMENTAL

### 1. Vesicle Preparation

Polymer-complexed phosphatidylcholine vesicles with diameters in the micrometer range were prepared by a two-step emulsification technique. A volume of an aqueous solution containing a fluorescent marker or of human hemolyste was dispersed as fine droplets into an equal volume of 5% dipalmitoylphosphatidylcholine [DPPC] solution in chloroform or in dichloromethane to yield a W/O emulsion. The emulsion obtained was quickly added with stirring into 5 volumes of 0.1% aqueous poly(acrylic acid)-[PAA] solution or of 0.2% aqueous carboxymethylchitin [CM chitin] solution to give a W/O/W complex emulsion. Aqueous vesicle dispersion was obtained after evaporating the organic solvent. The use of [PAA] gave paucilamellar vesicles while multilamellar vesicles were formed when [CM chitin] was used.

### 2. Estimation of Vesicle Stability

The stability of [PAA]-complexed [DPPC] vesicles was estimated by measuring the amount of 5(6)-carboxyfluorescein released from the vesicles into the surrounding medium under various conditions. A 0.5% vesicle suspension in buffer solutions with various pH values and ionic strengths was incubated at different temperatures, and sampling was done at appropriate time intervals. The samples were immediately centrifuged at 15,000 rpm for 20 minutes. After dilution of a volume of the supernatant with an appropriate volume of the buffer, the concentration of the fluorescent marker was determined spectrofluorometrically.

The release tests were also carried out in solutions of different glucose concentrations. Glucose was dissolved in a phosphate buffer solution (pH 5.5, ionic strength 0.0154) at various concentrations and the vesicles were redispersed in the glucose solutions. The resultant suspension was incubated at 30°C for 2 hr., followed by centrifugation to remove the vesicles. The concentration of the fluorescent marker in the supernatant was determined by the same procedure as described above.

Britton-Robinson buffer solutions were used to examine the effect of pH. The ionic strength of the buffer solutions was adjusted to 0.01 by addition of NaCl, into which the vesicles were redispersed. The release tests were made in the manner described above.

The effect of ionic strength was studied by employing phosphate buffer solution (pH 5.5) with different ionic strengths as the suspension media of the vesicles. The experimental procedures were the same as those used in the release tests in glucose solutions.

The degree of disintegration of [CM chitin]-complexed phosphatidylcholine vesicles loaded with hemolystae by the action of phospholipase C was evaluated by determining spectrophotometrically the hemoglobin concentration leaked from the vesicles. A volume of vesicle suspension at 0.5% in the phosphate buffer solution was mixed with an equal volume of solution of the enzyme in the same buffer. After incubation for 2 hr. at 37°C, the mixture was centrifuged at 15,000 rpm for 30 min. to remove the unbroken vesicles and the supernatant containing leaked hemoglobin was collected as the sample.

### 3. Estimation of Vesicle Phagocytosis

As an index of phagocytosis, oxygen consumed by guinea-pig polymorphonuclear leucocytes was measured when the cells phagocytosed [CM chitin]-complexed phosphatidylcholine vesicles. To 3 mL of leucocyte suspension was added 0.5 mL of the vesicle suspension and the rate of decrease of oxygen in the medium was recorded with a Clark type oxygen electrode. The initial rate of oxygen consumption was calculated and used as the index of phagocytosis.

## RESULTS AND DISCUSSION

[PAA]-complexed [DPPC] vesicles were found to be quite stable to the action of serum albumin, serum globulins, and fluorescent marker, whereas [CM chitin]-complexed phosphatidylcholine vesicles were slightly disintegrated by low concentrations of serum albumin even though they were highly resistant to serum globulins, fibrinogen, and high concentrations of serum albumin.

Release of the fluorescent marker from the vesicles was rather inappreciable at temperatures below 30°C for up to 5 hr. incubation, while the incubation of the vesicles at 37°C caused marked release in the same time period. This would be related to the phase transition of [DPPC] layers, which takes place around 41°C. Complexation with [PAA] made the phase transition of the phospholipid layers somewhat ambiguous and the peak of heat absorption on DSC chart shifted to a higher temperature.

The addition of glucose to a phosphate buffer (pH 5.5, ionic strength 0.0154) as the dispersion medium of the vesicles, reduced the percentage marker release at 30°C as shown in Figure 1. This should demonstrate that the osmotic pressure of glucose solution acts on the vesicle surface to suppress marker release.

Changes in pH of the medium also affected the stability of the vesicles as indicated in Figure 2, either in the absence or in the presence of glucose. Marker release increased with increasing pH of the medium, suggesting an important contribution of hydrogen bonding between undisso-

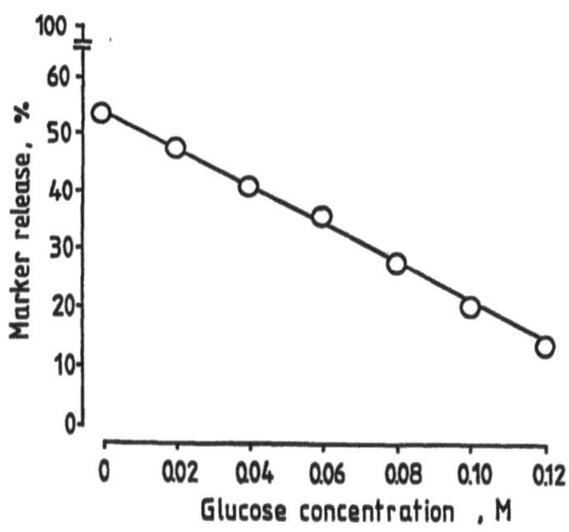


Figure 1. Effect of glucose concentration on marker release from [PAA]-complexed [DPPC] vesicles in phosphate buffer (pH 5.5, ionic strength 0.0154).

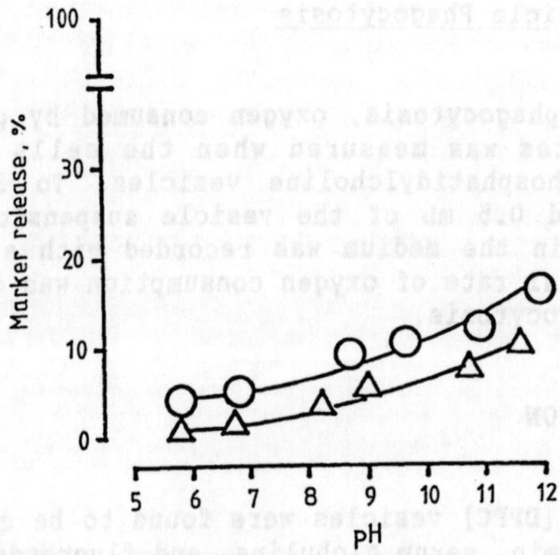


Figure 2. Effect of pH on marker release from [PAA]-complexed [DPPC] vesicles in the presence and absence of glucose. Glucose concentration: o, 0 M; Δ, 0.028 M.

ciated carboxylic acid groups of [PAA] and phosphodiester head groups of [DPPC] to the interaction of these vesicle membrane components.

The ionic strength of the dispersion medium was found to play an important role in stabilizing the vesicles. An increase in the ionic strength of the medium gave rise to a decrease in marker release as shown in Figure 3. The percentage marker release decreased drastically as the ionic strength increased in the absence of glucose, whereas the presence of glucose, at a concentration of 0.12 M, caused only slight changes in vesicle stability. This shows the importance of the electrostatic interaction between the vesicle membrane components.

Hemoglobin release from [CM chitin]-complexed phosphatidylcholine vesicles increased almost linearly with increasing concentration of phos-

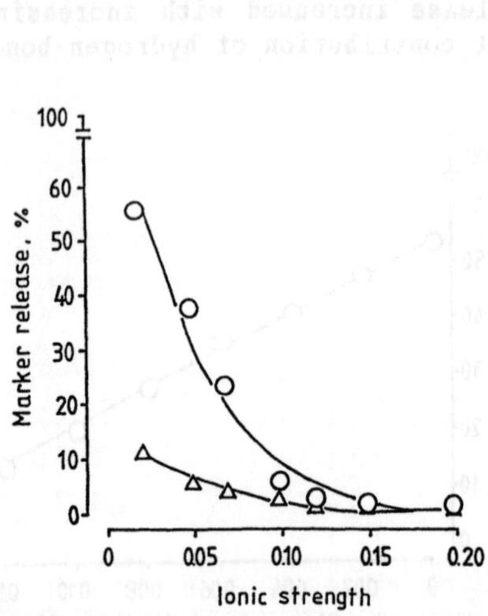


Figure 3. Effect of ionic strength on marker release from [PAA]-complexed [DPPC] vesicles in phosphate buffer (pH 5.5) in the presence and absence of glucose. Glucose concentration: o, 0 M; Δ, 0.12 M.

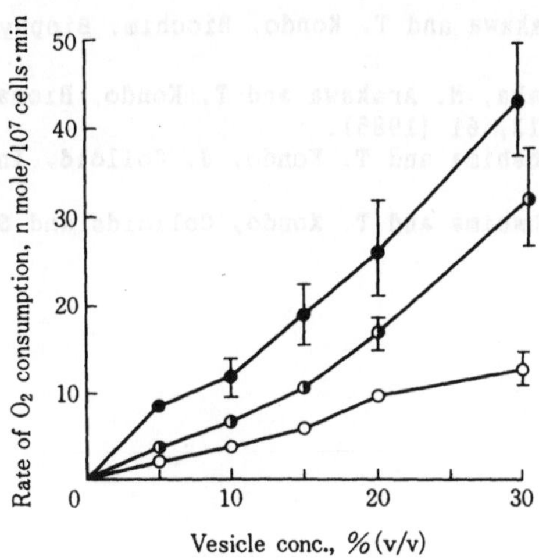


**Figure 4.** Oxygen consumption by guinea-pig polymorphonuclear leucocytes versus polymer-coated vesicle concentration added. Each plot shows the mean value of 6 runs and the bar means the standard error. Polymer: o, carboxymethyl chitin; ◑, carboxymethyl cellulose; •, carboxymethyl chitosan.

pholipase C, which digests the phospholipid, when the enzyme was added to the vesicle suspension. This suggests that **[CM chitin]** molecules cover the phospholipid layers of the vesicles rather coarsely forming a mesh-like structure with openings large enough to allow phospholipid C molecules to make easy access to and act on the phospholipid molecules on the vesicle surface.

Figure 4 shows the degree of phagocytosis, as measured by oxygen consumption by guinea-pig polymorphonuclear leucocytes, when the phosphatidylcholine vesicles coated with **[CM chitin]**, **[CM cellulose]**, and **[CM chitosan]** were introduced into suspensions of the leucocytes. It is clear that **[CM chitin]** suppressed most vesicle phagocytosis by the leucocytes. This fact will be useful in designing artificial red blood cells and drug delivery systems using phospholipid vesicles.

## REFERENCES

1. M. Takada, T. Yuzuriha, K. Katayama, K. Iwamoto and J. Sunamoto, Biochem. Biophys. Acta , **802**, 237 (1984).
2. K. Aliev, H. Ringsdorf, B. Schlarb and K. H. Leister, Makromol. Chem. Rapid Commun., **5**, 345 (1984).
3. K. Seki and D. A. Tirrell, Macromolecules, **17**, 1692 (1984).
4. S. L. Regen, J. S. Shin and K. Yamaguchi, J. Amer. Chem. Soc., **106**, 2446 (1984).
5. H. Bader, K. Dorn, B. Hupfer and H. Ringsdorf, Adv. Poly. Sci., **64**, 53 (1985).
6. D. A. Tirrell, D. Y. Takigawa and K. Seki, Ann. N. Y. Acad. Sci., **446**, 237 (1985).
7. A. Kato, M. Arakawa and T. Knodo, J. Microencap., **1**, 105 (1984).
8. E. Wehrli, T. Kondo, A. Kato, H. Izawa and J. Kreuter, J. Microencap., **1**, 329 (1984).

9. H. Izawa, M. Arakawa and T. Kondo, Biochim. Biophys. Acta , 885, 243 (1986).
10. A. Kato, I. Tanaka, M. Arakawa and T. Kondo, Biomater. Med. Dev. Artif. Organs, 13, 61 (1985).
11. K. Fukuda, H. Ohshima and T. Kondo, J. Colloid. Interface Sci., 123, 442 (1988).
12. K. Fukuda, H. Ohshima and T. Kondo, Colloids and Surfaces, 32, 211 (1988).

# POLYMERIC CONTROLLED RELEASE OF CARDIOVASCULAR DRUGS

Robert J. Levy, M.D.,[a] Steven F. Bolling, M.D.,[a] Rivka Siden, M.Sc.,[a] Alan Kadish, M.D.,[b] Yashwant Pathak, Ph.D.,[c] Parvin Dorostkar, M.D.[a] Amnon Sintov, Ph.D.,[d] Gershon Golomb, Ph.D.,[e] Thomas P. Johnston, Ph.D.[f]

(a) University of Michigan
R5014 Kresge II
Ann Arbor, MI 48109-0576
(b) Northwestern University Medical Center
250 E. Superior
Chicago, IL 60611
(c) South Dakota State University
Box 2202C
Brookings, SD 57007-0197
(d) Perio-Products Ltd.
Box 23950
Jerusalem, Israel, 91232
(e) Hebrew University of Jerusalem
Jerusalem, Israel, 91120
(f) University of Illinois at Chicago
833 S. Wood Street
Chicago, IL 60612

Address for reprints: Robert J. Levy, M.D. University of Michigan Medical Center Department of Pediatrics R5014 Kresge II 0576 Ann Arbor, MI 48109 (313 936-2850, FAX 313 936-9470)

Controlled release drug-polymer implants have been useful for a variety of cardiovascular disease processes. Implanting a controlled release system at the site of a pathologic focus results in regional drug administration being optimized and the chance for systemic side effects greatly reduced. The general formulations, which have been useful, include nondegradable polymers such as silicone rubbers, polyurethanes, and ethylene:(vinyl acetate). In addition, biodegradable matrices composed of compounds such as high molecular weight polyanhydrides have also been valuable. These polymers may be combined with dispersed or dissolved drugs to form matrix systems, which release contained agents based on either diffusion or biodegradation or both. Furthermore, reservoir configured controlled release systems, with rate limiting membranes and refilling capability, also represent an important therapeutic advantage. Controlled release systems have been demonstrated by our group to be advantageous for the

*Cosmetic and Pharmaceutical Applications of Polymers*
Edited by C.G. Gebelein *et al.*, Plenum Press, New York, 1991

therapy of ventricular arrhythmias, cardiovascular calcification, and prevention of cardiac transplantation rejection. In addition, the work of others has demonstrated controlled release implants to be valuable for preventing prosthetic valve endocarditis and for preserving pacemaker lead electrical function. Future controlled release systems will be based on unique agents, advanced reservoir systems, and feedback-modulation capabilities.

## INTRODUCTION

This paper will examine the use of controlled release drug delivery matrices for cardiovascular diseases. A controlled release matrix consists of a sustained release preparation composed of a base polymer containing a dispersed or dissolved drug. Release of the drug from this polymer matrix can be based either on simple diffusion or hydrolysis due to biological degradation. The advantages of using controlled release implants for cardiovascular disease are that local therapy with a controlled release matrix implant can result in very high regional levels of the desired drug, while systemic exposure can be held to minimally low levels, thereby reducing the possibility of side effects.

A number of specific cardiovascular disease processes have been investigated by our group in terms of therapy with controlled release systems. These include cardiac arrhythmias,[1-3] cardiovascular calcification,[4-6] and cardiac transplant rejection.[7] In addition, the work of others has demonstrated the efficacy of controlled release therapy for preventing bacterial endocarditis on cardiac valve prostheses,[8,9] and prevention of the increase in the electrical pacing threshold of cardiac pacemaker leads.[10,11] This paper will update results and review progress in this field.

## EXPERIMENTAL

### 1. Materials

Polymers used to formulate controlled release matrices consisted of the following: Silicone rubbers; Dow Corning Silastics, were obtained from Dow Corning (Midland, MI). The following polyurethanes were used: Mitralthane MPU5 (Symbion, Denver, CO), Biomer (Ethicon, Summerville, NJ), TecoFlex 80A (Thermedics, Woburn, MASS). Ethylene:(vinyl acetate) was obtained from Dupont (Wilmington, DE). A high molecular weight polyanhydride, poly(sebacic acid):(carboxyphenoxypropane), was obtained from Nova (Baltimore, MD) per Dr. Abraham Domb. Other matrices were formulated from purified rat tail tendon collagen.[12]

Ethanehydroxydiphosphonate was provided by Procter & Gamble (Cincinnati, OH). Lidocaine was obtained from Abbott Laboratories (Chicago, IL). Verapamil, propranolol, and dexamethasone were purchased from Sigma (St. Louis, MO). D-sotalol was provided by Bristol-Myers Squibb (Wallingford, CT). Cyclosporin-A was provided by Sandoz (Basel, Switzerland).

Aluminum salts and ferric salts, as well as all other chemicals, were all American Chemical Society reagent grade purity and were obtained from Mallinkrodt Chemicals (St. Louis, MO).

Glutaraldehyde pretreated bovine pericardium was prepared according to the established methodology of our laboratory,[5] and was stored for at least two weeks prior to use in 0.2% glutaraldehyde, pH 7.4 (0.05 M HEPES).

## 2 Controlled Release Methodology

Silastic controlled release matrices were formulated by levigating the drug of interest in prepolymerized silicone rubber, which was then polymerized according to standard methods of catalysis.[5] Solvent casting procedures were carried out by dispersing or dissolving the drug of interest in a solvent compatible with the base polymer and then casting thin films, which were dried in a vacuum oven at 50°C. Ethylene:(vinyl acetate) was dissolved in dichloroethane, and polyurethanes were dissolved in dimethylacetamide. The high molecular weight polyanhydride was also dissolved in methylene dichloride. Collagen films were cast by dissolving the Type I rat tail tendon collagen in 0.1 M acetic acid. In addition, primary polyurethane polymerization of TecoFlex 80A was carried out using a two component system with a $FeCl_3$ catalyst as previously published.[1] Typical matrix formulations ranged in drug loadings from 20-30%. *In vitro* drug release was carried out under perfect sink conditions at pH 7.4, as previously published.[1-6] *In vivo* dosage was estimated based upon the *in vitro* release data, and the well documented agreement, in general, between *in vitro* and *in vivo* release results.[5,13,14]

## 3 Animal Models

Arrhythmia animal model systems utilized dogs in standard preparations. Several models have been used including ouabain induced ventricular tachycardia,[1] rapid ventricular pacing,[15] and repeated hourly ten-minute occlusions of the left anterior descending coronary artery.[16] Ventricular tachycardia (VT) was defined as the occurrence of 3 or more sequential ventricular premature beats. The efficacy of controlled release therapy of ventricular tachycardia was documented by continuous recordings of the electrophysiologic data on a Hewlett Packard Physiologic Recorder, and 8-channel Analogue Tape Deck (Hewlett Packard, Philadelphia, PA).

Calcification and the efficacy of anticalcification agents were assessed, using either the rat subdermal model of bioprosthetic heart valve calcification,[4-6] or tricuspid valve replacements in sheep.[17] Explant calcification was evaluated both by atomic absorption spectroscopy for calcium,[4-6] or specific elemental localization of calcium and phosphorous.[8]

Heart transplantation was carried out using the rat heterotopic model of cardiac transplantation,[7] transplanting the heart of a Brown Norway rat to a White Lewis rat, and using all male animals weighing 250-300 grams. Surgical techniques have been previously published.[19]

# RESULTS AND DISCUSSION

## 1 Controlled Release Therapy of Cardiac Arrhythmias

The general format for these studies involved placing the controlled release matrix directly on the surface of the heart in an open chest

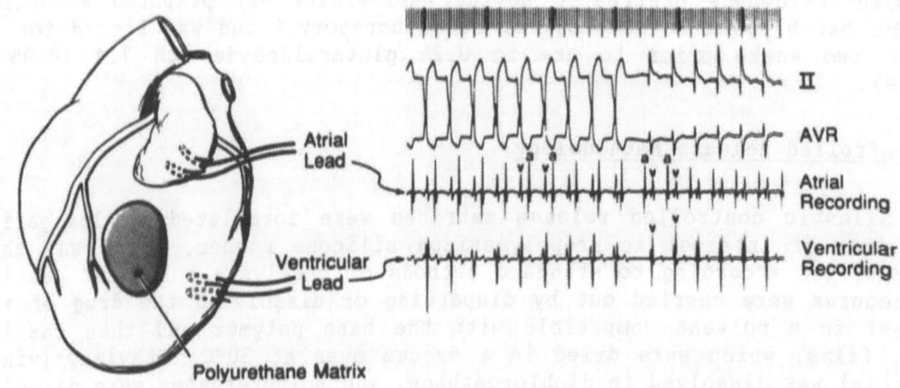


Figure 1. An example of the placement of a controlled release matrix on the left ventricular cardiac surface to treat ouabain induced ventricular tachycardia successfully.[1] Reproduced with permission (Elsevier, North Holland) from Sintov, et al.[1]

canine preparation (Figure 1). The significant advantage of this approach with respect to higher coronary venous blood levels compared to systemic blood levels is apparent in Figure 2. In addition, the animals administered lidocaine by this route of administration experienced no significant effects on cardiac function.[3]

Controlled release antiarrhythmia therapy for preventing ischemia-induced ventricular tachycardia (VT) has been successfully achieved with virtually all of the classical types of antiarrhythmic agents according to the Vaughan-Williams Classification (Table 1).[20] As can be seen (Table 1), thus far, the calcium channel blockers (verapamil) and Type III

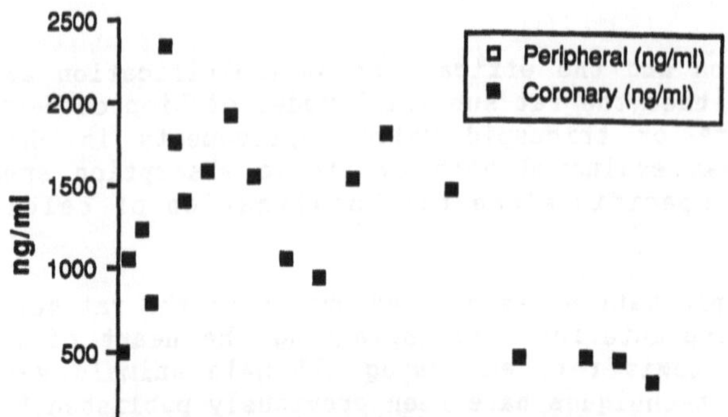


Figure 2. Cardiac controlled release drug delivery of lidocaine resulting in a markedly enhanced coronary lidocaine concentration compared to simultaneous peripheral sampling.[2] Figure reproduced from Sintov, et al., with permission of the publisher (Elsevier, North Holland).[2]

Table 1. Antiarrhythmia controlled release therapy: results of acute ischemia-ventricular tachycardia (VT) studies in dogs

| Drug | Type* | Mechanism | Polymer | Loading | Dose (mg/kg/2hr) | N | VT (per/min.) |
|---|---|---|---|---|---|---|---|
| Lidocaine | I | Na Channel Blockade | Tecoflex | 28% | 0.23 | 5 | 0.6 ±0.2 |
| Propranolol | II | Beta Blocker | MPU-5 | 30% | 0.14 | 6 | 1.22±0.12 |
| D-Sotalol | III | Delayed Repolarization | MPU-5 | 30% | 0.20 | 9 | 0.46±0.11 |
| Verapamil | IV | Calcium Channel Blocker | MPU-5 | 30% | 0.30 | 11 | 0.10±0.03 |
| No Therapy | - | - | Control | - | - | 9 | 1.10±0.30 |

*Vaughan-Williams Classification.[20]

agents (D-sotalol) were the most effective for arrhythmias due to acute myocardial ischemia. Type I agents, such as lidocaine, are less effective. In addition, the β-blocking agent, propranolol, had no significant effect for preventing ventricular arrhythmias due to acute ischemia.

Future considerations, for controlled release arrhythmia therapy, will include the development of feedback electrocardiogram-sensing systems which eventually will be coupled with modulation schemes to regulate antiarrhythmia therapy according to disease activity. In addition, it is expected that customized rather than conventional agents, will be most desirable for direct cardiac antiarrhythmia therapy.

## 2. Prevention of Cardiovascular Calcification with Controlled Release

Calcification of native heart valves, blood vessels, and cardiovascular prostheses represents an important disease process for which there is presently no means for preventing or treating. In particular, calcification is the most frequent cause of the clinical failure of cardiac valve bioprostheses composed of stent-mounted, glutaraldehyde pretreated porcine aortic valves or bovine pericardium.[21] Several useful animal models of bioprosthetic heart valve calcification, including both subdermal [4-6] and circulatory implants[17] (see above), have been used in our studies of controlled release therapy for preventing cardiovascular calcification.

The advantage of controlled release therapy for preventing bioprosthetic heart valve calcification is that the side effects of the agents used, such as diphosphonates, can be avoided, particularly with respect to bone toxicity. A number of other strategies are being looked at with respect to preventing cardiovascular implant calcification, most notably biomaterial pretreatments with either detergents or specific anticalcification agents. It is conceivable that controlled release could be advantageously coupled with any of these other approaches.

As can be seen in Table 2, a number of successful drug-polymer matrix combinations have been demonstrated to be efficacious for preventing bioprosthetic heart valve calcification in the rat subdermal model.[4-6,22] Of note in these results, is the potential long term duration of controlled release possible with the relatively less soluble calcium salt of

Table 2. Controlled release anticalcification therapy: rat subdermal model results (21 days)

| Agent | Polymer | Loading | Dose (mg/kg/24hr) | N | Dur.* (days) | Calcium (ug/mg) |
|---|---|---|---|---|---|---|
| $Na_2EHDP$* | Ethylene Vinyl-Acetate | 20% | $10^{-3}$ | 3 | 21 | 2.5±0.04 |
| $Na_2EHDP$ | Silastic 382 | 30% | $10^{-3}$ | 10 | 6 | 2.3±0.05 |
| CaEHDP | Silastic 382 | 30% | $10^{-6}$ | 10 | 60 | 34.7±14.5 |
| $FeCl_3$ | Silastic 6605 | 10% | 0.63 | 10 | 21 | 14.3± 2.9 |
| $AlCl_3$ | Silastic 6605 | 10% | 0.56 | 10 | 21 | 14.3± 5.7 |
| Unimplanted | Control | - | - | 3 | 21 | 88.0± 3.1 |
| | Control | - | - | 10 | 60 | 160.0± 9.1 |
| No Therapy | - | - | - | 10 | - | 1.0± 0.1 |

* EHDP, ethanehydroxydiphosphonate

ethanehydroxydiphosphonate (EHDP). The estimated extrapolated release duration of this matrix system would be more than fifty years. In addition, a preliminary report has demonstrated that combined Na-Ca EHDP controlled release matrices were effective for inhibiting calcification in sheep tricuspid bioprosthetic heart valve replacements.[17]

Future studies will focus on the efficacy of preloaded bioprosthetic tissue, coupled with controlled release. In addition, recent results from our laboratory have demonstrated synergism between several of the agents, most notably between ferric chloride[22] and EHDP.[5] Finally, reservoir configured controlled release systems have been shown to be effective with EHDP[6] and offer the advantage of reloading with drug, therefore will also be the subject of further research.

## 3. Cardiac Transplantation: Prevention of Rejection with Regional Controlled Release

Conventional immunosuppressive therapy for organ transplants with systemic administration of agents such as cyclosporin-A, has significant side effects,[23-25] including increased susceptibility to overwhelming

Table 3. Controlled release local cardiac immunosuppression for prevention of rat heterotopic heart transplant rejection

| Drug | Polymer | Loading | Dose (mg/kg) | N | Time to Rejection (days) |
|---|---|---|---|---|---|
| Cyclosporine A (cardiac) | Type I Collagen | 20% | 1.0 | 6 | 17.1±0.6 |
| Cyclosporine A (subcutaneous) | Type I | 20% | 1.0 | 6 | 8.2±0.6 |
| Dexamethasone (cardiac) | Polyanhydride | 0.2% | 0.07 | 6 | 10.2±0.06 |
| No Therapy | Control | - | - | 10 | 6.9±0.03 |

infection as well as toxicity to other organs such as the liver and kidney. The results of our studies, shown in Table 3, have demonstrated that local cardiac immunosuppression, with either cyclosporin-A or dexamethasone, successfully prevented cardiac transplant rejection in a rat model at doses which would be ineffective when given systemically.

Ideal agents, for use in regional cardiac transplant immunosuppression, should be agents which cannot be given by any other route. These might include antibodies to important cytokines such as tumor necrosis factor[26] or other potent agents, such as prostaglandin $E_2$, which cannot be given systemically due to side effects. In addition, the reservoir approach just mentioned with respect to preventing calcification with controlled release therapy, could also be relevant for continued immunosuppression of cardiac or other organ transplants.

## 4. Other Uses of Cardiovascular Controlled Release

Prevention of cardiac valve prosthesis bacterial endocarditis with controlled release antibiotic therapy has been demonstrated experimentally[8] and this approach has been successful clinically as well.[9] Furthermore, a dexamethasone eluting cardiac pacing lead has been shown to successfully prevent the untoward rise in cardiac electrical pacing thresholds due to the local inflammatory response after implantation.[10,11] Furthermore, Edelman and his colleagues have recently demonstrated transarterial controlled release for preventing the progression in the events in experimental atherogenesis.[27]

## CONCLUSION

Cardiovascular controlled release through the use of drug-polymer matrices implanted at the site of various disease processes, has been shown to be a powerful general approach. This route of administration has been demonstrated to be valuable for treating life threatening cardiac arrhythmias, preventing cardiovascular calcification, enhancing cardiac transplant survival, preventing bacterial endocarditis, and optimizing the function of cardiac pacing leads.

## ACKNOWLEDGMENT

The authors thank Mrs. Catherine Wongstrom for preparing this manuscript. This work was supported, in part, by an NHLBI grant, HL 41663. Dr. Levy is an Established Investigator of the American Heart Association.

## REFERENCES

1. A. Sintov, W. Scott, M. Dick & R. J. Levy, J. Cont. Rel., 8, 157 (1988).
2. A. Sintov, R. Siden & R. J. Levy, J. Chromatog. Biomed. App., **496**, 335 (1989).
3. A. Sintov, W. A. Scott, K. P. Gallagher & R. J. Levy, Pharm. Res, **7**, 28 (1990).

4. R. J. Levy, J. Wolfrum, F. J. Schoen, M. A. Hawley, S. A. Lund & R. Langer, Science, 228, 190 (1985).
5. G. Golomb, M. Dixon, M. S. Smith, F. J. Schoen & R. J. Levy, J. Pharm. Sci, 76, 271 (1987).
6. T. P. Johnston, J. A. Boyd, B. L. Ciesliga, F. J. Schoen, G. Amidon & R. J. Levy, Int. J. Pharm, 59, 95 (1990).
7. S. F. Bolling, H. Lin, J. A. Boyd, K. P. Gallagher & R. J. Levy, J. Heart Transpl. In Press.
8. L. S. Olanoff, J. M. Anderson & R. D. Jones, Trans. Am. Soc. Artif. Intern. Organs, 25, 334 (1979).
9. M. Karck, F. Siclari, H. Wahligh, U. Sperling, C. Schmid & A. Haverich, Eur. J. Cardiothorac. Surg., 4, 142 (1990).
10. H. Mond, K. Stokes, J. Holland, L. Griggs, P. Kertes, B. Pate & K. D. Hunt, PACE, 11, 214 (1988).
11. A. S. Radovsky, J. F. Van Vleet, K. B. Stokes & W. A. Tacker, PACE, 11, 1085 (1988).
12. G. Chandrakasan, D. A. Torchia & K. Piez, J. Biol. Chem., 251, 6062 (1976).
13. L. R. Brown, C. L. Wei & R. Langer, J. Pharm. Sci, 72, 1181, (1983).
14. G. Golomb, R. Langer, F. J. Schoen, M. S. Smith, Y. M. Choi & R. J. Levy, J. Cont. Rel., 4, 181 (1986).
15. A. Sintov, W. A. Scott, R. Siden & R. J. Levy, J. Cardiovasc. Pharm., 16, 812-817 (1990).
16. V. Elharrar, W. E. Gaum & D. P. Zipes, Am. J. Cardiol., 39, 544 (1977).
17. T. P. Johnston, E. L. Bove, S. F. Bolling, J. A. Boyd, B. L. Ciesliga, G. L. Amidon, F. J. Schoen & R. J. Levy, Int. J. Pharm., 52, 139 (1989).
18. F. J. Schoen, R. J. Levy, A. C. Nelson, W. F. Bernhard, A. Nashef & M. Hawley. Lab. Invest., 52, 523 (1985).
19. N. C. Tilney, W. M. Padberg, R. H. A. Lord, D. Araneda, T. B. Strom & J. W. Kupiec-Weglinski, Transplantation, 46, 122S (1988).
20. E. M. Vaughan Williams, J. Mol. Cell. Cardiol. 19, Suppl. 5, 5 (1987).
21. F. J. Schoen, J. L. Kujovich, R. J. Levy & M. S. J. Sutton, in: "*Contemporary Issues in Cardiac Pathology,*" B. F. Waller, Ed., in: "*Cardiovascular Clinics,*" A. N. Brest, Ed. in Chief, F. A. Davis, Philadelphia, 1988, Chapter 13, p. 289.
22. Y. Pathak, J. Boyd, F. J. Schoen & R. J.Levy, Biomaterials, In Press.
23. P. C. DeGroen, A. J. Askamit, J. Pakela, G. S. Forbes & R. A. Krom, N. Engl. J. Med., 317, 361 (1987).
24. B. Ryffel, B. M. Foxwell, A. Gee, B. Greiner, G. Woerly & M. J. Mihatsch, Transplantation, 46, 90S (1988).
25. R. C. Starling & R. J. Cody, Am. J. Cardiology, 65, 106 (1990).
26. H. Lin, S. W. Chensue, R. M. Streiter, D. G. Remick, K. Gallagher & S. F. Bolling, Transplantation. In Press.
27. E. R. Edelman, D. H. Adams & M. J. Karnovksy, Proc. Natl. Acad. Sci. U.S.A., 87, 3773 (1990).

# CONTROLLED RELEASE OF PROTEIN AND VACCINES FROM POLY(ESTER) MICROSPHERES *IN VITRO*

H.T. Wang, H. Palmer, R.J. Linhardt*, D.R. Flanagan and E. Schmitt

Division of Medicinal and Natural Products Chemistry
and
Division of Pharmaceutics
College of Pharmacy
University of Iowa
Iowa City, Iowa 52242

Biodegradable microspheres of poly(L-lactide) and copolymers of lactide and glycolide have been prepared by spray drying. Degradation studies of these microspheres using residual mass measurements, viscometry and gel permeation chromatography indicated the entire mass of polyester matrices was maintained for 10, 30 days and 6-10 months for 50:50 and 85:15 copolymers and poly(L-lactide), respectively. The continuous drop in polymer intrinsic viscosity and molecular weight during hydrolysis suggested that matrix degradation began as soon as these microspheres were placed in the buffer and that their degradation proceeded through random-chain scission. Protein release, using bovine serum albumin microspheres, showed that release from 50:50 copolymer was independent of polymer molecular weight over a range from 31,000 to 93,000. The release was, however, dependent on the polymer composition and BSA loading in the microspheres. A burst-effect was found in the release study for microspheres prepared from copolymers. The identity and integrity of the released protein was confirmed by sodium dodecyl sulfate polyacrylamide gel electrophoresis on the release products of polymeric microspheres containing BSA, FPL-K and EcoBac$^{R}$-Plus vaccines. These results suggested that the release BSA (or vaccine) from polymeric microspheres could be sustained for up to one month.

## INTRODUCTION

Biodegradable polymers have been used in delivery of drugs[1-9] and biologicals.[10-18] The major advantages of biodegradable polymers used in controlled release applications are that they do not require surgical removal from the host after delivery of bioactive agent is complete. Additionally, adverse tissue reactions from implanted polymer may be ameliorated.[19-20]

*Cosmetic and Pharmaceutical Applications of Polymers*
Edited by C.G. Gebelein *et al.*, Plenum Press, New York, 1991

There has been considerable interest in developing new methods to deliver both vaccines and adjuvants.[21-23] A polymeric delivery device incorporating an immunogen in the presence or absence of an adjuvant could sustain the release of a small amount of antigen over a prolonged period of time eliminating the need for boosters. Appropriate presentation of the vaccine to the immune system may also be required, particularly for highly purified antigens prepared by recombinant technology.

Poly(esters) were chosen for study as a vaccine delivery vehicle because they are currently used in suture materials and have demonstrated low toxicity.[24-27] In recent studies we have prepared polyester microspheres by a spray-drying method.[28-30] These microspheres undergo chemical hydrolysis to soluble monomeric or oligomeric units and degrade in four major stages: (1) Polymer hydration, (2) Strength loss caused by breakage of backbone bonds in polymer, (3) Loss of mass integrity, and (4) Solubilization.[30] Since poly(ester) matrix biodegradation generally involves bulk erosion, release takes place prior to matrix biodegradation under diffusion-control. This has created problems in the delivery of drugs where zero-order release kinetics is desirable.[1,4] Precise control of release kinetics, however, may not be as important as delayed or sustained release in the application of these polymers to vaccine delivery.

This research focuses on microspheres prepared from three biodegradable poly(esters) as potential carriers of vaccines. The matrix characteristics during degradation show matrix lifetimes compatible with vaccine applications. *In vitro* bovine serum albumin and vaccine release studies demonstrate sustained release is possible, making these formulations potentially useful for vaccine applications.

## EXPERIMENTAL

### 1. Materials

Poly(DL-lactide:glycolide, 50:50 and 85:15) (Medisorb®, bioresorbable polymers) were obtained from Dupont (Wilmington, DE). Other poly(DL-lactide:glycolide, 50:50) copolymers were obtained from Birmingham Polymers, Inc. (Birmingham, AL) and Henley Chemicals, Inc. (Montvale, NJ). Poly(L-lactide) polymers were obtained from Birmingham Polymers, Inc. and Polysciences, Inc. (Warrington, PA), respectively. Bovine serum albumin (BSA) and Coomassie Brilliant Blue R were from Sigma (St. Louis, MO). Acrylamide was purchased from Boehringer Mannheim Biochemicals (Indianapolis, IN). N,N'-Methylenebisacrylamide (BIS) was from Fisher Chemical Company (Pittsburgh, PA). Sodium dodecyl sulfate (SDS) was from BDH Chemicals Ltd (Poole, England). *Escherichia coli* Bacterin (EcoBac®-Plus) and killed panleukopenia virus (FPL-K) were used as vaccines and were obtained from Solvay Animal Health, Inc. (Mendota Heights, MN). All other chemicals and solvents were reagent grade and used without further purification.

### 2. Preparation of Microspheres

Polymeric microspheres, including poly(DL-lactide:glycolide, 50:50 or 85:15), were prepared by dissolving 10 g of polymer in 500 mL of methylene chloride, and were spray-dried at 37°C according to the methods described previously.[30] Spray-dried BSA (100-1000 mg), with a particle size distribution; <0.7 µm (69%), 0.7-1.0 µm (12%), 1.0-2.0 µm (4%), 2.0-3.0 µm (4%), 3.0-4.0 µm (4%) and >4 µm (7%), was suspended in a methylene chloride solution of polymer (2%, w/v), 500 mL) using sonification, and this BSA suspension was subsequently spray-dried at 37°C. Vaccine containing polymer microspheres were prepared by dissolving poly(DL-lactide: glycolide, 85:15, 4.5 g) in methylene chloride (250 mL). Lyophilized vaccine (0.5 g) was then added and homogenized using a handmixer (ESGE Handmixer M122, Biospec Products, Bartlesville, OK) for 5 minutes and then spray-dried at 37°C. The FPL-K vaccine was prepared for formulation by dialyzing (molecular weight cut-off, 10-15,000) against 5 mM $Na_2HPO_4$ buffer (pH = 7) to remove most of the salt before lyophilization. The EcoBac$^R$-Plus vaccine was lyophilized in the presence of L-lactose (5 mg/mL).

## 3. Characterization of Polymers and Microspheres

### 3A. Molecular Weight and Intrinsic Viscosity

Polymer molecular weights were determined by gel permeation chromatography (GPC) with 0.5% w/v solution of polymer in methylene chloride. The intrinsic viscosities of polymer methylene chloride solution, at 25°C, were measured by a Cannon-Ubbelode viscometer (Cannon, State College, PA). The details of these measurements were published previously.[30]

### 3B. BSA Content in Microspheres

Microspheres (50-100 mg) were dissolved in $CH_2Cl_2$ (5 mL) and passed through a glass syringe equipped with a solvent resistant filter (2.5 cm in diameter, 0.45 µm in pore size, Millex$^R$-HV, Millipore Products Division, Bedford, MA). Three 5 mL volumes of $CH_2Cl_2$ were used to wash the residual polymer from the filter. Four 5 mL volumes of phosphate buffered saline (PBS, containing 150 mM $Na_2HPO_4$ and 0.9 w/v% NaCl) were passed through the filter to dissolve and remove the retained BSA particles. The buffer washes were collected and the BSA content was determined by Bradford's method.[31] Elemental analysis was also used to determine the BSA or vaccine content of the microspheres. The nitrogen content of BSA (or vaccine), polymer microspheres and protein loaded microspheres were each determined and used to calculate protein loading.

## 4. Degradation of Microspheres

Twelve to fifteen screw-cap test tubes containing microspheres (200 mg) dispersed in PBS (2.5 mL) were prepared and agitated at 37°C and 100 rpm in a G-24 environmental incubator shaker (New Brunswick Scientific Co., Edison, NJ). At various time intervals, one sample tube was removed, and the residual mass was separated, washed with double distilled water and dried under vacuum. The residual mass, polymer molecular weight and intrinsic viscosity were determined.

## 5. *In Vitro* Release Studies

Protein release studies were conducted using poly(DL-lactide: glycolide, 50:50 or 85:15) or poly(L-lactide) microspheres containing BSA, EcoBac$^R$-Plus. FPL-K vaccine was prepared using poly(DL-lactide: glycolide, 85:15). Microspheres (200 mg) were placed in 2.5 mL of PBS (pH = 7.2). Each day the buffer solution was completely removed and the protein content was measured using Bradford's method.[31] Fresh buffer (2.5 mL) was added to replenish that which had been removed.

The protein recovered in the release studies was identified using sodium dodecyl sulfate (SDS) polyacrylamide gel electrophoresis (PAGE). The recovered buffer (1 mL) in the release studies was dialyzed (molecular weight cut-off, 10-15,000) against double distilled water and then frozen and freeze-dried. The sample was dissolved in 50 µL SDS reducing buffer,[32] and heated at 95°C for 5 minutes. A sample of the above buffer (10 µL) was loaded into a SDS-PAGE system, containing stacking gel (4% acrylamide, 0.1% bisacrylamide, 0.01% SDS and 0.05% ammonium persulfate) and separating gel (7.5% acrylamide, 0.2% bisacrylamide, 0.01% SDS, and 0.05% ammonium persulfate).[32] Electrophoresis was performed for 40 minutes at 200 volts in running buffer.[32] The gel was, then, removed from glass plates and stained with a solution of 0.1% Coomassie blue R in fixative (40% methanol/10% acetic acid). Destaining was carried out twice with 50 mL of 40% methanol and 10% acetic acid in water.

## RESULTS AND DISCUSSION

### 1. Characterization of Polymers and Microspheres

The molecular weights and the intrinsic viscosities of a number of poly(DL-lactide-co-glycolide) and poly(L-lactide) polymers obtained from differently commercial sources are shown in Table 1. The poly(esters) used in this study had molecular weights between 31,000 and 93,000.

Microspheres of poly(esters) prepared by spray-drying gave a particle size around 1-2 µm and had a tendency to aggregate.[30] The BSA loadings of microspheres prepared by spray-drying were determined by the Coomassie dye binding assay.[31] The experimentally measured loadings (Table 2) were comparable to the theoretically calculated loadings based on the amount of BSA put into the methylene chloride solution of poly(ester). Nitrogen composition by elemental analysis of BSA loaded microspheres confirmed

Table 1. Properties of various biodegradable polymers

| Polymer[a] | Composition and Manufacturer | | Molecular Weight $M_N$, Da | Intrinsic Viscosity (dL/g) |
|---|---|---|---|---|
| 1. P(DL-LA/GA) | (50/50) | Dupont | 60,000 | 0.34[c] |
| 2. P(DL-LA/GA) | (50/50) | Birmingham | 67,000 | 0.46[c] |
| 3. P(DL-LA/GA) | (50/50) | Henley RG503[b] | 31,000 | 0.4[d] |
| 4. P(DL-LA/GA) | (50/50) | Henley RG504[b] | 42,000 | 0.5[d] |
| 5. P(DL-LA/GA) | (50/50) | Henley RG505[b] | 74,000 | 0.7[d] |
| 6. P(DL-LA/GA) | (50/50) | Henley RG506[b] | 93,000 | 0.8[d] |
| 7. P(DL-LA/GA) | (85/15) | Dupont | 63,000 | 0.43[c] |
| 8. P(L-LA) | | Polysciences | 88,000 | 0.51[c] |
| 9. P(L-LA) | | Birmingham | 95,000 | 1.28[d] |

(a) LA, Lactide; GA, Glycolide.
(b) Resomer[R].
(c) Determined in $CH_2Cl_2$.
(d) Determined in $CHCl_3$ and specified by manufacturer.

| Table 2. Protein loading of spray-dried microspheres | | | |
|---|---|---|---|
| Spray-dried Microspheres | Composition and Manufacturer | Loading (%) Theoretical | Experimental |
| 1. P(DL-LA/GA) | (50/50) Dupont | 10 | 10.37, 8.3[a] |
| 2. P(DL-LA/GA) | (50/50) Dupont | 5 | 5.50 |
| 3. P(DL-LA/GA) | (50/50) Dupont | 1 | 1.02 |
| 4. P(DL-LA/GA) | (50/50) Henley RG503[b] | 5 | 5.44 |
| 5. P(DL-LA/GA) | (50/50) Henley RG504[b] | 5 | 5.44 |
| 6. P(DL-LA/GA) | (50/50) Henley RG505[b] | 5 | 5.50 |
| 7. P(DL-LA/GA) | (50/50) Henley RG506[b] | 5 | 5.18 |
| 8. P(DL-LA/GA) | (50/50) Birmingham | 5 | 6.08 |
| 9. P(DL-LA/GA) | (85/15) Dupont | 10 | 10.75 |
| 10. P(DL-LA/GA) | (85/15) Dupont | 5 | 5.18 |
| 11. P(DL-LA/GA) | (85/15) Dupont | 1 | 1.60 |
| 12. P(L-LA) | Polysciences | 5 | 6.08, 5.7[a] |
| 13. P(L-LA) | Birmingham | 5 | 5.95 |

(a) Determined from elemental analysis.
(b) Resomer[R].

these theoretical and experimental loadings as shown for two systems in Table 2.

## 2. Degradation of Microspheres

Degradation studies *in vitro* were carried out with three types of polymer microspheres, poly(DL-lactide:glycolide, 50:50, (Birmingham Polymers), poly(DL-lactide:glycolide, 85:15, (Dupont) and poly(L-lactide, (Polysciences) possessing intrinsic viscosities of 0.46, 0.43 and 0.51 dl/g corresponding to number average molecular weights of 67, 63 and 88 x $10^3$ Daltons, respectively. Poly(ester) microsphere degradation in PBS was followed by measuring residual mass, polymer intrinsic viscosity and polymer molecular weight (Figures 1-3).

No mass loss could be detected from the 50:50 and 85:15 copolymer microspheres for 10 and 30 days, respectively. However, polymer intrinsic viscosities and molecular weights decreased continuously during this period. The relative molecular weight ratio ($Mn/Mn_o$) dropped faster than the relative intrinsic viscosity ratio($[\eta]/[\eta]_o$). For example, it took 27 days to decrease the intrinsic viscosity of the 85:15 copolymer to half of its initial value, but it only took 17 days to decrease its initial molecular weight by half. After a lag period during which residual mass remained constant, weight loss began and then accelerated. Only 40 and 90 days were required for nearly complete mass loss of the 50:50 and 85:15 copolymers, respectively. In addition, the degradation profiles of residual mass, intrinsic viscosity and molecular weight for poly(DL-lactide: glycolide, 50:50) (Birmingham Polymers) microspheres (number average molecular weight 67,000) were similar to those profiles for microspheres of 50:50 copolymer (Dupont, with a number average molecular weight of 60,000 Daltons).[30]

It should be noted that even though no mass loss was observed in poly(L-lactide) over the time frame of these experiments (Figure 3a), significant changes in both polymer viscosity and molecular weight were

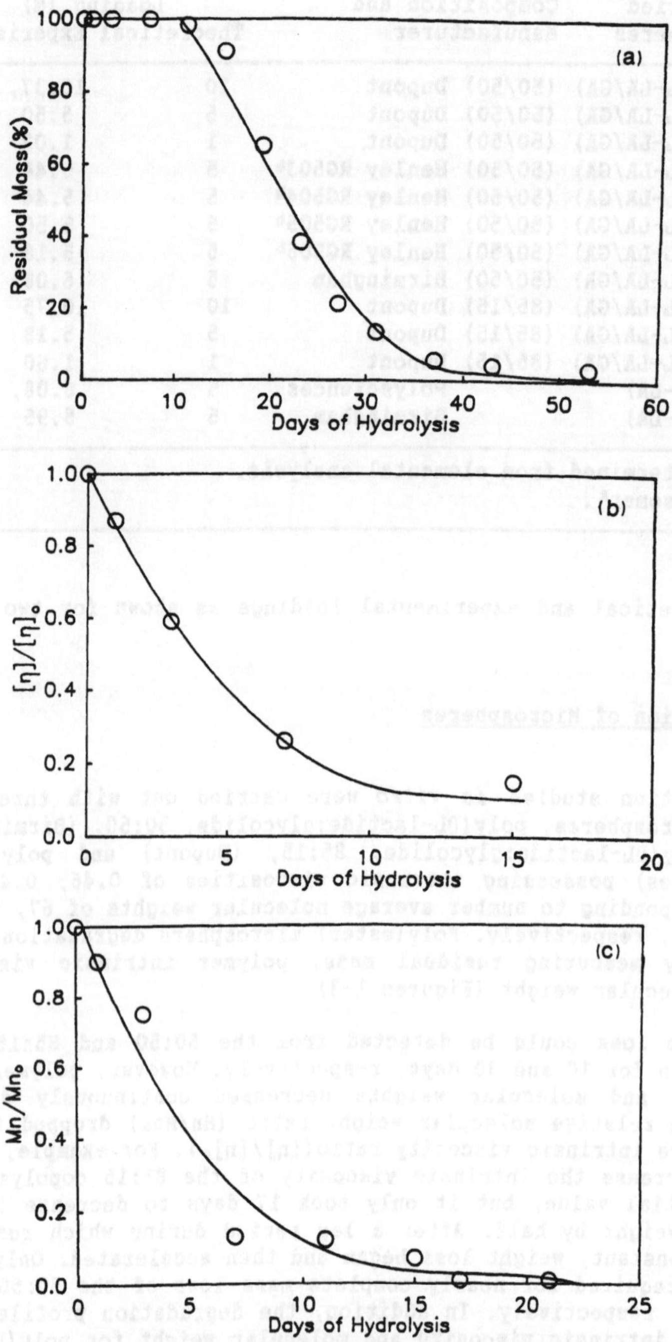


**Figure 1. Degradation studies on poly(DL-lactide:glycolide, 50:50), (Birmingham Polymers) microspheres by various methods: (a) Residual mass (%); (b) ratio of intrinsic viscosity; (c) molecular weight ratio (degraded vs. initial) are shown as a function of time.**

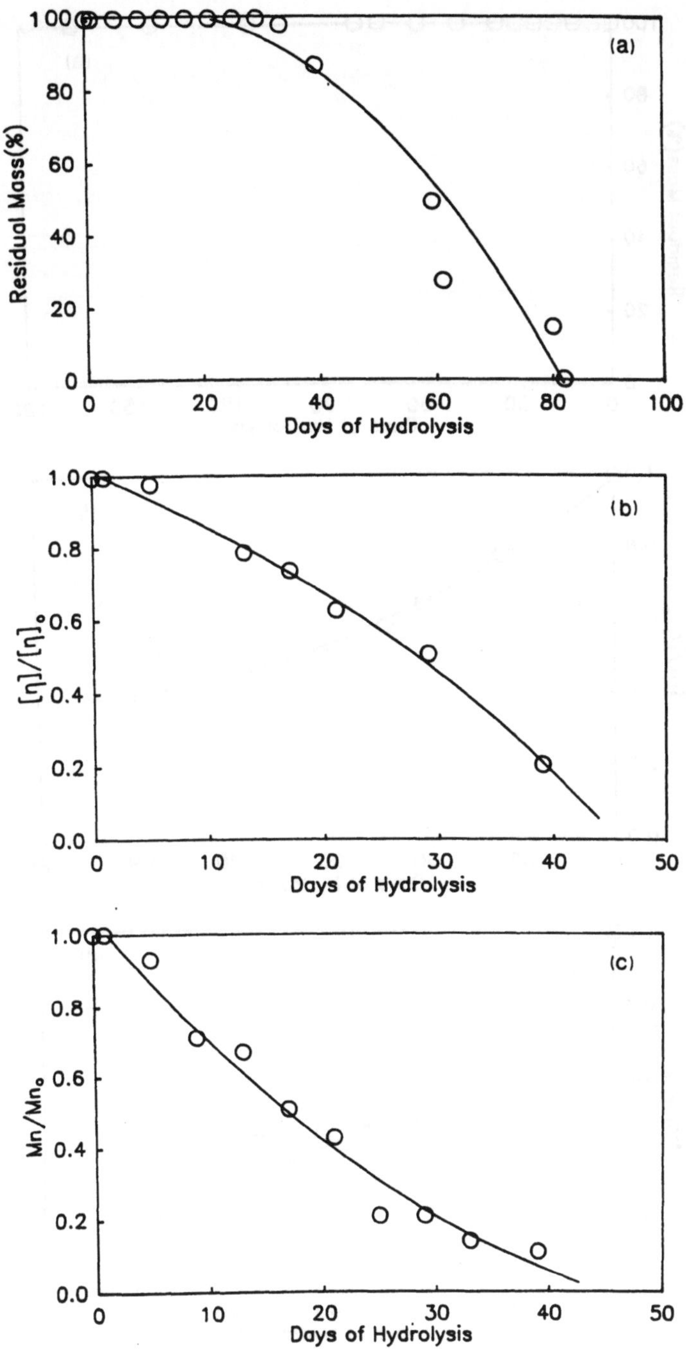


**Figure 2.** Degradation studies on poly(DL-lactide:glycolide, 85:15), (Dupont) microspheres by various methods: (a) Residual mass (%); (b) ratio of intrinsic viscosity; (c) molecular weight ratio (degraded vs initial) are shown as a function of time.

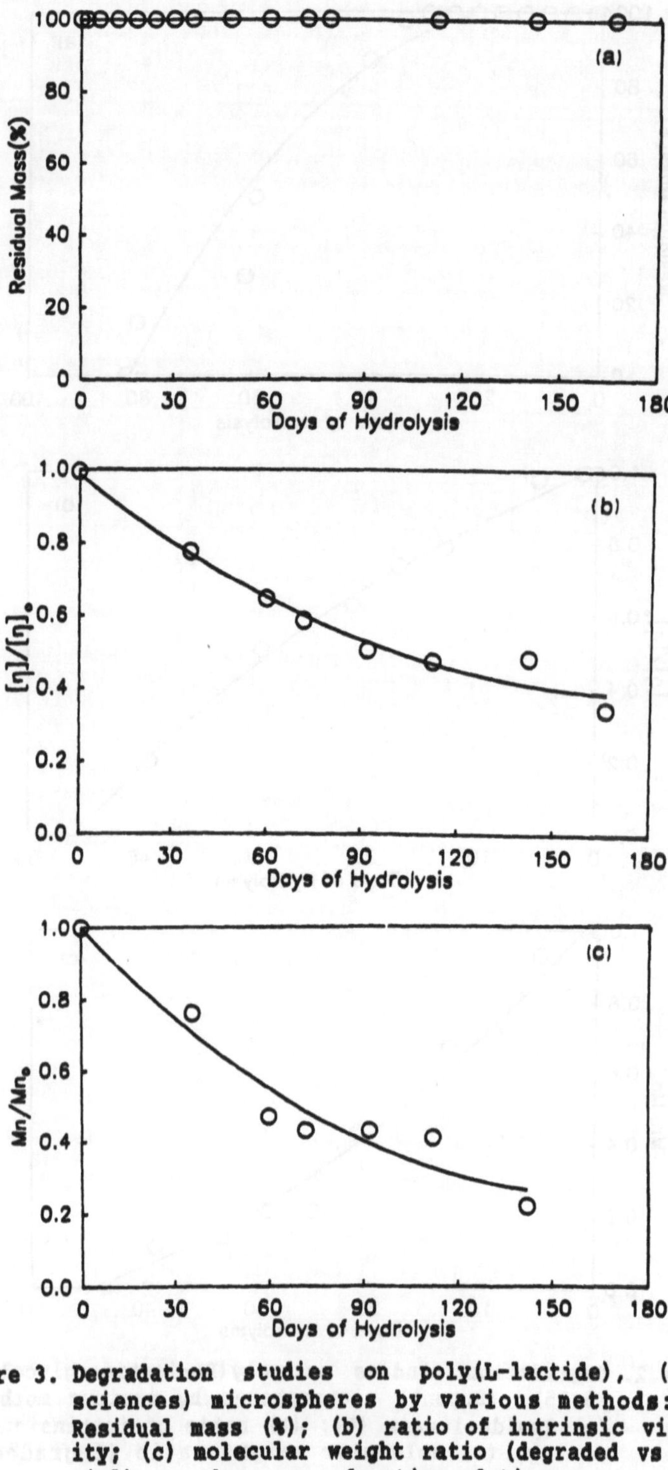


**Figure 3.** Degradation studies on poly(L-lactide), (Polysciences) microspheres by various methods: (a) Residual mass (%); (b) ratio of intrinsic viscosity; (c) molecular weight ratio (degraded vs initial) are shown as a function of time.

detected (Figure 3b & 3c). These results suggest that polymer chain scission begins as soon as the polymer is exposed to PBS, but that even over a period of 180 days, the polymer molecular weight is not reduced sufficiently to produce water-soluble oligomers. Other experiments in our laboratory have shown that poly(L-lactide) microspheres maintained >60% of their initial mass after exposure to PBS at 37°C for 10 months.

These results indicate that the degradation profiles can be separated into three phases. Initially there is a constant residual mass with no formation of soluble monomeric or oligomeric products but a rapid decrease in polymer intrinsic viscosity and molecular weight is observed. This implies a random-chain scission process throughout the bulk of the polymer microspheres. Secondly there is simultaneous loss in residual mass, decrease in polymer intrinsic viscosity and molecular weight, and production of soluble monomeric and oligomeric products.[30] The final phase involve a reduced rate of residual mass loss, probably due to rate-determining dissolution of monomeric and oligomeric products.

## 3. *In vitro* Release Studies

*In vitro* release studies of BSA-containing microspheres of poly(DL-lactide:glycolide, 50:50) of molecular weights varying from 31,000 to 93,000 daltons showed that the release of the BSA is not very sensitive to variation in polymer molecular weight (Figure 4). The time at which half of the loaded BSA loading had been released was less than 1 day, irrespective of the polymer molecular weight.

The release of BSA from poly(DL-lactide:glycolide, 50:50 or 85:15) microspheres with different BSA loadings (Figures 5a & 6b) indicated that an initial burst release of BSA was observed followed by a linear release period. A significant burst-effect was found for microspheres loaded with 5 and 10% BSA in both polymers. In poly(DL-lactide:glycolide, 50:50) microspheres loaded with 5.5 and 10.37% BSA, half of the BSA dose was released within 1 day. The copolymer microspheres loaded with 5% BSA showed a slightly lower burst-effect than those loaded with 10% BSA. In

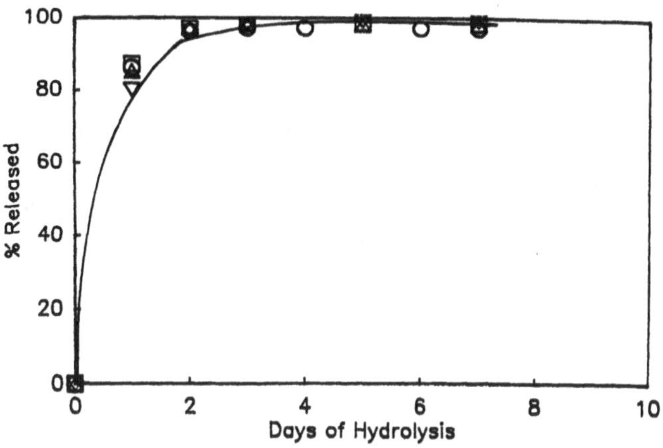


Figure 4. BSA release profiles from poly(DL-lactide:glycolide, 50:50), (Henley Chemicals) microspheres of various molecular weights. Molecular weights: 31,000 (o), 42,000 (△), 74,000 (□) and 93,000 (▽) are shown as a function of time.

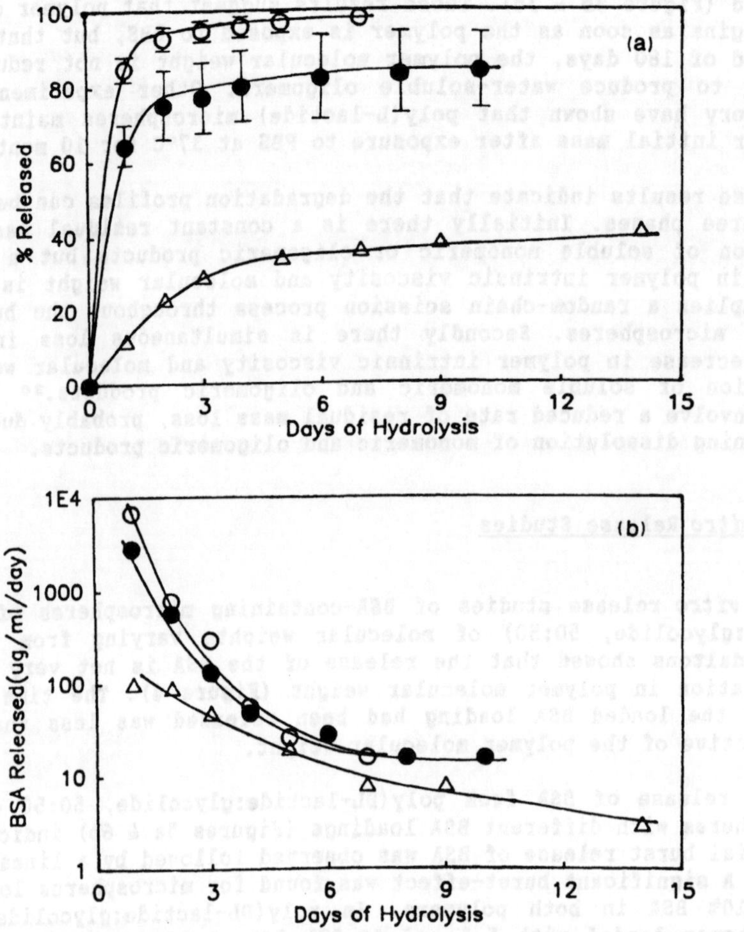


**Figure 5. Effect of BSA loading on protein release from poly-(DL-lactide:glycolide), 50:50, (Dupont): (a) cumulative percents released; (b) daily release from microspheres with 10.37% (o), 5.50% (•), and 1.02% (Δ) (w/w) BSA loading are shown as a function of time.**

**microspheres of the 85:15 copolymer, the difference in release profiles of microspheres with 5 and 10% loading was much less and well within the experimental error. At 1% loading the burst-effect was substantially lower in the 50:50 copolymer formulation and nearly absent in the 85:15 copolymer formulation. These loadings may, however, be too low for application in vaccine delivery. In these experiments, the BSA remaining was confirmed by drying the degraded copolymer microspheres and measuring their BSA content. The summation of the cumulative percent released and the remaining percent BSA resulted in 100% mass balance for the delivery system. The daily release of BSA rapidly decreased over the first 7 days for the 50:50 copolymer and over the first 15 days for the 85:15 copolymer (Figures 5b & 6c), then remained relatively constant at a low rate for the duration of the release studies. Because of the detection limit, of the Bradford assay (5 µg/mL),[31] the daily release of BSA could only be measured for 14 days for the 50:50 copolymer and 30 days for the 85:15 copolymer.**

**The *in vitro* profiles of BSA release from poly(L-lactide) micro-**

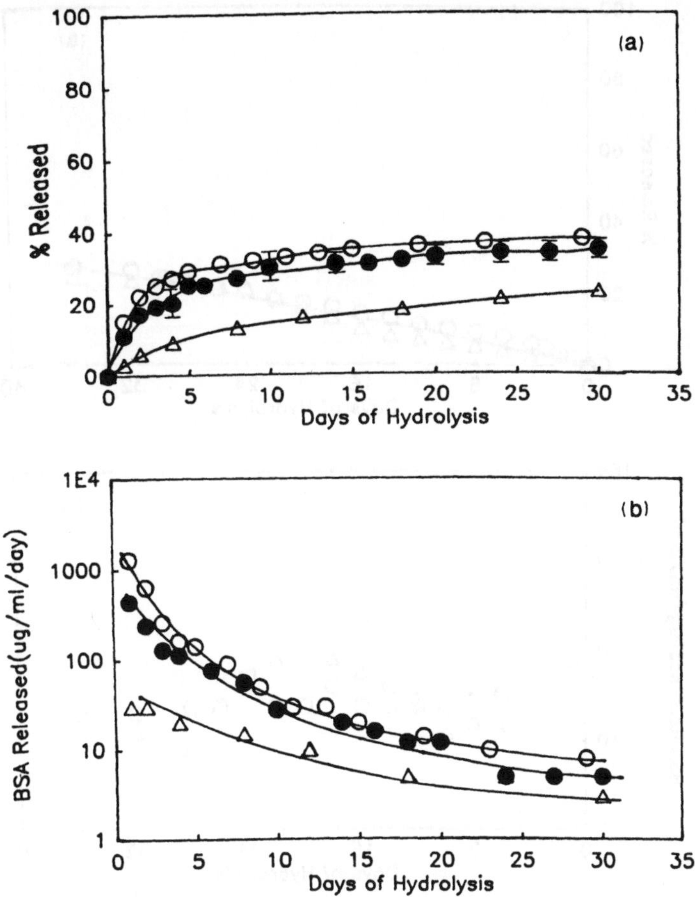


Figure 6. Effect of BSA loading on protein release from poly-(DL-lactide:glycolide, 85:15), (Dupont): (a) cumulative percent released; (b) daily release from microspheres with 10.75% (o), 5.18% (•), and 1.60% (▲) (w/w) BSA loading are shown as a function of time.

spheres (Figure 7a) were nearly linear over the time period studied for both microsphere preparations. A burst effect was not observed. The daily BSA release (Figure 7b) ranged between 10-100 µg level throughout a 36 day period. The different release behavior (i.e., the absence of a burst effect and a nearly linear release profile) between poly(L-lactide) and copolymer of lactide and glycolide may be due to the partially crystalline character of poly(L-lactide). The crystalline domain inside the BSA loaded poly(L-lactide) microspheres is probably slowly degraded compared to the amorphous domain. This may serve to block the diffusional release of BSA from the poly(L-lactide) microspheres.

To confirm the identity of the protein measured by Bradford assay,[31] *in vitro* release of BSA (5% loading) from poly(DL-lactide:glycolide, 50:50) microspheres was followed using SDS-PAGE (Figure 8a). The last two lanes corresponding to the first and third day's release, reflected the large burst-effect observed using the Bradford assay (Figure 5). The first lane was loaded with a 20 µg BSA standard. The BSA release could be confirmed throughout the entire 27 day period. Quantification of BSA using this method is probably not reliable because of the precipitation

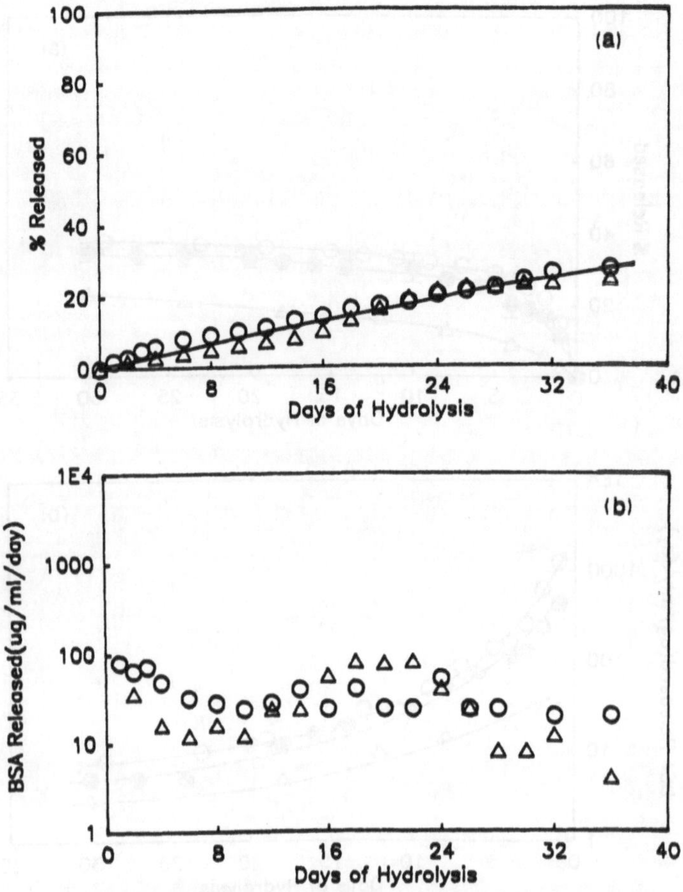


Figure 7. BSA released from poly(L-lactide) microspheres: (a) cumulative percent released; (b) daily release from microspheres prepared with polymer from Polysciences (o); Birmingham Polymers (Δ) as a function of time.

of protein that occurs during dialysis and reconstitution.

SDS-PAGE was also applied to the release studies of copolymer microspheres containing FPL-K and EcoBac®-Plus (Figures 8b & c). These two vaccines contained BSA, hence, the BSA band in the SDS-PAGE gel could be regarded as a marker to establish overall release of the vaccine. A similar burst effect was observed in the SDS-PAGE gel during the first three days of the release study on FPL-K (Figure 8b), while no burst effect was detected for the microspheres containing EcoBac®-Plus (Figure 8c). The reason for this difference might be due to the low concentration of BSA contained in the EcoBac®-Plus vaccine. The SDS-PAGE results (Figures 8b & c) suggested that release of vaccine from poly(DL-lactide: glycolide, 85:15) microspheres could be sustained for up to 33 days.

## CONCLUSIONS

Degradation profiles of poly(ester) microspheres prepared by spray drying showed a lag period during which the residual mass remained con-

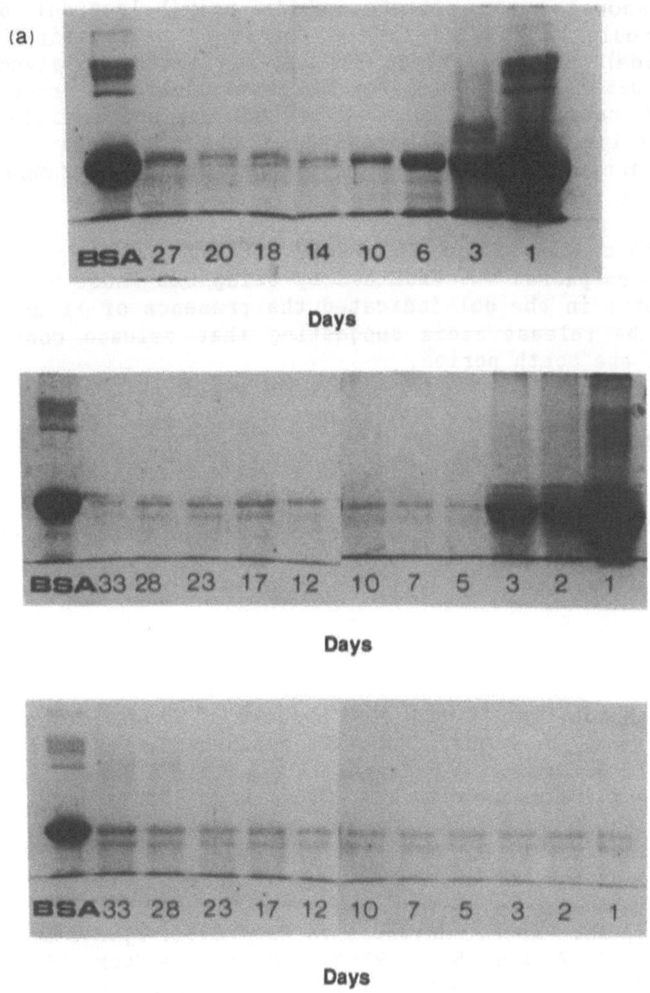


**Figure 8.** SDS-PAGE analysis of protein released from formulated microspheres: (a) BSA-poly(DL-lactide:glycolide, 50:50), (Dupont) (b) FPL-K-poly(DL-lactide: glycolide, 85:15), (Dupont) (c) EcoBac[R]-Plus-poly-(DL-lactide:glycolide, 85:15), (Dupont) microspheres are shown as a function of time.

stant. The length of this period was a function of monomer structure and copolymer ratio, ranging from 10 to 30 days for 50:50 and 85:15 copolymers to 6-10 months for poly(L-lactide). The decrease in polymer intrinsic viscosity and molecular weight during this period suggests that the hydrolysis of microspheres proceeds by random-chain scission.

The release profiles of BSA from microspheres prepared from poly(DL-lactide:glycolide, 50:50) having molecular weights, 31,000, 42,000, 74,000 and 93,000, were identical. This indicated that variation in molecular weight of polymers between 31,000 and 93,000 would not change the release characteristics of BSA loaded polymeric microspheres.

The monomer structure, copolymer ratio and BSA loading of the polymeric microspheres gave different release curves. The 50:50 and the 85:15 copolymers showed burst effects, while poly(L-lactide) did not. This difference could be related to the partially crystalline nature of the poly(L-lactide). The difference in poly(DL-lactide-co-glycolide) copolymer ratio (50:50 versus 85:15) for BSA incorporated microspheres resulted in different rates of degradation in PBS and hence different rates of release. The increase in BSA loading of the microspheres increased the volume fraction of BSA, decreasing the protective entrapment of BSA and increasing the rate of BSA release.

The protein (BSA, FPL-K and EcoBac$^R$-Plus) release from incorporated polymeric microspheres was examined by using SDS-PAGE. The BSA band (released protein) in the gel indicated the presence of >1 μg quantities of protein in the release media suggesting that release continued to take place over a one month period.

## ACKNOWLEDGMENT

The authors are grateful for Solvay Animal Health Inc. for their generous support of this research.

## REFERENCES


1. R. J. Linhardt, in: "*Controlled Release of Drugs: Polymers and Aggregate Systems,*" M. Rosoff, Ed., VCH Publishers, New York, 1989, Chapter 2, p. 53.
2. K. Juni & M. Nakano, Crit. Rev. Ther. Drug Carrier Sys., 3, 209 (1987).
3. J. P. Kitchell & D. L. Wise, in: "*Methods in Enzymology,*" Vol. 112, K. J. Widder & R. Green, Eds., Academic Press, Orlando, FL, 1985, p. 436.
4. J. Heller, in: "*Recent Advances in Drug Delivery Systems,*" J. M. Anderson & S. W. Kim, Eds., Plenum Press, New York, 1984, p. 101.
5. T. R. Tice & D. R. Cowsar, Pharm. Tech., 8(11), 26 (1984).
6. R. S. Langer & N. A. Peppas, Biomaterials, 2, 201 (1981).
7. D. A. Wood, Intel. J. Pharm., 7, 1 (1980).
8. S. Yolles & M. F. Sartori, in: "*Drug Delivery Systems,*" R. L. Juliano, Ed., Oxford University Press, New York, 1980, p. 84.
9. A. Schindler, R. Jeffcoat, G. L. Kimmel, C. G. Pitt, M. E. Wall, & R. Zweidinger, in: "*Contemporary Topics in Polymer Science,*" Vol. 2, E. M. Pearce & J. R. Schaeffgen, Eds., Plenum Press, New York, 1977, p. 251.
10. N. Marcotte, A. Polk & M. F. A. Goosen, J. Pharm. Sci., 79(5), 407 (1990).
11. Y. Tabata & Y. Ikada, Pharm. Res., 6(4), 296 (1989).
12. Y. Ogawa, H. Okada, M. Yamamoto & T. Shimamoto, Chem. Pharm. Bull., 36(7), 2576 (1988).
13. H. V. Maulding, J. Controlled Release, 6, 167 (1987).
14. J. Kohn, S. M. Niemi, E. C. Albert, J. C. Murphy, R. Langer & J. G. Fox, J. Immunol. Methods, 95, 31 (1986).
15. F. G. Hutchinson & B. J. A. Furr, Biochem. Soc. Trans., 13, 520 (1985).
16. L. M. Sanders, J. S. Kent, G. I. McRae, B. H. Vickery, T. R. Tice & D. H. Lewis, J. Pharm. Sci., 73(9), 1294 (1984).
17. T. M. S. Chang, J. Bioeng., 1, 25 (1976).
18. D. H. Carter, M. Luttinger & D. L. Gardner, J. Controlled Release, 8, 15 (1988).

19. M. Vert, P. Christel, F. Chabot & J. Leray, in: "*Macromolecular Biomaterials*," G. W. Hastings & P. Ducheyne, Eds., CRC Press, Boca Raton, FL, Chapter 6, 1984, p.119.
20. J. H. Ratcliffe, I. M. Hunneyball, A. Smith, C. G. Wilson & S. S. Davis, J. Pharm. Pharmacol., 36, 431 (1984).
21. J. Kreuter & E. Liehl, J. Pharm. Sci., 70(4), 367 (1981).
22. I. Preis & R. S. Langer, J. Immunol. Methods, **28**, 193 (1979).
23. J. Kreuter, R. Mauler, H. Gruschkau & P. Speiser, Expl. Cell Biol., **44**, 12 (1976).
24. D. Wasserman & A. J. Levy, Ger. Offen. 2,406,539, Aug. 14, (1975).
25. Anon, Chem. Ind., 905 (1970).
26. A. K. Schneider, Fr. Patent 1,478,694, April 28, (1976).
27. E. E. Schmitt & R. A. Polistina, US Patent 3,297,033, Jan. 10, 1967.
28. R. J. Linhardt, D. R. Flanagan, E. Schmitt & H. T. Wang, Polymer Preprints, 30(1), 464 (1989).
29. R. J. Linhardt, D. R. Flanagan, E. Schmitt & H. T. Wang, Polymer Preprints, 31(1), 249 (1990).
30. H. T. Wang, H. Palmer, R. J. Linhardt, D. R. Flanagan & E. Schmitt, Biomaterials, 11, 679 (990).
31. M. Bradford, Anal. Biochem., 72, 248 (1976).
32. Anon, in: "*Mini-protean$^R$ II Dual Slab Cell Instruction Manual*," Bio-Rad Laboratories, Richmond, CA, 1987, p.24.




# TREATMENT OF YEAST INFECTIONS EMPLOYING ORGANOTIN-CONTAINING MODIFIED POLY(VINYL ALCOHOL)

Charles E. Carraher, Jr., Cynthia W. Butler and Lisa Reckleben

Florida Atlantic University
Departments of Chemistry and Biological Sciences
Boca Raton, FL 33431

Organotin-containing modified poly(vinyl alcohol) effectively controls the growth of *Candida albicans*, the microorganism responsible for yeast infections. Other gram negative and gram positive bacteria are either unaffected or less severely inhibited. Thus, the inhibition of *Candida albicans* is selective.

## INTRODUCTION

Candida is a yeast that has several species, any one of which is capable of causing candidiasis. The organism is a component of the normal flora of skin and mucous membranes as well as the gastrointestinal tract. It gains its foothold either during the birth process or shortly after. Thus, the organism presents an ever present risk of becoming pathogenic when conditions are favorable. Infections are most often of the skin or mucous membranes. Occasionally, life-threatening systemic candidiasis occurs. Of all the species of Candida, *C. albicans* causes the most infection followed closely by *C. tropicalis*.

Those individuals most susceptible to this organisms pathogenicity are women on oral contraceptives and individuals taking broad spectrum antibiotics. Other conditions which increase susceptibility to candidiasis are: malignancies, pregnancy, diabetes mellitus, tight undergarments, obesity, corticosteroid therapy, blood dyscrasias and immunologic deficiencies. Newborn infants are also victims of this organism if vaginal candidiasis is present during the birth process. Infants often experience oral thrush or long and painful bouts of diaper rash.

The ability of cells of *C. albicans* to attach to epithelial cell membranes through a ligand-receptor interaction is associated with its pathogenicity. The yeast cells also exhibit a remarkable tenacity for plastic surfaces which enable them to colonize plastic catheters and prosthetic devices. These properties make Candida species an opportunistic organism whenever a patient's natural ability to control invasion is compromised. Patients with AIDS often contract mucosal candidiasis.

*Cosmetic and Pharmaceutical Applications of Polymers*
Edited by C.G. Gebelein *et al.*, Plenum Press, New York, 1991

Of the three categories of candidiasis: cutaneous, systemic and chronic mucocutaneous, we are most concerned with the first and last.

## EXPERIMENTAL

Commercial PVA (99% hydrolyzed, Aldrich, Milwaukee, WI) was employed along with the organostannane halides (also from Aldrich, Milwaukee, WI) as received.

Modification procedures and equipment are similar to those described elsewhere.[1-3] Briefly, an aqueous solution of PVA containing sodium hydroxide equal to the theoretical maximum mole amount of chloride available from the organostannane halide was added. Stirring (about 18,000 rpm, no load) was begun and the organic solution (generally carbon tetrachloride or chloroform) containing the organostannane halide was added. The product rapidly precipitated from the reaction mixture, was collected by suction filtration, washed repeatedly with water, transferred to a glass petri dish and dried. When plasticizers were added, they (2 mL) were added to the organic phase prior to the reaction.

Infrared spectra were obtained using films employing a Nicolet 5DX-FTIR. Mass spectral analysis was carried out by the Nebraska Mass Spectrometry Service Laboratory. Analysis was carried out using a direct insertion probe connected to a Kratos MS-50 mass spectrometer operating in the E.I. mode, 8 kr acceleration and 10 sec/decade scan rate with a probe temperature to 450°C.

Bacterial studies were conducted in the usual manner. Plates containing a suitable growth medium were seeded with suspensions of the test organism that produced a suitable lawn of test organism after 24 hours incubation at 37°C. Shortly after, the plates were seeded using 1-3 $mm^2$ sections of the test material. The plates were incubated and growth inhibition was noted.

## RESULTS AND DISCUSSION

### 1. Structural

We have been investigating the synthesis and properties of polymeric materials containing metal atoms.[1,2] Much of this work was based on the idea that organometallics, such as organostannane halides, act as electrophiles for attack by Lewis bases, such as hydroxyl-containing compounds.

Some time ago we reported the modification of poly(vinyl alcohol) through the reaction with various organostannane halides.[3] As expected, these products are crosslinked and insoluble when organostannane dihalides are used, but they are linear when the monohalide is employed. Originally only the dibutyl, tributyl, diphenyl and triphenyl stannanes were successfully reacted with the PVA. Here we report the synthesis of analogous products, except where the number of organostannane moieties successfully reacted is greatly increased.

Our initial synthetic procedure was not sufficient to allow the inclusion of many of the organostannane moieties into the PVA chain. Yields were low, with the percentage of tin also low. This procedure employed

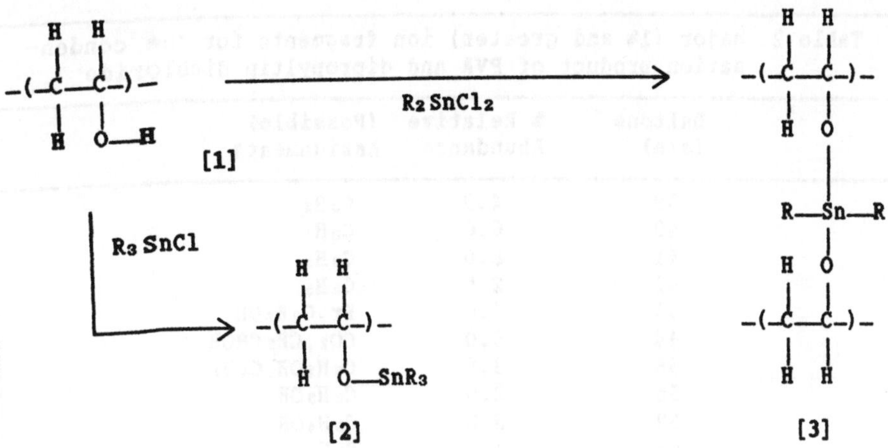

Structures 1 to 3.

diethyl ether as the organic liquid. Here we employed "poorer solvents" as the organic liquids and were able to effect the reaction of organostannane halides, where the R is methyl, ethyl, propyl, etc. Infrared study of the products is consistent with a product containing units such as [1], [2], and [3]. The Sn-O-R asymmetric stretch is found around 670 (all bands are given in $cm^{-1}$) while the Sn-O-R symmetric stretch is found about 600 and 570. For products derived from aliphatic organostannane bands characteristic of the aliphatic portion and C-H stretching in the PVA are present in the 2900 to 3000 region. The presence of unreacted OH groups is indicated by the presence of a broad band about 3300-3700.

The mass spectra of the products are consistent with structures as described by 1-3. Table 1 contains a listing of the natural abundances for isotopes of tin. Table 2 contains the major ions derived from the condensation product of PVA and dipropyltin dichloride. Smaller ion fragments are derived from the propyl moieties and the fragmented PVA backbone. Higher ion fragments (>150 Daltons) have tin present. Table 1 contains a listing of the percentages of the major isotopes naturally found along with absolute and relative ion abundances for the SnPr moiety. There is a good agreement between the percentage natural abundance of the various tin isotopes and that found in the fragmentation of the product. There is a noticeable lack of oxygen-containing low molecular weight ion fragments (<150 Daltons). Presumably the oxygen remains with the tin-containing fragments, though the most abundant ion fragments containing tin are oxygen free (such as SnPr (116-124 Daltons) and $SnPr_2$ (205-207 Daltons)). This is consistent with the observation that for many tin-containing polymers bound through oxygen, tin-containing volatile moieties are absent with the tin remaining in the residue. Additional tin-

Table 1. Percentage natural abundance of tin isotopes

| Tin isotope | %-Natural Abundance | Tin Isotope | %-Natural Abundance |
|---|---|---|---|
| 116 | 14 | 120 | 33 |
| 117 | 7.7 | 122 | 4.7 |
| 118 | 24 | 124 | 6.0 |
| 119 | 8.7 | | |

**Table 2. Major (1% and greater) ion fragments for the condensation product of PVA and dipropyltin dichloride**

| Daltons (m/e) | % Relative Abundance | (Possible) Assignments |
|---|---|---|
| 39 | 4.9 | $C_3H_3$ |
| 40 | 6.6 | $C_3H_4$ |
| 41 | 8.0 | $C_3H_5$ |
| 42 | 2.5 | $C_3H_6$ |
| 43 | 7.0 | Pr, $C_2H_2OH$ |
| 44 | 5.0 | $CO_2$, $CH_2CHOH$ |
| 55 | 3.8 | $C_3H_2OH$, $C_4H_7$ |
| 56 | 2.0 | $C_3H_3OH$ |
| 57 | 3.0 | $C_3H_4OH$ |
| 65 | 1.3 | $C_5H_5$ |
| 66 | 1.4 | $C_5H_6$ |
| 69 | 2.8 | $C_4H_4OH$ |
| 70 | 1.9 | $C_4H_5OH$ |
| 94 | 4.0 | $C_5H_{10}(OH)_2$ |
| 159 | 1.1 | |
| 161 | 1.5 | $OSnMe_2$ |
| 163 | 2.0 | |
| 223 | 1.2 | $C_3H_4OSnPr$ |
| 225 | 2.1 | |
| 241 | 1.1 | $C_4H_6OSnPr$ |

containing ion fragments are found at 221-229 ($CHCH_2CHOSnPr$), 237-245 ($CH_2CHCH_2CHOSnPr$), 260 ($CH_2CHCH_2CHCH_2CHOSnPr$) and 282-287 ($CH_2CHCH_2CHCH_2$-$CHOSnPr$) Daltons. The absence of large amounts of chloride is indicated by the absence of ion fragments at 35 and 37 and the presence of ion fragments at 36 and 38 (HCl), except not in the correct (isotopic) abundance.

Similar results are found for the condensation product of PVA and dimethyltin dichloride. Table 3 contains assignments for the most abundant ion fragments. Again, there are numerous ion fragments containing tin present in the predicted isotope abundance. Tables 4 and 5 contain the results for the SnMe and $OSnMe_2$ units. Higher molecular weight ion fragments are also found at 385-387 (2 units), 403 (2 units + $CH_2$), 429-431 (2 units + $C_3H_5$) and 453-458 (2 units + $C_5H_6$) Daltons. The absence of ion fragments at 35-38 is consistent with the absence of the Cl atom.

## 2. Biological Results

As noted before, yeast infections are among the most common infections of mankind. Some descriptions regarding the nature of yeast infections were given in the introductory section.

Tables 6 and 7 contain biological results emphasizing common microorganisms. In general, almost all of the organotin-containing PVA modified compounds inhibit *C. albicans* to a greater extent than either gram positive (such as *S. aureus*) and gram negative (*P. aeruginosa* and *E. coli*) organisms. Further, in general the monomeric organotin chlorides do not show selectivity in the inhibition of gram negative and gram positive organisms compared with *C. albicans*. A number of the modified PVA com-

Table 3. Major (10% and greater) ion fragments derived from the condensation of dimethyltin dichloride and PVA

| Daltons (m/e) | %-Relative Abundance | (Possible) Assignment |
|---|---|---|
| 39 | 11 | $C_3H_3$ |
| 41 | 16 | $C_3H_5$ |
| 44 | 30 | $CO_2$, $CH_2CHOH$ |
| 91 | 13 | |
| 120 | 12 | Sn |
| 131 | 15 | SnMe |
| 132 | 10 | SnMe |
| 133 | 28 | SnMe |
| 134 | 15 | SnMe |
| 135 | 35 | SnMe |
| 137 | 15 | SnMe |
| 141 | 10 | SnMe |
| 161 | 35 | $OSnMe_2$ |
| 163 | 67 | $OSnMe_2$ |
| 164 | 30 | $OSnMe_2$ |
| 165 | 100 | $OSnMe_2$ |
| 166 | 10 | $OSnMe_2$ |
| 167 | 35 | $OSnMe_2$ |
| 297 | 22 | $U+C_4O_3H_x$ |
| 299 | 21 | $U+C_4O_3H_x$ |
| 326 | 11 | 2U -3Me |
| 327 | 26 | 2U -3Me |
| 328 | 19 | 2U -3Me |
| 330 | 11 | 2U -3Me |
| 333 | 10 | 2U -3Me |

Table 4. Relative abundance of $OSnMe_2$ associated isotopes

| | | | | | | | |
|---|---|---|---|---|---|---|---|
| Sn- | 116 | 117 | 118 | 119 | 120 | 122 | 124 |
| $OSnMe_2$ | 159 | 160 | 161 | 162 | 163 | 165 | 167 |
| %-Relative Abundance | 18 | 7.6 | 24 | 10 | 33 | 3.4 | 4.2 |
| Predicted | 14 | 7.7 | 24 | 8.7 | 33 | 4.7 | 6.0 |

Table 5. Relative abundance of SnMe associated ion fragments

| | | | | | | | |
|---|---|---|---|---|---|---|---|
| Sn- | 116 | 117 | 118 | 119 | 120 | 122 | 124 |
| SnMe | 131 | 132 | 133 | 134 | 135 | 137 | 139 |
| %-Relative Abundance | 12 | 8 | 23 | 12 | 29 | 11 | 5 |
| Predicted | 14 | 7.7 | 24 | 8.7 | 33 | 4.7 | 6 |

Table 6. Zones of inhibition for selected poly(vinyl alcohol) materials containing organostannane

| Tested Material | Zone of Inhibition (cm) *P. auer.*(-) | *S. aureus*(+) | *C. albicans* |
|---|---|---|---|
| PVA/$Pr_3Sn$(1) | 0.4 | 2.5 | NG |
| PVA/$Pr_3Sn$ | 0.3 | 3 | NG |
| PVA/$Pr_3Sn$(3) | 0.2 | 1.7 | 3.4 |
| PVA/$Pr_3Sn$(2) | 0.2 | 0.7 | NG |
| PVA/$Et_3Sn$ | 0.7 | 1.6 | 3.7 |
| PVA/$Me_3Sn$(1) | NI | 0.4 | 3 |
| PVA/$Me_2Sn$(1) | NI | 0.4 | 2 |
| PVA/$Bu_3Sn$ | 0.2 | 4.7 | 3 |
| PVA/$Me_2Sn$ | 0.2 | 1.0 | 1.5 |
| PVA/$Me_3Sn$ | NI | 0.7 | 2.5 |

where: (1) = dialkyl phthalate; (2) = isodocyl diphenylphosphate; (3) = dioctyl sebocate and NI = no inhibition, NG = no growth.

Table 7. Selected biological activities of PVA modified through reaction with organostannane halides

| Tested material | Zone of inhibition (cm) *E. coli*(-) | *P. auer.*(-) | *S. aureus* (+) | *C. albicans* |
|---|---|---|---|---|
| $Me_3Sn$/PVA(3) | NI | NI | NI | 0.7 |
| $Me_3Sn$/PVA(1) | NI | NI | NI | NI |
| $Me_3Sn$/PVA | NI | NI | 0.3 | 1.2 |
| $Me_3Sn$/PVA(2) | NI | NI | 0.6 | T |
| $Me_3SnCl$ | 1.7 | 1.3 | 1.8 | T |
| $Pr_3Sn$/PVA(1) | 0.9 | 0.7 | 3.0 | T |
| $Pr_3Sn$/PVA(3) | 0.7 | 0.8 | 1.7 | T |
| $Pr_3Sn$/PVA(2) | 0.9 | 0.1 | 3.0 | T |
| $Pr_3Sn$/PVA | 0.8 | 0.7 | 3.0 | T |
| $Et_3Sn$/PVA(2) | 1.0 | 0.8 | 1.9 | T |
| $Et_3Sn$/PVA(3) | 0.8 | 0.7 | 0.8 | T |
| $Et_3Sn$/PVA(1) | 0.9 | 0.4 | 0.8 | T |
| $Et_3Sn$/PVA | 2.3 | 0.4 | 2.4 | T |
| $Et_3Sn$/Br | ST | NI | 1.1 | T |
| $Me_2Sn$/PVA | 0.5 | 2.7 | 0.7 | T |
| $Me_2SnCl_2$ | 1.4 | 1.4 | 1.5 | T |
| $Ph_3Sn$/PVA | NI | NI | 0.7 | T |
| $Ph_3SnCl$ | NI | NI | 1.9 | T |
| $Bu_3Sn$/PVA | ST | 0.5 | 2.6 | T |
| $Bu_3SnCl$ | 0.1 | 1.0 | 2.0 | T |

Where: ST = Slight inhibition; NI = No inhibition; T = No growth; (1) = dialkylphthalate, (2) = isodocyl diphenylphosphate, and (3) = dioctyl sebocate.

pounds exhibited total inhibition of *C. albicans* while exhibiting only moderate inhibition of the other organisms.

The value of these results lie in the possible manufacture of a cream or powder to eliminate a vaginal yeast infection without destroying the remaining normal flora of the vaginal vault. Further, the treatment may be one step as compared to the present treatments that are multistep.

## ACKNOWLEDGMENT

We are pleased to acknowledge partial support from ACS-PRF 15508-87.

## REFERENCES

1. Y. Naoshima and C. Carraher, J. Macromol Sci.-Chem., **A25**(8), 895 (1988).
2. D. Siegmann, C. Carraher and A. Friend, J. Polymer Materials, **8**, 19 (1987).
3. C. Carraher and J. Piersman, Angew. Makromole. Chemie, **28**, 153 (1973).

# DRUG CARRIER SYSTEMS BASED ON RESORBABLE POLYESTER COLLAGEN AND/OR BIOMATERIAL COMBINATIONS

A. Stemberger, M. Unkauf, D. E. Arnold and G. Blümel

Institut für Experimentelle Chirurgie
Technische Universität München
Ismaninger Str. 22, D-8000
München 80, Germany

In the treatment of bone and soft tissue infections only minor active agent concentration can be attained at the site of infection with the use of conventional therapy procedures, after oral or parenteral administration of antibiotics. The possibility of administering drugs locally exists during the course of surgery, after the infection focus has been removed. This is also true for the targeted application of drugs by modern surgical techniques such as endoscopy. Beginning with a gentamicin collagen compound, developed for treatment of bone infection osteomyelitis, a sustained release system was contrived. Through shell-like coating of the collagen fibrils of this carrier system with resorbable poly(esters), a sustained liberation system of the incorporated aminoglycoside could be achieved. The liberation kinetics were dependent on the nature and concentration of the aliphatic poly(esters) used as links to the poly(glycolic), poly(lactic) or poly(butyric acid). During the course of this research it could further be shown that drugs can also be placed on medical devices such as artificial joints through use of these carriers in the form of thin films.

## INTRODUCTION

Research in sustained drug release systems used to be focused, as a rule, on orally ingested drugs and rectal or vaginal application systems. To achieve the goal of prolonged serum levels in parenteral administration, low soluble salts or micronized powder were applied.

Within the framework of so-called perioperative infection prophylaxis, antibiotics are administered intraoperatively, but the tissue levels are usually not sufficient to provide effective bacterial concentration. The treatment of infection resulting from the implantation of vascular prostheses is especially problematic. In manifest infections the necessary high tissue levels cannot be achieved through increased antibiotic dosage. Exactly this problematic nature, which exists in soft tissue and bone infection, motivated our research group to take new paths[1]. This especially concerns research to improve the release system of a locally applied gentamicin collagen compound by means of resorbable aliphatic poly(esters).

Additional problems are presented by orthopedic implants, especially the so-called "cementless" prostheses. Contrary to fixation with bone cement containing antibiotics, no local antibiotic prophylaxis exists in these cases. The questions concerning the local application of antitumor drugs in soft tissue and bones are similar. Targeted local administration is possible in this case, especially through the modern surgical endoscopy intervention.

## MATERIALS AND METHODS

For the following studies polymers from poly(lactic acid) and poly(glycolic acid) (Boehringer Ingelheim, Germany) as well as poly(hydroxybutyric acid) (Chemie AG Linz, Austria) were used (see Tables 1 and 2).

These resorbable poly(esters) were dissolved in organic solvents, as described by the manufacturer, prior to mixing with the following drugs and dyes: methylene blue as a model substance (E. Merck, Darmstadt, Germany), mitoxanthron (Novanthrone[R], Lederle, Wolfratshausen, Germany), gentamicin for labaratory use (Sigma Chemie, Germany), collagen sponge (Collatamp[R], Ruhland, Germany) gentamicin-collagen (Sulmicin Implant[R], Essex Chemie, Munich, Germany). Collagen materials, in petri dishes, were coated with the dissolved polymers, evacuated and stored over silica gel.

So-called "cementless prostheses" were coated in a further series of experiments. This type of implant was originally designed for dogs, and coating on the surface areas was calculated as depicted in Table 3.

Elution profiles were performed with isotonic saline or serum. The drugs released were determined by photometrical techniques or with the TDX assay for gentamicin (Abbott diagnostic products, Wiesbaden, Germany).

## RESULTS

The porous structure of the various drug collagen compounds was coated with resorbable or degradable poly(ester) compounds and eluted as a function of time. Gentamicin-collagen sponges, which had been sealed with PHB polymers, showed the best results concerning delayed release. See Figure 1.

Table 1. Poly(lactide)/poly(glycolide) polymers

| Product name of polymer | R104 poly(D,L-lactide) | R203 poly(D,L-lactide) | L104 poly(L-lactide) | RG504 poly(D,L-lactide-co-glycolide) |
|---|---|---|---|---|
| Molecular weight (daltons) | 2,000 | 25,000 | 2,000 | 50,000 |

| Table 2. Poly(hydroxybutyrate)/poly(hydroxyvalerate) polymers | | | | |
|---|---|---|---|---|
| Product name of polymer | PHB-C1 poly(hydroxy-butyrate) | PHB/19%PHV | PHB/10%PHV poly(hydroxybutyrate) with % poly(hydroxyvalerate) | PHB/5%PHV |
| Molecular weight (daltons) | 400,000 | 800,000 | | |

After coating the surface with methylene blue, which had been incorporated into a collagen sponge, a significantly improved retention rate could be observed in the poly(ester)-bound drug, as compared to the corresponding gentamicin compound. These experiments were repeated with mitoxanthrone, and comparable results were obtained. The elution profiles presented, in Figures 2 and 3, show delayed release after coating. Here the mitoxanthrone-collagen sponges, treated with poly(lactide) and poly-(glycolide), result in an efficient retarding effect.

Following the coating of the surfaces of artificial joints, such as the so-called "cementless prosthesis," comparable results were obtained, dependent on the drug and the poly(ester). The retarding effect was especially evident in the use of substances such as methylene blue which are soluble in organic solvents (see Figure 4). Even after three weeks, complete release was not achieved in elution tests, *in vitro*.

In a further series of experiments, gentamicin powder was dispersed in chloroform poly(lactide)/poly(glycolide) solutions and used to coat steel devices. In comparison to the non-sealed gentamicin powder device combinations, where the antibiotic crystals dissolved quickly, delayed release could be achieved through a low molecular (R104) poly(lactide) polymer and a high molecular (R203) poly(lactide) polymer (see Figure 5). In comparison to coatings with soluble drugs, thicker layers of coating are necessary.

## DISCUSSION

As these experiments have shown, systems can be developed which, depending on the coating technique and the polymer used, upon combination with resorbable drug carrier systems such as the collagen-gentamicin compound with resorbable poly(esters), can result in delayed release of the aminoglycosides *in vitro*. In general collagen sponges with incorpo-

| Table 3. Coating scheme | | | |
|---|---|---|---|
| Thickness polymer/surface mg/cm² | Polymer % | Methylene blue % | Gentamicin sulfate % |
| 1.5 | 93.5 | 6.5 | - |
| 4.0 | 97.4 | 2.6 | - |
| 8.0 | 98.7 | 1.3 | - |
| 11.8 | 90.0 | - | 10 |

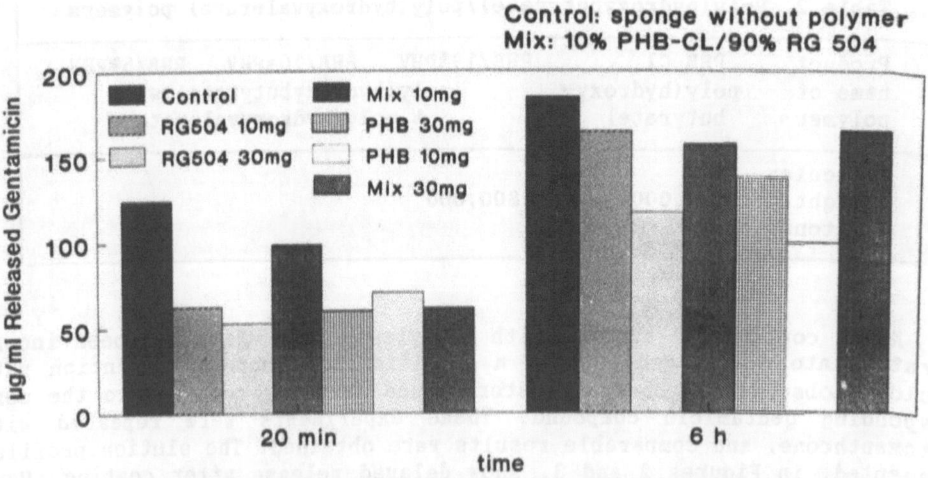


**Figure 1. Gentamicin liberation from gentamicin-collagen sponges (200 mg gentamicin sulfate per 10 x 10 cm) coated with resorbable poly(esters).**

**rated drugs, and subsequently coated, revealed a delayed liberation of the resorbable poly(ester) used. Scanning electron microscopy demonstrated no evidence of change in the collagen fibrils in the sense of embedding.**

**It could then be shown that methylene blue, as opposed to gentamicin, distributes itself homogeneously in the poly(ester) layers. The collagen fibrils were clearly visible, revealing no direct encapsulation. This phenomenon is due to the solubility of methylene blue in the solvent of resorbable poly(esters).**

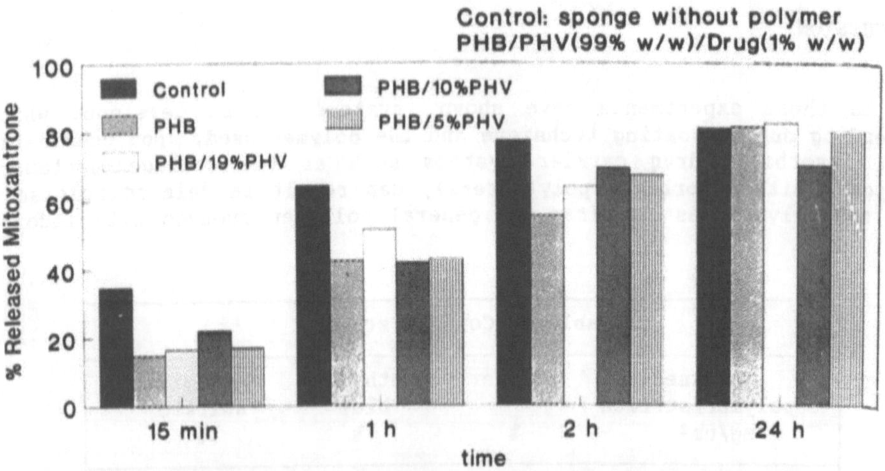


**Figure 2. Mitoxanthrone liberation from mitoxanthrone-collagen sponges (for details see Materials and Methods section) coated with resorbable poly(esters). Coating performed with PHB and/or PHV.**

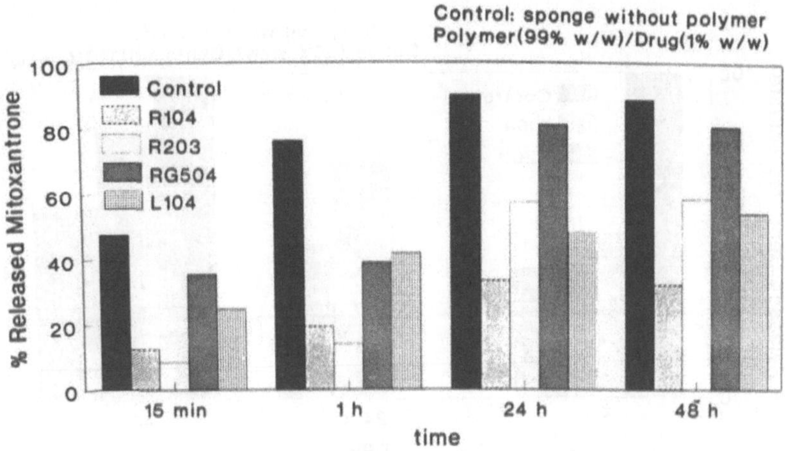


Figure 3. Mitoxanthrone liberation from mitoxanthrone-collagen sponges (for details see Materials and Methods section) coated with resorbable poly(esters). Coating performed with poly(lactide) and/or poly(glycolide).

Release of the drugs can be modified by surface properties such as the thickness of the coating, the type of poly(ester) and parameters such as good or low solubility of the drug in the solvent poly(ester) mixture. Sustained release was also provable in the use of the water-soluble antibiotic gentamicin.

During the course of these studies it could also be shown that the results are transferable to biomaterials. A coating of biomaterials, especially the metal surfaces of orthopedic implants, can lead to targeted drug application with high local effective substance levels. The layer

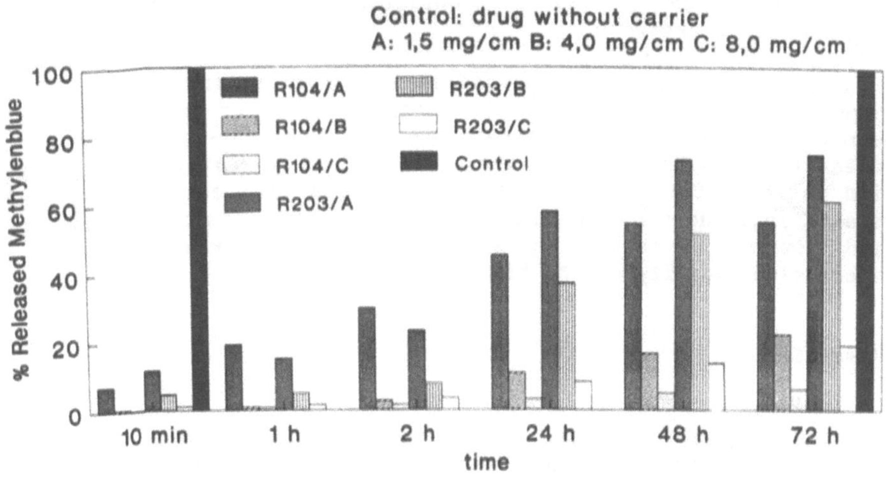


Figure 4. Dye liberation from a stainless-steel device coated with methylene blue and various resorbable poly(esters).

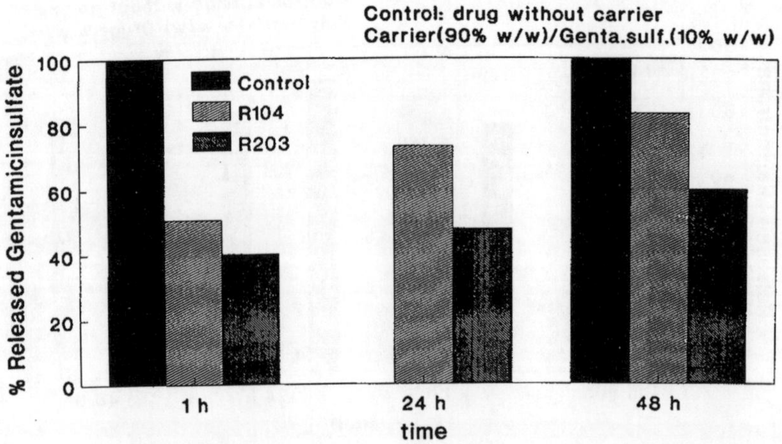


Figure 5. Gentamicin sulfate liberation from a stainless-steel device coated with gentamicin in combination with resorbable poly(esters).

density of the polymer used and and the concentration of the effective substance are variable in broad areas and must be adapted according to the clinical demands. Generally, drugs which are soluble in the polymer solutions can be used well. When polymers which show a rough cast-like structure after evaporation are applied, even fine drug crystals can be used on biomaterials.

Under the conditions of elution *in vitro*, the aminoglycoside is released quickly from the gentamicin-collagen compound. Animal studies have revealed high gentamicin levels in the cortical bone for a period of 14 days following implantation of an uncoated gentamicin-collagen sponge in a rabbit's femur. After instillation of corresponding doses of an injection solution only low bone levels could be observed.[1] From these investigations we have learned that results obtained by *in vitro* release study do not necessarily reflect the *in vivo* situation. Similar animal studies with resorbable poly(ester) drug-collagen compounds coated with this technique are in progress.

This research may indicate that even substances with good solubility in aqueous systems have better prolonged release controllability after placement on biomaterials with subsequent coating. The necessary *in vivo* studies will soon be able to reveal information on sustained drug distribution in soft tissue and bone. Delivery of a drug coated on a medical device is influenced by factors such as solubility, rate of dissolution, degradation of the carrier by hydrolysis and/or enzymatic cleavage and the perfusion capacity of the tissue.

## REFERENCES

1. A. Stemberger, R. Ascherl, F. Lechner & G. Blümel, Eds., "*Kollagen als Wirkstoffträger*," Schattauer, Stuttgart-New York, 1989.

# ANTICANCER DRUGS BASED ON ANALOGUES OF PLATINOL DERIVED FROM HISTIDINE

Charles E. Carraher, Jr. and Melanie M. Williams

Department of Chemistry
Florida Atlantic University
Boca Raton, FL 33431

The reaction products of tetrachloroplatinate (II) and histidine are studied as a function of pH employing elemental analysis, NMR, MS and IR in preparation of the use of site-specific proteins as carriers for the cis-dichloroplatinum II moiety. Unlike the analogous products derived from simple diamines, the products are a mixture of units with the structures dependent on the pH of the reaction system.

## INTRODUCTION

### 1. General

We have previously reported the synthesis of a number of polymeric derivatives of the widely employed anticancer drug cisdiaminodichloroplatinum II, cis-DDP.[1-4]

One objective is the delivery of drugs to the specific site in need of that drug. A number of approaches to achieve such "magic bullets" have been offered, but few have been tested. One approach is to employ site specific proteins as carrier agents to deliver the desired drug. The composition and geometry of proteins is quite varied, but most contain a variety of units derived from many of the typical amino acids. The geometry and uptake of specific drugs is dependent on the specific composition and reaction conditions. Here we report the synthesis of complexes from the reaction of the amino acid histidine with tetrachloroplatinate as a function of the particular form of histidine, i.e., pH of the reaction system.

The ability of amino acids to form complexes with metal ions has been known since the 1800s when the preparation of a copper salt of leucine was found by Gossman to be an adequate means of purifying amino acids.[5] In 1847, Weidmann discovered the biuret reaction, in which a deep red complex is produced in the reaction between Cu(II) salts and biuret ($H_2N$-CO-NH-CO-$NH_2$).[6] In the early 1900s, Ley described the synthesis of metal complexes of glycine and other amino acids, and pioneered studies on the determination of formation and stability constants, as well as the bind-

*Cosmetic and Pharmaceutical Applications of Polymers*
Edited by C.G. Gebelein *et al.*, Plenum Press, New York, 1991

ing sites, geometry, and structures of the complexes.[7,8] Most of the early studies of metal complexes of amino acids were conducted on aqueous complexes, and were concerned with the determination of stability constants. Studies of solid state complexes have only recently been undertaken to any significant extent.

The interactions between metal ions and amino acids is of considerable interest in coordination chemistry as well as biochemistry. Metal ions may react with amino acids in proteins to sustain or modify protein structure and configuration and in doing so affect the physical and biological properties of the protein.[9] On the other hand, metalloenzymes contain metal ions which form an integral part of the protein structure.

Most of the naturally occurring transition metals are bound through amino acid residues in enzymes.[10] Since enzymes are principally composed of proteins, the nature and extent of metal ion-protein interactions is an important topic in biochemistry.

While numerous studies have been reported on the metal complexes of simpler bifunctional amino acids such as glycine, considerably less research has been focused on amino acids with higher functionality. This is especially true for the basic trifunctional amino acids.

The simple bifunctional alpha amino acids generally act as bidentate ligands, coordinating through the alpha amino and carboxylate groups, whereas, higher amino acids containing side chain residues have multiple binding sites.[11] Depending on the structure of the polydentate amino acid and the properties of the metal ion, some or all of the available donor atoms may coordinate.[12] As a result, many structural possibilities often exist for metal complexes of polydentate amino acids, and their study is of great theoretical interest in coordination chemistry.

The nature of the metal ion, pH, metal to ligand ratio, and total concentration are important factors that determine which complex is formed.[13] While many metal complexes of amino acids have been reported to be monomeric, it has been suggested that others may be polymeric.[14] For example, Block has suggested that the copper complexes of HOOC-$CH_2$-NH-$(CH_2)_n$-NH-$CH_2$-COOH, prepared by Bailar and co-workers may be polymeric as a result of their insolubility and inertness as well as the indication by molecular models that these amino acids can not coordinate quadritentately. Additionally, when the carboxyl group is not part of a chelate, simple amino acids sometimes act as tridentate ligands in a bridging polymeric system as shown below in Structure 1.[15]

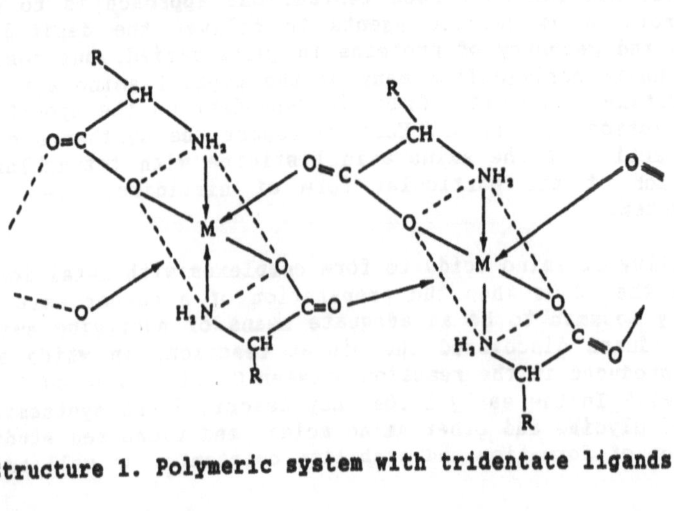

Structure 1. Polymeric system with tridentate ligands.

## 2. Platinum-complexes of Amino Acids as Chemotherapeutic Agents

Considerable interest in platinum complexes has resulted from the discovery of the antitumor activity of cis-DDP by Rosenburg and Van Camp in 1965.[16] The often severe side effects of the drug, especially renal toxicity, have prompted efforts to develop platinum derivatives which have not only reduced toxicity, but also increased solubility and enhanced activity.[17] Amino acids appear to be ideal ligands to achieve such properties. There are several advantages in using amino acids as ligands in platinum containing antitumor agents. First, amino acids are naturally occurring biological ligands and are generally nontoxic; toxic effects should therefore not result from their release in cells or in the bloodstream. Second, amino acids have previously been employed to direct chemotherapeutic agents into tumor cells which have reduced blood supplies.[18-20] Transportation of the drug is accomplished by an active transport mechanism which specifically absorbs amino acids into the cell. It has been suggested that as a result of this transport mechanism, platinum complexes of amino acids may be readily uptaken by cancer cells.[18-20] The ability of the amino acids to transport drugs into the cell would then alleviate problems with drug uptake and solubility. Finally, certain cancer cells are unable to synthesize a specific essential amino acid. As a result, these cells possess special mechanisms for the intracellular transport of the particular amino acids. For example, lymphoblastic leukemia cells require sparginine, while myeloblastic leukemia cells require glutamine,[18-21] and each of these types of cancer cells possess a special mechanism for obtaining these amino acids. Use of platinum derivatives of specific amino acids such as these may introduce the possibility of selectively attacking cancer cells which must obtain these amino acids exogenously.[19,20] Thus, the use of amino acids as ligands in platinum containing chemotherapeutic agents introduces the possibility of not only cancer cell specificity, but also specificity for certain types of tumors.

Allcock, Carraher and co-workers have reasoned that the use of polymeric platinum derivatives may lead to decreased renal toxicity, as high polymers cannot penetrate the semi-permeable membranes of the kidneys, and should, therefore, retard their rate of excretion. Polymeric platinum derivatives synthesized by these groups have shown significant activity against a wide variety of tumor systems.[1-4,22]

## 3. Metal Complexes as Models for Metal-Protein Interactions

Metal protein interactions have become a topic of immense importance in biochemistry. Throughout all biological systems, metal ions are associated with peptides and proteins. The nature of metal-protein interactions may take several forms. Metalloproteins contain a metal atom which forms an integral part of the protein structure. Other proteins, such as transferrin and conalbumin, bind specific metal ions strongly yet reversibly.[23] Yet, other proteins associate with metals but these reactions are readily reversible.

Metal ions are involved in a number of different types of biological processes. The role of the essential transition metal ions (Co, Cr, Zn, Mo, Mn) in biological processes is generally that of the catalysis of enzymes.[6] However, there are a number of transition metal metalloproteins which act as storage proteins, for example, hemoglobin and ferritin.[9]

In 1956, Gurd and Wilcox recognized that "it had become both feasible and desirable to compare in detail the formation of metal-protein complexes with the formation of simpler complexes".[23] In more recent years, much research has been directed towards understanding the reactions between metal ions and simpler organic molecules including amino acids, and peptides. As enzymes are principally proteins composed of specific sequences of amino acids, the study of metal complexes can lend insight into the mechanism of action, binding sites and structure of the sites of catalysis of enzymes reactions. In this sense, interactions between metal ions and amino acids can serve as models for metal-protein interactions.[9,11]

Amino acid complexes of the first row transition metals have been investigated to a large extent to mimic and understand enzyme reactions. An important step towards understanding the function of the metalloproteins involves determining the nature of the metal-protein interaction in terms of the binding group and the structure of the interaction with the metal.[24] However, identification of metal binding sites in metalloproteins is a difficult problem owing to the large number of binding sites. Synthetic polyamino acids have been shown to be instructive protein models, and it is reasonable that their interaction with metal ions should be comparable to that of metalloproteins.[25] A logical step towards the use of these bipolymers is to study metal complexes of tridentate amino acids which have exposed residues in proteins.[26]

The study of protein interactions with the heavy metals (such as the second and third row transition metals) also have relevance in enzymology, but are important in areas such as toxicology and chemotherapy also. The purpose of such studies in all of these areas is that of identifying the effect of the foreign metal on natural biological processes. For example, when a metal-containing chemotherapeutic drug is administered, the drug can enter normal cells as well as cancerous cells, causing anti-proliferation or death of both cell types. It is of importance to establish the nature of these interactions.

The presence of foreign metals in biological systems can also have detrimental effects on enzyme functions. These metals often displace the normal catalytic metal ion from its binding site on a protein, resulting in disrupted enzyme activity.[9] The displacement of an essential metal by a foreign one is dependent on thermodynamic, kinetic, and steric factors, as well as other factors.[9] Mercury exhibits a strong affinity for sulfur-containing ligands and its binding to cysteine residues results in the inhibition of sulfhydryl enzymes. On the other hand, Co(II) can replace Zn(II) in many enzymes with the retention of enzyme activity.[27] It appears that when the substituted metal has properties that are similar to those of the metal replaced, retention of enzyme activity can often be retained. Otherwise, substitution of the metal results in inhibition or disruption of enzyme activity. As with the first row metals, complexes of the second and third row metals with amino acids and peptides can be instructive models in determining the adverse effects these metals have on enzyme activity.

## 4. Histidine Structure as a Function of pH

A knowledge of the acid-base properties of amino acids is important in determining which functional groups will be available for metal binding, and the relative reactivities of these groups in complexing with metal ions. Thus, the form of the amino acid used in complexation is important in determining which products are formed. At the isotonic point, amino acids exist in aqueous solutions as zwitterions with the amino group protonated and the carboxyl group ionized. The structure of histidine as a zwitterion is given below in Structure 2.

Competition between metal and hydrogen ions frequently determines which of several donor atoms on a given ligand will preferentially bind to a metal ion at a given pH.[23] The concentration of hydrogen ions affects complex formation by competition with metal ions for the ligand. The ligand A is involved in a successive equilibria with the metal ion as well as reacting with available protons according to the relationship shown in Equation 1.[23] The pKa', or apparent dissociation constant (the dissociation constant measured under specified conditions) is the equilibrium constant for the reaction: $HA \rightleftharpoons H^+ + A^-$. Table 1 gives the pKa' values for selected multifunctional amino acids.

$$pKa' = pH + \log [HA]/[A] \qquad \text{(Equation 1)}$$

The equilibrium of the basic amino acids in aqueous solution can be illustrated by the following sequence, shown in Figure 1, where the order for deprotonation of the R-group and the amino group is reversed for histidine. In neutral solutions, the free amino acids exist in the zwitterion form. Addition of an equivalent of base to a solution of the basic amino acid deprotonates the zwitterion to the fully depronated form. The existence of the basic amino acids in the fully depronated form facilitates coordination of a metal ion to the amino acid and side chain donors.[29] At lower pH's, chelation through the alpha amino and carboxylate groups is favored. The pKa' value is a measure of the basicity of a ligand $A^-$; the higher the pKa', the stronger the base is towards hydrogen ions. The binding of a metal ion, also a Lewis acid, parallels to some extent, the affinity of hydrogen ions to a given site.[11]

## 5. Nature of Metal in Determining the Binding Site

Binding of a metal ion to one or more of the functional groups of an amino acid depends upon several factors including the basicity of a given functional group, the nature of the metal ion, and the stability of the complex formed. Based upon its existence in the deprotonated state at all but the lowest pH values, it would appear that the carboxylate ion would be the most common binding site. However, under conditions of moderate pH, binding is favored at this site only when conditions exist for chelate formation.[11] Non-chelated bonding to the carboxlate group is found in some complexes prepared at low pH.[30,31]

Coordination of amino acids through the carboxylate oxygens is known for metals which are "hard" or "class a" metals, but is rare in complexes of the "soft" or "class b" metals. Appleton and Hall reported the first Pt(II) complex in which an amino acid was bound through the carboxylate group.[32] At low pH (1.5-4.5), the mono and disubstituted complexes are

Table 1. Ionization constants of amino acids in aqueous solution at 25°C [28]

| Amino Acid | pKa'(COOH) | pKa'($^+NH_3$) | pKa'(R' group) |
|---|---|---|---|
| l-lysine | 2.18 | 8.95 | 10.53 |
| l-arginine | 2.01 | 9.04 | 12.48 |
| l-histidine | 1.77 | 6.10 | 9.18 |

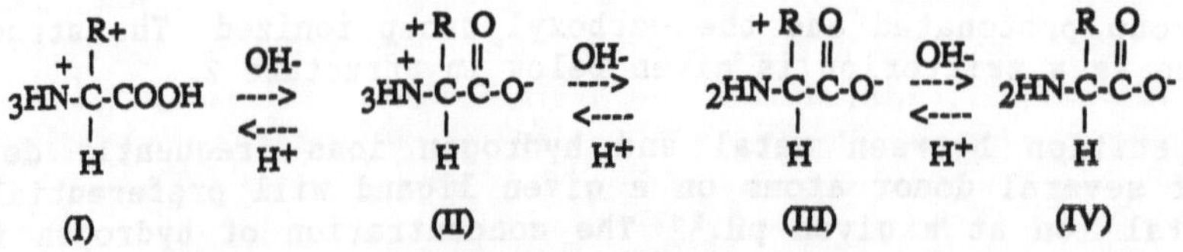

**Figure 1. Equilibrium of basic amino acids.**

formed from the reaction of equimolar quantities of glycine and cis-$[Pt(NH_3)(H_2O)]_2(NO_3)_2$. In these solutions, conversion to the (N,O) chelate begins to occur, requiring about 24 hours for complete conversion. Formation of the (N,O) chelate is more rapid upon the addition of base. This conversion was found to be irreversible as the (N,O) chelate is the thermodynamically more stable product. The conversion is thought to occur only under conditions where the leaving group is labile enough to be displaced and under mild enough conditions to prevent rearrangement.

The terminal amino groups of amino acids are the most common sites for metal binding. Coordination to the nitrogen donors of amino acids is especially favorable for "soft" or "class b" transition metals, such as Pt(II) and Pd(II), and the order of binding preference for the common donor atoms in amino acids is S >> N > O. These soft transition metals have a decreased shielding of the nucleus, resulting in a higher effective charge. This leads to a greater polarization of the lone pairs of electrons on the coordinating donor atoms of the ligand, and results in a substantial degree of covalent bonding in these complexes.[33]

Preferential coordination to nitrogen donors in amino acids is also a result of their strong basic character and the ligand field stabilization energy imparted to complexes coordinated through these donor atoms.[11] For example, a square planar geometry is favored by Pt(II) and Pd(II) complexes because these metals contain eight d-electrons and gain considerable ligand field stabilization in this geometry where the highest energy d-orbital is left empty.[34] The filled-d orbitals are then further stabilized, or lowered in energy, when coordinated to nitrogen donors that are at the strong field end of the spectrochemical series. The effect of ligand field stabilization energy is especially pronounced in square planar complexes of Pt(II) and Pd(II). In fact, this effect is so pronounced that Pt(II) has been shown to bind to the alpha amino groups of glycine in both monodentate and bidentate fashions even at pH values where the carboxyl groups are protonated.[11] It is the high polarizing power of soft metals such as Pt(II) that brings about the ionization of the protonated amino groups.[35]

The alpha amino acids are all capable of forming five-membered chelate rings with a metal ion through the alpha amino and carboxylate groups, and this is often the mode of amino acid coordination to transition metal ions. The tendency for these amino acids to form chelates rather than coordinating to one or another of these donor groups in a monodentate fashion is a result of the increased stability imparted to the complex as a result of chelate formation.[23]

Of the twenty alpha amino acids naturally occurring in proteins, approximately half contain side chain donor atoms which are potentially capable of forming a chelate ring with a metal ion through a side chain donor atom and the alpha amino nitrogen.[36] Consideration of the amino acids of interest in this study reveals that bidentate chelation through the side chain nitrogen donor atom and the alpha amino nitrogen would give six membered chelates with histidine, whereas lysine and arginine would form eight and nine membered chelates, respectively. Other factors being equal, five membered chelates are the most stable, and the stability of the chelate decreases as the size of the chelate ring increases.[33] It would then be expected that tendency towards side chain chelation would decrease as the size of the chelate ring increases. However, when other factors come into play, formation of larger membered chelates can become favorable. For example, reaction of ornithine with Pd(II) ions yields equivalent amounts of both the five membered (N,O) chelate and the seven membered (N,N) chelate. Formation of the larger chelate competes with the formation of the five membered chelate as a result of the preference of palladium for nitrogen donors over oxygen donors.[37] The structure of the seven membered chelate was recently confirmed by X-ray studies.[38]

When the side chain functional groups are non-chelating in the complex, these groups can either stabilize or destabilize the chelate. Functional groups which contribute an inductive effect, such as the guanidium group in arginine, have been found to destabilize the chelate.[26] On the other hand, the presence of a carboxylate group may stabilize the chelate by increasing the basicity of the donor group.

## 6. Metal Complexes of Histidine

Metal complexes of histidine are among the most extensively studied complexes of all the alpha amino acids. The imidazole group in histidine is an important residue in peptides and proteins due to its ability to act as a buffer at physiological pH.

The imidazole group in the side chain of histidine is approximately half ionized at pH 6.1. Thus the physiological form of histidine is a mixture of zwitterionic and cationic forms.[29] The unique ability of histidine to function as both an acid and a base at physiological pH is an important feature for many biological functions.[29] The imidazole group functions not only as a buffer, but also as a binding site for metal ions in biological systems. For example, histidyl side chains function as metal binding groups in metalloproteins, vitamin B12, myoglobin and hemoglobin.[9,24] In fact, histidine is one of the strongest metal binders among the amino acids and provides one of the primary means of metal-protein binding.[23]

The imidazole ring in histidine is generally regarded as an aromatic heterocycle with the electron pair from the N-1 "pyrrole-like" nitrogen contributing to the aromatic sextet, as shown in Structure 2.[13] Thus, the electron pair on the N-3 are the only electrons available for coordination with metal ions without compromising the aromaticity of the ring.

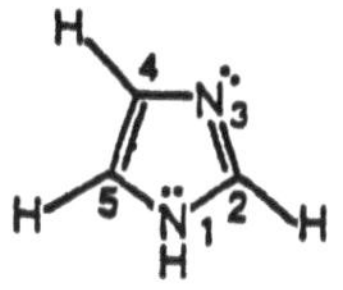

Structure 2. Imidazole ring with "pyrrole-like" nitrogen.

Although bonding structures through the "pyrrole" nitrogen of histidine have been proposed,[30] it is generally accepted that such bonding is unlikely, and there are no authenticated cases of such bonding.[13] Over the normal pH range, histidine possesses three potential donor sites in aqueous solutions.[39] This topic will be dealt with in the Results and Discussion section.

The results of crystallographic studies have shown that histidine can coordinate through each of its potential donor sites in metal binding, in monodentate, bidentate and tridentate fashions. With metal ions that prefer to coordinate in an octahedral geometry, histidine often acts as a tridentate ligand, Structure 3.[13,40-42] For example, $Ni(His)_2\text{-}H_2O$ contains histidine coordinated as the tridentate anionic ligand.[18-20] Since three donor atoms of histidine cannot occupy planar positions simultaneously, the histidinate ligand generally forms a facial arrangement, with one ligand occupying an axial position.[26,43,44]

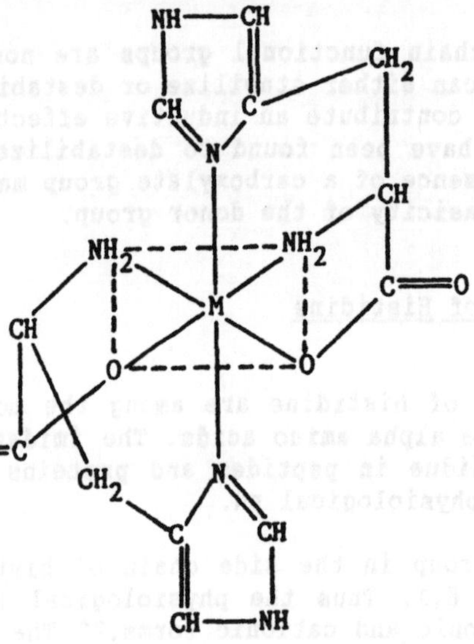

Structure 3. Histidine as a tridentate ligand.

Monodentate coordination through the carboxyl oxygens has been observed in 1:1 histidine complexes of Mn(II), Fe(II), Co(II), Ni(II), Cd(II), and Cu(II) prepared in solutions of pH < 4.[30,45] Additionally, the reaction of $Hg(II)Cl_2$ with histidine hydrochloride at low pH yields the following product, Structure 4, in which Hg(II) is coordinated to histidine only through the carboxylate oxygens.[46]

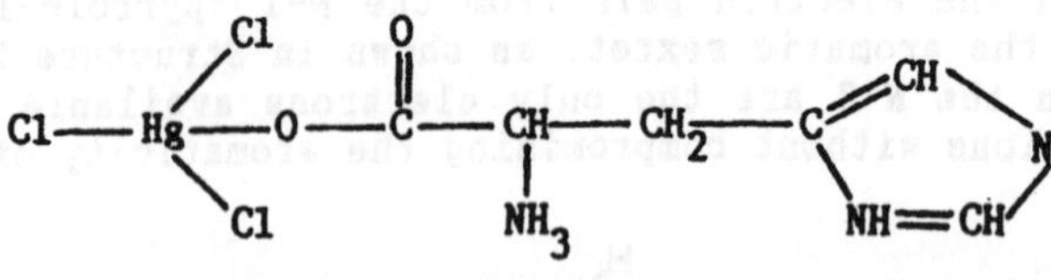

Structure 4. Monodentate coordination with histidine.

In solution, complexes of histidine with Cu(II) have been the most extensively studied, but the equilibria with Cu(II) is among the most complicated. Because Cu(II) complexes of histidine have been extensively studied throughout a wide pH range and for a variety of metal to ligand radios, these complexes will be presented as an illustrative example of the types of complexes histidine forms in solution with metal ions.

Cu(II) coordination compounds typically consist of four nearby donor atoms in a tetragonal plane with one or two more distant axial ligands.[34] Results of titration studies indicate that 1:1 and 2:1 solutions of Cu(II) and histidine contain $Cu(His)^{+}$, $Cu(HisH)^{+2}$, $Cu(His)(HisH)^{+}$, and $Cu(His)_2$ as the predominate species. Complexes of tridentate anionic histidine dominate above pH 7, while at lower pH several types of bidentate chelation may also exist. The structures of many of these complexes have been matters of uncertainty.[13]

With 1:1 metal to ligand ratios, the complex $[Cu(His)]^{+}$ predominates in solution at pH 6, being converted to the hydroxo complex Cu(His)(OH), above this pH. In $[Cu(His)]^{+}$, histidine acts as a tridentate ligand with the two nitrogen donors in the coordination plane, and a more weakly bound axial carboxylate group. Below pH 5, several species coexist in solution, with the complex $[Cu(HisH]^{2+}$ comprising the greatest amount but accounting for less than 40% of the total complexes present.[38]

Solutions containing 2:1 or higher molar ratios of histidine to Cu(II) have similar complex distributions for most pH values.[47] From pH 7-8, nearly all of the Cu(II) present is bound as $Cu(His)_2$, with this complex persisting to about pH 11. The most widely accepted structure for this complex is one in which both histidine molecules are tridentate with trans amino groups; one histidine is coordinated through the alpha amino and carboxylate groups, with the imidazole nitrogen occupying an axial position, while the second histidine coordinates through the imidazole and amino groups in the plane, with an axial carboxylate.[36] However, this complex is in equilibrium with other structures, and is thus a dynamic one. At pH values higher than 8, complexes containing one less hydrogen, either due to hydroxyl formation, or pyrrole hydrogen ionization begin to appear. The complex $[Cu(HisH)(His)]^{+}$ represents about 90% of the total complexes present in solution from pH 4-4.5, but is converted to $Cu(His)_2$ at a pH of 5.5. $[Cu(HisH)His)]^{+}$ contains one anionic histidine bound as a tridentate ligand with an apical carboxylate, and one neutral histidine coordinated as an (N,O) chelate in the plane with a protonated imidazole ring. The only other species which approaches 50% of the total complexes present in solution is $[Cu(HisH)_2]^{2+}$, which occurs just below pH 3. The proportion of $[Cu(HisH)_2]^{2+}$ reaches a maximum percentage at about pH 3.3, and accounts for only 15% of the total species present. This complex was isolated from acid solutions as crystals for X-ray studies even though other complexes are more predominate. In this complex, below, the ligands are bound to the metal through the alpha amino and carboxylate groups, with the imidazole group remaining uncoordinated; two water molecules occupy the remaining coordination sites.[13] The monohistidine complex, $[Cu(His)]^{+}$, is also present in the 2:1 solutions, comprising a maximum of 25% of the total species present at just above pH 3, Structure 5.

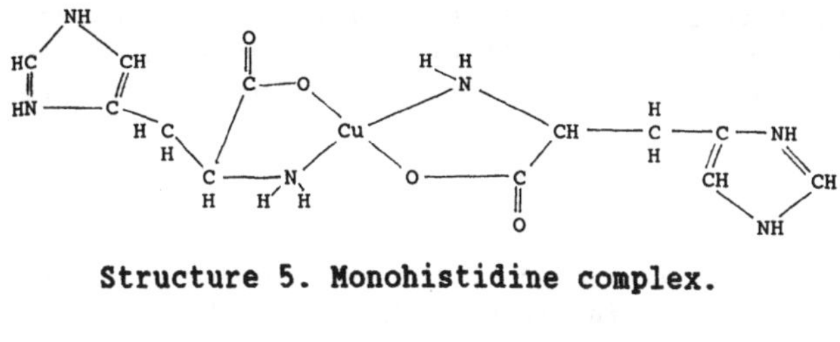

Structure 5. Monohistidine complex.

There have been several studies on solid complexes of Pt(II) with histidine.[48-54] No stability constant studies on aqueous complexes have been reported. This is probably due to the kinetic inertness of Pt(II), and therefore the slow reaction times to form complexes. Additionally, precipitates of the complexes are readily formed upon completion of the reaction. Using 2:1 ratios of histidine to $K_2PtCl_4$, with histidine reportedly used in the zwitterionic form, Nelson and Frye reported the isolation of $[Pt_2(His)_3(H_2O)]^+$ in which histidine, in the anionic form, is bound to the metal through the imidazole and alpha amino nitrogens; the negatively charged carboxylate groups remain uncoordinated.[55]

Volshtein reports the isolation of $Pt(His)^{-2}$ when 4:1 ligand to metal ratios were used with histidine in its anionic form. The protonated complex, $Pt(HisH)_2$-2Cl is formed when $Pt(His)_2$ is dissolved in HCl.[53] The two complexes differ only in the protonation and deprotonation of the carboxylate groups.

Using 3:1 ligand to metal ratios, the cis and trans isomers of $Pt(His)_2$ were reportedly obtained in solutions of pH 7.5. However, attempts to obtain consistent elemental analysis were not successful and the characterization of these complexes was accomplished solely on the basis of NMR studies.[48]

A particularly interesting study reports data on the purity of $Pt(His)_2$, and other similar Pt(II)-histidine complexes, by NMR, ion-exchange chromatography, and electrophoretic techniques. In this study, it was found that when cis-DDP reacts with histidine, a mixture containing up to three different complexes were formed.[54]

Although there have been several reports of 1:2 platinum complexes of histidine, there have been no studies on the 1:1 complexes. As histidine is a trifunctional ligand, it is possible that polymeric complexes will result when the reactants are used in equimolar ratios.

## EXPERIMENTAL

### 1. Synthesis

#### 1A. Method A

An aqueous solution of $K_2PtCl_4$ was added to a solution of the histidine dissolved in water to give a 1:1 molar ratio of amino acid to $K_2PtCl_4$. The amino acids used were either in the zwitterionic form (as the free base) or the cationic form (as the monohydrochloride). The solutions were stirred at room temperature for four to seven days. Generally, after 24-48 hours, the solution clouded and within several days a precipitate formed. The precipitate was separated from the solution by vacuum filtration and repeatedly washed with water, acetone, alcohol, and diethyl ether. Attempts to isolate soluble complexes by adjustment of the pH with NaOH and HCl, as well as by the addition of acetone or ethanol to the mother liquor, did not produce any additional precipitates.

#### 1B. Method B

The procedure used was analogous to Method A with the exception that an equivalent of NaOH was added to the solution of the amino acid to give a molar ratio of 1:1:1 amino acid to $K_2PtCl_4$ to base.

### 2. Synthesis of $Pt(HisH)_2\cdot 2Cl$ and $Pt(His)_2$

Synthesis of $Pt(HisH)_2\cdot 2Cl$ and $Pt(His)_2$ was accomplished according to

Volshtein.[53] An aqueous solution of $K_2PtCl_4$ was added to a solution of L-histidine containing one equivalent of NaOH to give a 1:4:4 molar ratio of $K_2PtCl_4$ to amino acid to NaOH. The solution was stirred at room temperature for seven days after which a white solid precipitate of $Pt(His)_2$ formed. The solid was isolated by vacuum filtration and washed with water, acetone, alcohol and diethyl ether. $Pt(HisH)_2 \cdot 2Cl$ was obtained by dissolving $Pt(His)_2$ in concentrated HCl. Following the initial dissolution of $Pt(His)_2$, a new white precipitate of $Pt(HisH)_2 \cdot 2Cl$ formed. Acetone was added to complete the precipitation. The precipitate was filtered, washed with acetone, and dried under vacuum at 80°C.

### 3. Physical Characterization

The pH of the reaction solutions using both synthetic methods was monitored as a function of time using an Orion analog pH meter (Model 301), and are good only to about plus or minus 0.2-0.4 pH units at intermediate pH values, and less accurate at extreme pH values.

Most elemental analyses were determined at Gailbraith Laboratories; Knoxville, TN. Some platinum analyses were conducted in our laboratory gravimetrically, by heating the product to metallic platinum, and determining the weight percentage platinum. Light scattering photometry was carried out in the usual manner. Infrared spectra were recorded using a Mattson Alpha-Centauri FTIR equipped with KBr and Mylar optics and a DTGS detector. Deuteration was accomplished by dissolving the product in $D_2O$ and adding enough NaOD to dissolve the sample.

$^1H$-NMR spectra were recorded on a General Electric-QE 300, 300 MHz FT-NMR spectrophotometer. Approximately 0.1% solutions of the polymer in $d_6$-DMSO were prepared and stirred overnight to dissolve the polymer. One hundred scans were acquired with a 3 microsecond pulse at 22° pulse angle and a spectral width of 6024 Hz. All spectra were recorded at ambient temperature.

High resolution electron impact (HREI) positive ion mass spectral analyses were carried out at the Midwest Center for Mass Spectrometry; Lincoln, Nebraska. A Kratos MS-50 Mass Spectrometer, operating at 8 KV acceleration, and a ten second per decade scan rate, was used for all measurements. HREI spectra were recorded by heating the sample ballistically to about 450°C; FAB spectra were also carried out using the same instrumentation, with NBA (nitrobenzyl alcohol) and ONPOE (2-nitrophenyl octyl ether) used as the matrices for the sample. Positive ion electron impact spectra were also recorded in our laboratory using a DuPont 24-961 B Double Focusing Mass Spectrometer with a resolution of 1/1500 at 0% valley. The sample was inserted as a solid in a glass ampule using a direct insertion probe.

## RESULTS AND DISCUSSION

### 1. General

Simple bi- and trifunctional amino acids can result in many structural possibilities. Here we will describe data related to the structure of the platinum-containing complexes of histidine. Coordination of amino acids through the carboxylate oxygens is known for metals that are "hard" or "class a" metals, but is rare for complexes of "soft" or "class b"

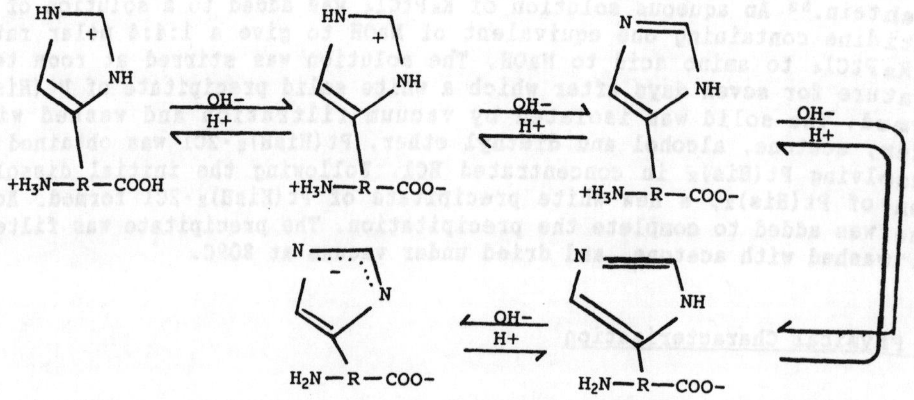


**Figure 2. Equilibrium forms of histidine in aqueous solution.**

metals, such as Pt(II) and Pd(II), with the order of binding preference for common donor atoms in amino acids being S > N > O. The "softness" is a result of a low shielding of the nucleus, resulting in a higher effective charge allowing a greater polarization of the lone electron pairs on the donor atoms of the ligand resulting in a substantial degree of covalent bonding.[33] Preferential coordination to nitrogen donors in amino acids is also a result of their strong basic character and the ligand field stabilization energy imparted to such complexes coordinated through these donor atoms.[11]

The alpha amino acids are all capable of forming five-membered chelate rings with the platinum. Such complexes will be noted as (N,O) complexes. Equilibrium forms of histidine in aqueous solution are shown in Figure 2. The pka values for proton abstraction are carboxylic acid, 1.9; imidazolium, 6.1; ammonium, 9.1; and imidazole (N-1), 14.4.[9]

Representative monomeric, dimeric and polymeric structures are given as Structures [1 - 8] in Figure 3. Over the normal pH range, histidine possesses three potential donor sites.[39] Thus, histidine can form chelates with Pt(II) through (1) the alpha amino and carboxyl forming five-membered (N,O) complexes [1, 5]; (2) the alpha amino and imidazole groups forming six membered (N,N) complexes [2, 8]; (3) the carboxylate and imidazole group forming seven-membered (N,O) complexes [3]; and (4) through all three donor groups as a tridentate ligand [4]. Structure [3] is the least likely and is possible only under acidic conditions where the nitrogen atoms are not readily available since formation of seven-membered chelates without other stabilizing factors is not favored.[23] Structures [2] and [8] are favored by metals that have large increases in crystal fild stabilization energy when bonded to nitrogen, particularly at high pH.[11,44] In fact an NMR study of aqueous 1:2 Pt(II)-histidine complexes suggests this as the sole mode of binding over the pH range of 1-12.[48]

## 2. pH of the Reaction Solution

The pH of the reaction solutions were monitored as a function of time. The results indicate that the reactions are generally complete after 3-4 days. When no base is added to the reaction solutions (Method A), the pH drops to a value between 2-3, indicating that precipitation of the products occurs in fairly strong acidic solutions. Precipitates ob-

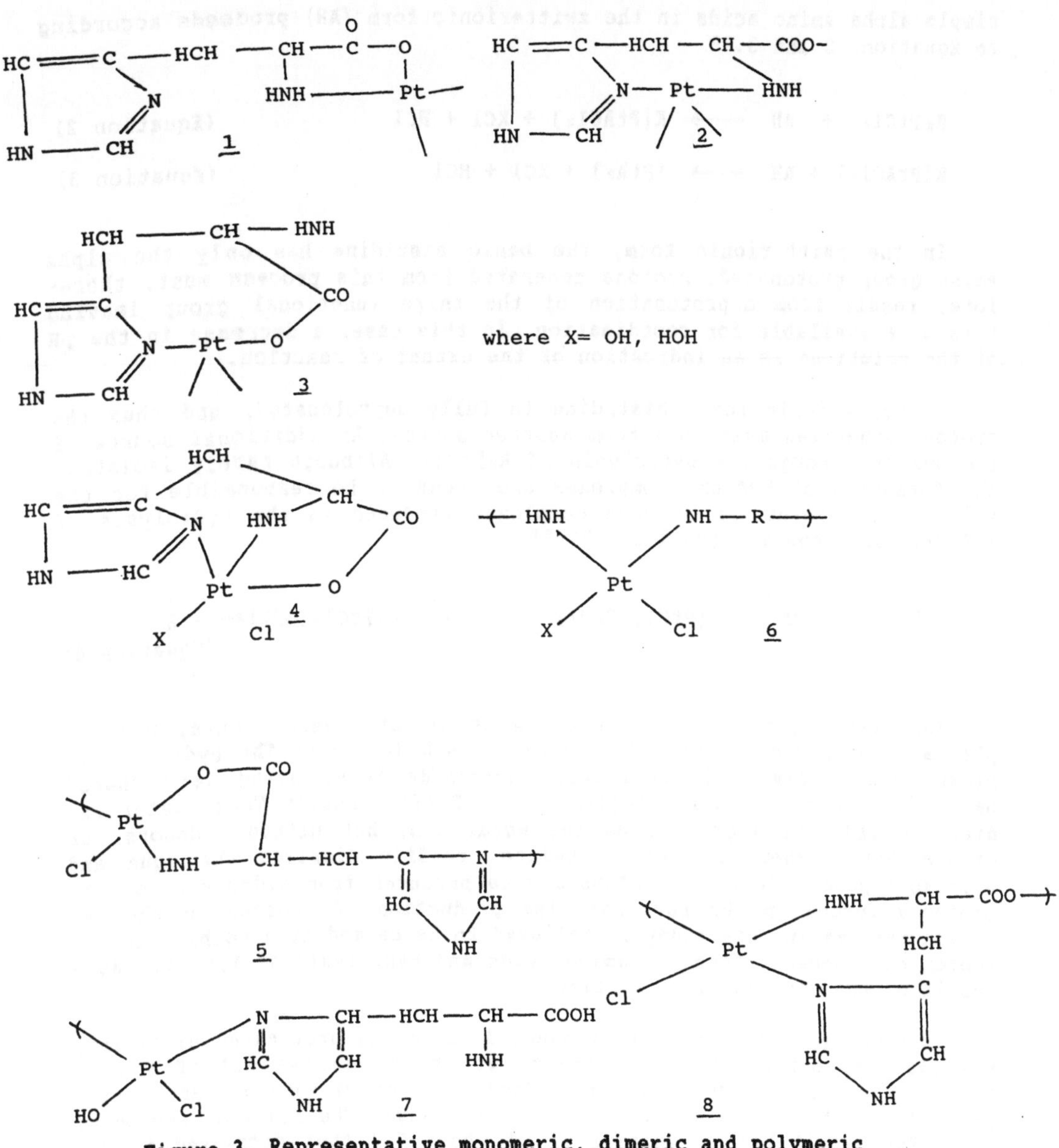

**Figure 3. Representative monomeric, dimeric and polymeric structures.**

tained in these solutions should have most free carboxyl groups protonated. When an equivalent of base is added to the amino acid solution (Method B), the pH drops to about 4. In these solutions the products are precipitated in neutral to somewhat acidic solutions indicating that most free carboxyl groups on the amino acids should be deprotonated at these pH values (pKa, COOH, 2.0 for the free amino acids).

The large decrease in the pH of the reaction solutions indicates that protons are being produced in the reaction systems. There are several possible sources of protons in these reaction systems. First, displacement of a proton from one of the protonated functional groups in the amino acids by a metal ion is necessary for coordination at these sites. The reactions of $K_2PtCl_4$ with the amino acids have been extensively studied by Volshtein and others. According to Volshtein, reaction of the simple alpha amino acids in the zwitterionic form (AH) proceeds according to Equations 2 and 3.[56]

$$K_2PtCl_4 + AH \longrightarrow K[PtACl_2] + KCl + HCl \qquad \text{(Equation 2)}$$

$$K[PtACl_2] + AH \longrightarrow [PtA_2] + KCl + HCl \qquad \text{(Equation 3)}$$

In the zwitterionic form, the basic histidine has only the alpha amino group protonated. Protons generated from this process must, therefore, result from deprotonation of the third functional group leaving this site available for coordination. In this case, a decrease in the pH of the solutions is an indication of the extent of reaction.

In the anionic form, histidine is fully deprotonated, and thus the protons generated must come from another source. An additional source of protons is through the hydrolysis of $K_2PtCl_4$. Although rarely isolated, the formation of hydroxy complexes are known to be responsible for the acidities of solutions of aqua complexes produced in the hydrolysis of $K_2PtCl_4$, as shown in Equation 4.[34,57]

$$[PtCl_4]^{-2} + H_2O \longrightarrow [PtCl_3(H_2O)]^- + Cl^- \longrightarrow [PtCl_3(OH)]^{2-} + H^+ \qquad \text{(Equation 4)}$$

The first aquation begins after a period of several hours, but replacement of a second chloride ion takes much longer.[57] The hydroxo complexes are rarely isolated because hydroxide ions, being very "hard" bases, do not form stable complexes with Pt(II) ions.[34] Thus, water can displace chlorine ligands from the metal ion, but nitrogen donors, or oxygen donors when part of a chelate, readily displace the aqua and hydroxo ligands. Finally, protons may be produced from hydrolysis of the products formed in the reaction. The production of protons in the reaction systems in this study is believed to be caused by a combination of proton displacement from the amino acids and hydrolysis of $K_2PtCl_4$, as a result of the complexation reactions.

In summary, reactions were conducted employing protonated histidine, histidine free base and histidine to which one equivalent of sodium hydroxide has been added. The base added reaction mixture had an initial pH of 10.2, that decreased to 4.2 after two days. The reaction employing histidine itself had an initial pH of about 7.2 that decreased to 2.9 after one day. Thus the major form of histidine for the reaction employing added base is probably the ammonium form while for histidine itself it is probably largely in the imidazolium form.

## 3. Elemental Analysis

The elemental analysis of the histidine products prepared in Method A (Table 2) is consistent with the products having one histidine unit and one chlorine atom per platinum atom, corresponding to the empirical formula Pt(His)Cl. Addition of $AgNO_3$ to a freshly prepared solution of the histidine products prepared in Method A does not produce a precipitate of AgCl, consistent with the chlorine in these products being bound to the metal and not simply present as a counter ion. The elemental analysis of the histidine product prepared by Method B is essentially the same as that prepared in Method A, with the exception that this product apparent-

Table 2. Elemental analysis for products derived from histidine

| | Percent Element | | | | Atom Number Average[c] | | |
|---|---|---|---|---|---|---|---|
| | Found[a] | | Expected[b] | | Found[d] | | Expected |
| Product | 3 | 4 | | | 3 | 4 | L-Histidine |
| C | 17.6, 17.7 | 17.9 | 17.9 | | 6 | 6 | 6 |
| H | 2.6 | 3.1 | 2.7 | | 11 | 13 | 9 |
| N | 10.1, 10.2 | 10.3 | 10.4 | | 3 | 3 | 3 |
| Pt | 47.7, 45.3 | 46.5 | 48.4 | | 1 | 1 | - |
| Cl | 8.6, * | 5.42 | 8.7 | | 1 | 0.5 | - |
| O[e] | 13.5, * | 16.78 | 7.9 | | 3.5 | 4 | 2 |

(a) Data from two analyses.
(b) Based on the empirical formula Pt(His)Cl·$H_2O$.
(c) Rounded to nearest integral of half-integral.
(d) Adjusted to carbon.
(e) Oxygen by difference.
* Analyses not done.

ly contains more water or hydroxyl ligands, and less chlorine. Thus, the elemental analysis of the histidine products is consistent with a polymeric product such as 6 or 7, a monomeric complex having the empirical formula Pt(His)Cl, or other structure.

## 4. Solubility

The histidine products prepared by Method A were soluble upon the addition of base to aqueous solutions. The solubility of these products only in basic solutions is an indication of the acidic nature of the products. The products prepared by Method B were insoluble in all solvents tested, even upon heating (boiling water bath). The histidine products were yellow to brown in color.

## 5. Infrared Spectroscopy

Infrared spectroscopy is one of the most widely used techniques for elucidating the structure of metal complexes of amino acids. This is due to the sensitivity of the amino and carboxyl group stretching modes to coordination by metals.[26] However, the infrared spectra of most of the amino acids have not been well characterized; with the exception of the amino and carboxyl group stretching and bending vibrations, assignment of the bands in the spectra is often difficult. An additional problem arises in the spectra of amino acids in the solid state where the effect of hydrogen bonding on the amino and carboxyl stretching frequencies is appreciable.[58] However, several thorough studies have been made on simpler amino acids such as glycine and alanine, and assignments of some bands can be made by comparison with these studies.

The assignment of the infrared bands in the following discussion are made by comparison with a number of general references on amino acids and amino acid-metal complexes,[58-72] as well as a number of studies of related complexes. The most prominent bands in the spectra of the amino acid and the corresponding complexes are assigned with particular emphasis on the carboxylate and amino group stretching and bending vibrations and the Pt-N stretching vibrations.

As a result of various spectroscopic studies, it is generally accepted that free amino acids exist in the solid state almost exclusively as zwitterions.[58] As a result of this zwitterionic structure, the infrared spectra of the free amino acids all show characteristic absorptions due to stretching of the carboxylate group near 1600 $cm^{-1}$ (asym.) and 1400 $cm^{-1}$ (sym.), and the N-H stretching modes of the $^{+}NH_3$ group in the 3100-2600 $cm^{-1}$ region. Multiple combination and overtone bands extend the $^{+}NH_3$ absorptions to about 2000 $cm^{-1}$. Two bending modes are usually observed for the $^{+}NH_3$ group near 1600 $cm^{-1}$ (asym.) and 1550-1480 $cm^{-1}$ (sym.).[71] These bands are common to all of the amino acids.

A study of over 53 pairs of D- and L-amino acid has shown that no differences exist in their solid state spectra between 5000 $cm^{-1}$ and 600 $cm^{-1}$.[63] Therefore, no significant spectral differences are expected for the D- and L-isomers of a given amino acid or their complexes in the mid-IR region of the spectrum, and assignment of only one isomer is presented. Whereas the D- and L-isomers of a given amino acid shown no differences in the mid-IR region, marked differences are noted between the spectra of a given isomer and the DL-racemate. However, these differences do not greatly affect interpretation of the spectra.

The D- and L-isomers of histidine have essentially identical infrared spectra in the mid-IR region. However, the spectrum of the DL-racemate contains more bands and is in general more complicated.

The infrared spectra of the products prepared from D-histidine, L-histidine, and DL-histidine prepared in Method A in the mid-IR region are identical. However, the histidine product prepared by Method B differs from those prepared in Method A in the carboxylate and carboxylic acid stretching regions. Volshtein has synthesized the (N,N) chelates of histidine with platinum[53] and the spectra of these products, $Pt(His)_2$ and $Pt(HisH)_2 \cdot Cl$ have been previously reported.[73] These complexes were synthesized and their infrared spectra recorded for comparison with the histidine products obtained in this study. A comparison of the spectra of the histidine products [3] and [4] prepared by Methods A (Products) and B and the spectra of $Pt(His)_2$ and $Pt(HisH)_2 \cdot 2Cl$ was made and selected assignments given in Table 3. Interpretation of the histidine spectra was aided by comparison with previous studies on imidazole[74] and solution studies of histidine.[31,45]

In the solid state, L-histidine exists in the zwitterion form.[63] On the other hand, L-histidine monohydrochloride, which exists in the cationic form, has the imidazole "pyridine-like" nitrogen protonated. This is in contrast to the hydrochlorides of other amino acids in which the proton resides on the carboxyl group, generating the carboxylic acid. The proton resides on the pyridine nitrogen because this site is more basic than the carboxylate group and attracts protons more readily.[66] The strong broad absorptions of histidine in the N-H stretching region is a result of strong hydrogen bonding of both the imidazole ring and the alpha amino and carboxylate groups.

A strong band is observed at 3429 $cm^{-1}$ in the spectrum of L-histidine·HCl which is absent from the spectrum L-histidine. Previous studies have shown that imidazole exhibits a strong band between 3470 and 3440 $cm^{-1}$, attributed to a non-hydrogen bonded N-H stretching of the imidazole ring;[75] this band is absent from the solid state spectrum of imidazole.[74]

| Table 3. Infrared frequencies and assignments for histidine, Pt(His)₂, Pt(HisH)₂·2Cl, and Product 3 | | | | |
|---|---|---|---|---|
| 1-Histidine | Pt(His)$_2$ | Pt(HisH)$_2$·2Cl | Product 3 | Assignment |
| | | | 3195 | N-H(H-bonded) stretch |
| 3126,3073 | 3105 | 3124,3097 | 3114 | Im. C-H stretch |
| - | 3044 | 3053 | 3031 | NH$_2$ stretch |
| 2877 | 2899 | 2912 | 2912 | CH$_2$ stretch |
| 2711 | - | - | - | $^+$NH$_3$ stretch |
| 2644 | 2546,2678 | 2520,2651 | broad | Im. N-H stretch |
| - | - | 1736 | 1726 | C=O stretch(COOH) |
| 1633 | 1618 | - | 1619 | COO-asym. stretch |
| 1569 | 1541 | 1555 | hidden | Im. C=C stretch |
| 1590 | - | - | - | $^+$NH$_3$ asym. bend |
| - | sh | 1598 | 1597 | NH$_2$ bend |
| 1497 | 1510 | 1511,1502 | 1503 | C=N stretch |
| 1461 | - | - | - | $^+$NH$_3$ bend |
| sh | 1423 | 1426 | 1429 | CH$_2$ scissor |
| 1415 | 1406 | - | 1392 | COO-sym. stretch |
| 1343 | 1342 | 1339 | 1333 | CH$_2$ wag |
| 1315 | 1305 | - | - | Im. C-H bend |
| 1271 | 1272 | 1272 | 1271 | Im. Ring Breathing |
| - | - | 1201 | 1214 | C-O stretch(COOH) |
| 1062 | 1051 | 1046 | 1054 | Im. C-H bend |
| 597 | 597 | 595 | 597 | Pt-N stretch |

In the same study, a band at 3097 cm$^{-1}$ was attributed to the hydrogen-bonded N-H stretch.

The imidazole C-H stretching frequencies of histidine occur at higher frequencies than those of the aliphatic CH and CH$_2$ frequencies due to ring strain. Accordingly, the imidazole C-H stretching frequencies of histidine and the histidine products are assigned to bands between 3126 cm$^{-1}$ and 3105 cm$^{-1}$, while the aliphatic C-H stretching modes are assigned to the bands occurring between 2877 cm$^{-1}$ and 2919 cm$^{-1}$.

The N-H stretching frequencies of the $^+NH_3$ group in histidine occur near 2700 cm$^{-1}$ and are absent from the histidine complexes and products. Bands occurring between 3030-3045 cm$^{-1}$, which appear in the complexes and products but are absent in spectrum of histidine, are assigned to coordinated NH$_2$ stretching modes. The N-H stretching vibrations of the imidazole ring are assigned to weak bands occurring between 2700 cm$^{-1}$ and 2500 cm$^{-1}$ in both histidine and the histidine products and complexes.

The histidine products prepared using histidine in the zwitterionic form (Method A) exhibit a strong band at 1726 cm$^{-1}$ characteristic of a COOH stretch and a band at 1619 cm$^{-1}$ characteristic of the COO-asymmetric stretch.[59,61,70,73] The band at 1726 cm$^{-1}$ is absent from the product prepared in Method B, but the carboxylate ion asymmetric stretch is again present at 1615 cm$^{-1}$. The histidine products prepared using zwitterionic histidine (Method A) therefore contain both protonated and ionized carboxyl groups whereas the product of anionic histidine (Method B) contains only ionized carboxylate groups. In the spectrum of Pt(His)$_2$, in which the carboxyl groups are ionized, a band is also observed at 1618 cm$^{-1}$, but no band near 1720 cm$^{-1}$ is present. On the other hand, the spectrum of Pt(HisH)$_2$·2Cl exhibits a band at 1726 cm$^{-1}$, characteristic of the COOH C=O stretching mode, but no band near 1620 cm$^{-1}$.

The absorptions between 1650 $cm^{-1}$ and 1450 $cm^{-1}$ in the histidine complexes are somewhat difficult to assign because the carboxylate, C=C, and C=N stretching modes and the $^{+}NH_3$ and $NH_2$ bending modes occur in this region. However, deuteration studies help in distinguishing these vibrations. Two $^{+}NH_3$ bending modes are expected for 1-histidine between 1660-1590 $cm^{-1}$ and 1550-1480 $cm^{-1}$. Balice and Theophanides[73] assigned the bands at 1473 $cm^{-1}$ and 1603 $cm^{-1}$ to the $^{+}NH_3$ symmetric and asymmetric bending modes respectively, since these bands disappear upon deuteration. The band at 1598 $cm^{-1}$ in $Pt(HisH)_2 \cdot 2Cl$ was assigned by them[73] to the coordinated $NH_2$ asymmetric bend (this band shifted upon deuteration), whereas in the spectrum of $Pt(His)_2$, it was assigned to the band at 1555 $cm^{-1}$ (1541 in our spectra). However, this band would not be expected to shift 45 $cm^{-1}$ from that in $Pt(HisH)_2 \cdot HCl$, since these complexes differ only in the protonation of the carboxyl group. Since the position of this band seems unreasonable, and deuteration of $Pt(His)_2$ was not undertaken by Balice and Theophanides, (only the spectrum of deuterated $Pt(His)_2.2Cl$ was reported) this band is believed to be incorrectly assigned. In our spectra, a shoulder is found in the spectrum of $Pt(His)_2$ near 1590 $cm^{-1}$ and this band is assigned to the coordinated $NH_2$ asymmetric bend. The band at 1541 $cm^{-1}$ in $Pt(His)_2$ can then be assigned to an imidazole C=C stretch, in agreement with the frequency found for this vibration in imidazole.[74] Another assignment of Balice and Theophanides in question is that of the $NH_2$ symmetric bend at 1435 $cm^{-1}$ ($Pt(His)_2$) and 1496 $cm^{-1}$ ($Pt(HisH)_2 \cdot 2Cl$). While the $^{+}NH_3$ group generally exhibits both symmetric and asymmetric bending modes, only one bending mode is observed for primary amines (substituted or otherwise). Therefore, a second $NH_2$ bending vibration is not expected for the complexes. These bands have therefore been reassigned and are given in Table 3. Assignment of the $NH_2$ bending modes in the histidine products synthesized in this study was accomplished by comparison with the deuterated product. Upon deuteration, the band at 1597 $cm^{-1}$ in Product 3 disappeared and this band is therefore assigned to the N-H bending mode of the coordinated $NH_2$ group.

The carboxylate symmetric stretch near 1400 $cm^{-1}$ appears in the spectra of histidine, $Pt(His)_2$, and all of the histidine products. This band is absent from the spectrum of $Pt(HisH)_2 \cdot 2Cl$, in which the carboxyl groups are protonated. On the other hand, the spectrum of $Pt(HisH)_2 \cdot 2Cl$ contains a strong band at 1201 $cm^{-1}$ due to the C-O stretch of the COOH group. This band is absent from $Pt(His)_2$ and histidine Product 4. These observations are consistent with the presence of deprotonated carboxylate groups in $Pt(His)_2$ and all of the histidine products, and with protonated carboxylic acid groups in $Pt(HisH)_2 \cdot 2Cl$ and the histidine products only from Method A.

A weak band is present in the histidine complexes and products near 596 $cm^{-1}$, which is absent in the spectrum of histidine, and is therefore assigned to a Pt-N stretching mode. The remaining assignments in the spectra of histidine, the histidine complexes, and the histidine products appear in Table 3.

The infrared data of the histidine products lends evidence for the binding sites in these products. The presence of coordinated N-H stretching and bending modes, and Pt-N stretching vibrations are consistent with coordination through the amino and imidazole nitrogens. The presence of both ionized and protonated carboxyl groups in the products of Method A and only ionized carboxyl groups in Method B suggests that coordination through the carboxylate groups occurs in all of the histidine products. If the carboxylate groups were simply deprotonated (and not coordinated) in the histidine products, no bands would be expected in the 1600 $cm^{-1}$ range of the products prepared in Method A, as these products were precipitated in acid solutions, and dissolve upon the addition of base. To ensure that the carboxylate stretching band observed in these products is not due solely to the presence of some free carboxylate groups, a sample of Product 3 was dissolved in NaOH, concentrated HCl was added to the solution until the pH was about 0.5, and the product reprecipitated. At this pH, any free carboxyl groups which may have originally been deprotonated would become protonated, and the band near 1620 $cm^{-1}$ would disappear, whereas the coordinated carboxylate stretching band would remain. The product which precipitated from this highly acidic solution still showed a strong band near 1620 $cm^{-1}$, and the intensity of the C=O stretch (1726 $cm^{-1}$) increased relative to that of the carboxylate stretching band. From this information it can be concluded that all of the histidine products contain coordinated carboxylate groups and the histidine products prepared in Method A also contains free carboxylic acid groups. The histidine products prepared in Method A may additionally contain some free carboxylate groups, but differentiation between free and coordinated carboxylate groups can not be made on the basis of the IR data.

The product prepared from histidine monohydrochloride, which is in the cationic form, shows the same bands in the infrared as the other products prepared in Method A. This is evidence that bonding through the same functional groups occurs in this product.

## 6. NMR Spectral Results

Erickson, et al., have reported the NMR spectra of complexes which were presumed to be cis and trans isomers of the neutral and anionic forms of $Pt(His)_2$.[48] (It should, however, be noted that reproducible elemental analyses or neutralization equivalents on these complexes could not be obtained). The imidazole ring protons are flanked by 195 Pt side bands at both high and low pHs, and the position of these bands varies with pH and the particular isomer. This was taken to indicate that coordination to the imidazole nitrogen occurs at both high and low pHs. The upfield portion of the spectrum exhibits peaks due to the CH and $CH_2$ protons. Based on coupling constants, the change in the position of these peaks with pH was interpreted in terms of the various conformations of the complexes at different pHs.

The spectrum of Product 3 is similar to that of the $Pt(His)_2$ complexes at pH > 12, but the peaks are substantially broader and less resolved; as a result, the 195Pt side bands are not resolved. The broad peak centered around $\delta = 3.0$ is due to the CH and $CH_2$ protons, whereas the downfield peaks arise from the imidazole C-H protons. The position of the peaks in the spectrum of Product 3 are near those of the cis $Pt(His)_2$ complex at pH 12, but since these peaks shift with pH and the particular isomer, no conclusions can be drawn from the peak positions.

## 7. Mass Spectral Results

The mass spectra of the products were obtained using high resolution electron impact ionization (HREI) and Fast Atom Bombardment (FAB) techniques, in the mass range of 28-600. Under EI conditions, product decomposition occurs with the evolution of $CO_2$, Cl ($Cl,^{35}$ $HCl,^{36}$ $Cl^{37}$ and $HCl^{38}$) all masses given in Dalton units) and ion fragments characteristic of the histidine moiety. Sample results are given in Table 4.

**Table 4. Significant (greater than 5%) ion fragments generated for product of tetrachloroplatinate and DL-histidine using high resolution EI (m/e greater than 32)**

| m/e | Relative Intensity (%) | Assignment | m/e | Relative Intensity (%) | Assignment |
|---|---|---|---|---|---|
| 35.9795 | 11 | HCl(35) | 68.0620 | 6 | RH |
| 37.9785 | 3 | HCl(37) | 69.0708 | 31 | $RH_2$ |
| 41.0434 | 27 | CH $CNH_2$ | 71.0868 | 19 | $CH_2CHCO_2$ |
| 42.0505 | 6 | $CH_2CN_2$ | 73.0305 | 21 | $NH_2CHCO_2$ |
| 43.0181 | 29 | $CH_2CH\ NH_2$ | 81.0722 | 15 | $RCH_2$ |
| 43.0584 | 30 | $CH_2CH_2NH_2$ | 82.0796 | 7 | $RHCH_2$ |
| 44.0191 | 26 | $NHCH_2NH$ | 85.1031 | 12 | $CHCNH_2CO_2$ |
| 45.0009 | 7 |  | 95.0860 | 7 | $RHCH_2CH$ |
| 45.0373 | 1000 | $CO_2H$ | 97.1016 | 10 | $RH_2CH_2CH_2$ |
| 55.0568 | 27 | $NHCH_2NC$ | 111.1174 | 5 | $RHCH_2CHNH_2$ |
| 56.0646 | 9 | $CCO_2$ | 129.0912 | 11 |  |
| 57.0730 | 32 | $CHCO_2$ | 149.0241 | 11 | His-$XH_5$ |
| 60.0302 | 45 | $H_2NCO_2$ | 256.2393 | 5 | PtClNH |
| 61.0316 | 7 | $NH_2CO_2H$ |  |  |  |
| 67.0555 | 9 | R (where R = Ring) |  |  |  |

## 8. Molecular Weight

The molecular weight of the histidine product prepared by Method A were determined by light scattering photometry to be 1.25 x $10^5$ Daltons, corresponding to an average DP of about 300. Measurements were made to determine the rate of degradation of the product in solution through

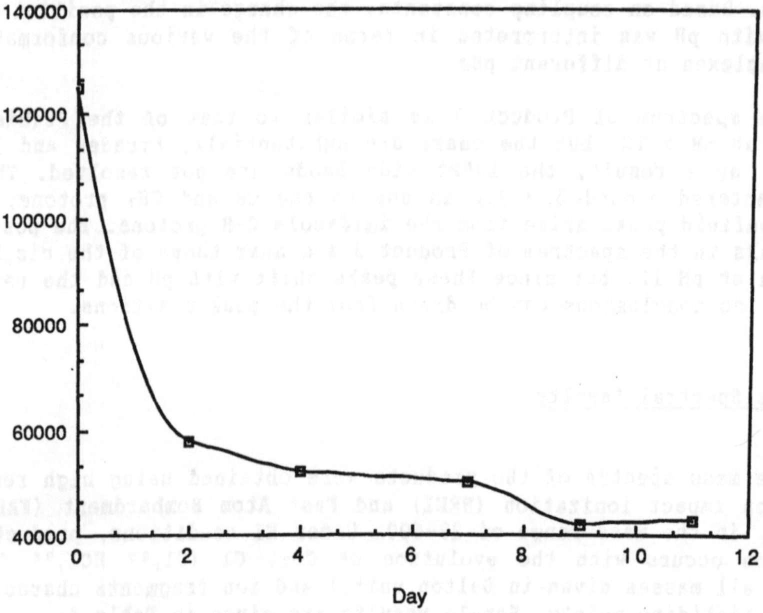


Figure 4. Apparent molecular weight versus time for Product 3.

determining the apparent molecular weight (0.1% solutions). As shown in Figure 4, the molecular weight decreased to about one half of its original value within the first twenty-four hours and then decreased very little over the next eleven days. Because of possible molecular association, etc., molecular weight values should be considered as approximate.

## SUMMARY

The present study is consistent with histidine coordinating platinum through the carboxylate groups, in contrast to the results of previous studies. Previous studies have reported histidine complexes coordinated to platinum only through the amino and imidazole (N-3) nitrogens. Coordination through the carboxylate oxygen would be expected to occur if both of the nitrogen donors of histidine are also coordinated, thus (presumably) forming a polymeric structure. The products synthesized in Method A contain both protonated carboxylic acid groups and coordinated carboxylate groups, according to the results of the infrared data. The values obtained for the molecular weight of these products from light scattering are consistent with a polymeric product but may be high due to molecular association.

## REFERENCES

1. C. Carraher and R. Strothers, in: "*Applied Bioactive Polymeric Systems*," C. G. Gebelein, C. E. Carraher & V. Foster, Eds., Plenum, NY, 1989, pp. 139-150.
2. D. Siegmann, D. Brenner and C. Carraher, PMSE, 59, 535 (1988).
3. C. Carraher, R. Strothers and D. Brenner, PMSE, 57,173 (1987).
4. D. Siegmann, C. Carraher and A. Friend, J. Polymer Materials, **4**, 19 and 29 (1987).
5. J. P. Greenstein and M. Winitz, "*Chemistry of the Amino Acids*," Vol. 3, John Wiley and Sons, New York, 1961, Chapter 6.
6. G. Wiedemann, J. Prakt. Chem., **42**, 255 (1847); **43**, 271 (1848).
7. H. Ley, Z. Electrochem., **10**, 954 (1904).
8. H. Ley, Ber., **42**, 354 (1909); B. Kirson and Barsily, Bull. Chim. Soc. France, 1336 (1957).
9. M. N. Hughes, "*The Inorganic Chemistry of Biological Processes*," John Wiley and Sons, New York, 1972, Chapter 1.
10. D. R. Williams, Chem Rev., **72**(3), 203 (1972).
11. R. J. Angelici, in: "*Inorganic Biochemistry*, Vol. 1, G. L. Eichorn, Ed., Elsevier, New York, 1973.
12. H. Sigel, "*Metal Ions in Biological Systems*," Vol. 9, Marcel Dekker, New York, 1979, Preface and Chapter 1.
13. R. J. Sundberg and R. B. Martin, Chem. Rev., **74**(4), 470 1974.
14. H. P. Block, in: "*Inorganic Polymers*," F. G. A. Stone and W. A. G. Graham, Eds., Academic Press, New York, 1962, Chapter 5, p. 189-307.
15. C. A. McAuliffe, J. V. Quagliano and L. M. Vallarino, Inorg. Chem., 5, 1996 (1966).
16. B. Rosenburg, in: "*Nucleic Acid-Metal Ion Interactions*," T. G.Spiro, Ed., Wiley-Interscience, New York, 1980, pp. 1-29.
17. C. F. J. Barnard, M. J. Cleare, P. C. Hydes, Chemistry in Britain, 1001 (1986).
18. D. R. Williams, Inorg. Chem. Acta Rev., **123**, (1972).
19. A. J. Carlson, K. E. Trainor and E. C. Walton, J. Proc. Roy. Soc. New South Wales, **108**, 6 (1975).
20. A. J. Carlson, R. J. Banner, R. P. Gale, N. T. McArdle, K. E. Trainor and E. C. Walton. J. Clin. Hemat. Oncol., **7**, 294 (1977).

21. R. D. Graham and D. R. Williams, J. Inorg. Nucl. Chem, 41, 1245 (1979).
22. H. R. Allcock, R. W. Allen and J. P. O'Brien, J. Am. Chem. Soc., 99 (12), 3984 (1977).
23. F. R. N. Gurd and P. E. Wilcox, in: *"Advances in Protein Chemistry,"* Vol. II, M. L. Anson, K. Baqiley and J. T. Edsall, Eds., Academic Press, New York, 1956, p. 311.
24. Ei-Ichiro Ochiai, *"Bio-Inorganic Chemistry, An Introduction,"* Allyn-Bacon, Massachusetts, 1977.
25. M. Hatano and T. Nozawa, in: *"Metal Ions in Biological Systems,"* Vol. 5, H. Sigel, Ed., Marcel Dekker, New York, 1976.
26. S. T. Chow and C. A. McAuliffe, in: *"Progress in Inorganic Chemistry,"* Vol. 19, S. J. Lippard, Ed., Interscience, New York, 1975, p. 51.
27. B. L. Vallee, in: *"Metal Ions in Biological Systems,"* Advances in Experimental Medicine and Biology, Vol. 40, S. K. Dhar, Ed., Plenum Press, New York, 1973, p. 1.
28. *"Handbook of Chemistry and Physics,"* 65th edition, CRC Publishing Co., Boca Raton, Florida, 1985.
29. R. Barker, *"Organic Chemistry of Biological Compounds,"* Prentice-Hall, New Jersey, 1971.
30. C. C. McDonald and W. D. Phillips, J. Am. Chem. Soc., 90, 6371 (1968).
31. (a) R. A. Carlson and T. L. Brown, Inorg. Chem., 5, 268 (1966).
32. T. G. Appleton and J. R. Hall, J. Chem. Soc., Chem Commun., 911, 1983.
33. D. P. Craig and R. S. Nyholm, in: *"Chelating Agents and Metal Chelates,"* F. P. Dywer and D. P. Mellor, Eds., Academic Press, New York, 1964.
34. F. R. Hartley, *"The Chemistry of Platinum and Palladium,"* Applied Science, London, 1973.
35. A. J. Thompson, R. J. P. Williams and S. Reslova, in: *"Structure and Bonding,"* Vol. 2, Springer-Verlag, Berlin, 1972.
36. R. B. Martin, in: *"Metal Ions in Biological Systems,"* Vol. 9, H. Sigel, (Ed.), Marcel Dekker Inc., NY, 1979, Chapter 1.
37. E. W. Wilson and B. Martin, Inorg. Chem., 9(3), 528 (1970).
38. Y. Nakayama, K. Matsumoto, S. Ooi, H. Koroya, J.Chem. Soc. Chem. Comm., 170 (1973).
39. G. R. Lenz and A. E. Martell, Biochemistry, 3, 750 (1964).
40. R. Leberman and B. R. Rabin, Trans. Faraday Soc., 55, 1660 (1959).
41. N. C. Li, B. E. Doody and J. M. White, J. Am. Chem. Soc., 79, 5859 (1957).
42. D. D. Perrin and V. S. Sharma, J. Chem. Soc, A. 724 (1967).
43. J. H. Ritsma, J. C. Vander Grampel and F. Jellinek, Rec. Trav. Chim., 88, 411 (1969).
44. H. Sigel, R. Criesser and D. B. McCormick, Arch. Biochem. Biophys., 134, 217 (1969).
45. C. B. Acland and H. C. Freeman, Chem. Commum., 1016 (1971).
46. M. J. Adams, D. C. Hodgkin and V. A. Racburn. J. Chem. Soc., (A), 2632 (1970).
47. H. C. Freeman and R. P. Martin, J. Biol. Chem., 224, 4823 (1969).
48. L. E. Erickson, J. W. McDonald, J. K. Howie, R. P. Chow, J. Am. Chem. Soc., 90, 6371 (1968).
49. L. M. Volshtein and I. G. Lukyanova, Neorg. Khim, 11(6), 708, (1966).
50. L. M. Volshtein and L. D. Dikanskaya, Russ J. Inorg. Chem., 19(1), 81, (1974).
51. L. M. Volshtein and L. D. Dikanskaya, Russ. J. Inorg. Chem., 13(9), 1304 (1968).
52. G. R. Brubaker and D. H.Busch, Inorg. Chem, 2110 (1966).
53. L. M. Volshtein and I. G. Luk'yanova, Russ. J. Inorg. Chem., 11(6), 708 (1966).

54. V. Saudek, H. Pivcova, D. Noskova and J. Drobnik, J. Inorg. Biochem., 23, 55 (1985).
55. D. D.Nelson and H. Frye, Z. Naturforsch., 21, 630 (1966).
56. L. M. Volshtein and O. P. Slyudkin, Russ. J. Inorg. Chem., 17(8), 1168 (1972), and Reference 10; p. 169.
57. L. F. Grantham, T. S. Elleman and D. S. Martin, J. Am. Chem. Soc., 77, 2966 (1955).
58. K. Nakamoto, Y. Morimoto and A. E. Martell, J. Am. Chem. Soc., 83, 4528 (1961).
59. M.Avram and G. H. Mateescu, "*Infrared Spectroscopy,*" Wiley-Interscience, NY, 1972.
60. K. Nakamoto and P. J. McCarthy, "*Spectroscopy and Structure of Metal Chelates,*" John Wiley, NY, 1968.
61. K. Nakamoto, "*Infrared and Raman Spectra of Inorganic and Coordination Compounds,*" 4th edition, John Wiley, N Y, 1968.
62. F. S. Parker, "*Applications of Infrared Spectroscopy in Biochemistry, Biology, and Medicine,*" Plenum Press, NY, 1971.
63. R. J. Koegal, J. P. Greenstein, M. Winitz, S. N. Birnbaum and R. A. McCallum, J. Am. Chem. Soc., 77, 5708 (1955).
64. M. Tsuboi, T. Onischi, I. Nakagawa, T. Shimanouchi and S. Mizushima, Spectrochim. Acta., 12, 253 (1958).
65. J. R. Kincaid and K. Nakamoto, Spectrochim. Acta , 32A, 277 (1976).
66. L. Larson, Acta Chem. Scand., 4, 27 (1950).
67. J. A. Kieft and K. Nakamoto, J. Inorg. Nucl. Chem., 29, 2561 (1967).
68. K. Fukushima, T. Onishi, T. Shimanouchi and S. Mizushima, Spectrochim. Acta , 13, 236 (1959).
69. R. A. Condrate and K. Nakamoto, J. Chem. Phys., 42(7), 2590 (1965).
70. C. N. R. Rao, "*Chemical Applications of Infrared Spectroscopy,*" Academic Press, NY (1963).
71. R. M. Silverstein and G. C. Bassler, "*Spectroscopic Identification of Organic Compounds,*" John Wiley, New York, 1981.
72. G. C. Perry and H. S. Stenton, J. Chem. Soc. Dalton Trans., 1446, 2429 (1976); Spectrochim. Acta , 32(A), 1287 (1967).
73. V. Balice and T. Theophanides, J. Inorg. Nucl. Chem., 32, 1237 (1970).
74. M. Cordes and J. L. Walker, Spectrochim. Acta , 24A, 237 (1968).
75. C. N. C. Drey and J. S. Fruton, Biochemistry, 4, 1 (1965).



# MEDICAL APPLICATIONS OF HYALURONAN AND ITS DERIVATIVES

Endre A. Balazs

Matrix Biology Institute
Ridgefield, NJ
and
Biomatrix, Incorporated
Ridgefield, NJ

**Hyaluronan (hyaluronic acid) is a natural polyanionic polysaccharide (glycosaminoglycan) present in the intercellular matrix of most vertebrate connective tissues. The unique properties of hyaluronan result in a molecular network which, in highly hydrated conditions (>99% water content), is extremely elastoviscous and pseudoplastic. These rheological properties are responsible for the biological function of hyaluronan and its medical applications. During the last decade, a highly purified fraction of hyaluronan was introduced in human and veterinary medicine as a therapeutic device for arthritis and eye surgery. The use of elastoviscous hyaluronan solutions in surgery was called viscosurgery. The medical use of viscous hyaluronan solutions for the treatment of arthritis is called viscosupplementation. The use of the natural, highly purified, high molecular size hyaluronan did not fulfill all the medical needs for elastoviscous fluids. Therefore, in the mid-1980s the development of crosslinked derivatives of hyaluronan began. Fluids, gels and solids were made from the native high molecular mass hyaluronan by creating permanent attachment between the polysaccharide chains. These crosslinked derivatives of hyaluronan are called hylans (generic name). Hylans have the same tissue compatibility as native hyaluronan. They, however, have a considerably longer residence time in tissues, which offers a major advantage in their medical efficacy compared to hyaluronan. Most importantly, hylan fluids and gels have enhanced rheological (elastic, viscous, pseudoplastic) properties and therefore more suitable for viscosurgery use and for viscosupplementation.**

## INTRODUCTION

Hyaluronan, a glycosaminoglycan, is present in nearly all tissues of the vertebrate body. As an elastoviscous fluid, it fills the space between cells and collagen fibrils of all connective tissues and covers the

*Cosmetic and Pharmaceutical Applications of Polymers*
Edited by C.G. Gebelein *et al.*, Plenum Press, New York, 1991

surface of some endothelial tissues. Its function is to protect and lubricate cells, and regulate their movement in the elastoviscous matrix. This elastoviscous matrix also stabilizes the collagen network and protects it from mechanical dislocation. Highly elastoviscous solutions of the non-inflammatory fraction of hyaluronan were introduced into medicine as viscosurgical tools in ophthalmology and as a treatment to decrease arthritic pain and restore normal joint function by viscosupplementation of the joint fluid.

Hylans are crosslinked forms of hyaluronan in which the crosslinking process does not change the polyanionic character of the polysaccharide chain but creates fluids, gels and solids with a variety of viscoelastic properties. Hylans were developed to broaden the spectrum of the medical use of the natural hyaluronan. This is achieved by varying the rheological properties of hylan devices and by their longer residence time in the tissues compared with that of native hyaluronan. Hylans, just as hyaluronan, are highly biocompatible. They do not cause immunological reactions, are compatible with all components of blood and do not elicit foreign body reaction or capsule formation.

Hylan devices are being used in clinical studies for a great variety of medical applications. Hylan gel slurries and hylan fluids are used as viscosurgical tools and implants in ophthalmology. Hylans are used as viscosupplementation devices to treat arthritic pain and stiffness. Hylan gels and solids are used in matrix engineering to influence and control the healing and regenerative process in connective tissue. Hylan devices are also used as drug delivery systems. The development of the highly purified, non-inflammatory fraction of hyaluronan and the subsequent development of hylans has opened new chapters in medicine, with the introduction of such techniques as viscosurgery, viscosupplementation and matrix engineering.

## HYALURONAN

Hyaluronan (hyaluronic acid, HA, or hyaluronate) is a glycosaminoglycan whose disaccharide repeating unit is N-acetylglucosamino $\beta1\rightarrow4$ Na-glucuronate. These units are linked with $\beta1\rightarrow3$ glucoside bonds to an unbranched chain of 10 to 15 thousand disaccharides.[1] This long polysaccharide chain, with a molecular mass of 4-5 million in physiological salt solution (ionic strength 0.15, pH ~7.0), forms a random coil and occupies a large spheroidal hydrated volume whose largest diameter is in the magnitude of 400-500 nm. This means that the hydrated specific volume of this hyaluronan molecule can be as high as four (4) liters per gram. In other words, in one liter of solvent, 0.25 grams of hyaluronan occupies all of the solvent space. Thus one can say that a hyaluronan solution, with a concentration of 0.25 g/L (0.025% or 0.25 mg/mL), is saturated.[2] In this solution, all hyaluronan molecules touch each other and leave no space for other solute molecules in the solvent. Other solutes, small or large, must occupy the solvent space inside the molecular domain of the hyaluronan or be "sandwiched" in between the individual hyaluronan molecules. The important chemical properties of this polysaccharide are that it contains carboxylic acid groups and that the amino group of the glucosamine is acetylated. Thus hyaluronan is a polyanion. As a polyanion with an extremely large hydrated volume the cationic counterions can be "bound" to the anionic site to various degrees. This coulombic interaction depends on the size and nature of the cation and the concentration and molecular volume of the hyaluronan.[3]

Hyaluronan is synthesized by enzymes located in the cell membrane.

During synthesis the long polysaccharide chain is extruded from the cell membrane into the extracellular matrix.[4,5] Many different types of cells possess the hyaluronan synthesizing enzyme system,[6] including cells of the connective tissue such as fibroblasts and mononuclear phagocytic cells, as well as epithelial and endothelial cells. This does not mean that under physiological conditions in the normal adult organism these enzymes are active. It is known, however, that hyaluronan is present in the intercellular matrix of many tissues. It is present in all connective tissue matrices at relatively high concentrations (0.10 to 10 mg/mL). It is also found in tissue fluids such as the blood, aqueous, pleural fluid, cerebrospinal fluid and synovial fluid. In these fluid compartments, however, the hyaluronan concentration is extremely low ($10^{-5}$ to $10^{-3}$ mg/mL).[7,9] The turnover of the hyaluronan in the body fluids is very fast (half-life time: a few minutes to 8 hours)[8,9] compared with that in the connective tissue matrices, which is relatively long (half-life time: 2-30 days).[10]

The biological function of hyaluronan in the extracellular matrix is based on the rheological properties of its "concentrated" (0.1 - 10.0 mg/mL) solutions.[11,12,13] Hyaluronan, in physiological salt solution at this concentration, forms pseudoplastic solutions with extremely shear- and frequency-dependent viscosity and elasticity. These rheological properties are dependent on the molecular volume and concentration of hyaluronan.[13] Greater molecular volume and higher concentration yield more elasticity and viscosity. The naturally-occurring hyaluronan solutions in the body such as the liquid vitreous of the eye and the joint fluid (synovial fluid) show high elastoviscous properties. The biological functions of these solutions are based on their rheological properties.[14] They have protective, shock-absorbing and structure stabilizing effects. Since highly elastoviscous hyaluronan solutions also fill the space between collagen fibrils and filaments in such tissues as the gel vitreous of the eye, the synovial and capsular tissues of the joint and the subcutaneous tissues, the role of this polysaccharide is to maintain the separation of the fibrils and, in general, to stabilize the extracellular structures.[15] The flow of water is considerably decreased in elastoviscous hyaluronan solutions; therefore, the collagen-hyaluronan matrix stabilizes tissue structures by preventing the movement of fluid in tissues under mechanical impact.[11,12]

Another biological function of the molecular network of hyaluronan is the regulation of movement and function of solutes in the extracellular space. The density and negative charge of the molecular network determine the flow or diffusion of molecules. The network also acts as a charged molecular sieve for metabolites with various sizes and charges.[16]

Finally, the elastoviscous solution of hyaluronan acts as a regulator of movement, function and mitosis of white blood cells and cells of the lyphomyeloid system in connective tissue when they migrate to and from the lymph and blood vessels.[17,21] It was shown using *in vitro* systems that the phagocytic activity and prostaglandin-releasing function of mononuclear phagocytes are inhibited by the presence of elastoviscous solutions of hyaluronan in the medium. The viscosity-dependent regulation of cellular activity of lymphomyeloid cells is an important aspect of the biological function of hyaluronan.[17-22]

An entirely different type of biological function was found in tissue culture models. When cells grow on plastic surfaces in media that is free of exogenous hyaluronan, some cell lines produce hyaluronan. It was suggested that this hyaluronan, which is added to the immediate environment of the cell, influences and regulates the aggregation, movement and chemotactic activity of some cells.[23-28] It must be emphasized that the concentration of the hyaluronan in the culture media of these cells is very low and not comparable to the elastoviscous medium surrounding cells in the intercellular matrix of connective tissue.

Another biological activity was proposed for hyaluronan during embryonic development, tissue repair and tumor development. Based mostly on *in vitro* studies it was suggested that hyaluronan synthesis by certain cells is connected with the regulation of cell proliferation, cell-cell interaction, cell migration and regulation of growth factors.[29-32] Under such experimental conditions the concentration of hyaluronan is not sufficient to produce an elastoviscous environment for cells. One must conclude that there are two entirely different hyaluronan environments for cells: one in which the hyaluronan concentration in the microenvironment of the cell is too low to form an elastoviscous matrix, and another in which the cells are embedded in an elastoviscous matrix. Under either *in vitro* or *in vivo* conditions, cells apparently respond differently to these distinctively different hyaluronan environments. The regulatory mechanism of cell activity ascribed to hyaluronan must be evaluated in view of the type of cell (embryonic fibroblasts, adult fibrocytes, lymphomyeloid cells, cells of endothelial origin, etc.) and the amount (concentration) and rheological properties (elastoviscosity) of the hyaluronan or the hyaluronan-collagen matrix surrounding the cells.

## HYLANS

Hylan is the generic name of a crosslinked hyaluronan molecule in which the carboxylic acid and the N-acetylglucosamine groups are not affected by the crosslinking process. Thus, the characteristic chemical nature of the polymeric chain, its polyanionic character, and the acetamido group are not altered. These unaltered, long polysaccharide chains are crosslinked by two different methods. In one procedure, the hyaluronan molecule is treated with aldehydes in the presence of certain specific protein molecules. A methylene crosslink is created through a protein bridge between the hydroxyl groups of two neighboring hyaluronan chains and the amino or imino groups of the bridge protein. Thus, several hyaluronan molecules are permanently coupled into one larger molecular enity. The amount of crosslinked protein in this molecule is less than 0.5% (per weight) of the polysaccharide content. The molecular mass of this hylan can be several times that of the native hyaluronan molecule (8 to 25 million). This hylan is water soluble and the elastoviscous properties of this fluid are enhanced over those of the solutions of natural hyaluronan.[33,34]

In another crosslinking procedure, a sulfonyl-bis-ethyl crosslink is created between two hydroxyl groups of the polysaccharide chains. The resulting molecular network is water insoluble. Depending on its polymeric concentration and the extent of crosslinks, soft gels (0.2-1.0% polymer content) or solids (particles, membranes, tubes and other forms) can be created with a polymer content of 1-20%. The soft gels are mostly used as slurries of deformable particles of various sizes (0.1-1 mm). This gel slurry can be used as such or dispersed in a viscoelastic hylan fluid, forming a two phase system.[35] For example, such a two phase system can contain 20% per volume hylan gel slurry and 80% per volume hylan fluid. The solvent of both phases is the same physiological salt solution. The gel phase can contain 0.4% polymer and the liquid phase can contain 1% polymer. The two phases differ primarily because the gel phase is water-insoluble and is suspended in the more concentrated, highly elastoviscous hylan fluid. The water soluble hylan molecules can partially penetrate the gel particles. These solids are used in various shapes and forms as tissue implants.[33-36]

Thus, with two different crosslinking procedures, four types of hyaluronan derivatives were created.

1. Hylan fluid: a water soluble molecule with enhanced molecular size and rheological properties.
2. Hylan gel: a water insoluble, highly hydrated gel and its slurries (99.0-99.8% hydration fluid).
3. Hylan gel-fluid: two phase system, with varying gel-slurry and fluid volume ratios.
4. Hylan solids of various forms (80-95% hydration fluid).

The enhanced rheological properties of the hylan fluid and the special rheology of the hylan gel solids and fluid-gel mixtures considerably broaden their use in medicine.

## BIOCOMPATIBILITY OF HYALURONAN AND HYLANS

The prerequisite for the use of these polysaccharides in medicine as tools, tissue supplements, implants or delivery systems is their biological compatibility. The great advantage of using polysaccharides is that they are non-immunogenic. No antibody formation or other immunological reaction could be observed as a result of repeated use of hyaluronan or hylan in animal and human bodies.[37-40] The second most important aspect of biocompatibility of hyaluronan and hylan is that they do not cause inflammation or foreign body reaction (granulation, capsule formation). During the development of high molecular mass hyaluronan (molecular mass ≥2 million) for medical use, it was found that it was important to remove a fraction of hyaluronan that was closely associated with an inflammatory agent. The development of a highly purified hyaluronan fraction that did not cause inflammation when injected into various tissues (eye, joint, abdominal cavity) of the animal and human body made possible its use in medicine. This hyaluronan fraction was called the non-inflammatory fraction of hyaluronan (NIF-NaHA).[42-45]

Hylan fluid and hylan gel are also free from inflammatory agents. NIF-NaHA and hylans (fluid and gel and their mixtures) also are highly biocompatible because they do not initiate the so-called foreign body reaction which the implantation of all other polymeric materials produces to various degrees. The reason for the biological compatibility of NIF-NaHA is the fact that it is applied and used in the same tissue compartments where it is present as a natural component of the extracellular matrix. Hylans show the same biocompatibility as the native hyaluronan. This means that the crosslinking process did not change the molecule in such a manner that the body can differentiate it from the native molecule. This is further emphasized by the finding that the catabolic pathway for hylan molecules is the same as that of hyaluronan.[41]

## METABOLISM OF HYALURONAN AND HYLAN

The metabolic pathway of hyaluronan is fairly well known.[46] Briefly, it is as follows: the hyaluronan, produced by the cell membrane and extruded to the intercellular matrix, diffuses or flows with the extracellular fluids through the lymph vessels to the blood. There is evidence that during this process some of the hyaluronan enters the lymph nodes where it is metabolized. But the bulk of the hyaluronan passes unaltered from the lymph to the blood and is removed by the liver endothelium. The hyaluronan molecule is completely degraded in the liver to carbon dioxide and water. A small amount of this molecule is also taken up from the blood by the kidney where it is degraded and secreted. No evidence has been found for degradation of this molecule in the extracellular matrix of normal tissues.

Exogenous hyaluronan (NIF-NaHA) introduced into such tissue spaces as the vitreous and aqueous of the eye and the joint fluid cleared these spaces at the same speed as the endogenous material.[10,47]

The catabolism of hylan fluid injected into various tissue spaces follows the path of endogenous or exogenous hyaluronan. Because the molecular size and volume of the soluble hylan is greater than that of the native hyaluronan, its half-life time in various tissues is somewhat longer. From the blood, however, diluted solutions of hylan fluid are removed by the liver in the same manner and at the same speed as hyaluronan. The half-life of hylan gel in various tissues depends on the movement of fluids and solid tissues in that compartment. The vitreous of the eye represents one extreme condition where the tissue is not exposed to any mechanical movement or fluid flow. In this tissue the implanted hylan gel does not degrade and remains unchanged for years, as we can tell from experiments in monkey eyes. The joint space (synovial space) represents another extreme condition where the tissue is exposed to constant movement that provides mixing and flow of the joint fluid. Hylan gel slurry injected into the fluid space of the joint is driven by mechanical forces to the synovial tissue and from there to the lymph vessels. During this process, the size of the gel particles decreases, and finally the particles completely disappear from the joint tissues. The half-life time of the hylan gel in the joint space is 8-9 days. The residence time of hylan gel in the anterior chamber of the eye and the skin was also studied, and it was found to be very long (months), depending on the mechanical stress to which the tissue is exposed. Partially degraded hylan gel injected into the blood is completely taken up by the liver and degraded and excreted through the same pathway as hyaluronan. No accumulation of any part of the hylan gel could be observed in any tissue.[41]

## MEDICAL APPLICATION OF HYALURONAN AND HYLAN

All medical applications of hyaluronan and its crosslinked derivative, hylan, are based on the physicochemical and rheological properties of the molecular matrices they form when hydrated in physiological salt or buffer solution. The physical and rheological properties of these hydrated matrices depend on the volume of the individual molecules, their interaction (entanglement) and the amount (substitution) and nature of crosslinks. Their biological and medical function depends on the rheological properties and the residence time of these polymers in the tissue where they are applied.[42,44,47-50]

### 1. Viscosupplementation

The first medical use of hyaluronan was conceived in the 1960s. Its intended use was for the replacement of pathological joint (synovial) fluid in various painful arthritic conditions. The objective was to replace the synovial fluid, which in pathological conditions had lost its normal elastoviscosity, with a highly elastoviscous hyaluronan solution (NIF-NaHA).[48,49,51] The replacement constitutes a supplementation of the rheological properties of the fluid. Thus, this medical application was later called viscosupplementation.[42] Animal model studies[52] and clinical studies in arthritic race horses[51,53] and arthritic human patients showed that viscosupplementation is an efficacious treatment of painful arthritis.[54-57] The benefit of this treatment is the decrease of pain in the joint and consequent improvement of joint mobility and function. In the late 1970s, based on the concept of viscosupplementation, the treatment of arthritic race horses with elastoviscous hyaluronan solution (NIF-NaHA) became a standard medical treatment world-wide.[58-67] During the late 1980s NIF-NaHA was marketed in Japan and Italy for the treatment of painful osteoarthritis in humans.[68]

During the past years hylan fluid-gel mixtures have also been used in clinical studies for viscosupplementation in arthritic joints in both horses and humans. These studies showed that the more elastoviscous hylans with a longer residence time in the joints are more efficacious than NIF-NaHA for the treatment of painful arthritis. The mechanism of the beneficiary effect of viscosupplementation in the joint is physical. In arthritic joints the elastoviscosity of the synovial fluid that fills the space between the surfaces of moving tissues is considerably lower than in normal tissues. The high elastoviscosity of the synovial fluid is essential for the lubrication and protection of cells and tissues, and one of the causes of joint pain and the subsequent decrease of joint mobility is the loss of this protective elastoviscous mechanism. With viscosupplementation this protection is restored until the joint can regenerate its own elastoviscous fluid and restore normal joint function.[69,70].

## 2. Viscosurgery in Ophthalmology

Viscosurgery is defined as the use of elastoviscous substances during or after surgery to facilitate surgical procedures and to aid the healing and regenerative processes in the surgical wound.[71] Thus, in viscosurgery the elastoviscous substance is used to separate and move tissues, create space for surgery, and protect tissue surfaces that are sensitive to mechanical damage or drying out during the surgical procedure.[72-74] The elastoviscous substances are used in these cases as surgical tools to carry out the objectives of the surgery.[44,75-77]

Viscosurgical implants are used postsurgically to decrease bleeding and to fill tissue spaces to prevent postoperative adhesions, excessive scar formation, and to provide a passage flow of tissue fluids.

The first application of viscosurgical implants and viscosurgery was in retinal detachment surgery and corneal transplantation. Elastoviscous NIF-NaHA solutions were used to facilitate the repositioning of the detached retina and fill the vitreous space after the pathological vitreous was removed. In corneal transplantation, NIF-NaHA was used to protect the corneal endothelium of the transplanted tissue from mechanical damage during surgery.[78-90] Later NIF-NaHA, as a viscosurgical tool, became widely used in surgery of the anterior segment and especially in cataract surgery when the cataractous lens was replaced with an artificial intraocular lens.[91-97] Today viscosurgery is routinely used in cataract surgery to manipulate tissues, create space and to protect sensitive tissues during the implantation of intraocular lenses. NIF-NaHA was the first viscosurgical tool and implant marketed world-wide under the trade name Healon® in the early 1980s (Pharmacia AB, Uppsala, Sweden).[45] Following the success of Healon, several other hyaluronan-based viscosurgical products have been marketed, but none of them has as high an elastoviscosity as Healon. Therefore, Healon remains today the most widely used viscosurgical tool and implant in ophthalmic surgery.[94,98-100]

The crosslinked derivatives of hyaluronan, the hylans, offer additional advantages as viscosurgical tools and implants. The greater elastoviscosity of hylan fluids and the unique rheological properties of the water-insoluble hylan gel slurries give a new dimension to their surgical use in ophthalmology. Studies in primates showed both the safety and efficacy of viscosurgical applications of hylans in glaucoma and retinal detachment surgery.[101-108] Clinical studies in which hylan fluid is used for surgery where greater elastoviscosity is needed than can be provided by hyaluronan-based products are currently in progress. Hylan gel slurry is used in clinical studies for retinal and glaucoma surgery and as a viscosurgical implant. Hylan fluid is used in trauma surgery as a viscosurgical tool. In all of these surgical procedures, the use of Healon did not completely fulfill the medical need.

## 3. Viscosurgery in Other Surgical Specialties

It was suggested that the viscosurgical principles so successfully used in ophthalmology be extended to other surgical and microsurgical areas.[76] The first such application was in orthopedic surgery.[109] Clinical studies are in process on the use of hylan fluid in arthroscopic surgery of the knee and other joints including the temporomandibular joint. During arthroscopy, the joint surfaces (the cartilage and synovial tissues) are exposed to mechanical damage (scuffing) by the instrument used. The use of an elastoviscous solution like hylan fluid provides protection to these tissue surfaces. This protection is extremely important, because it is known that some of these tissues do not heal after mechanical damage. Viscosurgery in arthroscopy can provide better handling of tissues and tissue debris and, in general, facilities the surgical procedure.[109-112]

Viscosurgical implants (Healon®) were used in the early 1970s in animal models to demonstrate that highly elastoviscous NIF-NaHA, when used to fill the cartilage wound, stimulated its healing.[49,52] It also decreased scar formation in subcutaneous wounds and reduced adhesion formation between traumatized tendon and tendon sheaths in monkey hands.[113-115] Capsule formation around plastic implants was also reduced when Healon was applied as a protective shield around the implant.[52]

The development of hylan gel triggered a new interest in the use of viscosurgical implants. Animal experiments with hylan gel slurries clearly demonstrated that hylan gel prevents the formation of undesirable adhesions and excessive internal scars. Clinical studies are in progress to evaluate the efficacy of hylan gel slurry in preventing adhesions between tendon and tendon sheaths, decreasing scar formation around peripheral nerves (neutral entrapment) and around nerve roots (laminectomy) and in other postsurgical applications in orthopedic surgery.[116-119]

It has been suggested that viscosurgical tools and implements be used in otosurgical procedures. Pilot clinical studies were carried out in middle ear surgery to investigate the effect of Healon on the healing of the tympanic membrane and the reepithelization of mucous membranes.[120-124]

## 4. Matrix Engineering

**Matrix engineering is defined as the therapeutic use of various biological macromolecules from the intercellular matrix for promoting, directing or modifying the regenerative and healing processes in the various tissues in the body.[48] Historically, the first material used for this purpose was collagen as surgical suture material. The first real matrix engineering was proposed in the mid 1960s with the use of so-called "reconstituted vitreous".[125-127] This was a human tropocollagen (salt soluble collagen) solution mixed with high molecular mass hyaluronan. The mixture of these two biological macromolecules extracted from the intercellular matrix of human umbilical cords was injected into the eye to replace the liquid vitreous. As the cold solution warmed to body temperature, a gel formed which was made of a random network of non-crosslinked collagen fibrils filled with hyaluronan molecules. Thus, a reconstituted vitreous, the simplest intercellular matrix, was engineered to fill the vitreous space inside the eye. The rheological properties of this matrix implant were comparable to the natural vitreous gel which is also made of a random collagen fibrillar network filled with hyaluronan. While the "reconstituted vitreous" never became a usable matrix implant, because of its lack of optical purity, the concept was used later for matrix augmentation in the skin to smooth wrinkles and scars.[128]**

**Matrix engineering with high molecular mass hyaluronan (NIF-NaHA) was not very successful because of the short residence time of this natural substance in the tissues. Water-insoluble hylan gel, however, is an ideal substance for this purpose. When implanted in the extracellular matrix of connective tissues, it does not cause inflammation, immunological or foreign body reaction, or capsule formation. It remains in the tissue indefinitely. When applied as a gel slurry, the fibrocytes of the surrounding tissue slowly fill the crevasses between the gel particles with newly-formed collagen fibrils.[129]**

**Hylan gel slurry was used in mouse and guinea pig dermis for tissue augmentation. Radio-labeled material remained in the tissue for more than 28 days, with no evidence of tissue capsule formation or adverse reaction.[129-130] Clinical studies are in progress to use hylan gel slurry in matrix augmentation in the dermis to fill wrinkles and smooth scars.**

**Hylan gel slurry, as a matrix engineering implant, was placed in between the two ends of severed peripheral nerves. Nerve elements (axons) regenerated in between the gel particles. Most importantly this regeneration was faster and more complete in the presence of the gel than in its absence, and the presence of the gel also prevented the invasion of connective tissue cells into the regenerating area of the nerve.[131]**

**Another use of hylan gel in matrix engineering was shown using a guinea pig urinary sphincter model. Hylan gel slurry introduced into the connective tissue matrix of the sphincter muscle improves sphincter function by filling the space between the muscles, thus making it an ideal substance to be used for the treatment of urinary incontinence.[130,132]**

**A modified hylan gel was developed for percutaneous embolization.[133] This gel contains blood clotting factors and radio-opaque material. When introduced through small lumen catheters to the appropriate vascular site, permanent vascular occlusion is effected. The embolus permanently blocks blood circulation without causing imflammation or damage to neighboring tissues, and the presence of radio-opaque material permits the localization of the embolus radiologically. Animal studies with this hylan gel indicate that it will be an important tool for treatment of**

aneurysms, arteriovenous malformations and other vascular lesions.[133-136]

The extraordinary blood compatibility of hylan has made it possible to use it as a coating or filling material for vascular grafts and prostheses and for blood handling equipment. Numerous *in vitro* and *in vivo* studies have shown that hylan or hylan coated surfaces extend partial thromboplastin times, prevent protein adsorption, do not trigger activation of coagulate factor XII, and prevent platelet activation and adhesion.[130] Thus, materials used for blood handling (dacron, cellulose, polyurethane), when coated or filled with hylan, have enhanced tissue and blood compatibility. Thus blood coagulation, denaturation of blood proteins, complement activation, fibrinolysis and, in general, inflammatory reactions can be prevented by hylan or hylan coated surfaces. When certain polymeric surfaces are coated with hylan, endothelization can be achieved. Thus, matrix engineering with hylan can be applied to promote healing and regeneration of vascular tissues.[137-139]

## 5. Drug Delivery Compartments

Hylan gels, microparticles and membranes were developed for the controlled and localized delivery of biologically active molecules. *In vitro* and *in vivo* studies showed that hylan substances provide sustained controlled release of specific drug molecules. At the same time, the extraordinary biocompatibility of hylan prevents inflammation or capsule formation which could interfere with the mechanism of delivery of drugs. Studies are in progress to use hylan delivery compartments with such pharmacologically active molecules as antibiotics, growth factors, anti-inflammatory and anti-neoplastic agents.[140-142] The hylan gel developed for percutaneous embolization was also tested as a drug delivery component through the formed embolus.[143]

The development of the hylans significantly expanded the applications made possible using elastoviscous solutions of ultrapure hyaluronan by adding these hyaluronan-based derivatives to the medical compendium. Additionally, the availability of various forms of hylans such as fluids, gel slurry, gel-fluid mixtures, membranes, particles and tubes opened new potentials for the medical use of this highly biocompatibility, polysaccharide polymer. The availability of this new class of polymers triggered such new medical treatments as viscosupplementation, viscosurgery and matrix engineering, and also opened a new chapter in drug delivery technology.

## REFERENCES


1. J. E. Scott, in: "*Ciba Foundation Symposium 143: Biology of Hyaluronan*," J. Wiley and Sons, New York, 1989, pp. 6-15. "Secondary structures in hyaluronan solutions: chemical and biological implications."
2. E. A. Balazs, Fed. Proc., 17, 1086-1093 (1958). "Physical chemistry of hyaluronic acid."
3. E. A. Balazs, J. V. Davies, G. O. Phillips, and D. S. Scheufele, J. Chem. Soc., 12, 1424-1428 (1968)."Polyanions and their complexes. IV. A pulse radiolysis study of the interaction between methylene blue and heparin in aqueous solution."
4. P. Prehm, Biochem. J., 220, 597-600 (1984). "Hyaluronate is synthesized at plasma membranes."
5. P. Prehm, Biochem. J., 267, 185-189 (1990). "Release of hyaluronate from eukaryotic cells.

6. J. R. E. Fraser and T. C. Laurent, in: *Ciba Foundation Symposium #143: Biology of Hyaluronan,* D. Evered and J. Whelan, Eds., J. Wiley & Sons, New York, 1989. "Turnover and metabolism of hyaluronan."
7. A. Engström-Laurent, in: *Ciba Foundation Symposium #143: Biology of Hyaluronan,* J. Wiley & Sons, New York, 1989, pp. 233-247. "Changes in hyaluronan concentration in tissues and body fluids in disease states."
8. J. R. E. Fraser and T. C. Laurent, in: *Ciba Foundation Symposium #143: Biology of Hyaluronan,* J. Wiley & Sons, New York, 1989, pp. 41-59. "Turnover and metabolism of hyaluronan."
9. H. D. Schubert, J. L. Denlinger, and E. A. Balazs, Exp. Eye Res., 39, 137-152 (1984). "Exogenous Na-hyaluronate in the anterior chamber of the owl monkey and its effect on the intraocular pressure."
10. J. L. Denlinger, Ph.D. Thesis, Universite des Sciences et Techniques de Lille, Lille, France, 1982. "Metabolism of sodium hyaluronanate in articular and ocular tissues."
11. E. A. Balazs, in: "*New and Controversial Aspects of Retinal Detachment,*" A. McPherson, Ed., Harper & Row, New York, 1968, pp. 3-15. "The molecular biology of the vitreous."
12. E. A. Balazs, in: "*Nordiska Bokhandelns Forlag.,*" A. Engel and T. Larsson, Eds., Stockholm, 1969, pp. 107-122. "Some aspects of the aging and radiation sensitivity of the intercellular matrix with special regard to hyaluronic acid in synovial fluid and vitreous."
13. D. A. Gibbs, E. W. Merrill, K. A. Smith and E. A. Balazs, Biopolymers, 6, 777-791 (1968). "The rheology of hyaluronic acid."
14. E. A. Balazs, "*Symposium: Prognosis for Arthritis: Rheumatology Research Today and Prospects for Tomorrow,*" University Mich. Med. Ctr. J., 1968, pp. 255-259. "Viscoelastic properties of hyaluronic acid and biological lubrication."
15. E. A. Balazs and D. A. Gibbs, in: "*Chemistry and Molecular Biology of the Intercellular Matrix,*" Vol. III, E. A. Balazs, Ed., Academic Press, New York, 1970, pp. 1241-1254. "The rheological properties and biological function of hyaluronic acid."
16. W. D. Comper, T. C. Laurent, Physio. Rev., 58, 255-315 (1978). "Physiological function of connective tissue polysaccharides."
17. Z. Darzynkiewicz and E. A. Balazs, Exp. Cell. Res., 66, 113-123 (1971). "Effect of connective tissue intercellular matrix on lymphocyte stimulation, I, Suppression of lymphocyte stimulation by hyaluronic acid."
18. E. A. Balazs, "*Handbook of the Biology of Aging,*" C. E. Finch and L. Hayflick, Eds., Van Nostrand Reinhold Co., New York, pp. 222-240 (1977). "Intercellular matrix of connective tissue."
19. J. V. Forrester and E. A. Balazs, Trans. Ophth. Soc., U. K., 97, 554-557 (1977). "Effect of hyaluronic acid and vitreous on macrophage phagocytosis."
20. J. V. Forrester and E. A. Balazs, Immunology, 40, 435-446 (1980). "Inhibition of phagocytosis by high molecular weight hyaluronate."
21. E. A. Balazs and Z. Darzynkiewicz, in: "*Biology of the Fibroblast,*" E. Kulonen and J. Pikkarainen, Eds., Academic Press, London, pp. 237-252 (1973). "The effect of hyaluronic acid on fibroblasts, mononuclear phagocytes and lymphocytes."
22. E. A. Balazs, S. Briller, and J. L. Denlinger, in" "*Seminars in Arthritis and Rheumatism,*" J. H. Talbott, Ed., Grune and Stratton, New York, pp. 141-143 (1981). "Na-hyaluronate molecule size variations in equine and human arthritic synovial fluid and effect on phagocytic cells."
23. "*The biology of hyaluronan,*" "*Ciba Foundation Symposium #143,*" J. Wiley & Sons, New York, 1989.
24. M. Brecht, U. Mayer, E. Schlosser, and P. Prehm, Biochem. J., 239, 445-450 (1986). "Increased hyaluronate synthesis is required for fibroblast detachment and mitosis."

25. C. B. Underhill, J. Cell. Sci., 131, 177-189 (1982). "Interaction of hyaluronate with the surface of simian virus 40-transformed 3T3 cells."
26. C. B. Underhill, and B. P. Toole, J. Cell. Physiol., 110, 123-128 (1982). "Transformation-dependent loss of the hyaluronate-containing coats of cultured cells."
27. M. W. Lark and L. A. Culp, J. Biol. Chem., 257, 14073-14080 (1982). "Selective solubilization of hyaluronic acid from fibroblast substratum adhesion sites."
28. J. Laterra and L. A. Culp, J. Biol. Chem., 257, 719-726 (1982). "Differences in hyaluronate binding to plasma and cell surface fibronectins."
29. B. P. Toole, in: "*Cell Biology of Extracellular Matrix,*" E. D. Hay, Ed., Plenum Press, New York, pp. 259-294 (1981). "Glycosaminoglycans in morphogenesis."
30. B. P. Toole, R. L. Goldberg, G. Chi-Rossio, C. B. Underhill and R. W. Orkin, in: "*The Role of Extracellular Matrix in Development,*" R. L. Trelsted, Ed., Alan R. Liss Inc., New York, pp. 43-66 (1984). "Hyaluronate-cell interactions."
31. B. P. Toole, S. I. Munaim, S. Welles, and C. B. Knudson, "*Ciba Foundation Symposium #143: Biology of Hyaluronan,*" J. Wiley & Sons, New York, 1989, pp. 138-149. "Hyaluronate-cell interactions and growth factor regulation of hyaluronate synthesis during limb development."
32. C. Biswas and B. P. Toole, in: "*Cell Membranes,*" E. Elson, et al., Eds., Plenum Press, New York, 3, 341-363 (1987). "Modulation of the extracellular matrix by tumor cell-fibroblast interactions."
33. E. A. Balazs and A. Leshchiner, in, "*Cellulosics Utilization: Research and Rewards in Cellulosics,*" H. Inagaki and G. O. Phillips, Eds., Elsevier, Applied Science, London and New York, pp. 233-241 (1989). "Hyaluronan, its crosslinked derivative hylan, and their medical applications."
34. E. A. Balazs, A. Leshchiner and P. Band, U. S. Patent #4,713,448, 1987."Chemically modified hyaluronic acid preparation and method of recovery thereof from animal tissues."
35. E. Leshchiner, E. A. Balazs, N. E. Larsen and A. Leshchiner, U. S. Patent pending (#7/550,287), 1990."Biocompatible viscoelastic gel slurries, their preparation and use."
36. E. A. Balazs and A. Leshchiner, U. S. Patent #4,582,865, 1986. "Crosslinked gels of hyaluronic acid and products containing such gels."
37. W. Richter, Int. Arch. Allergy, 47, 221-217 (1974). "Non-immunogenicity of purified hyaluronic acid preparations tested by passive cutaneous anaphylaxis."
38. W. Richter, E. Ryde and E. O. Zetterström, Int. Arch. Allergy Appl. Immunology, 59, 45-48 (1979). "Non-immunogenicity of purified sodium hyaluronate preparations in man."
39. J. L. Denlinger, A. A. M. El-Mofty and E. A. Balazs, Exp. Eye Res., 31, 101-117 (1980). "Replacement of the liquid vitreus with sodium hyaluronate in monkeys, II. Long term evaluation."
40. E. A. Balazs, N. Larsen, B. Morales, J. L. Denlinger, H. Dursema and M. Kling, "*Biocompatibility of hylans,*" Biomatrix Report #102, 1988.
41. E. A. Balazs, P. Band, N. E. Larsen, and A. Goldman, "*Clearance of hylan devices from the joint and the blood,*" Biomatrix Report #105, 1990.
42. E. A. Balazs and J. L. Denlinger, *Ciba Foundation Symposium #143: Biology of Hyaluronan, Chinchester,* D. Evered, J. Whelan, Eds., Wiley & Sons, pp. 265-280 (1989). "Clinical uses of hyaluronan."
43. J. L. Denlinger and E. A. Balazs, Exp. Eye Res., 3, 81-99 (1980). "Replacement of the liquid vitreus with sodium hyaluronate in monkeys, I. Short term evaluation."

44. E. A. Balazs, in: "*Healon (Sodium Hyaluronate), A Guide to its Use in Ophthalmic Surgery*," D. Miller and R. Stegmann, Eds., John Wiley & Sons, New York, 1983 pp. 5-28. "Sodium hyaluronate and viscosurgery."
45. E. A. Balazs, U. S. Patent #4,141,973, 1979. "Ultrapure hyaluronic acid and the use thereof."
46. L. Roden, P. Campbell, J. R. E. Fraser, T. C. Laurent, H. Pertoft and J. N. Thompson, "*Ciba Foundation Symposium #143: Biology of Hyaluronan*, D. Evered and J. Whelan, Eds., Wiley, Chinchester, 1980. "Enzymatic pathways of hyaluronan catabolism."
47. J. L. Denlinger and E. A. Balazs, in: "*Viscoelastic Materials: Basic Science and Clinical Applications*," E. S. Rosen, Ed., Pergamon Press, New York, pp. 185-199 (1989). "The fate of exogenous viscoelastic hyaluronan solutions in the primate eye."
48. E. A. Balazs, "Hyaluronic acid and Matrix implantation," Arlington, Biotrics Inc., pp. 131-147, 1971.
49. E. A. Balazs, in: "*Disorders of the Knee*," A. J. Helfet, Ed., Lippincott Corp., Philadelphia, 2nd ed., 1982, pp. 61-74. "The physical properties of synovial fluid and the special role of hyaluronic acid."
50. E. A. Balazs, in: "*Viscoelastic Materials: Basic Science and Clinical Applications*," E. S. Rosen, Ed., Pergamon Press, New York, 1989, pp. 167-183. "The introduction of elastoviscous hyaluronan for viscosurgery."
51. E. A. Balazs and J. L. Denlinger, J. Equine. Vet. Sci., 5, 217-228 (1985). "Sodium hyaluronate and joint function."
52. N. W. Rydell and E. A. Balazs, Clin. Orthop., 80, 25-32 (1970). "Effect of intraarticular injection of hyaluronic acid on the clinical symptoms of osteoarthritis and on granulation tissue formation."
53. N. W. Rydell, J. Butler and E. A. Balazs, Acta. Vet. Scand., 11, 139-155 (1970). "Hyaluronic acid in synovial fluid, VI, Effect of intraarticular injection of hyaluronic acid on the clinical symptoms of arthritis in track horses."
54. J. G. Peyron and E. A. Balazs, Pathol. Biol., 22, 732-736 (1974). "Preliminary clinical assessment of Na-hyaluronate injection into human arthritic joints."
55. C. Weiss, E. A. Balazs, R. St. Onge and J. L. Denlinger, in: "*Seminars in Arthritis and Rheumatism*," Vol. 11, J. H. Talbott, Ed., Grune & Stratton, New York, 1981, pp. 143-144. "Clinical studies of the intraarticular injection of Healon[R] (sodium hyaluronate) in the treatment of osteoarthritis of human knees."
56. A. J. Helfet, "*Disorders of the Knee*," J. B. Lippincott, Philadelphia, 1st ed. 1974, 2nd ed. 1982, 194 pages.
57. E. A. Balazs and J. L. Denlinger, in: "*Osteoarthritis, Current Clinical and Fundamental Problems*," J. G. Peyron, Ed., Ciba Geigy, Paris, 1984, pp. 165-174. "The role of hyaluronic acid in arthritis and its therapeutic use."
58. O. G. Swanstrom, Proc. Amer. Assoc. Equine Pract., 24th Annual Convention, 1978, pp. 345-348. "Hyaluronate (hyaluronic acid) and its use."
59. R. J. Rose, J. NZ. Vet., 27, 5-7 (1979). "The intra-articular use of sodium hyaluronate for the treatment of osteo-arthritis in the horse."
60. J. A. Auer, G. E. Fackelman, D. A. Gingerich and A. W. Fetter, AMJ. Vet. Res., 41, 568-574 (1980). "Effect of hyaluronic acid in naturally occurring and experimentally induced osteoarthritis."
61. D. H. G. Irwin, JS. African Vet. Assn., 50, 231-233 (1980). "Sodium hyaluronate in equine traumatic arthritis."
62. M. W. Phillips, Proc. Am. Assn. Equine Pract. 26th Ann. Convention, 1980, pp. 389-394. "Intra-articular sodium hyaluronate in the horse: a clinical trial."
63. D. A. Gingerich, J. A. Auer and G. E. Fackelman, Res. Vet. Sci., 30,

192-197 (1981). "Effect of exogenous hyaluronic acid on joint function in experimentally induced equine osteoarthritis: dosage titration studies."
64. A. Churchill, J. Eq. Vet. Sci., 5, 240-242 (1985). "Treating tendonitis with sodium hyaluronate."
65. J. Selway, J. Eq. Vet. Sci., 5, 236-238 (1985). "Notes on the use of sodium hyaluronate in thoroughbreds."
66. J. Selway, J. Eq. Vet. Sci., 5, 238-240 (1985). "Post-surgical use of sodium hyaluronate."
67. R. Steele, J. Eq. Vet. Sci., 5, 235-236 (1985). "Notes on hyaluronate use in the standardbred."
68. O. Namiki, H. Toyoshima and N. Morisaki, Int. J. Clin. Pharm. Ther. Toxicol., 20, 501-507 (1982). "Therapeutic effect of intraarticular injection of high molecular weight hyaluronic acid on osteoarthritis of the knee."
69. E. A. Balazs, *"The development of the concept of viscosupplementation and a review of clinical studies with hyaluronan,"* Biomatrix Report #106, 1989.
70. E. A. Balazs and J. L. Denlinger, in: *"Modern aging research 4. Comparative Pathology of Major Age-Related Dieseases, Current Status and Research Frontiers,"* D. G. Scarpelli and G. Migaki, Eds., Alan R. Liss, New York, 1984, pp. 129-143. "The synovial cell."
71. E. A. Balazs, D. Miller and R. Stegmann, Presented at the International Congress and First Film Festival on Intraocular Implantation, Cannes, France, 1979. "Viscosurgery and the use of Na-hyaluronate in intraocular lens implantation."
72. E. L. Graue, F. M. Polack and E. A. Balazs, Exp. Eye Res., 31, 119-127 (1979). "The protective effect on Na-hyaluronate to corneal endothelium."
73. D. Miller, P. O'Connor and J. Williams, Ophthalmic Surg., 11, 19 (1980). "Use of Na-hyaluronate during intraocular lens implantation in rabbits."
74. H. Bothner, in: *"Viscoelastic Materials, Basic Science and Clinical Applications,"* E. S. Rosen, Ed., Pergamon Press, New York, 1989, pp. 3-22. "Rheology of intraocular solutions."
75. E. A. Balazs, Curr. Canadian Ophthalmol. Prac., 2, 24-28 (1984). "What is viscosurgery?"
76. E. A. Balazs, in: Transplant Today, 1985, pp. 62-64. "Viscosurgery."
77. E. A. Balazs, in: *"Viscoelastic Materials, Basic Science and Clinical Applications,"* E. Rosen, Ed., 1988, pp. 149-165. "The introduction of elastoviscous hyaluronan for viscosurgery."
78. E. A. Balazs, in: *"Importance of the Vitreous Body in Retina Surgery with Special Emphasis on Reoperations,"* Proceedings of the 11th Conference of the Retina Foundation, Ipswich, May 30-31, C.V. Mosby, St. Louis, 1958, pp. 29-48. "Physiology of the vitreous body."
79. E. A. Balazs, M. I. Freeman, R. Klöti, G. Meyer-Schwickerath, F. Regnault and D. B. Sweeny, *"Modern Problems in Ophthalmology (Secondary Detachment of the Retina),"* E. B. Streiff, Ed., S. Karger, Basel, 10, 3-21 (1972). "Hyaluronic acid and the replacement of vitreous and aqueous humor."
80. P. Algvere, Acta Ophthalmol., 49, 975 (1971). "Intravitreal injection of high molecular weight hyaluronic acid in retina detachment surgery, a preliminary report."
81. J. Edmund, Mod. Prob. Ophthalmol., 12, 370-377 (1974). "Vitreous substitute in the treatment of retinal detachment."
82. F. M. Polack, T. Demong and H. Santaella, Presented as a scientific exhibit, American Academy of Ophthalmology Meetings, Chicago, Illinois 1981, pp. 1-6. "Sodium hyaluronate (Healon®) in keratoplasty."
83. E. A. Balazs, Beiträge des ersten Nationalen Healon®-Symposiums, October 15-16, Ferdinand Enke Verlag, Stuttgart, 1982, pp. 1-13. "Pharmakologische Eigenschaften von Natrium-Hyaluronat im Auge."

84. F. Renault, Bull. Soc. Ophthalmol. Fr., 1971. "Acide hyaluronique intravitreen et cryocoagulation dans le traitment des formes graves de decollement de la retine."
85. R. Klöti, Schweiz, Ophthalm. Ges., 165, 351-359 (1972). "Hyaluronsäure als Glaskörpersubstituent."
86. F. Regnault and P. Bregeat, Mod. Prob. Ophthalmol., 12, 378-383 (1974). "Treatment of severe cases of retinal detachment with highly viscous hyaluronic acid. Limitations and prospects for retinal surgery."
87. J. Edmund, in: "*Advances in Vitreous Surgery*," S. R. Irvine and E. R. O'Malley, Eds., Charles C. Thomas, Springfield, Illinois, pp. 624-625 (1976). "Comments on the clinical use of Healon[R] and a short survey of the use of intraocular injection of hyaluronic acid."
88. R. C. Pruett, C. L. Schepens and D. A. Swann, Arch. Ophthalmol., 97, 2325-2330 (1979). "Hyaluronic acid vitreous substitute. A six-year evaluation."
89. S. Stenkula, L. Ivert, I. L. Gislason, R. Tornquist and L. Weijdegard, Ophthalmic Surg., 12, 435-437 (1981). "The use of sodium hyaluronate (Healon[R]) in the treatment of retinal detachment."
90. F. M. Polack, Tr. Am. Ophth. Soc., 80, 248-261 (1982). "Penetrating keratoplasty using MK stored corneas and Na-hyaluronate (Healon[R])."
91. L. G. Pape and E. A. Balazs, J. Am. Acad. Ophth., 87, 699-705 (1980). "The use of sodium hyaluronate (Healon[R]) in human anterior segment surgery."
92. D. Miller and R. Stegmann, J. Am. Intraocular Implant Soc., 6, 13-15 (1980). "Use of Na-hyaluronate in anterior segment eye surgery."
93. D. Miller and R. Stegmann, Am. Ophthalmol., 13, 811-815 (1981). "Use of sodium hyaluronate in human IOL implantation."
94. G. Meyer-Schwickerath, Ed., "*Viskochirurgie des Auges, Beitrage des ersten nationalen Healon[R],*" Symposium, October 15-16, Ferdinand Enke Verlag, Stuttgart, 1982.
95. E. A. Balazs, Ocular Inflammation Ther., 1, 91-92 (1983). "Viscosurgery in the eye."
96. J. J. Alpar, Am. Ophthalmol., 19, 350-353 (1987). "Viscoelastic surgery."
97. R. Stegmann and D. Miller, Ann. Ophthalmol., 14, 813-815 (1982). "Extra-capsular cataract extraction with hyaluronate sodium."
98. D. Miller and R. Stegmann, Eds., "*Healon (sodium hyaluronate) a guide to its use in ophthalmic surgery,*" J. Wiley & Sons, New York, 1983.
99. G. Eisner, Ed., "*Ophthalmic viscosurgery, A review of standards, techniques and applications,*" Medicopea, Montreal, 1986.
100. E. Rosen, Ed., "*Viscoelastic materials, Basic science and clinical applications,*" Pergamon Press, Oxford, 1989.
101. E. A. Balazs, Deutsche Ophthalmologische Gesellschaft 86, Tagung, September 1988, Berlin, p. 87. "Hyaluronan molecular matrices (hylans) for ophthalmic surgery."
102. A. Vadasz, E. A. Balazs, M. F. Ghilardi, J. Bodis-Wollner, A. Glover and A. Goldman, Proc. Int. Soc. Eye Res., Vol. V, 1988, p. 67. "Hylan gel biocompatibility and residence time of a new vitreous substitute."
103. A. Vadasz, A. Goldman and E. Balazs, Invest. Ophthal. Vis. Sci., Vol. 29, 1988, p. 440. "The use of hylan gel in ophthalmic surgery."
104. A. Vadasz, M. F. Ghilardi, A. Goldman, A. Glover, E. A. Balazs and I. Bodis-Wollner, Invest. Ophthal. Vis. Sci., 30 (3), 70 (1989). "The intensity-response function of the owl monkey ERG; Normative data and the evalution of a vitreous substitute, hylan gel."
105. A. Goldman, R. Akiyama, A. Vadasz, J. P. Koniarek, J. Fischbarg, L. S. Liebovitch and E. A. Balazs, Invest. Ophthal. Vis. Sci., 30 (3), 260 (1989). "The effect of crosslinked hyaluronan (hylans) on corneal endothelial function."
106. E. A. Balazs, A. Vadasz and A. I. Goldman, Proc. Int. Soc. Eye Res.,

Vol. VI, 1990, p. 165. "The use and fate of cross-linked hyaluronan in the eye."
107. I. Goldman, A. Vadasz, E. A. Balazs and I. Bodis-Wollner, Invest. Ophthal. Vis. Sci., 31 (4), 392 (1990). "Further investigations on the non-monotonic intensity-response function of the owl monkey ERG."
108. E. A. Balazs, Proc. Int. Soc. Eye Res., Vol. VI, 1990, p. 186. "Viscosurgery and its tools."
109. C. Weiss and E. A. Balazs, "*Mediguide to Orthopedics*," Vol. 4, Issue 4, W. L. Jaffe and M. G. Neuwirth, Eds., Dellacorte Publications, Inc., 1984, pp. 1-4. "The physiologic function of hyaluronic acid and the therapeutic role of viscosurgery in the skeletal system."
110. C. Weiss and E. Balazs, Arthroscopy, 3 (2), 138 (1987). "Arthroscopic viscosurgery."
111. C. Weiss and E. Balazs, Third Int. Symp. on Arthros. of the TMJ, New York, December 1988. "Viscosurgery and viscosupplication in joints."
112. J. P. McCain, E. A. Balazs and H. de la Rua, J. Oral Maxillofac Surgery, 47, 1161-1168, (1989). "Preliminary studies on the use of a viscoelastic solution in arthroscopic surgery of the temporomandibular joint."
113. R. St. Onge, C. Weiss, J. L. Denlinger and E. A. Balazs, Clin. Orthop. and Related Res., 146, 269-275 (1980). "A preliminary clinical assessment of Na-hyaluronate injection into "No-Man's Land" for primary flexor tendon repair."
114. C. Weiss, H. J. Levy, J. Denlinger, J. M. Suros and H. E. Weiss, Bulletin of the Hospital for Joint Diseases Orthopaedic Institute, 46, 9-15 (1986). "The role of Na-hylan in reducing postsurgical tendon adhesions."
115. C. Weiss, J. M. Suros, A. Michalow, J. Denlinger, M. Moore and W. Tejeiro, Bulletin of the Hospital for Joint Diseases Orthopaedic Institute, 47, 31-39 (1987). "The role of Na-hylan in reducing postsurgical tendon adhesions, Part 2."
116. N. E. Larsen, K. M. Lombard and E. A. Balazs, 35th Ann. Meet. Ortho. Res. Soc., 1989, p. 151. "The effect of hyaluronan on cartilage and chondrocyte response to mechanical and biochemical perturbation."
117. C. Weiss, J. M. Suros, J. Dennis, J. Denlinger, A. Badia, J. Gross and S. Eremenco, 35th Ann. Meet. Ortho. Res. Soc., 1989, p. 539. "Effect of Na-hylan on articular cartilage."
118. C. Weiss, J. Dennis, J. M. Suros, J. Denlinger, A. Badia and I. Montane, Orthopaedic Research Society, Febuary 6-9, Las Vegas, Nevada, 1989, p. 44. "Sodium hylan for the prevention of postlaminectomy scar formation."
119. N. E. Larsen, K. M. Lombard, E. G. Parent and E. A. Balazs, J. Orthop. Res., (in press). "Effect of hylan on cartilage and chondrocyte cultures."
120. C. Laurent, S. Hellström and L. E. Stenfors, Am. J. Otolaryngol, 7, 181-186 (1986). "Hyaluronic acid reduces connective tissue formation in middle ears filled with absorbable gelatin sponge, an experiment study."
121. S. Hellström and C. Laurent, Acta Otolaryngol (Stockholm), 442, 54-61 (1987). "Hyaluronan and healing of tympanic membrane perforations, an experimental study."
122. L. E. Stenfors, Acta Otolaryngol (Stockholm), 442, 81-87 (1987). "Treatment of typanic membrane perforations with hyaluronan in an open pilot study of unselected patients."
123. L. E. Stenfors, Acta Otolaryngol (Stockholm), 442, 88-91 (1987). "Repair of traumatically ruptured tympanic membrane using hyaluronan."
124. C. Laurent, Umeå University Medical Dissertations New Series Number 211, 1988. "Hyaluronan in the middle ear."
125. E. A. Balazs and D. B. Sweeney, "*Controversial Aspects of the*

*Management of Retinal Detachment*," C. L. Schepens and C. D. J. Regan, Eds., Little, Brown & Company, Boston, Vol. 3, 1965, pp. 200-202. "The use of hyaluronic acid and collagen preparations in eye surgery."

126. E. A. Balazs and D. B. Sweeney, "*Modern Problems in Ophthalmology (Surgery of Retinal Vascular Diseases and Prophylactic Treatment of Retinal Detachment)*," E. B. Streiff, Ed., Amersfoort, S. Karger, Basel, Vol. 4, 1966, pp. 230-232. "Replacement of the vitreous body of monkeys with reconstituted vitreous and hyaluronic acid."

127. E. A. Balazs and D. B. Sweeney, "*New and Controversial Aspects of Retinal Detachment*," A. McPherson, Ed., Harper & Row, New York, 1968, pp. 371-376. "The injection of hyaluronic acid and reconstituted vitreous in to the vitreous cavity."

128. D. P. Clark, W. Hanke and M. D. Swanson, J. Am. Acad. Derm., 21, 992-998 (1989). "Dermal implants, safety of products injected for soft tissue augmentation."

129. N. E. Larsen, M. B. Kling, E. A. Balazs and E. A. Leshchiner, Proc. 16th Annual Meeting Society for Biomaterials, 1990, p. 302. "Hylan gel for soft tissue augmentation."

130. E. A. Balazs, J. L. Denlinger, E. Leshchiner, P. Band, N. Larsen, A. Leshchiner and B. Morales, Biotech USA 1988, Proceedings of the Conference held in San Francisco, California, November 14-16, 1988. Conference Management Corp., 1988, pp. 442-451. "Hylan: hyaluronan derivatives for soft tissue repair and augmentation."

131. A. Gutierrez-Miguelez, N. Cuenca, J. de Juan and C. Belmonte, Trabajos del Instituto Cajal 86, 141 (1987). Proc. 2nd Cong. Sociedad Española de Neurociencia Barcelona, Spain. "El ácido hialurónico no constituye un soporte adecuado para la regeneración de fibras nerviosas periféricas."

132. M. Ramer, J. Wilson, D. Walker and A. E. Clark, Proc. 16th Annual Meeting Society for Biomaterials, 1990, p. 195. "Bioglass[R]-hylan suspension for treatment of urinary incontinence."

133. S. K. Hilal, E. A. Leshchiner, N. E. Larsen, A. G. Khandji, F. G. Moser and E. A. Balazs, American Society of Neuroradiology, Twenty-Fifth Annual Meeting, 1987, p. 80. "A novel viscoelastic composition for percutaneous embolization."

134. D. E. Gunson, S. K. Hilal, E. A. Leshchiner, N. E. Larsen and E. A. Balazs, Twenty-Third Annual Meeting, American College of Clinical Pathology, 1988, p. 43. "Percutaneous embolization of rabbit auricular arteries using hylan gel."

135. A. Leshchiner, N. E. Larsen, E. A. Balazs and S. K. Hilal, U. S. Patent 4,795,741, 1988. "Novel compositions for therapeutic percutaneous embolization and the use thereof."

136. N. E. Larsen, E. A. Leshchiner, E. G. Parent, J. Hendrickson-Aho and E. A. Balazs, J. Biomedical Materials Res., 1991, (in press). "Hylan gel composition for percutaneous embolization."

137. N. E. Larsen, E. A. Leshchiner and E. A. Balazs, 1st Atlantic Congress, "The Future of Collagen," Lyon, France, Conference, Tome 1, 1985, pp. 116-123. "Use of hyaluronic acid as a biocompatible interface in vascular prosthesis."

138. E. A. Balazs, D. Wedlock and G. Phillips, U. S. Patent 4,487,865, 1984. "Polymeric articles modified with hyaluronate."

139. E. A. Balazs and A. Leshchiner, U. S. Patent 4,500,676, 1985. "Hyaluronate modified polymeric articles."

140. E. A. Balazs, A. Leshchiner and N. Larsen, U. S. Patent, filed 1985 (pending). "Drug delivery systems based on hyaluronan, derivatives thereof and their salts and methods of producing same."

141. N. E. Larsen, E. A. Leshchiner, E. G. Parent and E. A. Balazs, First International Symposium on the Use of Polymers in Pharmaceuticals and Cosmetics," American Cosmetic Society, Polymers Symposium Series, 1991, in press. "Hylan and hylan derivatives in drug delivery."

142. N. E. Larsen, E. G. Parent and E. A. Balazs, 17th Annual Meeting Society for Biomaterials, Scottsdale, Arizona, May, 1991, in press. "Release of the antibiotic gentamicin from hylan vehicles."
143. N. E. Larsen, E. A. Leshchiner, E. G. Parent, J. Hendrickson-Aho and E. A. Balazs, 15th Annual Meeting Society for Biomaterials, Lake Buena Vista, Florida, 1989, p. 141. "Local intravascular drug delivery via percutaneous embolization with hylan gel."

# HYDROPHILIC LUBRICIOUS COATINGS FOR MEDICAL APPLICATIONS

Y. L. Fan

Specialty Chemicals Division
Union Carbide Corporation
Bound Brook, NJ, 08805

The need for a hydrophilic lubricious surface for medical devices intended for either a transient or permanent implant in the human body has generated a great deal of research and development activities among industrial and academic communities. The goal is to develop a biocompatible, low-friction surface that gives a normal feel when dry, but would become instantly slippery upon exposure to an aqueous fluid, such as a body fluid. The different approaches undertaken in the literature have been reviewed and grouped into five categories by virtue of the chemical nature of the resultant surfaces. Comparisons are made on the performance characteristics of each of these technologies including a new hydrophilic lubricious coating system commercialized by Union Carbide. A mechanistic analysis of the hydrophilic lubricity is also presented.

## INTRODUCTION

Synthetic and natural-occurring polymeric materials, such as poly(urethane), acrylic resin, vinyl resin, poly(olefin), nylon and rubber, are widely used for making catheters, contact lenses, implant devices, heart valves, intra-uterine devices, peristaltic pump chambers, endotracheal tubes, gastroenteric feed tubes and arteriovenous shunts. It is highly desirable that these materials can be fabricated to provide a surface that is not only hydrophilic, but also have a low coefficient of friction when in contact with an aqueous fluid, such as a body fluid. Such a lubricious surface would facilitate insertion or removal of a medical device into or out of a patient, and would minimize injury or inflammation of mucous membranes, as well as aid to patient comfort. For ease of handling, it is even more desirable if the surface exhibits a normal feel when dry, but becomes slippery only upon exposure to an aqueous fluid.

Earlier approaches to provide a low-friction surface include the application of a non-permanent coating such as silicone oil, olive oil, glycerine, or xylocain jelly, and the use of conventional, low-friction materials such as fluoroplastics, silicone and poly(ethylene).[1-4]

*Cosmetic and Pharmaceutical Applications of Polymers*
Edited by C.G. Gebelein *et al.*, Plenum Press, New York, 1991

These earlier techniques, however, are unsatisfactory in practice. In the case of oils being applied, they tend to run off and the lubricity does not last. In the case of low-friction materials, such as Teflon, the finished medical devices are difficult to handle because they are slippery at all times.

A great deal of research and developmental activities has been generated during recent years to address this need at both industrial and academic institutions. As a leading manufacturer of water-soluble polymers, Union Carbide has been working actively on the development of such a technology. This paper reviews the technological advances in the field of hydrophilic lubricious surfaces. The author will provide comparisons of the performance characteristics of these technologies including a hydrophilic lubricious coating system recently commercialized by Union Carbide. A mechanistic analysis of the hydrophilic lubricity is be presented.

## TYPES OF HYDROPHILIC LUBRICIOUS SURFACES

A variety of approaches have been undertaken to develop a hydrophilic lubricious surface for medical devices that would provide good handling characteristics when dry, but become instantaneously slippery upon exposure to a body fluid.[5-12] These different techniques may be grouped into five categories by virtue of their chemical distinctions.

### 1. Lubricating Coatings Derived from a Hydrophilic Polymer

A most commonly used technique for imparting a hydrophilically slippery surface to a medical device, such as surgical gloves, catheters, and sutures, is the application of an exterior coating containing a suitable hydrophilic polymer. The latter may be either water-soluble or water-swellable. The coating may be applied using any suitable coating process, but a dip-coating process is probably the most commonly used.[13]

Poly(vinyl pyrrolidone) or poly(ethylene oxide) is an effective hydrophilic polymer for providing a coating surface having a low coefficient of friction when wet while retaining a normal feel when dry.[13-17]

Coatings made from these polymers have been accepted by the medical communities because of their proven processability, biocompatibility, some antithrombogenicity, protein repellent properties, and commercial availability.[18,19] The polymer is dissolved in an organic solvent, such as dimethylformamide or tetrahydrofuran, and applied by either a dip- or spray-coating process. An acid pretreatment of substrates made from latex rubber may be employed to improve adhesion. A similar effect was described on hydrophilic poly(urethane) surfaces[20] and non-hydrophilic acrylic surfaces.[21]

Poly(ethylene oxide) has been modified in many ways either to offer an extra niche for the intended application or being compositionally different from the homopolymer. Poly(ethylene oxide) having a free-radical curable unsaturated end group(s) has been made, which permitted simultaneous sterilization and curing for medical applications.[22] The finished coating was lubricious in water, but otherwise water resistant. A poly(oxyethylene):polyoxypropylene) block copolymer coating was found to impart an improved run-down characteristic and reduced tissue drag on some absorbable surgical sutures.[23] The lubricity was further improved by formulating with a humectant such as glycerol.[24] Similar coatings have been prepared using poly(alkylene oxide glycolic acid ester) block or random copolymers.[25,26]

Hydrophilic poly(urethanes) constitute another class of synthetic polymer used for this purpose. Poly(urethanes) offer a wide range of physical and mechanical properties as well as processabilities and biocompatibilities.[27] Poly(urethanes) and urethane copolymers have been used for friction reduction in pharmaceutical closures,[28] as orthopedic casting materials,[29] in hydrogel carrier systems,[30] and as burn dressings and friction-reduction coatings.[72]

Other synthetic polymers useful for affording a friction-reduction coating include poly(alkylene oxalate),[30] poly(vinyl alcohol),[73] ionene (ionic amine) polymers,[32] and caprolactone homo- and co-polymers. The latter constitutes a class of biodegradable lubricating polymers.[33]

Chitin, poly(N-acetyl-D-glucosamine), and its derivatives have been employed for imparting lubricity on surgical elements; e.g., tubing, catheters, drains and gloves.[34] This natural-occurring polymer and its derivatives are sufficiently thermally stable to endure autoclaving, and are biodegradable. Many cellulose derivatives such as carboxymethyl cellulose, hydropropyl cellulose, and methyl cellulose have been used as a hydrophilic lubricant.[35-37] A slippery coating based on gelatin, alcohol, and formaldehyde was also claimed.[38]

Certain hydrophilic lubricious coatings are formulated to be intentionally leachable or absorbable so that the hydrophilic polymer is removed from the device in a controlled fashion.[26,33,39] This type of coating is often used on surgical sutures and shaving devices among others.

The hydrophilic lubricity of the coating compositions described in this group, while good initially, is often neither lasting nor durable. The hydrophilic polymers are readily removed from the substrate surface through either leaching or mechanical dislocation due to a lack of adhesion.

## 2. Hydrophilic Polymer Blends or Complexes

Hydrophilic coatings having a somewhat improved durability or adhesion, or both, have been obtained by either forming a polymeric blend or complex with the substrate material. Thermoplastic poly(urethane) is blended with poly(N-vinyl lactam) to yield a durable lubricious coating when wet.[16,40] Slippery poly(urethane) surfaces having improved abrasion resistance have been obtained by mixing with a dialkyl sulfonate surfactant. The poly(urethane) may be produced *in-situ* from a diol and a diisocyanate in the presence of stannous octoate.[41] A stable hydrophilic polymer blend of poly(vinyl butyral) and poly(vinyl lactam) was claimed to retain slippery properties in an aqueous environment.[42] High molecular weight poly(ethylene oxide) was incorporated in a cured poly(urethane) casting tape to give a controlled slipperiness.[43] Flexible tube for introducing catheters has been made using a poly(vinyl pyrrolidone)-poly(urethane) blend to impart lubricity and thrombo-resistance.[44] Such a hydrophilic polymer coating may be rendered more stable by treating with non-toxic organic or inorganic salt, such as sodium citrate and sodium chloride, respectively.[45] The substrate surface may be cured with a diisocyanate.

Poly(acrylates), such as poly(2-hydroxyethylmethacrylate) or poly(2-ethylhexl acrylate) and their copolymers, have been used for forming a lubricating layer, bonded to a rubber article, with respect to damp skin.[46] Guidelines were made antithrombotic and slippery by coating with a poly(urethane)-poly(siloxane) composition.[47] Organosiloxane copolymers containing amino-alkyl-siloxane and methyl siloxane have been used for a similar purpose.[48]

Generally speaking, the hydrophilic coatings described in this group exhibit a somewhat better performance than the simple hydrophilic polymer coating mentioned earlier. They are, however, still deficient in abrasion resistance and coating durability.

## 3. Formation of Interpenetrating Polymeric Networks

The need for a greater adhesion between the lubricious coating and the substrate has led to the development of techniques for forming a surface layer of interpenetrating polymeric network (IPN) where the hydrophilic polymer is bonded permanently with the substrate molecules. In many instances, however, since the hydrophilic polymer is a thermoplastic rather than a thermoset polymer, the compositions are not true IPNs, but are mixtures where the long-chain hydrophilic polymers intertwine physically within the network of a poly(urea) or poly(urea)-poly(urethane) or poly(urea)-poly(acrylate).

The hydrophilic polymer most often cited for this application is poly(vinyl pyrrolidone). It is usually used in conjunction with a multifunctional isocyanate, with or without the presence of a poly(urethane).[49-51] Poly(ethylene oxide) has been applied in pretty much the similar fashion, with a poly(isocyanate), to produce a network structure mostly through physical entangement.[52] When relatively low molecular weight hydroxy-terminated poly(alkylene oxide) and/or poly(alkylene glycol) molecules are present in the coating composition, the hydrophilic polymers may become chemically attached to the network structure through the formation of an IPN.[53]

Hyaluronic acid and its salts, chondroitin sulfate and agarose are other classes of biocopolymers used for producing a hydrophilic lubricious coating on medical devices. The dimensional stability of such a coating was significantly improved by the incorporation of albumin.[54] The latter was claimed to facilitate the wetting of hyaluronic acid on the more hydrophobic surface of an acrylic polymer anchor coat, such as poly(hydroxyethyl methacrylate). This lubricious poly(saccharide) coating was reported to be particularly useful for providing a long-lasting comfort to wearers of body implants or prostheses. The poly(saccharide) may be crosslinked to afford a more permanent lubricity. A biocompatible, blood compatible structure with low-friction crosslinked hydrophilic polymer coating was reported.[55]

Since there are probably relatively few true IPNs formed by the technique described in this group, the hydrophilic polymer molecules are often not permanently anchored on the substrate surface. As a result, these coatings will show a good initial lubricity, but only a moderate retention of lubricity upon exposure to body fluid(s) or subjection to mechanical abrasion or both.

## 4. Chemically Reactive Hydrophilic Polymer Coatings

While good adhesion does not necessarily involve chemical bonding, the presence of chemical bonding in an adhesive joint usually insures a durable bond. Following this thinking, a number of systems have been devised where the hydrophilic polymer layer is capable of chemically bonding to the substrate, usually through a reactive primer system.

Water-soluble polymers such as cellulose derivatives, poly(maleic anhydride), poly(acrylamide) or water-soluble nylon have been claimed to chemically bond to a reactive primer such as aldehyde, epoxy, or isocyanate.[56] Depending on the chemical nature of the substrate material, however, the reactive primer may or may not bond chemically to the substrate. Poly(saccharides), such as hyaluronic acid, have been bonded to acrylic polymers using a poly(isocyanate) to afford an improved hydrophilicity and lubricity.[57,58] Vinyl pyrrolidone copolymers containing active hydrogen, such as a (vinyl pyrrolidone-2-hydroxyethyl methacrylate) copolymer, have been described to bond to substrate through a poly(isocyanate) primer.[59] In all these cases, the hydrophilic lubricious coatings have been claimed to be durable and not leachable. A common feature claimed by the patentees of this group was that the hydrated hydrophilic coating was abrasion resistant and would survive either repeated abrasions during a surgical procedure or a long period of implanting in the human body.

A hydrophilic lubricious coating system, designated as POLYSLIP™, has been developed recently at Union Carbide.[60] This coating system, composed of a reactive water-soluble polymer and a reactive primer, is applied by a two-step coating process. The resultant surface exhibits a normal feel when dry, but a fast rate of hydration upon exposure to water, high lubricity, and good durability even in the hydrated state. The hydrophilic polymer used for the topcoat meets the standards of the National Formulary.

## 5. Surface Grafting of Hydrophilic Monomers

Instead of solution coating of a hydrophilic polymer, the latter can be polymerized *in situ* from the corresponding monomer(s) directly onto the surface of a medical device. A free-radical polymerization process is usually employed. An advantage of this process is that a thin and uniform coating, independent of the contour and shape of the medical device to be coated, can be readily obtained. On the other hand, this process may require a higher capital investment on equipment, and a more elaborate process control, than those of a conventional solution coating operation. Furthermore, one needs to be concerned with residual monomers, such as acrylamide, in the coated article which may be highly toxic.

A radiation grafting process of hydrophilic monomers onto organic polymeric substrates in the presence of cupric or ferric ions was described by Ratner, et al.[71] Maleic anhydride, n-vinyl pyrrolidone, hydroxy acrylate, hydroxy methacrylate, methacrylic and acrylic acids have been grafted onto the surface of gloves, sheets, and tubes made from natural rubber by immersing in a polymerization bath at an elevated temperature in the presence of a free-radical initiator.[61] According to this technique, the treated rubber articles were provided with a chemically bonded hydrophilic surface which had a lower friction coefficient to human skin.

Acrylamide and other suitable unsaturated monomers have been grafted using gamma-radiation to afford self-lubricating fill tubes which would insert easily in retention valves in medical appliances.[62] A similar technique using either a high energy radiation source or a free-radical initiator was employed to provide a gastrointestinal catheter having excellent surface slideability and compatibility with living tissues.[63] Good hydrophilic lubricity was also obtained by a plasma-initiated copolymerization of acrylamide and methacrylate. No toxic side effect to humans was reported.[64] Other monomers or monomer-hydrophilic polymer mixtures have been radiation polymerized to produce slippery surfaces.[74,75] The process variables in a photo-induced graft polymerization have been studied by Uyama, et al.[65] They report that a graft amount of 50-100 μg/cm² of poly(acrylamide) on a poly(propylene) substrate was required to afford optimum hydrophilicity.

Besides surface modification, lubricious medical devices may be fabricated directly from a hydrophilic polymer. The latter is produced by either copolymerization or chemical modification of a base resin. (Ethylene-vinyl acetate) copolymers have been radiation-graft copolymerized with acrylic acid to impart a greater hydrophilicity. The neutralized resin was suitable for making surgical devices having high water-swelling characteristics and slipperiness.[66] Low-friction plastic catheters were made from a copolymer of acrylonitrile with either a controlled hydrolysis of poly(acrylonitrile) or through copolymerization.[65] Thermosetting allyl ester copolymers containing maleic lubricity were claimed to form optical parts having excellent abrasion resistance and lubricity.[67]

Organosilicone compounds have been copolymerized with methyl methacrylate to yield an oxygen permeable contact lens that possesses reduced surface friction and improved wettability.[68] A hydrophilic polymer composition has been rendered bacteriostatic by incorporating a quaternary ammonium salt-containing comonomer. For instance, a terpolymer exhibiting such behavior was made from an ester of an unsaturated carboxylic acid and an alcohol having a terminal quaterary ammonium group, a comonomer, and a crosslinking agent.[69]

While all the techniques discussed in this section are expected to produce a lubricious surface, the rate of lubricity development upon exposure to water may vary considerably. Generally speaking, devices having a hydrophilically modified surface are expected to hydrate at a rapid rate because of a high concentration of hydrophilic moieties on the surface. On the other hand, devices fabricated directly from a hydrophilic resin may hydrate considerably slower because of a lack of sufficient polar moieties on the surface initially. In the latter case, the hydrophilicity of the resin may have to be limited to afford an acceptable water resistance and/or other physical properties. Nevertheless, this technique holds promise for producing intricate medical devices which might not be easily coated with a conventional coating process.

## A MECHANISTIC ANALYSIS OF HYDROPHILIC LUBRICIOUS SURFACES

Hydrophilic surface lubricity is a common phenomenon found in marine life and living tissues in animal and human bodies. This surface lubricity serves to reduce friction in motion, or protect against tissue damage, or both. It is therefore not surprising that such a hydrophilic lubricious surface has become highly desirable for medical devices intended for either transient or permanent implant in the human body.

According to the modern theory of friction, 90% of the total frictional force is due to adhesion between a surface atoms.[70] Thus, the weaker the bond between the surface atoms, the lower the friction and the higher the lubricity.

Without modification or treatment, most of the medical devices made from either synthetic resins or rubber latex would exhibit a hydrophobic surface. Since water molecules would not wet the surface during application, direct contact between the medical device and tissues would result in friction when the surfaces are being abraded against each other. When the surface of a medical device is covered by a highly hydrophilic polymer layer, the situation changes dramatically. Such a surface is expected to hydrate rapidly forming a layer of water-swollen gel, i. e., hydrogel, upon exposure to a body fluid. This water-swollen layer, which is mostly water, forms a fluid barrier between the medical device and the tissues. Since fluids form only weak adhesive bonds,[70] friction between the surfaces is markedly reduced by eliminating any direct solid-to-solid surface contact during motion.

A satisfactory hydrophilic lubricious surface must be able to hydrate rapidly during application, while also exhibit good adhesion to the substrate in the hydrated state. It is, therefore, not surprising that the top layer of all hydrophilic lubricious coating formulations is composed of a water-soluble or swellable polymeric material. The molecular weight of the polymer should be sufficiently high such that it would not be leached out of the surface rapidly.

Even a high molecular weight water-soluble polymer will be leached or abraded away from the substrate surface unless it is firmly anchored. This may be accomplished by either a chemical bonding to the substrate or the formation of an adhesive layer which bonds simultaneously to both the hydrated water-soluble polymer and the hydrophilic substrate. The former is illustrated by the plasma and a high energy radiation grafting process, as well as the use of a reactive coating system where the primer is capable of chemically bonding to both the water-soluble polymer and the substrate. The latter is illustrated by the process of forming poly(urethane)-poly(urea) IPN at the interphase. The IPN functions as an adhesive joint anchoring the water-soluble polymer layer to the substrate.

Thus, the water-soluble (swellable) polymer layer in a good hydrophilic lubricious coating may be visualized to consist of two sub-layers. The exterior layer consists of free-flowing polymer chains or segments which are highly hydrated in water. The interior layer consists of polymer chains or segments which are either chemically bonded or physically entangled within a network of the primer molecules. When there is an insufficient population of hydrated molecules in the exterior layer, an inadequate lubricity would result. On the other hand, when the interior layer is insufficiently cured or reacted, a coating of poor abrasion resistance would develop. Thus, for a satisfactory hydrophilic lubricious coating, a proper balance between the two sub-layers is believed to be essential to achieve both a high degree of lubricity and good coating durability.

## REFERENCES


1. M. S. Heilmain, M. Sied and S. W. Maddell, U. S. Patent 4,080,706 (August 23, 1978).
2. Fuji Systems K. K., Japan Patent 63,164,956 (July 8, 1988).
3. H. E. Judd and R. D. Talty, U. S. Patent 4,169,163 (September 25, 1979).

4. E. W. Wong and D. G. Ballan, U. S. Patent 4.838,876 (June 13, 1989).
5. D. R. Shook, J. L. Doppman, E. L. Cattau and S. R. Goldstein, J. Biomech. Eng., 108 (2), p. 168-174 (May 1986).
6. P. M. Triolo & J. D. Andrade, J. Biomed. Mater. Res., 17(1), 149-165 (1983).
7. A. Cohen, J. Hosp. Infect., 6, Suppl. A, 155-161 (March 1985).
8. R. S. Pearce, L. R. West, G. T. rodeheaver and R. F. Edlich, Am. J. Surg., 148 (5), 687-691 (1984).
9. L. H. Harrison, J. Urol., 124(3), 347-349 (1980).
10. Y. Kikuchi, et al., Cardiovasc. Inervent. Radiol., 12(2), 107-109 (1989).
11. G. T. Rodeheaver, R. A. Foresman, M. T. Brazda and R. F. Edlich, Surg. Gynecol. Obstet., 164 (1), 17-21 (1987).
12. H. K. Buter and C. M. Kunin, J. Urol., 100 (4), 560-566 (1965).
13. A. Schwartz, J. Graper and J. Williams, U. S. Patent 4,589,873 (June 20, 1986).
14. Poly(vinyl pyrrolidone) USP grades are available from GAF and BASF, and poly(ethylene oxide) NF grades are available from Union Carbide.
15. W. Althana and J. Throne, European Patent 321,679 (June 28, 1989).
16. L. F. Becker, D. G. Laurin and J. A. Palomo, U. S. Patent 4,835,003 (May 30, 1989).
17. D. L. Podell, U. S. Patent 3,813,695 (June 4, 1974).
18. A. Goldstein, U. S. Patent 4,482,577 (November 13, 1984).
19. S. W. Kim, and J. Frijen, "*Critical Reviews in Biocompatibility,*" Vol. 1, No. 3, CRC Press, Boca Raton, 1985, pp. 229-260
20. F. E. Gould, C. K. Kliment and G. E. Seems, U. S. Patent 4,810,543 (March 7, 1989).
21. J. Sulc and Z. Krcova, German Patent 3,841,380 (June 22, 1989).
22. C. Golauder, E. Jonsson and T. Vladkova, European Patent 229,066 (July 22, 1987).
23. V. A. Perciaccante and H. P. Landi, U. S. Patent 4,047,533 (September 13, 1977): ibid., U. S. Patent 4,043,344 (August 23, 1977).
24. A. W. Hunter and S. Bogdansky, European Patent 128,043 (December 12, 1984).
25. L. T. Lehmann, D. W. Wang, D. J. Casey, L. Rosati and F. K. Jarrett, European Patent 258,749 (March 9, 1988).
26. D. J. Casey, F. K. Jarrett and L. Rosati, U. S. Patent 4,716,203 (December 29, 1987).
27. D. K. Hans, S. Y. Jeong and Y. H. Kim, J. Biomed. Mater. Res.; Applied Biomaterials, 23 (A2), p. 211-228 (1989).
28. V. G. Romberg, U. S. Patent 4,756,974 (July 12, 1988).
29. M. T. Scholz, D. C. Bartizal, K. E. Reed, W. K. Larson, T. C. Sandvis, R. S. Buskanin, D. A. Ersfeld and P. E. Hansen, European Patent 290,207 (November 11, 1988).
30. D. E. Hudgins, U. S. Patent 3,975,350 (August 17, 1976).
31. S. W. Shalaby and D. Jamiolkows, U. S. Patent 4,105,034 (August 8, 1978).
32. R. J. Schaper, U. S. Patent 4,166,894 (September 4, 1979); ibid., U. S. Patent 4,075,136 (February 21, 1978).
33. K. A. Messier & J. D. Rhum, U. S. Patent 4,624,256 (November 25, 1986).
34. D. J. Casey, U. S. Patent 4,068,757 (January 17, 1978).
35. F. R. Smith, U. S. Patent 4,199,367 (April 22, 1980).
36. D. E. Hanke, U. S. Patent 3,756,238 (September 4, 1973).
37. N. I. Trubitsina, O. A. Novikova and N. N. Gavrilyuk, U. S. S. R. Patent 1,265,226 (October 23, 1986).
38. P. Iwatschenk, U. S. Patent 4,798,593 (January 17, 1989).
39. Gillette Co., NetherlandS Patent 7,904,061 (December 4, 1979).
40. W. S. Creasy, D. H. Lorenz, R. G. Lacasse and S. A. Malagreca, U. S. Patent 4,642,267 (February 10, 1987).
41. V. L. Vailancourt, U. S. Patent 4,705,709 (November 11, 1987).

42. W. S. Creasy, U. S. Patent 4,847,324 (July 11, 1989).
43. H. K. Yoon AND R. L. J. Sun, European Patent 266,892 (May 11, 1988).
44. N. Kocak, U. S. Patent 4,705,511 (November 10, 1987).
45. E. G. Johansson AND J. M. R. Utassjoeberg, European Patent 217,771 (April 8, 1987).
46. H. I. Pudell, U. S. Patent 4,575,476 (March 11, 1986).
47. Nippon Zeon K. K., Japan Patent 60,012,069 (January 22, 1985).
48. P. Gold, U. S. Patent 4,534,363 (August 13, 1985).
49. H. R. Lambert, U. S. Patent 4,666,437 (May 19, 1987).
50. M. J. Micklus and D. T. On-Yang, U. S. Patent 4,100,309 (July 11, 1978).
51. Ibid., U. S. Patent 4,119,094 (October 10, 1978).
52. H. R. Lambert, U. S. Patent 4,487,808 (December 11, 1984).
53. J. M. Teffenhart, U. S. Patent 4,789,720 (December 6, 1988).
54. G. Halpern and J. U. Gould, U. S. Patent 4,657,820 (April 14, 1987); Ibid., U. S. Patent 4,810,586 (March 7, 1989); Ibid., U. S. Patent 4,722,867 (February 2, 1987).
55. E. G. Howard, W. O. Patent 8,909,246 (October 5, 1989).
56. N. Takemura and S. Tanabe, European Patent 166,998 (January 1, 1986).
57. E. M. Beavers, U. S. Patent 4,663,233 (May 5, 1987).
58. G. Halpern, C. Campbell, E. M. Beavers and H. V. Chen, U. S. Patent 4,801,475 (January 31, 1989).
59. C. K. Kliment and G. E. Seems, U. S. Patent 4,729,914 (March 8, 1988).
60. A Trademark of Union Carbide Corporations; U. S. and Foreign Patents pending.
61. F. R. Ainpour, U. S. Patent 4,526,579 (July 2, 1985).
62. T. E. Hyans, U. S. Patent 4,459,318 (July 10, 1984).
63, Nippon Medical, Japan Patent 61,209,670 (September 17, 1986).
64. Sumitomo Bakelite K. K., Japan Patent 63,033,866 (January 8, 1988).
65. A. Stoy, V. Stoy and J. Zima, U. S. Patent 4,026,296 (May 31, 1977).
66. P. Fydelor, R. A. Miller, B. T. Ringrose and T. W. Ramsay, U. S. Patent 4,785,059 (November 15, 1988); Ibid., European Patent 179,839 (May 7, 1986).
67. ITO Kogaku Kogyo K. K., Japan Patent 54,110,289 (August 29, 1979).
68. C. W. Neefe, U. S. Patent 4,280,759 (July 28, 1981).
69. C. H. Beede, H. L. Walch and T. Bluming, U. S. Patent 4,248,685 (February 3, 1981); Ibid., U. S. Patent 4,248,685 (February 3, 1981); Ibid., U. S. Patent 4,111,922 (September 5, 1978).
70. "*McGraw-Hill Encyclopedia of Science & Technology*," Vol. 1, p. 429-431 (1987).
71. R. D. Ratner and A. S. Hoffman, U. S. Patent 3,909,049 (February 10, 1976).
72. F. E. Gould and C. W. Johnston, German Patent 3,344,001 (June 13, 1985).
73. R. A. Miller, British Patent 2,179,258 (March 4, 1987).
74. Nippon Med. Supply K., Japan Patent 1,079,211 (March 24, 1989).
75. W. Althans, J. Thone and H. Ritter, European Patent 289,996 (November 9, 1988).

42. W. S. Creasy, U. S. Patent 4,347,324 (July 11, 1959).
43. H. K. Yoon AND E. L. J. Sun, European Patent 266,892 (May 11, 1988).
44. H. Kocak, U. S. Patent 4,705,511 (November 10, 1987).
45. R. G. Johansson AND J. M. R. Utassjoeberg, European Patent 217,771 (April 8, 1987).
46. H. I. Pudell, U. S. Patent 4,575,476 (March 11, 1986).
47. Nippon Zeon K. K., Japan Patent 60,012,069 (January 22, 1985).
48. E. Gold, U. S. Patent 4,534,365 (August 13, 1985).
49. H. R. Lambert, U. S. Patent 4,666,437 (May 19, 1987).
50. M. J. Michlus and D. T. On-Yang, U. S. Patent 4,100,309 (July 11, 1978).
51. Ibid., U. S. Patent 4,119,094 (October 10, 1978).
52. H. R. Lambert, U. S. Patent 4,487,808 (December 11, 1984).
53. J. H. Teffenhart, U. S. Patent 4,789,720 (December 6, 1988).
54. G. Halpern and J. U. Gould, U. S. Patent 4,657,820 (April 14, 1987); Ibid., U. S. Patent 4,810,586 (March 7, 1989); Ibid., U. S. Patent 4,722,867 (February 2, 1988).
55. E. G. Howard, W. O. Patent 4,909,246 (October 5, 1989).
56. N. Takemura and S. Tanabe, European Patent 166,998 (January 1, 1986).
57. E. M. Beavers, U. S. Patent 4,663,233 (May 5, 1987).
58. G. Halpern, C. Campbell, E. M. Beavers and E. Y. Chen, U. S. Patent 4,801,475 (January 31, 1989).
59. C. K. Kliment and G. E. Seems, U. S. Patent 4,729,914 (March 8, 1988).
60. A Trademark of Union Carbide Corporations; U. S. and Foreign Patents pending.
61. Y. R. Ainpour, U. S. Patent 4,526,579 (July 2, 1985).
62. T. E. Hyans, U. S. Patent 4,459,318 (July 10, 1984).
63. Nippon Medical, Japan Patent 61,209,670 (September 17, 1986).
64. Sumitomo Bakelite K. K., Japan Patent 63,033,866 (January 8, 1988).
65. A. Stoy, V. Stoy and J. Sing, U. S. Patent 4,026,296 (May 31, 1977).
66. P. Fydelor, R. A. Miller, J. T. Kingsnad and T. V. Ramsay, U. S. Patent 4,725,039 (November 15, 1988); Ibid., European Patent 179,839 (May 7, 1986).
67. ITO Kogaku Kogyo K. K., Japan Patent 54,130,289 (August 29, 1979).
68. C. W. Noele, U. S. Patent 4,280,759 (July 28, 1981).
69. C. E. Reeds, P. L. Walsh and T. Simoing, U. S. Patent 4,248,685 (February 3, 1981); Ibid., U. S. Patent 4,248,685 (February 3, 1981); Ibid., U. S. Patent 4,111,922 (September 5, 1978).
70. "McGraw-Hill Encyclopedia of Science & Technology," Vol. 1, p. 429-431 (1987).
71. R. D. Mather and A. C. Hoffman, U. S. Patent 3,903,049 (February 10, 1976).
72. Y. E. Gould and C. W. Johnston, German Patent 3,344,001 (June 13, 1985).
73. R. A. Miller, British Patent 2,179,258 (March 4, 1987).
74. Nippon Med. Supply K., Japan Patent 1,079,211 (March 24, 1989).
75. W. Ritzens, J. Teene and A. Mitter, European Patent 289,996 (November 9, 1988).

# NEW BIOMATERIALS: THE PREPARATION OF POLYESTERS DERIVED FROM HYDROXY AMINO ACIDS

Joachim Kohn

Department of Chemistry
Rutgers - The State University of New Jersey
New Brunswick, New Jersey 08903

In order to minimize the possible toxicity of degradable polymers, polymeric biomaterials were synthesized using exclusively natural amino acids as monomeric starting materials. Using new reaction schemes, the amino acids were linked by non-amide bonds, resulting in the formation of "pseudopoly(amino acids)." In particular, the preparation of poly(N-acyl-L-serine ester) from derivatives of N-acyl-L-serine was investigated in detail. N-benzyloxycarbonyl-L-serine was selected as a model monomer since the benzyloxycarbonyl groups can be removed, possibly with little degradation of the polymer backbone. Best results were obtained by the ring opening polymerization of N-benzyloxycarbonyl-L-serine-β-lactone. Under optimized conditions, poly(N-benzyloxycarbonyl-L-serine ester) with a molecular weight (MW) of about 40,000 (GPC, relative to polystyrene standards) was obtained. Such serine derived polyesters can be regarded as structural analogs of poly(lactic acid) and may find numerous applications as biomaterials.

## INTRODUCTION

After 1950, advances in metallurgy, ceramics and polymer chemistry resulted in the identification of numerous materials with improved biological and engineering properties. Attempts to utilize these new materials in medical applications led to the rapid growth of the field of biomaterials research. At this point, perhaps the most significant trend in biomaterials research is the gradual shift from inert to degradable implant materials. The design of sutures and the formulation of implantable drug delivery devices are currently well established applications for degradable implant materials. In the future, degradable implants will undoubtedly be deployed in a number of additional applications where the implant is required for a limited period of time only.

Since degradable implant materials must fulfill stringent requirements, related to their degradability, biocompatibility, and their physicomechanical properties, only a very small number of promising candidates for medical applications have so far been identified. Although some bio-

*Cosmetic and Pharmaceutical Applications of Polymers*
Edited by C.G. Gebelein *et al.*, Plenum Press, New York, 1991

degradable ceramics are known, the most intensely investigated, degradable implant materials are usually polymers.

It is a widely held assumption that the problems associated with the potential toxicity of a polymer and its degradation products can be alleviated by designing the degradable implant material in such a way that its degradation products are naturally occurring nutrients or metabolites. For example, the development of poly(glycolic acid) and poly-(lactic acid) had been based on this rationale.[1] Likewise, various poly-(amino acids) have been exclusively investigated as biomaterials[2] since the degradation of poly(amino acids) should not give rise to toxic degradation products. Unfortunately, however, most poly(amino acids) have highly unfavorable engineering properties and are insoluble, high-melting materials that cannot be processed into shaped objects by conventional fabrication techniques.[3] The failure of most synthetic poly(amino acids) to find useful applications as biomaterials illustrates the difficulty in balancing the often contradicting requirements for nontoxicity, degradability and favorable material properties.

We have recently suggested a new strategy for the improvement of the material properties of synthetic poly(amino acids).[4] Our approach is based on the introduction of non-amide bonds into the backbone of synthetic poly(amino acids): Whereas in conventional poly(amino acids), the amino acids are linked together by amide bonds via their N and C termini, we designed reaction schemes in which the amino acids are linked together via their side chains by non-amide bonds. This synthetic approach can give rise to poly(amino acids) with modified backbone structures. In analogy to widely used terminology in peptide chemistry,[5] the name "pseudo-poly(amino acid)" has been suggested for this new class of polymers.[6]

Our investigation of tryosine derived pseudo-poly(amino acids) revealed that these polymers do not suffer from the disadvantages that have limited the applicability of conventional poly(amino acids).[7-10] We are therefore currently extending our investigation of pseudo-poly(amino acids) to new polyesters derived exclusively from the natural hydroxy amino acids such as L-serine or hydroxy-L-proline.

## EXPERIMENTAL

### 1. Preparation of N-Z-L-Serine-β-Lactone[11]

The Verderas procedure[12,13] was modified as follows: To a solution of triphenylphosphine (13.12 g, 50 mmol) in 200 mL of a mixture of acetonitrile and THF (85:15) at room temperature was added diethyl azodicarboxylate (8.71 g, 50 mmol) under an atmosphere of argon. After stirring for 30 minutes, the reaction mixture was cooled to just above its freezing point (approximately -48 to -45°C). Then N-Z-L-ser (11.91 g, 49.8 mmol) in 200 mL of a mixture of acetontrile and THF (85:15) was slowly added over a period of 1 h. Thereafter, stirring was continued for an additional 30 min. The reaction mixture was slowly warmed to room temperature over a period of 3 h. After evaporation of the solvents under reduced pressure at a bath temperature of about 25°C, a white solid material was obtained. The crude product was dissolved in 20 mL of methylene chloride/ethyl acetate (85/15). Crystals of 1.2-dicarbethoxy-hydrazine precipitated and were removed by filtration. The filtrate was further purified by flash chromatography on silica gel 60, using methylene chloride/ethyl acetate (85/15) as the mobile phase: yield 8.92 g, 81%; m.p. 131-133°C.

After recrystallization from ethyl acetate/hexane the melting point increased to 133-134°C (lit. 133-134°C).[13]

The following analytical data was obtained. IR (KBr, $cm^{-1}$) 3366 (NH), 1835 (lactone C=O), 1688 (urethane C=O), 1532 (C-N), 1267 (C-O-C). 1109, 1019, 884; $^1H$ NMR ($CDCl_3$) δ 4.42 (2H, d, $CH_2$), 5.04 (1H, m, CH), 5.12 (2H, s, $CH_2$-Ph), 5.57 (1H m, NH), 734 (5H, s, Ph); $^{13}C$ NMR ($CDCl_3$) δ 168.7, 155.2, 135.4, 128.6, 128.5, 128.4, 128.2, 67.8, 66.3, 59.6;

Anal. Calcd for $C_{11}H_{11}NO_4$: C, 59.73; H, 5.01; N, 6.33.
Found: C, 59.46; H, 4.95; N, 6.21.

2. Ring Opening Polymerization of N-Z-L-Serine-β-Lactone[11]

Under argon, Z-L-serine-β-lactone (2.7 g, 12.2 mmol) was dissolved in 12 mL of THF. Tetraethylammonium benzoate (15 mg, 0.060 mmol) was added. The mixture was kept at 30°C for 7 days. At that time, TLC and GPC analysis showed that virtually all Z-L-serine-β-lactone had been consumed. The reaction mixture solidified in the course of the polymerization. The polymerization was terminated by the addition of 2 mL of methanol, and the polymer was fully precipitated by the addition of 50 mL of ethyl ether. After filtration and extensive washings with ethyl ether, the polymer was obtained as a white powder: yield 2.46 g, 90%. For purification, the crude polymer was dissolved in DMF and reprecipitated in ethyl ether.

## RESULTS AND DISCUSSION

### 1. The Use of Side-chain Hydroxyl Groups for the Synthesis of Amino Acid Derived Polyesters

Since there are no mild reactions that can be used to transform the amide linkages in the backbone of poly(amino acids) to ester bonds, amino acids derived polyesters must be prepared from scratch by suitably designed polymerization reactions. Our approach to the design of such polymerization reactions is based on the use of trifunctional hydroxy amino acids as monomeric starting materials.[4] This approach is, among others, potentially applicable to serine, hydroxyproline, threonine, and tyrosine and its only limited by the requirement that the ester backbone linkages give rise to polymers with desirable material properties.

One of the first pseudopoly(amino acids) prepared by us was a polyester derived from N-protected hydroxyproline. In these studies, we selected palmitic acid, a natural constituent of body fat, as the N-terminus protecting group.[4] The corresponding polymer, poly(N-hexadecanoyl-hydroxy-L-proline ester) [abbreviation: poly(Pal-Hpr ester)], was obtained by a melt transesterification procedure that yielded molecular weights (MW) of about 30,000 daltons (Scheme 1, R = $C_{15}H_{31}$).[4] Poly(Pal-Hpr ester) was soluble in organic solvents and readily processible by compression or injection molding. The polymer is highly hydrophobic and degraded very slowly both *in vitro* and *in vivo*. This polymer may thus be of value in the design of long term (1 to 2 years) implantable drug delivery systems.

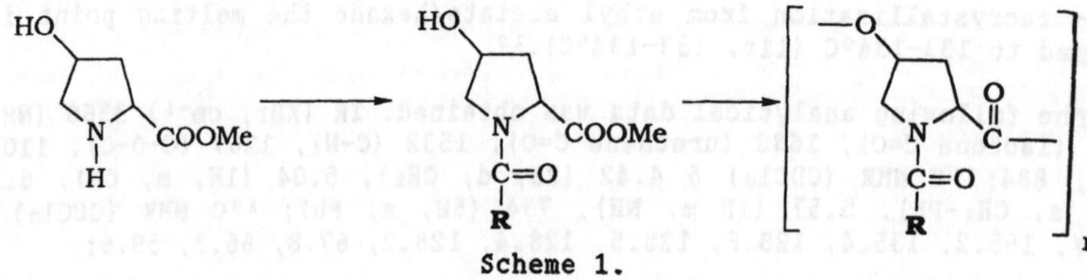

Scheme 1.

## 2. The Preparation of Poly(N-Acyl L-Serine Ester)

The direct esterification of N-protected serine in the presence of a suitable catalyst represents probably the most straightforward and most cost efficient approach to the preparation of poly(N-acyl-L-serine ester). We therefore started our synthetic exploration by suspending N-benzyloxycarbonyl-L-serine(N-Z-L-Ser) in toluene in a modified Dean-Stark distillation apparatus in the presence of 1% (w/w) of p-toluenesulfonic acid, using a previously published procedure.[14] Upon heating to reflux, the reaction mixture turned brown and gel permeation chromatography showed that the reaction product was a mixture of low oligomers and unreacted monomer. Although we repeated the above experiment under a large variety of different reaction conditions, no high polymers were obtained. We found evidence that the failure of the direct esterification technique was due to the β-elimination reaction at the serine side chain, leading to the formation of dehydroalanine (Scheme 2).[11]

Next, we investigated the transesterification of N-benzyloxycarbonyl-L-serine methyl ester (Z-Ser-O-Me). Transesterifications have the advantage of being catalyzed by a variety of catalysts such as bases, Lewis acids and numerous coordination compounds.[15] Furthermore, transesterification was successfully used in the preparation of hydroxyproline derived polyesters.

Surprisingly, the transesterification reaction failed in the case of serine. Although in most reaction mixtures some transesterification occurred (as indicated by the evolution of bubbles of methanol that were collected in a cold trap), even the most promising reaction conditions led only to the formation of low oligomers with an average molecular weight ($M_n$) of about 600 daltons.[11]

The difficulties observed in the preparation of polymers of high

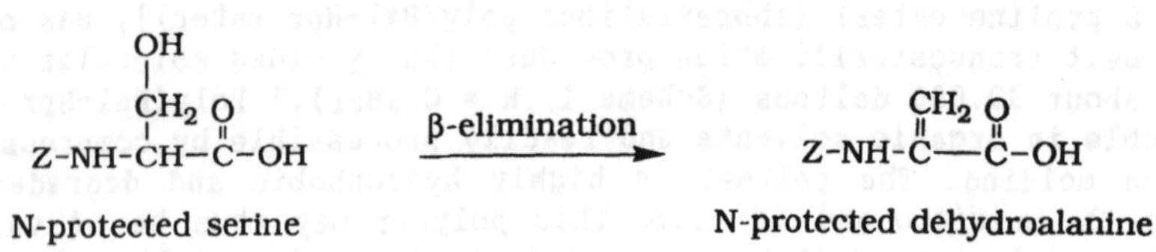


Scheme 2.

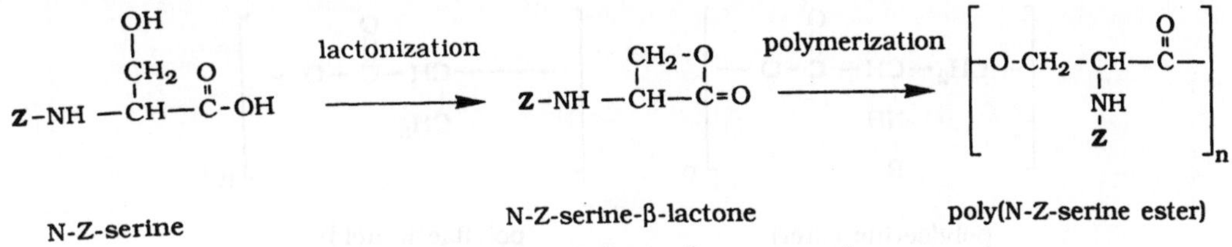

Scheme 3.

molecular weight by direct esterification or by transesterification prompted us to search for an alternative polymerization technique. The recent development by Vederas et. al., of a procedure for the synthesis of serine-β-lactones[12,13] made it possible to consider the ring-opening polymerization of N-protected L-serine-β-lactones (Scheme 3). This reaction had heretofore not been studied in detail.

In initial screening tests, tetraethylammonium benzoate (TEAB)[16] was identified as a particularly active initiator for the anionic solution polymerization of Z-serine-β-lactone. Next, we systematically optimized the reaction conditions. These studies resulted in the optimized procedure given in the experimental section. Using this procedure, poly(N-Z-serine ester) with a molecular weight (MW) of about 30,000 to 40,000 daltons could be obtained. Other N-terminus protecting groups could be used as well. For example, we found that N-Boc-L-serine polymerized even more rapidly than N-Z-L-serine. Using a very similar procedure, the polymerization of N-trityl-L-serine has, in the meantime, been reported by Spassky's group.[17]

There are two major limitations to the ring opening polymerization of serine-β-lactones: First of all, the preparation of the lactone itself is not trivial and could be a very costly procedure if the lactone will ever be required in pilot plant quantities. Second, a chain transfer reaction occurs during the ring opening polymerization and apparently limits the maximum molecular weight that can be obtained to about 40,000 daltons. For these reasons we are currently investigating improved synthetic procedures for the preparation of poly(N-acyl serine ester).

After completion of the polymerization reaction, the Z protecting group could be removed by catalytic transfer hydrogenation, a commonly used procedure in peptide synthesis.[18,19] The deprotection reaction was monitored by $^1$H-NMR. Fortunately, the Z group has a characteristic absorption (multiplet at 7.2 ppm) that makes $^1$H-NMR a sensitive method for the detection of residual protecting groups: In a quantitative test, as little as 0.005 mole% of residual Z protecting group could be detected in samples of "deprotected" polymer. Thus, we are confident that over 99.9% of the temporary protecting groups can be removed from the polymer. Unfortunately, we have not yet determined whether backbone cleavage occurs during the deprotection reaction. This question is still under investigation.

## CONCLUSION

Poly(serine ester) is a structurally new polyester whose backbone

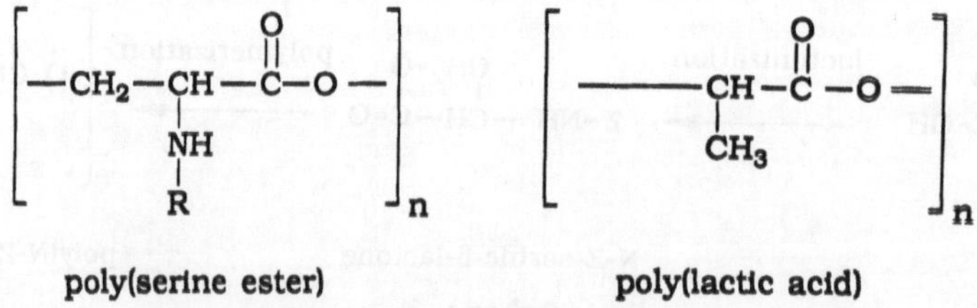


**Figure 1. Comparison of the chemical structures of poly(serine ester) and poly(lactic acid). One of the main differences between these two polymers is that poly(serine ester) has a reactive side chain (the amino group) which can serve as a convenient attachment point for drug molecules, crosslinkers, or other pendent chains.**

**resembles the backbone of poly(lactic acid) (Figure 1). However, contrary to poly(lactic acid), poly(serine ester) possesses a chemically accessible amino group that can be used as an attachment point for crosslinkers, drug molecules, or various blocking groups. This is potentially the most important advantage of poly(serine ester) over the well-established polyesters derived from lactic or glycolic acid. Thus we are currently exploring the use of poly(serine ester) for the preparation of fully biodegradable hydrogels and injectable drug carriers. No comparable synthetic schemes are possible for poly(lactic acid) (Scheme 4).**

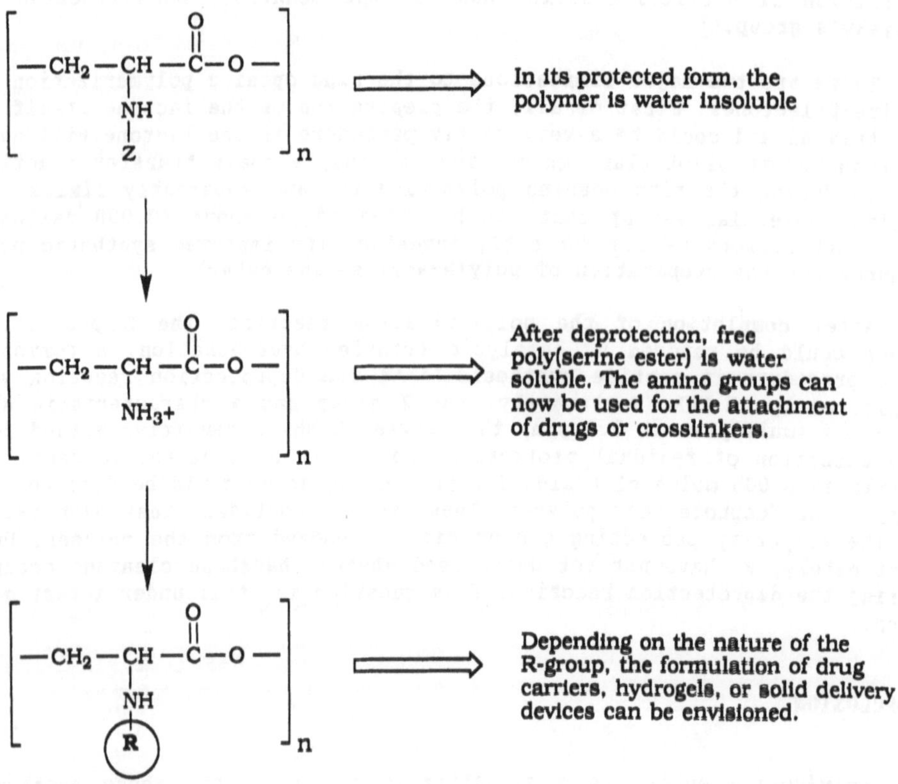


**Scheme 4.**

Furthermore, the freedom to choose any one of a large number of biocompatible moieties for attachment to the amino terminus should make it possible to modify the material properties of serine derived polyesters more readily than it is possible to modify the properties of polyesters derived from lactic or glycolic acid. We therefore expect serine derived polymers to be versatile materials whose properties can be custom-designed for many specific applications.

## ACKNOWLEDGMENTS

This work was supported by NIH grant GM39455.

## REFERENCES

1. E. J. Frazza and E. E. Schmidt, J. Biomed. Mat. Res. Symp., 1, 43 (1971).
2. J. M. Anderson, K. L. Spitizewski and A. Hiltner in: "*Biocompatibility of Tissue Analogs*," D. F. Williams, Ed., CRC Press, Boca Raton, Vol. 1, 1985, p. 67-88.
3. A thorough and still valid discussion of the basic physical properties of synthetic poly(amino acids) was written by C. H. Bamford, A. Elliot and W. E. Hanby, "*Synthetic Polypeptides*," Academic Press, New York, 1956.
4. J. Kohn and R. Langer, J. Am. Chem. Soc., 109, 817 (1987).
5. A. F. Spatola, in: "*Chemistry and Biochemistry of Amino Acids, Peptides, and Proteins*," B. Weinstein, Ed., Marcel Dekker, New York, 1983 p. 268.
6. J. Kohn, in: "*Biodegradable Polymers in Drug Delivery Systems*," R. Langer and M. Chasin, Eds., Marcel Dekker, New York, 1990, p.195-229.
7. J. Kohn and R. Langer, Polymeric Materials, Science and Engineering, American Chemical Society, 51 119 (1984).
8. J. Kohn, S. M. Niemi, E. C. Albert, J. C. Murphy, R. Langer, and J. G. Fox, J. Immunol. Meth., 95, 31 (1986).
9. S. Pulapura and J. Kohn, Polym. Preprints, 31(1), 233 (1990).
10. S. Pulapura, C. Li, and J. Kohn, Biomaterials, 11, 666 (1990).
11. Q. X. Zhou and J. Kohn, Macromolecules, 23, 3399 (1990).
12. L. D. Arnold, T. H. Kalantar, and J. C. Vederas, J. Am. Chem. Soc., 107, 7105 (1985).
13. L. D. Arnold, J. C. Drover, and J. C. Vederas, J. Am. Chem. Soc., 109, 4649 (1987).
14. H. Baltzer, H. Holtschmidt, F. Wiloth, and B. Mohr, Macromol. Chem., 7, 82 (1951).
15. S. R. Sandler and W. Karo, in: "*Polymer Synthesis*," Vol. 1, Academic Press, New York, 1974, p. 62.
16. See, for example, R. W. Lenz, Pure and Appl. Chem., 53, 1729 (1981): or D. Duchesne and R. E. Prud'homme, Polymer, 20, 1199 (1979).
17. I. Fietier, A. Le Borgne and N. Spassky, Polymer Bulletin 24, 349 (1990).
18. M. K. Anwar, R. A. Porter and A. Spatola, Int. J. Protein Res., 30, 489 (1987).
19. B. ElAmin, G. M. Anantharamaiah, G. P. Royer and G. E. Means, J. Org. Chem., 44, 3442 (1979).


# THE DEVELOPMENT OF A CELLULOSIC MATERIAL BASED METHOD FOR HEPARIN THERAPY MONITORING

Weiliam Chen and Victor C. Yang*

College of Pharmacy
The University of Michigan
Ann Arbor, MI 48109 -1065

We have developed a non-clotting based heparin assay. It is established by the attachment of protamine (a heparin antidote) to a porous filter paper strip; and the subsequent migration of heparin sample through the filter paper in an ascending or descending manner. The area of paper to which heparin adsorbed will be proportional to the heparin level in the sample. This region is visualized by spraying methylene blue NNX (a dye that interacts with heparin and the interaction causes a metachromatic shift of the dye's absorption maximum from blue to purple) solution onto the paper strip upon the exhaustion of the sample reservoir. This approach has proven to be capable of detecting and differentiating plasma heparin level within clinical range in a relatively time efficient and accurate manner.

## INTRODUCTION

Heparin, the most efficacious and widely utilized intravenous anticoagulant, for clinical use was developed by research groups in Canada and Sweden over six decades ago.[1] Since the introduction of heparin into modern internal medicine, its role as an anticoagulant has never diminished. The most important application of heparin is in clinical procedures in which extracorporeal blood circulation (ECBC) is required[2]: such as cytapheresis,[3] plasmapheresis,[4] cardiopulmonary bypass,[5] blood oxygenation,[6] vascular reconstructive surgery,[7,8] intraoperative autotransfusion,[9,10] hemodialysis[2] and so forth. For these procedures the therapeutic range of heparin in plasma is generally between 2 to 10 units/mL.[11]

Due to the potential morbidity of heparin therapy, continuous and careful monitoring of its therapeutic blood level during the entire course of ECBC treatment is essential in providing the clinicians with sufficient information for taking pertinent actions to mitigate the anticoagulant activity (e.g., moderating the infusion rate of heparin, administering of antidote, etc.) should an overdose be confirmed.[12] Hitherto,

* To whom correspondence should be addressed.

*Cosmetic and Pharmaceutical Applications of Polymers*
Edited by C.G. Gebelein *et al.*, Plenum Press, New York, 1991

the most effective and clinically well proven heparin antidote is protamine sulfate.[13] Protamine is routinely used to titrate and thus abate heparin's *in vivo* activity,[13] whenever necessary, during any medical procedure involving extracorporeal blood circulation. Unambiguous information regarding heparin level and its anticoagulant activity is crucial for clinicians to administer the minimal and effective dose of protamine required for heparin neutralization.

Numerous clotting assays are available for assessing the therapeutic efficacy of heparin. The most commonly employed methods include: Lee - White Clotting Time (WBCT),[14] Activated Coagulation Time (ACT),[15,16] Activated Partial Thromboplastin Time (APTT) and Plasma Recalcification Time (PRT).[16,17] They are widely utilized by medical laboratories. Since these heparin monitoring techniques are all based on the measurement of clotting time, external calibration, precise timing, manipulation of reagents and numerous sample handlings are the inevitable routines. Performing these assays can therefore be very labor intensive and time consuming. The physical conditions of samples have to be strictly controlled for the assurance of accuracy. In addition, most of these clotting based procedures rely heavily on relatively sophisticated instrumentation, therefore, these assays are confined to the medical laboratories with the appropriate instruments available on site.

We have developed a fundamentally different heparin sensing approach in which the heparin concentration is exemplified by its presence on filter paper strip surface, rather than anticoagulant activity (i.e., clotting time). The heparin level can be visually quantitated by the position of the color front on the paper strip. The new approach proven to be capable of abridging the aforementioned disadvantages of the clotting assays currently available for clinical laboratory use. This chapter focuses primarily on the development of this versatile method for determining heparin levels.

This new heparin sensing device is founded on some well characterized methodologies: (a) immobilization of protamine to cellulosic support, (b) the capillary migration of liquid in porous material, (c) the electrostatic binding of heparin to its antidote, protamine and (d) the metachromatic interaction between heparin and dyes belonging to the thiazine family.[1,12,18] The core of this new assay is a filter (or chromatographic) paper strip containing immobilized protamine. The concept of operation is schematically illustrated in Figure 1A (ascending mode) and Figure 1B (descending mode). To perform an assay in the ascending mode, the paper strip is settled in a reservoir containing a finite volume of heparin sample which will be wicked up by the paper strip through capillary migration. Assuming that the protamine distribution throughout the paper strip is homogeneous, each unit area of the filter paper will bind a specific amount of heparin. As the heparin sample migrates towards the top of the paper strip, heparin molecules will be electrostatically bound to the paper by the immobilized protamine molecules; while the non-heparin components of the sample (e.g., the liquid portion) will continue to move along that paper strip. Upon the exhaustion of the heparin sample in the reservoir, the paper strip will be sprayed with methylene blue NNX dye (a thiazine dye that interacts with heparin and the interaction results in a metachromatic shift of the dye's absorption maximum from blue to purple) solution.[19] Since both the dimension of the paper strip and sample volume are standardized, upon the completion of dye treatment, the length of the purple area on the paper strips will be proportional to the heparin concentration of the original sample. Heparin levels of the samples can be estimated by referencing the length of the purple area with an appropriate pre-constructed standard curve. Figure 1B depicted the descending mode of operation. A fixed volume of heparin sample is trans-

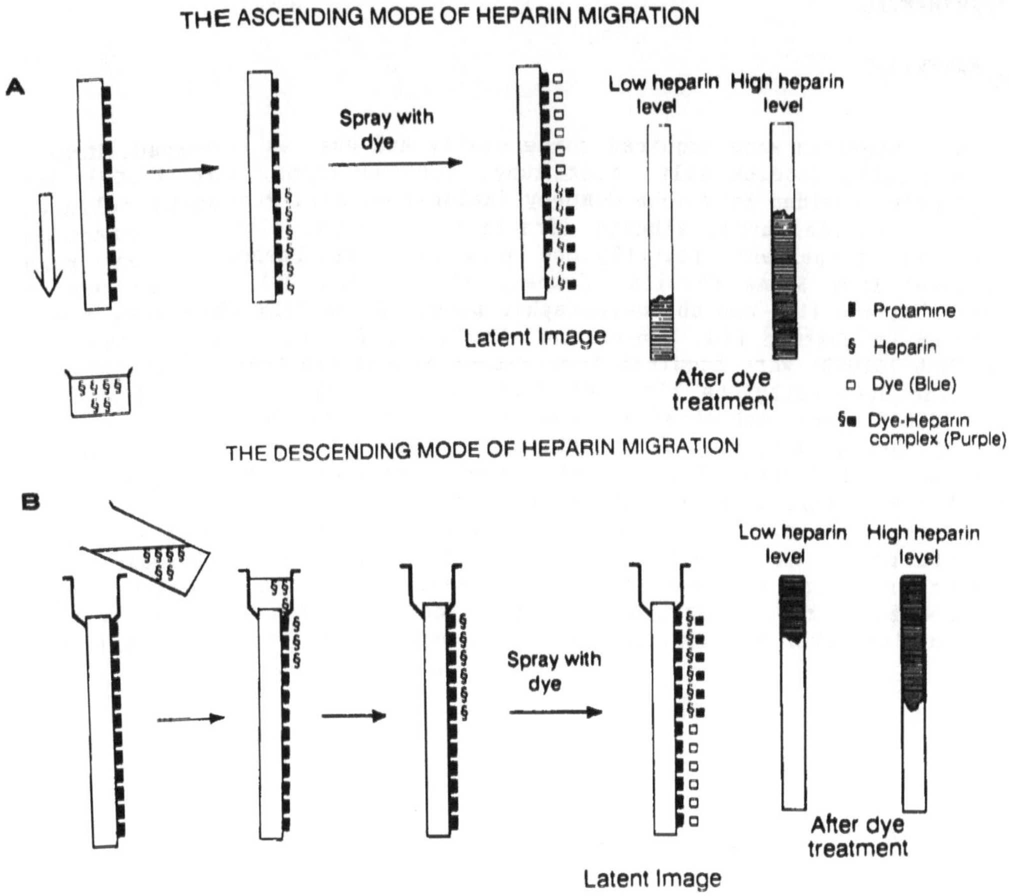


**Figure 1.** (A) The schematic illustration of the ascending mode of operation. (B) The schematic illustration of the descending mode of operation.

ferred to the fibrin cup which serves as sample reservoir. The sample will start to migrate down the paper strip (with bound protamine). Upon the exhaustion of the sample reservoir, methylene blue NNX dye solution will be sprayed onto the paper strip. The dye-heparin metachromatic interaction will render the area with bound heparin purple. Similar to the ascending mode of operation, the heparin level of the sample can be estimated in reference with an appropriate pre-constructed standard plot.

The results obtained using the new heparin monitoring method described in this chapter compared favorably with their counterparts obtained by the Azure A colorimetric assay (in aqueous sample) and the APTT clotting assay (in plasma sample). The average analytical recovery of heparin in aqueous and plasma samples were 100.1 ± 5.9% and 97.4 ± 7.7%, respectively.

**EXPERIMENTAL**

## 1. MATERIALS

All materials were acquired commercially and used as received. Heparin injectable (sodium salt, beef lung, 1000 units/mL, Lot 470CW) was generously provided by Upjohn Company (Kalamazoo, MI). Protamine sulfate, cyanogen bromide, human albumin (fraction V), human globulin (fraction IV-4), CFC propellant, polyethylene jars and Folin-Phenol reagent were purchased from Sigma Chemical Company (St. Louis, MO). Filter papers (Grade 541 and 113) and chromatographic paper (Grade 31ET Chr) were manufactured by Whatman Ltd. (Maidstone, England). Fibrin cups and Azure A dye (Lot 741002) were acquired from Fisher Scientific Company (Fairlawn, NJ). The dyes methylene blue NNX (Lot 277056-1087) and toluidine blue (Lot 405046) were both obtained from Fluka AG Chemical Corp. (Ronkonkoma, NY). The Actin$^R$ activated cephaloplastin reagent was purchased from American Dade del Caribe, Inc. (Puerto Rico). Silicone rubber sealant and Corex$^R$ Bottle were manufactured by Dow Corning (Midland, MI) and Corning Glasswork (Corning, NY) respectively. Freshly frozen citrated plasma was supplied by The American Red Cross (Detroit, MI). Other chemicals were reagent grade and deionized water was used. The Orbitron Rotator$^R$ was produced by Boekel Industries, Inc. (Philadelphia, PA). The Fibrometer$^R$ (Fibrosystem$^R$ BBL) was acquired from Becton-Dickinson Inc. (Hunt Valley, MD).

## 2. METHODS

### 2A. Immobilization of protamine on paper

2Aa. Activation of the paper: Five grams of cyanogen bromide were dissolved in 5 mL of acetonitrile and transferred to a 125 mL Erylenmeyer flask containing 25 mL of 1 M sodium carbonate solution. The content in the flask was promptly swirled and mildly agitated to form a homogeneous solution and poured into a Corex$^R$ Bottle containing 5 grams of paper strips (standard dimensions: 9 mm x 110 mm). After the bottle was tightly capped and shaken vigorously for exactly 5 minutes, the liquid phase was discarded. The filter paper strips were subsequently washed with 50 mL of chilled distilled water and 50 mL of 1 mM hydrochloric acid.

2Ab. Coupling of protamine sulfate to the activated paper strips: The activated paper strips were immediately transferred to a polyethylene jar and washed with 25 mL of 0.10 M sodium bicarbonate/0.50 N sodium chloride buffer (pH 8.3). Fifty milliliters of protamine sulfate solution (concentrations: 1 to 10 mg/mL, prepared by dissolving protamine sulfate in the same buffer described above) were added to the jar and it was tightly capped and placed on an Orbitron Rotator$^R$. The contents of the jar were incubated for a certain period of time (0.25 to 2.5 hours) while the rotator remained swirling. After the incubation, the residual protamine sulfate solution was poured out and collected. The filter paper strips were subsequently washed with 100 mL of 0.25 M sodium phosphate/0.50 M sodium chloride buffer (pH 7.0) followed by 100 mL of 0.07 M sodium phosphate/0.13 M sodium chloride buffer (pH 7.0). The protamine-bound paper strips were then rinsed with two 50 mL portions of distilled water. The drainage collected would then be assayed for the protamine sulfate content.

2Ac. Quantitation of protamine immobilized on paper: The quantity of protamine in the drainage collected was determined by Lowry assay.[19] The amount of protamine immobilized on the paper strips was determined by measuring the total protamine contents in the solution prior to and following the protamine immobilization process. Dividing the amount of immobilized protamine by the total surface area of the paper strips yielded the protamine distribution density (PDD) value.

## 2B. Determination of the Maximum Absorption Magnitudes of the Heparin-Dye Complexes of Three Thiazine Dyes

Two milliliters of dye (Azure A, toluidine blue and methylene blue NNX) solution (20 µg/mL) were blended with 2 mL of heparin solution (1 unit/mL). The absorbances of these dye-heparin mixture were measured at the wavelengths of 535 nm (Azure A), 529 nm (toluidine blue) and 551 nm (methylene blue NNX), respectively.

## 2C. Heparin Sample Migration Studies

2Ca. Ascending mode: A multiple sample holder (see Figure 2) was constructed from microslides in order to avoid the possibility of an alteration of migration pattern resulting from the contact of filter paper strip with any surface during the process of sample wicking. Filter paper strips were inserted through the gap (a) between the two bridges and settled in the wedge (b) at the bottom of the compartment. Heparin test samples were then pipetted into each chamber and allowed to travel up the paper strips by capillary migration. The whole setup was installed in a vacant dessicator in order to minimize the evaporation of the test samples. Sufficient time was allowed for the samples to be completely wicked up by the filter paper strips. They were subsequently transferred

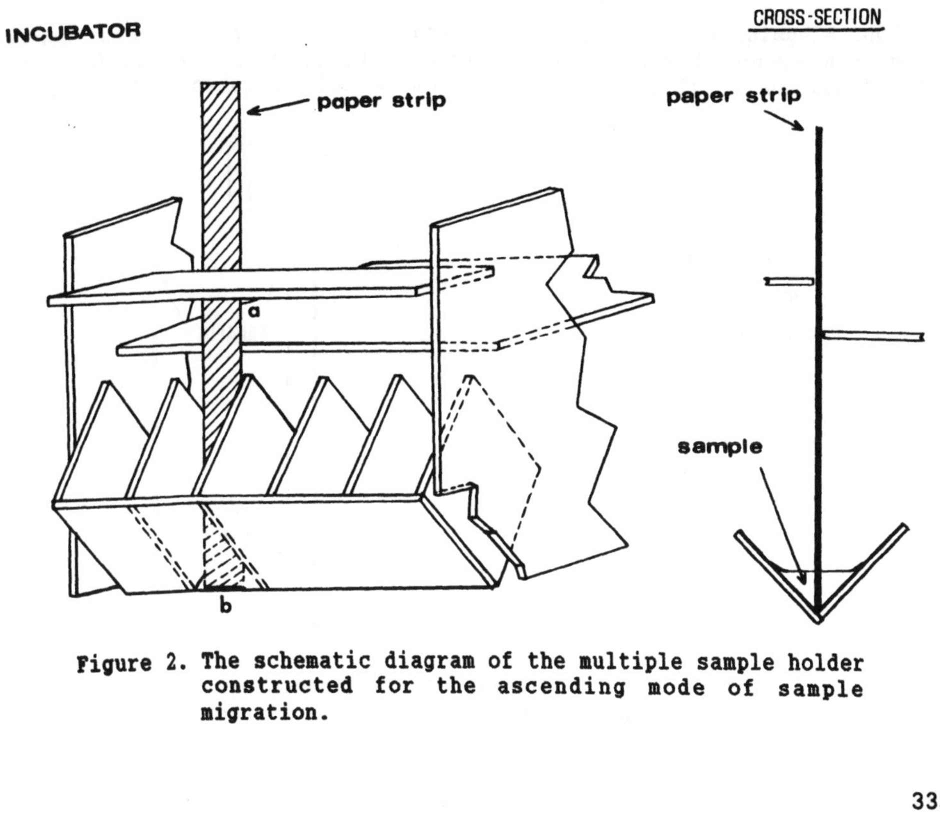


Figure 2. The schematic diagram of the multiple sample holder constructed for the ascending mode of sample migration.

and deployed in a fumehood. Methylene blue NNX dye solution (500 µg/mL) thrusted by a CFC propellant was sprayed onto the paper strips immediately. The heparin migration distance would be denoted by the length of the purple region on each paper strip.

2Cb. Descending mode: A fibrin cup was slashed in the middle with a razor blade. A filter paper strip, with its top laminated with two slides of parafilms (approximate dimensions: 10 mm x 2 mm), was inserted into the slat on the fibrin cup. All the edges were sealed with silicone rubber. The entire setup was then placed in a test tube (13 mm x 150 mm). This device is illustrated in Figure 3. To allow the application of a larger volume (beyond the liquid loading capacity of the paper strip) of heparin sample, a piece of benchtop mat (approximately 10 mm x 80 mm) was rolled and consolidated to the bottom of the paper strip with the aid of a plastic ring modified from 1 mL sized Eppendorf[R] pipette tips. Desired volume of heparin test sample was then pipetted into the fibrin cup and allowed to migrate down the paper strip. Upon the exhaustion of the entire sample volume, the device was transferred to a fumehood. Methylene blue NNX dye solution (500 µg/mL) thrusted by a CFC propellant was sprayed onto the paper strip immediately. The heparin migration distance would be denoted by the length of the purple region on the paper strip.

## 2D. The Effect of Several Major Plasma Components on the Migration Distance

2Da. Sodium chloride: Aqueous heparin solution (40 units/mL) were blended with aqueous sodium chloride solutions (0, 0.2, 0.4, 0.8, 1.2, 1.6, 2 and 2.4%, Wt/V) in a 1:1 proportion to form heparin samples (20 units/mL) with sodium chloride at various concentrations. Two hundred microliters of each sample was allowed to migrate through the type 541 paper strips (PDD = 45.4 µg/cm²) until the sample in the reservoir was exhausted. The influence of sodium chloride on the migration distance of heparin in the sensing system would then be assessed.

2Db. Albumin: Aqueous heparin solution (64 units/mL) were gently mixed with aqueous human albumin solutions (0, 1, 2.4, 3.2, 4, 6, 8 and 10%, Wt/Wt) in a 1:1 proportion to form heparin samples (32 units/mL)

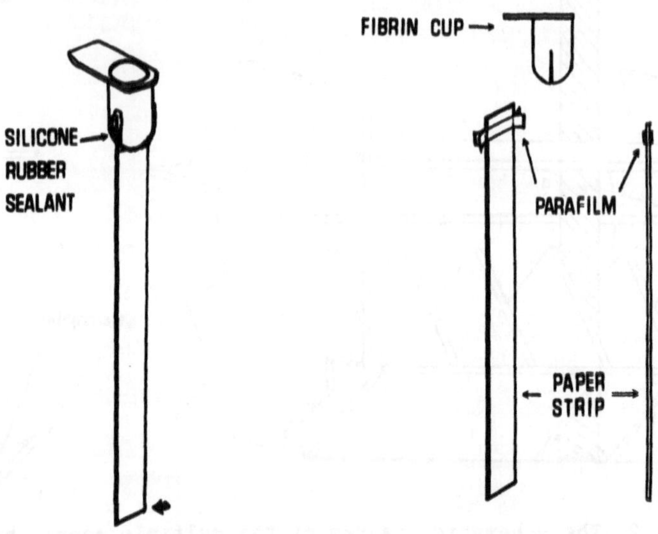


Figure 3. The schematic diagram of the heparin sensing device for the descending mode of sample migration.

with human albumin at various concentrations. Two hundred microliters of each sample was allowed to migrate through the type 541 paper strips (PDD = 64.4 μg/cm²) until the sample in the reservoir was exhausted. The effect of albumin on the migration distance of heparin in the sensing system would then be evaluated.

2Dc. Globulin: Aqueous heparin solution (60 units/mL) were gently mixed with aqueous human globulin solutions (0, 0.2, 0.4, 0.6, 0.8, 1, 2 and 3%, Wt/Wt) in a 1:1 proportion to form heparin solution (30 units/mL) with globulin at various concentrations. Two hundred microliters of each sample was allowed to migrate through the grade 541 paper strips (PDD = 79.8 μg/cm²) until the sample in the reservoir was exhausted. The leverage of globulin on the migration distance of heparin would then be studied.

### 2E. The Standard Addition Approach

Aqueous heparin solution (20 units/mL) was added to heparinized plasma samples of known concentrations (0, 2, 4, 6, 8 and 10 units/mL) in 1:1 (V/V) proportion. Three hundred microliters of each of the mixed samples was allowed to migrate down the 31ET Chr paper strips (PDD = 36.2 μg/cm²) according to the procedures described previously. The migration distance versus heparin concentration plot obtained would serve as the standard plot for determining the heparin concentrations in the plasma samples with unknown heparin levels.

### 2F. Validation of the Heparin Heparin Levels Deduced by Employing the New Sensing Device

2Fa. Aqueous heparin sample: The concentrations of aqueous heparin samples fabricated by random dilution for the migration studies were determined by the modified Azure A - Heparin colorimetric assay described by Jaques and Wollin.[20]

2Fb. Plasma heparin sample: The Activate Partial Thromboplastin Time (APTT) assay described by Jaques was utilized for determining the unknown heparin levels of plasma samples (prepared by random dilution).[20] Actin® activated cephaloplastin reagent and Fibrometer® (Fibrosystem® BBL) were employed for the assay.

## RESULTS AND DISCUSSION

Heparin therapy monitoring can be accomplished using assays measuring blood or plasma clotting time. The aforementioned WBCT, ACT, APTT and PRT assays are representative examples. They are widely employed in clinical laboratories. Most of these tests have been semi-automated through the introduction and integration of various mechanical gadgets; however, the reliability and accuracy of these clotting assays can easily be biased by numerous factors such as the physical conditions of samples (storage span, quality, etc.), temperature fluctuation, hemodilution, medications and so forth.[22] The new assay we described in this chapter proven to be capable of providing reliable and unbiased information regarding heparin therapy (i.e., the plasma heparin level) in a time efficient, inexpensive and sufficiently simple fashion. Since the assay is non-clotting based, the heparin level derived is not likely to be affected by the aforementioned factors that undermined the clotting based assays. Moreover, no sophisticated laboratory instrument is necessary; in conjunction with the fact that only minimal user expertise is needed; this new assay can thus be a potential remedy for the shortcomings of the currently available heparin monitoring techniques.

Clinically, protamine sulfate is routinely administered to patients for neutralizing the *in vivo* anticoagulant activity of heparin at the conclusion of most medical procedures involving extracorporeal blood circulation.[13] Protamine sulfate is a highly basic, low molecular weight protein that acts as a heparin antagonist by electrostatically complexing with the strongly acidic heparin to form a stable salt that has no anticoagulant activity.[13] The heparin-protamine interaction has been thoroughly investigated and well characterized. In addition, the ready availability of protamine commercially and the easiness of its detection (i.e., by Lowry assay) renders it to be the most convenient choice for the heparin binding role of this sensing system.

The porous material for use in designing the core of this new heparin sensing device should be easy for the attachment of protamine, have swift migration of liquid and possess sufficient mechanical strength to allow handling and chemical treatment without losing integrity. Since cyanogen bromide is utilized as the reagent for mediating protamine immobilization, cellulosic substances (e.g., filter paper), which are abundant in hydroxyl groups, are chosen as the support material. The hydroxyl groups exist on the surface of the filter paper can readily be activated with cyanogen bromide to form reactive residues which will subsequently react with the amine groups present on the surface of protamine molecule. Consequently, protamine can be coupled onto the filter paper through covalent linkage.[23,24] Porous filter paper also allows for rapid and uniform capillary migration to occur; thus, filter paper with sufficient rigidity and porosity (Whatman grade 113 and 541) was used as the support material for studying the feasibility of the heparin sensing device conceived. In addition, the possibility of applying chromatographic paper was also explored. Since the Whatman grade 31ET chromatographic paper also fulfilled the stipulated criteria for the choice of support material, that type of paper was likewise used.

The amount of protamine immobilized on paper (i.e., PDD value) could be controlled by adjusting three parameters: (a) the amount of activating agent (i.e., cyanogen bromide) used, (b) the concentration of protamine solution used for coupling, and (c) the incubation time for the coupling process. By utilizing different combinations of these variables, paper strips with different PDD values were prepared. These paper strips with diverse PDD values were employed in detecting and differentiating heparin levels at various concentration ranges.

The magnitudes of absorption of the dye-heparin complexes were measured at their corresponding maximum absorption wavelengths. The maximum

Table 1. Maximun absorption wavelengths of dyes and their heparin complexes. Dye concentration = 10 μg/mL; heparin concentration = 0.5 unit/mL.

| Dye | Absorption Maximum | Absorbance of Dye-Heparin complex |
|---|---|---|
| Methylene blue NNX | 551.2 nm | 0.675 |
| Toluidine blue | 534.8 nm | 0.505 |
| Azure A | 529.2 nm | 0.355 |

absorption wavelengths of the dyes and the maximum absorbances of their corresponding heparin complexes are depicted in Table 1. The absorption magnitudes of the dye-heparin complexes of methylene blue NNX, Azure A and toluidine blue were 0.675, 0.505 and 0.355 respectively. Evidently, methylene blue NNX was the most appropriate dye for detecting the presence of heparin since it provided the most intense color change that could greatly facilitate the visualization of the presence of heparin.

Initially, an ascending migrating approach was adopted and aqueous heparin solution was used for developing and elaborating the technique of a practical method. Heparin sample was wicked up by the paper strip solely through capillary migration. After the treatment with dye, the area to which heparin bound would appear to be purple whilst the area without heparin would remain blue. The correlation between the migration distance and heparin concentrations at various PDD values using the grade 541 paper is illustrated in Figure 4. A linear region that did not pass the origin was observed on each migration distance versus heparin concentration plot. Using 100 µL (which was the liquid loading capacity of the paper strip at the stipulated dimension) of aqueous heparin solution (15 units/mL); the migration distance were 15.9, 22.7, 32.4 and 47.1 mm for the paper strips of PDD values of 80.7, 61.2, 40.7 and 17.6 $\mu g/cm^2$, respectively. The difference in heparin migration distance at various PDD values reflected the difference in their degrees of sensitivity for heparin detection. The paper strips with higher PDD value (e.g. 80.7 $\mu g/cm^2$) were capable of detecting and differentiating a board range of heparin concentrations (10 to 50 units/mL). They were, however, not sufficiently sensitive to precisely differentiate a close range of heparin concentrations (10 to 15 units/mL). In contrast, the paper strips with low PDD value (e.g., 17.6 $\mu g/cm^2$) were only capable of serving comparable purpose within a narrow range of heparin concentrations (7.5 to 15 units/mL). These paper strips with low PDD values were far more sensitive to the change of heparin concentrations, as reflected by the increasing steepness of the slopes of the plots. The slopes of the linear regions on the plots were 3.3, 1.6, 1.1 and 0.8 mm-mL/unit for the PDD values of 17.6, 40.7, 61.2 and 80.7 $\mu g/cm^2$, respectively. The diminishing of slope with the increment of PDD value reflected the down shifting of sensitivity and thus the decrease of resolution of the paper strips to heparin.

Although the ascending migrating approach appeared to be feasible for accurately detecting heparin level in aqueous sample, it suffered from several insurmountable drawbacks. First, the heparin sample volume applied was restricted by the liquid loading capacity of the paper strip. Second, certain degree of user proficiency was necessary for proper operation. Lastly, since the heparin sample was migrating through the paper strip in an ascending manner, the speed of sample migration would inevitably be hindered by the gravitational force. This resulted in the excessive time requirement (greater than 40 minutes using the grade 541 paper) for the ascending mode of sample application. The exceedingly long time requirement thus undermined any further effort of developing the ascending mode of operation. Subsequent redesigning led to the descending migrating mode which virtually overcome all the intrinsic drawbacks associated with the ascending method.

By converting the paper strips' mode of wicking heparin sample from ascending to descending, it not only alleviated the technical complications encountered in the ascending mode, but also effectively converted the adverse effect of gravitational force on the ascending mode in favor of sample migration. The descending migrating approach required only two modest steps for proper operation: (a) transferring of sample into the reservoir (the fibrin cup) and (b) treating the paper strip with dye solution upon the exhaustion of heparin sample in the reservoir. Hence,

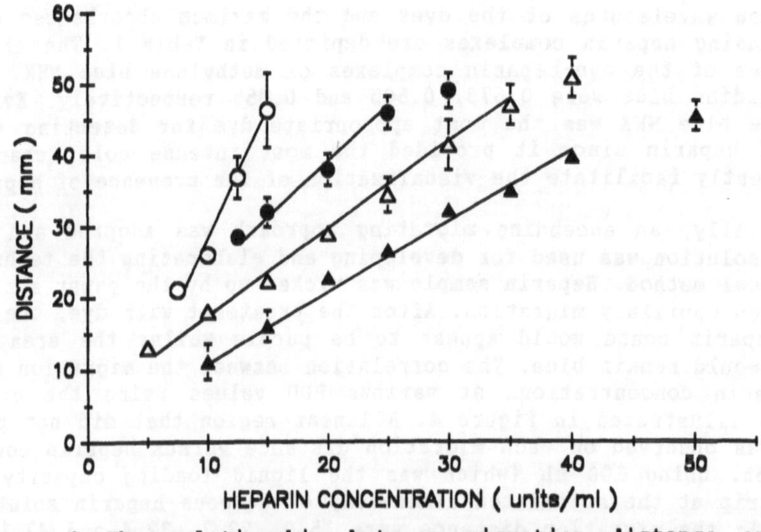


**Figure 4.** The relationship between migration distance and heparin concentration on the grade 541 paper strips of various PDD values in aqueous sample using the ascending mode of sample migration. Sample volume: 200 µL. PDD values: (o) 17.6 µg/cm², (●) 40.7 µg/cm², (△) 61.2 µg/cm² and (▲) 80.7 µg/cm².

it posed no difficulty, even for users without any experience with this system. In addition, with the presence of the liquid absorbent arrangement consolidated to the bottom of the paper strip, a much larger sample volume could be applied; thus, the amount of heparin available could be greatly increased. The appearance of the type 541 paper strips with the PDD value of 135.9 µg/cm² using the descending migrating approach and aqueous heparin samples following the treatment with methylene blue NNX dye is depicted in Figure 5. The purple area (the upper, darker region) signified the presence of heparin whilst the blue area (the lower, lighter region) indicated the absence of it. The heparin concentration of the sample was evidently proportional to the length of the purple area on the paper strip.

The correlation between the migration distance and heparin concentration at various PDDs using the grade 541 and 113 paper strips were illustrated in Figure 6A and 6B, respectively. The migration patterns were very similar to those observed in the ascending mode. As shown in Figure 6A, the paper strips with very high PDD values (e.g. 135.9 µg/cm²) were capable of detecting a board range of heparin concentration (the linear region covered the heparin concentration range of 10 to 50 units/mL). They were, however, not sufficiently sensitive for differentiating a close range of heparin levels (e.g. 10 to 20 units/mL). In contrast, the paper strips at low PDD value (e.g. 17.6 µg/cm²), although incapable of sensing a board heparin concentration range (the linear region covered the heparin concentration range of 5 to 15 units/mL), were adequate for differentiating heparin levels within close range. Due to the increased thickness (and thus its liquid loading capacity) of the grade 113 paper (0.42 mm) used, a significantly larger sample volume (500 µL) had to be applied. It should be noted that the scale for gauging the PDD values on the grade 113 paper was incongruous with that of the grade 541 paper. A PDD value of 92.9 µg/cm² on the grade 113 paper represented a low protamine content whilst the same number on the grade 541 paper depicted a relatively high PDD value. Using the minimum sample volume (i.e., the

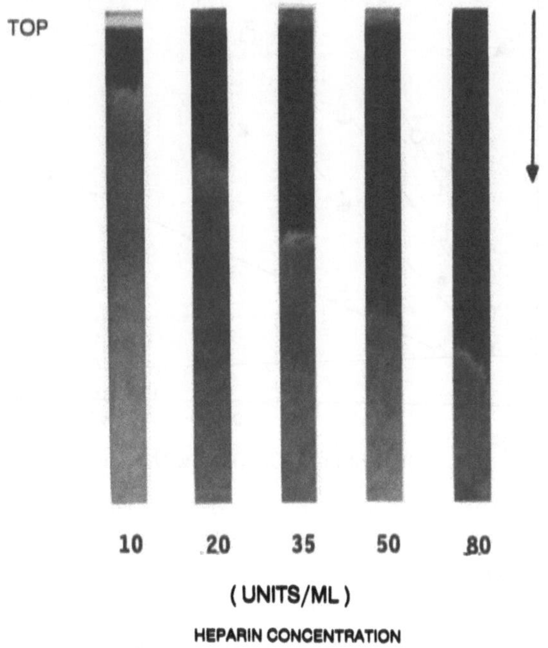


Figure 5. The appearance of the grade 541 paper strips (PDD: 135.9 µg/cm²) after the treatment with methylene blue NNX solution (500 µg/mL).

paper strip's liquid loading capacity), the time requirement for performing an assay in aqueous sample was 2.4 ± 0.2 minutes (using grade 113 paper, liquid loading capacity: 250 µL).

Figure 7 illustrates the relationship of the heparin levels of various samples (prepared by random dilution) determined by the new method and the Azure A colorimetric assay. The analytical recoveries of heparin in aqueous samples ranged from 92.6 to 110.2 %; the average analytical recovery attained was 100.1 ± 5.9 %. The correlation coefficient and slope were 0.995 and 0.941, respectively; and the intercept was 0.611 unit/mL; indicating that the new method compared favorably with the Azure A colorimetric assay.

With the new heparin sensing device achieved operational status in aqueous samples after surmounting all the technical problems; the satisfactory performance of this newly devised method led to the belief that a technically similar system could eventually be crafted for the non-clotting and non-instrumental monitoring of plasma heparin level in clinical laboratory in a very time efficient manner. Heparin spiked plasma samples were applied to the newly designed sensing system. The minimum time requirement for performing an assay was 5.9 ± 0.4 minutes (using grade 113 paper, sample volume: 250 µL). Figure 8 illustrates the relationship of the heparin levels of plasma samples (prepared by random dilution) determined by the new method and the APTT clotting assay. The analytical recoveries of heparin in plasma samples ranged from 85.1 to 109.9%; the average analytical recovery attained was 97.4 ± 7.7%. The correlation coefficient and slope were 0.983 and 1.065, respectively; and the inter-

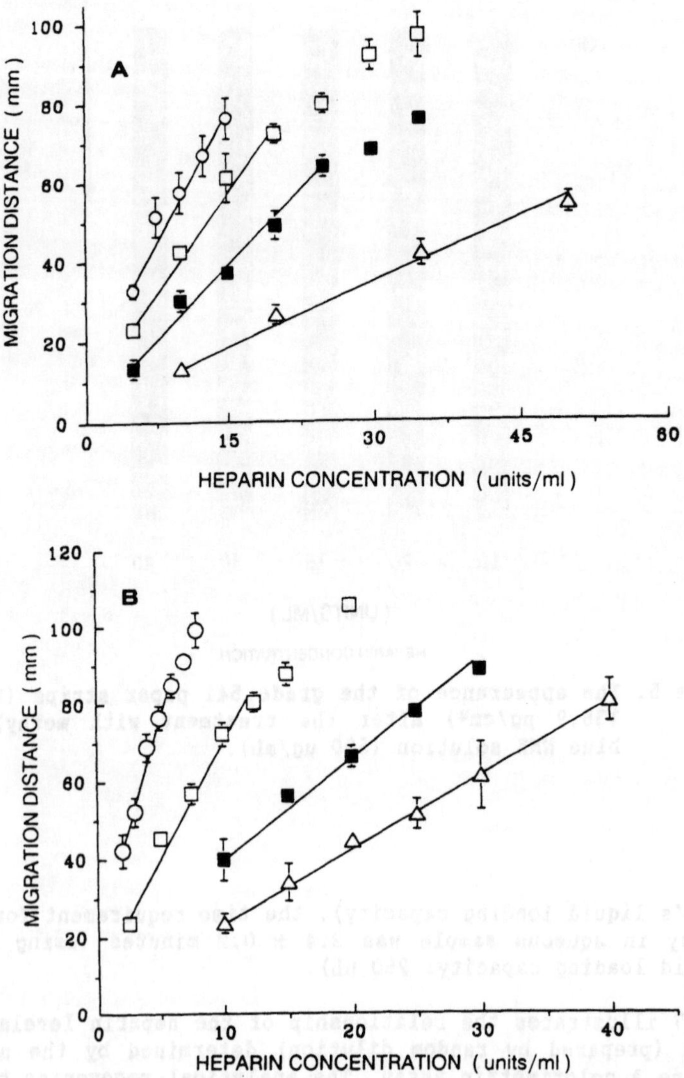


**Figure 6.** (A) The relationship between migration distance and heparin concentration on the grade 541 paper strips of various PDD values in aqueous sample using the descending mode of sample migration. Sample volume: 200 µL. PDD values: (o) 17.6 µg/cm², (□) 40.7 µg/cm², (■) 80.7 µg/cm² and (Δ) 135.9 µg/cm². (B) The relationship between migration distance and heparin concentration on the grade 113 paper strips of various PDD values in aqueous sample using the descending mode of sample migration. Sample volume: 500 µL. PDD values: (o) 92.9 µg/cm², (□) 123.0 µg/cm², (■) 219.8 µg/cm² and (Δ) 324.7 µg/cm².

cept was -0.729 unit/mL; which inferred that the heparin levels detected by the new method tended to be slightly lower than their counterparts determined by the APTT assay.

Despite the fact that this new sensing system's capability of measuring plasma heparin levels in a relatively accurate, time efficient and

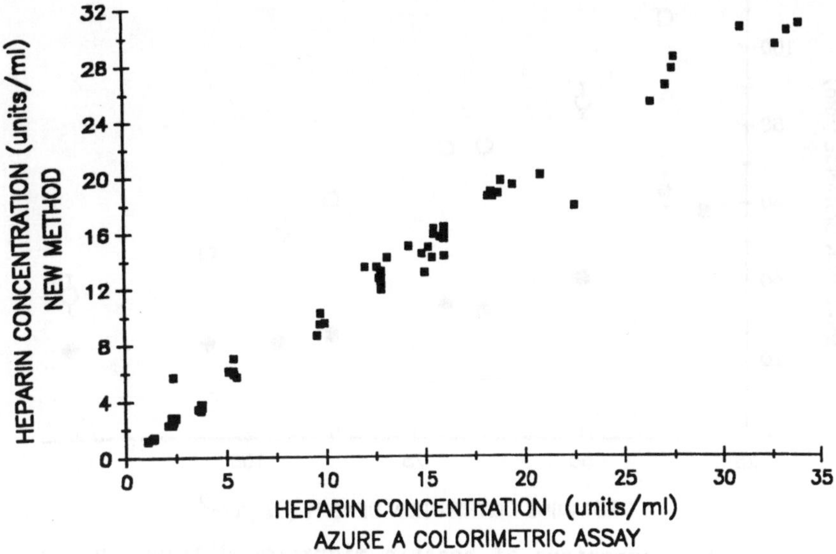


Figure 7. The correlation of aqueous heparin sample levels determined by the new method and the Azure A colorimetric assay.

technically simple manner, interference caused by various plasma components were encountered. Figure 9 illustrates the correlation between migration distance of heparin (using water and plasma as sample vehicles) and PDD values when both heparin concentration (25 units/mL) and sample volume (200 µL) were held constant. In general, the heparin migration distance using water and plasma as sample vehicles were very close at very high PDD value (e.g., 125.7 µg/cm²) whereas the heparin (in plasma)

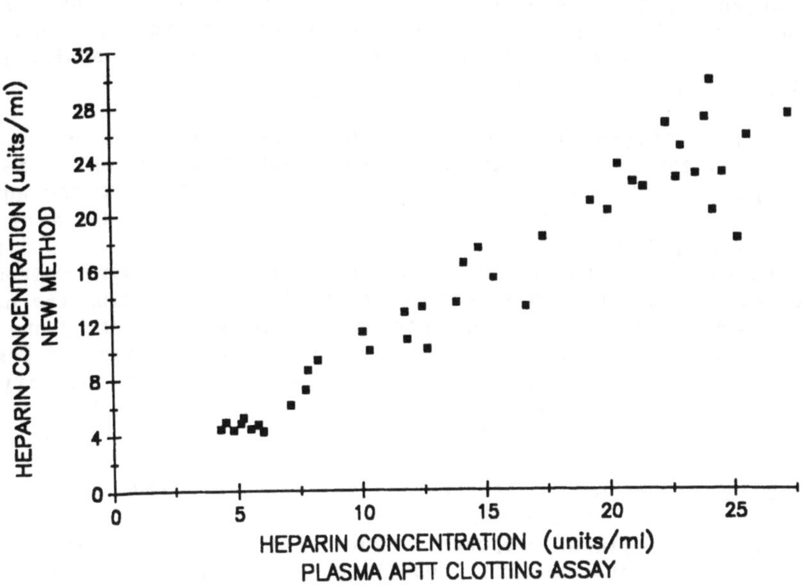


Figure 8. The correlation of plasma heparin sample levels determined by the new method and the APTT clotting assay.

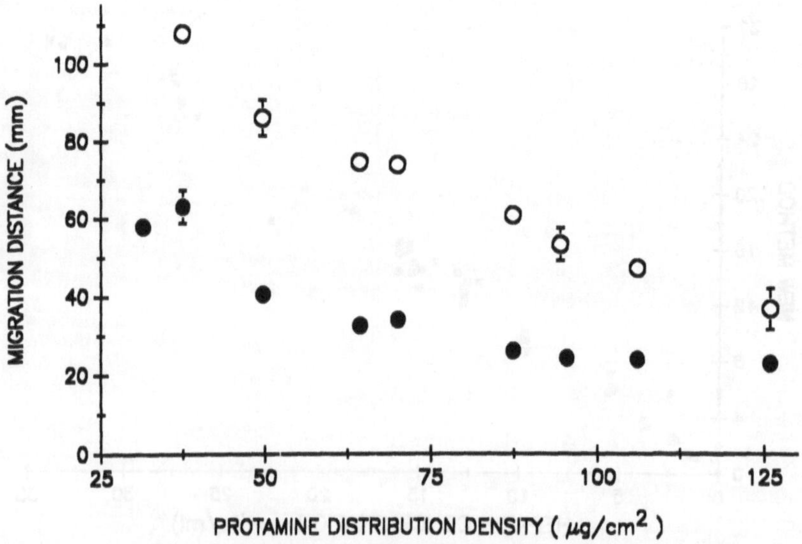


**Figure 9.** The comparison of heparin migration distance using plasma (●) and water (o) as sample vehicles at various PDD values. Heparin concentration: 25 units/mL, sample volume: 200 µL.

migration pattern started to deviate from that of water at medium PDD value (87.4 µg/cm²) with a gradual deterioration of color contrast. At a low PDD value (31.5 µg/cm², data not shown due to ambiguity), there was only a vague color change on the paper strip after the treatment with dye. Moreover, an ill-defined migration front was observed and the degree of ambiguity (or smearing effect) of the migration front elevated with the decrease of PDD value on the paper strips. The smearing effect in conjunction with the vague color change rendered this system to be incapable of detecting the presence of heparin in plasma sample at very low PDD value (e.g., 22.9 µg/cm² on grade 541 paper). Ironically, such a low PDD value was necessary for the application of this sensing system in the monitoring of heparin therapy.

Plasma is a complex biological system composed of various kinds of proteins, ions and organic solutes.[25] The observed deviation of heparin migration pattern using plasma sample as vehicle compared to the aqueous counterpart can be attributed to the presence of various kinds of ions and proteins in plasma interfering the binding of heparin to the immobilized protamine on the paper strips. In addition, these plasma components can also invoke a background interference and thus diminishing the magnitude of metachromatic interaction between heparin and the methylene blue NNX dye. At high PDD value (e.g., 125.7 µg/cm²), the abundance of immobilized protamine molecules per unit area imparts a strong affinity for heparin which can effectively offset the interference caused by various plasma components. At medium PDD values (e.g., 106.0 µg/cm²), the amount of immobilized protamine molecules per unit area is not sufficient to fully counteract the ionic interference and thus smearing of the migration front occurred. At low PDD values (e.g., 49.8 µg/cm²), the ionic interference of the plasma components overwhelms the metachromatic interaction between dye and heparin, resulting in a vague and inconclusive change of color from blue to very faint purple.

Water constituted more than 92% of the plasma weight. Albumin and globulin made up approximately 4.5 and 2.5% of the weight of plasma respectively, whilst the major inorganic electrolytes were sodium (0.33% w/v) and chloride (0.37% w/v) ions.[25] Table 2 summarized the effects of sodium chloride, albumin and globulin on the migration distance of heparin. Table 2A illustrates the correlation between the migration distance of heparin (20 units/mL, 200 µL) in the presence of sodium chloride at various concentrations when grade 541 paper strips (PDD = 45.4 µg/cm²) were used. The presence of sodium chloride significantly shortened the heparin migration distance. A gradual decline in migration distance was observed from 0 to 0.4% sodium chloride concentration and the declining of migration distance started to level off at 0.8% sodium chloride concentration. Table 2B summarizes the correlation between the migration distance of heparin (32 units/mL, 200 µL) in the presence of albumin at various concentrations when grade 541 paper strips (PDD = 64.4 µg/cm²) were used. A slight elevation of migration distance was observed at the initial phase and started to plateau off beyond the 1.2% concentration. The slight alteration of migration distance observed indicated that albumin was not the major factor of changing the heparin migration pattern. The observed color change (from blue to purple) after the dye treatment diminished gradually with the elevation of albumin level. Nevertheless, at 5% albumin level, the purple color observed in the region with heparin was still quite distinct. Table 2C depicts the correlation between the migration distance of heparin (30 units/mL, 200 µL) in the presence of globulin at various concentrations when grade 541 paper strips (PDD = 79.8 µg/cm²) were used. A nonlinear and dramatic surge in migration distance was observed at the initial phase (0 to 0.5 %) and seemed to plateau beyond 0.5 % globulin level. In addition, the observed color change (from blue to purple) deteriorated rapidly with only slight elevation in globulin level. It therefore suggested that sodium chloride and globulin were largely responsible for the alteration of migration distance of heparin spiked plasma in Figure 9.

Table 3 summarizes the time requirements for the exhaustion of the minimum sample volume (i.e., loading capacity) for each type of paper using water, 50% plasma and 100% plasma. The time requirement for grade 31ET and 113 papers compared favorably with that of the grade 541 paper. It should be noted that the viscosity albumin imparted on plasma could be

Table 2. Effect of sodium chloride, albumin and globulin on the migration distance of heparin

| A. Sodium Chloride (a) | | | | | | | | |
|---|---|---|---|---|---|---|---|---|
| % conc. | 0 | 0.1 | 0.2 | 0.4 | 0.6 | 0.8 | 1 | 1.2 |
| Distance ratio | 1 | 0.79 | 0.72 | 0.57 | 0.52 | 0.45 | 0.43 | 0.42 |
| **B. Albumin (b)** | | | | | | | | |
| % conc. | 0 | 0.5 | 1.2 | 1.6 | 2 | 3 | 4 | 5 |
| Distance ratio | 1 | 1.05 | 1.12 | 1.13 | 1.14 | 1.16 | 1.16 | 1.16 |
| **C. Globulin (c)** | | | | | | | | |
| % conc. | 0 | 0.1 | 0.2 | 0.3 | 0.4 | 0.5 | 1 | 1.5 |
| Distance ratio | 1 | 1.19 | 1.27 | 1.31 | 1.39 | 1.45 | 1.53 | 1.59 |

(a) PDD = 45.4 mcg/cm²; heparin concentration = 20 units/mL.
(b) PDD = 64.4 mcg/cm²; heparin concentration = 32 units/mL.
(c) PDD = 79.8 mcg/cm²; heparin concentration = 30 units/mL.

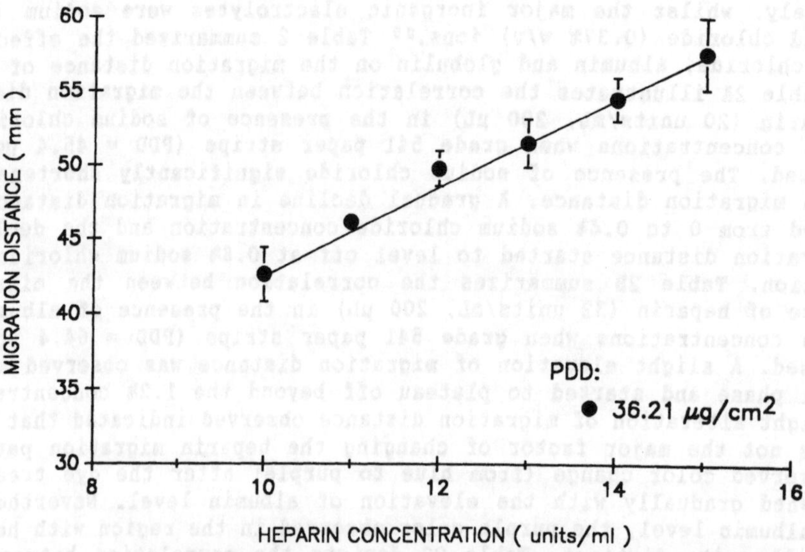


**Figure 10.** The relationship between migration distance and heparin concentration on the grade 31ET Chr paper strips at the PDD value of 36.2 μg/cm² in using the standard addition approach with the descending mode of sample migration.

accounted for the surge of time requirement. In general, using plasma as the sample vehicle at least doubled the migration time when compared to that of water in each type of paper. The concern for time efficiency rendered the grade 113 paper to be a reasonable selection as the support material for further development of the heparin sensing device for the application in plasma. Unfortunately, the grade 113 paper has a mild amber color which was similar to the tone of straw color of plasma. Using the type 113 paper would inevitably create an undesirable background hue which would eventually intensify the background contrast caused by the ingrained color of plasma upon the completion of sample migration. Due to the absence of any undesirable hue, the grade 31ET chromatographic paper was selected for further development of this sensing system even thought this had to be accomplished through the expense of time requirement.

A standard addition method was devised to counteract the interference of plasma components on the migration pattern of heparin and the lengthy time requirement caused by the relatively high viscosity of plasma. Fi-

Table 3. The time requirements for the exhaustion of the minimum sample volume (i.e., loading capacity) for each type of paper using water, 50% plasma and 100% plasma,

| Paper Grade | Loading Capacity (μL) | Water | 50% Plasma | 100% Plasma |
|---|---|---|---|---|
| 541 | 100 | 10.91±0.58 | 16.59±1.45 | 22.58±1.14 |
| 113 | 250 | 2.36±0.19 | 5.27±0.56 | 6.07±0.49 |
| 31ET | 300 | 6.52±0.66 | 10.93±0.72 | 14.75±1.16 |

gure 10 depicts the plot prepared by the standard addition approach. Aqueous heparin solution (20 units/mL) was added to the plasma sample in a 1:1 proportion. The added aqueous heparin solution served several purposes: (a) reduced the viscosity of the plasma sample and thus the migration time requirement; (b) halved the concentration of sodium chloride, globulin and albumin and thus significantly reduced the level of interference caused by these components; and (c) increased the heparin level in the plasma which would eventually result in a more intense color contrast after the dye treatment. As long as the added amount of heparin (and the volume) was maintained constant, the excess heparin introduced into the raw plasma sample would not likely to alter the performance of the method and the accuracy of the endpoint. The preliminary data obtained by adopting this technique indicated that the slope attained was 1.5 mm-mL/unit. Further investigation is necessary in order to improve the sensitivity (i.e., the steepness of the slope) and thus a more reliable resolution can be attained.

Further improvement of resolution can perhaps be achieved through the concurrent optimization of several factors including: (a) the PDD value on the paper strips, (b) the volume and concentration of the aqueous heparin solution used for the standard addition and (c) the proportion of the aqueous heparin solution and plasma sample. Investigation of the optimization is underway.

## ACKNOWLEDGMENT

This study was supported by the NIH Grant HL 38353 and the Whitaker Foundation of Biomedical Engineering Grant

## REFERENCES

1. L. B. Jaques, Pharmacol. Rev., 31, 99 (1980).
2. B. S. Bull, R. A. Korpman, W.M. Huse & D. B. Briggs, J. Thorac. Cardiovas. Surg., 69, 674 (1975).
3. M. R. Morales-Polanco, J. Pizzuto-Chavez, R. Ambriz-Fernandez, A. Aviles-Miranda and A. Sinco-Angeles, Transfusion. 22, 384 (1982).
4. M. Von Finck, J. Eulert, W. Heller & R. Sohorer, Der Anaesthesist. 34, 675 (1985).
5. J. Umlas. Human Path., 14, 271 (1985).
6. L. L. Sell, M. L. Cullen, G. C. Whittlesey, S. T. Yedlin, A. I. Philippart, M. P. Bedard & M. D. Klein, J. Pediatr. Surg., 21, 1087 (1986).
7. B. Y. Lee, W. R. Thoden, W. J. McCann Jr., L. T. Madden, L. R. Del-Guercio & P. Berkowitz, Surg. Gynecol. Obstet., 155, 809 (1982).
8. D. J. Effeney, J. Goldstone, D. Chin, E. S. Krupski & R. J. Ellis, Surgery, 90, 809 (1982).
9. A. J. Wilson, B. J. Cuddigan & A. P. Wyatt, R. Soc. Med. J., 81, 389 (1988)
10. N. Mummanei, M. Istanbouli, R. Pifarre & A. A. El-Etr, J. Thorac. Cardiovas. Surg., 86, 446 (1983).
11. A. M. Cumming, G. R. Jones, R. T. Wensley & R. B. Cundall. Thromb. Res., 41, 43(1986).
12. L Uziel, M Cugno, E Cacciabue, A Colombo, R Stabilini, D Mascheroni & A Agnostoni, Int. J. Artif. Organs , 9, 111 (1986).
13. L. B. Jaques, Can. Med. Assoc. J., 108, 1291 (1973).
14. J. N. Shanberge, Lab. Med., 9, 21 (1978).
15. G. C. Hoffman & A. Snyder, Cleveland Clin. Q., 33, 107, (1966).

16. P. J. Dauchot, L. Berzina-Moettus & A. Rabinovitch, Anesth. Analg., 62, 710, (1983).
17. B. Bain, T. Foster & B. Sleigh, Am. J. Clin. Pathol., 74, 668, (1980).
18. A. L. Stone, Fed. Proc., 36, 101, (1977).
19. O. H. Lowry, N. J. Rosebrough, A. L. Farr & R. J. Randall, J. Biol. Chem., 193, 265, (1951).
20. L. B. Jaques & A. Wollin, Can. J. Physiol. Pharmacol., 45, 787, (1967).
21. L. B. Jaques, in: "*Methods of Biochemical Analysis*," D. Glick, Ed., John Wiley & Son Inc., New York, 1977, p. 203.
22. B. S. Jacobs, B. L. Kasten, W. R. DeMott & W. L. Wolfson, 108, 1291, (1973).
23. P. A. Srere & K. Uyeda, in: "*Methods in Enzymology*," Vol XLIV, C. Mosbach, Ed., Academic Press Inc., New York, 1976, p. 11.
24. M. D. Lilly, in: "*Methods in Enzymology*," Vol XLIV, C. Mosbach, Ed., Academic Press Inc., New York, 1976, p. 46.
25. A. J. Vander, J. H. Sherman & D. S. Luciano, in: "*Human Physiology: The Mechanism of Body Function*," 3rd Edition, McGraw-Hill Inc., New York, 1980, p. 253.

# THE DEVELOPMENT OF PROTAMINE-COATED CELLULOSE HOLLOW FIBERS FOR HEPARIN REMOVAL: *IN VITRO* CHARACTERIZATION AND *IN VIVO* TESTING

Jae-Seung Kim, You-Yin Fu and Victor C. Yang

College of Pharmacy
University of Michigan
Ann Arbor, Michigan, 48109-1065

Nearly 20 million extracorporeal blood circulation (ECBC) procedures are conducted each year. They all require heparin anticoagulation to prevent the clotting process which is initiated when blood comes into contact with the extracorporeal device. However, heparin leads to a high incidence (8-30%) of bleeding complications. To avoid such complications, protamine, a heparin antidote, is generally administered at the conclusion of an ECBC procedure to neutralize the heparin anticoagulant activity. Unfortunately intravenous administration of protamine often causes serious adverse responses. We suggest an approach which could control both types of complications. The approach consists of placing a hollow fiber device containing immobilized protamine (i.e., a protamine filter) at the distal end of the ECBC procedure. The protamine filter would remove heparin before heparin is returned to the patient, thereby eliminating any heparin-induced bleeding complications. The protamine filter would also obviate the need of intravenous protamine administration and allow for an external protamine treatment, thereby eliminating protamine-induced complications. *In vivo* studies involving dogs have shown that the protamine filter is capable of removing heparin from the animal's blood circuit. The protamine filter also possesses certain anti-thrombogenic activity and induces significantly less thrombocytopenic responses when compared with the clinical procedure of intravenous protamine administration. Other preliminary *in vivo* data will also be presented.

## INTRODUCTION

Extracorporeal blood circulation (ECBC) has become one of the most widely frequently used medical procedures in recent years. It has been widely employed in many clinical situations such as renal dialysis, cardiothoracic surgery, organ transplantation, plasmapheresis, extracorporeal membrane oxygenation, and the implantation of artificial organs. It is estimated that approximately twenty million ECBC procedures are performed each year.[1,2]

*Cosmetic and Pharmaceutical Applications of Polymers*
Edited by C.G. Gebelein *et al.*, Plenum Press, New York, 1991

In all these applications, blood is drawn from the patient through an extracorporeal device. Upon contact with the synthetic materials forming the surface of the extracorporeal device, the blood's precisely regulated hemostasis is disturbed and it tends to clot within the device.[3] The thrombi formed occlude the perfusion channels in the machine. To prevent occulusion and maintain the fluidity of blood in the circulation, heparin, the most widely used anticoagulant, is systemically administered to the patient and the device prior to treatment. However, the high level of heparin required for this purpose poses a considerable hemorrhagic hazard to the patients. Gervin,[4] for instance, reports that 8-30% incidences of hemorrhagic complications occur during heparinization. Similarly, Fletcher and co-workers[5] report that 6-10% of patients develop coagulation abnormalities with excessive bleeding following open heart surgery. Swartz[6] also estimates that nearly 25% of all patients suffering from acute renal failure are subject to increased bleeding risk during and immediately following dialysis. The extent of these complications is enhanced for elderly patients, diabetic patients, patients with ulcers or other multiple trauma, and patients with current cardiac or vascular surgery.[7,8] In fact, the Boston Collaborative Drug Surveillance Program has cited heparin as "the drug responsible for the highest incidents of deaths in otherwise healthy patients".[9]

To avoid the life-threatening nature of the hemorrhage associated with systemic heparinization, considerable amounts of effort have been directed at solving this problem. The approaches taken include: (1) the administration of protamine sulfate, a heparin antagonist, to neutralize the anticoagulant activity of heparin;[10] (2) the development of heparin substitutes (e.g., prostacyclin)[11] and new antithrombotic agents (e.g., low molecular weight heparins);[12] (3) the use of low-dose heparinization[6] or regional heparin anticoagulation (i.e., by infusion of heparin into the blood entering the dialyzer, and neutralization by infusion of protamine into the heparinized blood as it returns to the patient);[13] (4) the use of regional citrate anticoagulation (this approach is similar to that of the regional heparin anticoagulation as described in (3), except that citrate is used as the anticoagulant agent and calcium is used as the neutralizing agent);[14] (5) the development of new blood compatible materials such as materials with surface-bound heparin for construction of the extracorporeal devices;[15] and (6) the development of an immobilized heparinase filter to degrade heparins at the termination of the ECBC procedure.[16-18]

Although these approaches have led to some improvement, they have not met much clinical success. For instance, prostacyclin infusion is associated with the unpredictable occurrence of sudden hypotension.[11] The use of low molecular weight heparins as new antithrombotic agents has been restricted due to the lack of neutralizing agents for these compounds.[19] The use of low-dose heparinization or regional heparin anticoagulation proves unsuccessful in preventing bleeding associated with dialysis.[6,13] In addition, regional heparin anticoagulation does not obviate the need of protamine as the neutralizing agent. Regional citrate anticoagulation is rarely used because of the technical difficulties in performing the procedures and in determining the adequate amount of calcium required for citrate neutralization. The use of heparin-bound materials for the construction of the extracorporeal device is associated with a major problem that unless the entire extracorporeal unit is made of these materials, clotting will still occur in the device. The heparinase-based heparin removal system under development looks most promising because it obviates the need of any heparin neutralizing agents. However, it is still in the experimental stage. Moreover, the approach may be associated with concerns such as the accumulation of heparin degradation products in the patient's body may pose some toxic effects, and the use of the microbial heparinase may trigger some immunological responses. To date, the administration of protamine as the conclusion of an ECBC procedure to reverse the anticoagulant activity of heparin still remains as the only and most accepted approach in clinical practice.

Unfortunately, the administration of protamine is frequently associated with important adverse hemodynamic side effects.[20-24] Protamine reactions range from mild hypotension[21-22] to severe or ultimately fatal cardiovascular collapse.[23-24] In a recent survey involving 1400 perfusionists, the most frequently cited perfusion accident in cardio-pulmonary bypass operations was the "protamine reaction".[25] However, as long as there are no alternatives to control the more serious and life-threatening bleeding problems associated with the systemic use of heparin, protamine will remain in use, in conjunction with heparin therapy, despite its toxic effects.

We propose a novel approach which would potentially control both heparin and protamine induced complications.[26] The approach consists of placing a blood compatible filter device containing immobilized protamine (defined as a protamine filter) at the distal end of the ECBC circuit (Figure 1). The protamine filter would bind and selectively remove heparin after heparin serves its anticoagulant purpose in the extracorporeal device and before heparin is returned to the patient. Meanwhile, the filter would restrict the protamine therapy to an external treatment. Since the majority of protamine toxicity results from the direct interaction of protamine with certain cells present in various organ sys-

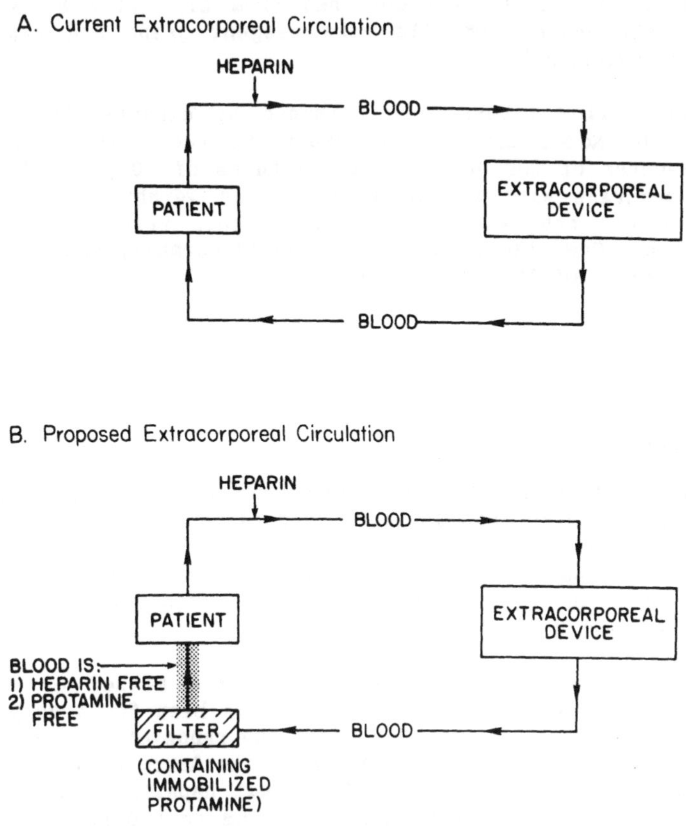


Figure 1. Schematic diagram of (A) current, and (B) proposed extracorporeal blood circulation.

tems,[27] the use of an external protamine treatment would theoretically minimize the adverse effects of protamine.

In this paper, we present results demonstrating the feasibility of the approach. Protamine was immobilized onto a cellulose hollow fiber bundle which was obtained from a clinically used hemodialyzer. An investigation of the parameters affecting protamine immobilization to cyanogen bromide (CNBr)-activated cellulose fibers was conducted in order to establish a protocol so that the immobilization process can be optimized and quantitatively controlled. Parameters affecting the adsorption pattern of heparin on the protamine-bound membrane was also examined. In addition, *in vivo* experiments involving dogs were conducted to demonstrate the effectiveness of the protamine filter in removing heparin and preventing protamine-induced toxicity.

## EXPERIMENTAL

### 1. Materials

Solid heparin (sodium salt from porcine intestinal mucosa, 150 USP units/mg) was purchased from Hepar Industries (Franklin, OH). Heparin injection (1000USP units/mL) was purchased from Elkins-Sinn (Cherry Hill, NJ). Bio-Rad dye reagent was from Bio-Rad Laboratories (Richmond, CA). Cyanogen bromide (CNBr) and solid protamine sulfate (grade III, Clupeine from herring) were purchased from Sigma (St. Louis, MO). Protamine sulfate injection (10 mg/mL) was obtained from Eli Lilly & Co. (Indianapolis, IN). All other chemicals were reagent grade and the water was distilled and deionized.

Cellulose hollow fibers were obtained by disassembling a Travenol (Deerfield, IL) Model 1500 CF hemodialyzer. Each hollow fiber has an internal diameter of 200 µm, a wall thickness of 10 µm, and is made of Cuprophane membrane with a nominal molecular weight cut-off of 10,000 daltons.[28] Unless otherwise stated, the fibers were cut to a uniform length of 15 cm. One gram of dry weight would normally consist of approximately 300 pieces of the cut hollow fibers.

### 2. Methods

#### 2A. Assays

The degree of activation was determined by measuring the cyanate ester content on the activated fibers, according to the method of Wilchek and Kohn.[28] The absorbance of the intense purple color generated by the reaction between the Konig's reagent and cyanate ester was measured at 588 nm, and the cyanate ester content was estimated using a molar absorptivity of 137 x $10^6$ $M^{-1}cm^{-1}$.[29]

Protamine concentration was assayed by the method of Bradford,[30] using the Bio-Rad dye reagent. This method has been found very sensitive for measuring protamine concentrations in the range of 1-5 µg/mL. Heparin concentration in solution was determined by the Azure A dye metachromasia assay.[31] The assay is based on the fact that heparin causes a shift of the absorption maximum of Azure A from 620 nm to 530 nm. The degree of shift is linearly related to the heparin concentration in the range of 1-10 µg/mL. Thus by measuring the absorbance at 620 nm, heparin concentration can be accurately determined from a standard curve. Heparin concentrations in plasma were determined by the activated partial thromboplastin time (aPTT) clotting assay,[32] using the activated Cephaloplastin reagent (Dade Diagnostics, Aguada, Puerto Rico). The assay was linear with heparin activities in the range 0.0 to 0.7 units/mL. A standard curve was constructed for each animal. Samples with heparin concentrations outside the linear range of the standard curve were diluted with plasma obtained from the same animal. Complement consumption was measured by the hemolytic complement assay (i.e., the $CH_{50}$ assay),[33] using the antibody-coated sheep erythrocytes.

## 2B. Investigation of Protamine Immobilization

Ground, fine particles (approximately 100 µm in diameter) were used for the investigation of protamine immobilization. To prepare fine fiber particles, intact hollow fibers were immersed in liquid nitrogen until frozen, and then ground in a crucible. Protamine sulfate was immobilized onto these fiber particles according to the method of March et al.[34] Variables affecting protamine immobilization such as the amount of CNBr used, the duration of incubation, and the protamine concentration in the coupling solution were examined. A general description of the immobilization procedures is listed below. One gram dry weight of the ground fibers was added to 20 mL of CNBr solution at concentrations ranging from 0.05 to 0.4 g/mL (prepared by combining adequate volumes of 1 g/mL of CNBr/acetonitrile solution with 1 M $Na_2CO_3$ solution). The mixture was vigorously agitated in a fume hood for 5 minutes, and the solution was filtered using a sintered glass funnel. The activated fiber particles were then washed subsequently with 70 mL each of distilled water, 1 mM HCl, and 0.1 M $NaHCO_3$ (pH 8.3) containing 0.5 M NaCl. Samples containing 0.10 g of the suction-dried fiber particles were withdrawn and assayed for the cyanate ester concentrations. The remaining suction-dried fiber particles were incubated at room temperature with 5 mL of the protamine solution prepared in the coupling solution containing 0.1 M $NaHCO_3$, and 0.5 M NaCl, at pH 8.3. After the incubation, the fibers were washed with 70 mL of 0.25 M phosphate buffer (pH 7.0), containing 0.5 M NaCl, to remove the nonspecifically adsorbed protamine. The amount of protamine immobilized on the fiber was calculated by measuring the protamine concentrations in the coupling solution before and after the immobilization procedure, as well as those in the washing solutions. The incubation time and protamine concentration were varied during the coupling process to examine their effects on protamine immobilization.

## 2C. Fabrication of the Hollow Fiber Bundle

The Travenol Model 1500 CF hemodialyzer was disassembled, and the fibers were cut down to a length of 15 cm. Approximately 800-1000 of the cut fibers (about 3 g in weight) were collected and tied together at each end with a teflon film to form a bundle. The bundle was then housed in Tygon tubing which was fitted with a molded connector at each end. The junctions were sealed with epoxy, and the epoxy was cured at room temperature until it hardened.

## 2D. Immobilization of Protamine onto the Hollow Fiber Bundle

The fabricated bundle prepared above was circulated with 200 mL of 1 M sodium carbonate solution for 5 minutes. After the circulation, the bundle was placed in a fume hood and circulated for 5 minutes with 60 mL of 0.1 g/mL CNBr solution. The CNBr activated bundle was subsequently washed with 200 mL each of distilled water, 1 mM HCl solution, and 0.1 M $NaHCO_3$ buffer (pH 8.3) containing 0.5 M NaCl. Immediately following the washing steps, 20 mL of the protamine solution were circulated through the bundle at a flow rate of 50 mL/min over a period of 2 hours at room temperature. The bundle was then washed with 200 mL of 0.25 M phosphate buffer (pH 7.0), containing 0.5 M NaCl, to remove the nonspecifically adsorbed protamine. The hollow fiber bundle thus prepared that contained immobilized protamine was defined as the "protamine filter".

## 2E. Characterization of Adsorption of Heparin on the Protamine Filter

Heparin solutions at concentrations ranging from 0.1 to 10 mg/mL were prepared by dissolving adequate amounts of heparin in normal saline (i.e., 0.9% in NaCl). Fifty mL of the solution was placed in a reservoir and recirculated with a peristatic pump through the protamine filter at a flow rate of 50 mL/min for one hour. Samples were withdrawn from the reservoir at various time intervals and assayed for the residual heparin concentrations. Plain hollow fiber bundles (i.e., not treated with protamine) were used as the control to determine the amount of heparin that was adsorbed on the bundle through the nonspecific, physical adsorption.

## 2F. Characterization of the Adsorption Isotherm of Heparin on the Protamine-Bound Fibers

Ground fiber particles containing immobilized protamine, prepared by freezing and grinding the intact hollow fibers obtained from a protamine filter, were used for the adsorption isotherm studies. A sequence of test tubes containing 5 mL of the heparin solution, at various heparin concentrations ranging from 0.01 to 15 mg/mL, was prepared. To each tube, 0.5 g of the protamine-bound fiber particles were added. The tubes were placed on an orbital rotor and rotated at 200 rpm for one hour at room temperature. The fiber particles were then removed by centrifugation and the equilibrium heparin concentration in the supernatant was measured by the Azure A assay.

The amount of heparin adsorbed (Q) per gram of the protamine-bound fibers was estimated using the equation, $Q = V(C_o - C)/W_t$, where $C_o$ and C were the initial and equilibrium (final) heparin concentrations, respectively, V was the volume of the heparin solution, and $W_t$ was the dry weight of the ground fibers.

## 2G. *In Vivo* Experiments

Twenty healthy female mongrel dogs, weighing 20-30 kg, were prepared for the study. The dogs were subdivided into three groups: in group I, dogs were utilized as the control in which no protamine was administered (n = 6); in group II, dogs were utilized for the testing of the protamine filter (n = 8); and in group III, dogs were utilized for the testing of the response resulting from direct intravenous administration of protamine (n = 6). All animals were anesthetized with sodium pentobarbital (30 mg/kg) and mechanically ventilated to maintain physiologic blood gases. Hydration was maintained by administration of lactated Ringer's solution (20 mL/kg bolus followed by a 10 mL/kg/hour infusion). The femoral artery and the femoral vein were cannulated, and to the cannules a hollow fiber bundle (5-7 grams in weight) was attached to allow blood to flow from the artery through the bundle and into the vein. For group I and II dogs, the bundles were not treated with protamine (i.e., sham filter), whereas for group III dogs, the bundles contained 60-70 mg of immobilized protamine. Heparin (1000 unit/mL) was administered intravenously to all animals at a dose of 150 units/kg of body weight. Twenty minutes after heparin administration, the two valves on the hollow fiber bundle were opened to allow blood to circulate through the bundle. The time of opening the valves was defined as "time zero". For the six dogs used for the testing of the protamine response (i.e., the group III dogs), protamine was administered intravenously at time zero at a dose of 1 mg/100 units of heparin. The blood activated clotting time (ACT) in all animals was periodically monitored with Hemocron (International Technidyne Corp., Metuchen, NJ). The plasma heparin levels were also determined using the ΔPTT clotting assay.

Hemodynamic changes in the animal were monitored every minute for 5 minutes, and then at 10, 20, 30 and 60 minutes after opening the fiber bundle. An oximetric Swan-Ganz catheter (Oximetrix, Mountain View, CA) was placed into the pulmonary artery for continuous determination of mixed venous oxygen saturation ($SvO_2$) and pulmonary artery systolic (PAS) and diastolic (PAD) pressures, and for intermittent measurements of cardiac output using thermodilution technique. Catheters were also placed into the left carotid artery and jugular vein for monitoring systemic arterial blood pressure and also for blood withdrawal.

Blood samples were collected for all dogs from the arterial line of the filter at time intervals of 3, 10, 20, 30 and 60 minutes after opening the fiber bundle. Samples were drawn into vacutainer tubes containing EDTA (Becton Dickinson, Rockville, MD) for measurements of hematocrit, total blood hemoglobin, and complete blood cell counts, and into vacutainer tubes containing sodium citrate for determinations of the heparin level and the activation of the complement system.

### 2H. Data Analysis

Data are presented as the mean plus or minus the standard deviation (SD). One way or two way analyses of variance were utilized to analyze the data.[35] Data obtained from the same animal group at various intervals were compared with that at time zero to identify if the change at each time interval was significant. Data obtained from the two experimental animal groups (i.e., group II and III in which animals were attached with a protamine filter or receiving intravenous protamine administration, respectively) were also compared separately with those obtained from the control animals (i.e., group I dogs) at the same time intervals to identify if the changes from the control group were statistically significant. Since the effects resulting from intravenous protamine administration were systemic and the effects from the use of the protamine filter were regional, data obtained from the two experimental groups (i.e., group II and III) at any time intervals could not be directly compared.

## RESULTS

Figure 2 shows the mechanism of protamine immobilization onto the CNBr activated cellulose hollow fibers. The overall immobilization process consists of two steps: (1) activation of the hydroxyl groups on the cellulose membrane with CNBr to produce reactive cyanate ester derivatives; and (2) coupling of protamine onto the activated cyanate ester groups through isourea linkages.

The effect of CNBr concentration on the extent of fiber activation was shown in Figure 3A. At low CNBr concentrations (<0.1 g/mL), the cya-

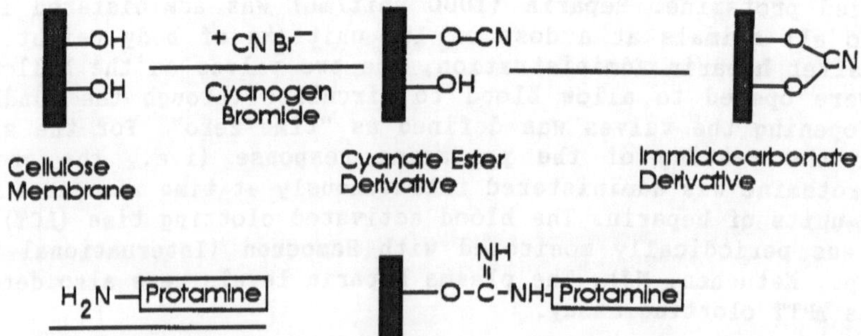


**Figure 2. Mechanism describing the activation and immobilization of protamine onto cellulose hollow fibers.**

**nate ester concentration on the fibers was increased in proportion to the increase in CNBr concentration. Doubling the CNBr concentration from 0.05 to 0.1 g/mL increased the cyanate ester content per gram of the fibers from 2.1 to 4.9 µmoles. This cyanate ester/CNBr concentration ratio (i.e., the slope of the plot), however, dropped significantly at higher CNBr concentrations. As the CNBr concentration was raised from 0.2 to 0.4 g/mL, the cyanate ester concentration on the fibers was increased slightly by only 1.2 fold (from 6.6 to 8.1 µmole/g).**

**When the protamine loading was plotted versus the cyanate ester content, a linear relationship was observed (Figure 3B). Since protamine was generally in excess compared to the cyanate ester content, the presence of a linear relationship in Figure 3B may thus suggest the existence of reaction stoichiometry between protamine and the cyanate ester groups.**

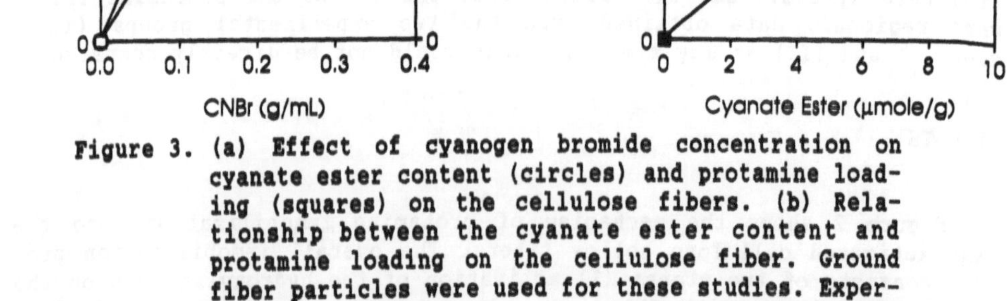


**Figure 3. (a) Effect of cyanogen bromide concentration on cyanate ester content (circles) and protamine loading (squares) on the cellulose fibers. (b) Relationship between the cyanate ester content and protamine loading on the cellulose fiber. Ground fiber particles were used for these studies. Experimental procedures were described in detail in the Methods. For each gram of the ground fibers, 5 mL of the protamine solution (10 mg/mL) were used during the coupling process. The coupling time was controlled to 2 hours.**

Table 1. Effect of protamine concentrations in the coupling solution on protamine loading on the cellulose fibers. Ground fiber particles were used for these studies. Experimental procedures were described in detail in the Methods. The incubation time was controlled to 2 hours.

| Protamine Concentration (mg/mL) | Protamine Loading (mg/g fiber) |
|---|---|
| 2.5 | 5.6 |
| 5.0 | 10.2 |
| 10.0 | 20.8 |
| 20.0 | 29.8 |

The effect of protamine concentration in the coupling solution on protamine loading is shown in Table 1. Up to 10 mg/mL, protamine loading on the fibers increased almost linearly with the change of protamine concentration in the coupling solution. Further increase in protamine concentration from 10 to 20 mg/mL, however, only imparted a 50% increase in the protamine loading on the fibers.

The effect of the duration of the immobilization process on protamine loading is shown in Figure 4. At room temperature, protamine loading increased with the prolongation of the incubation time, and reached to a plateau at about 30 minutes of incubation. All subsequent immobilization

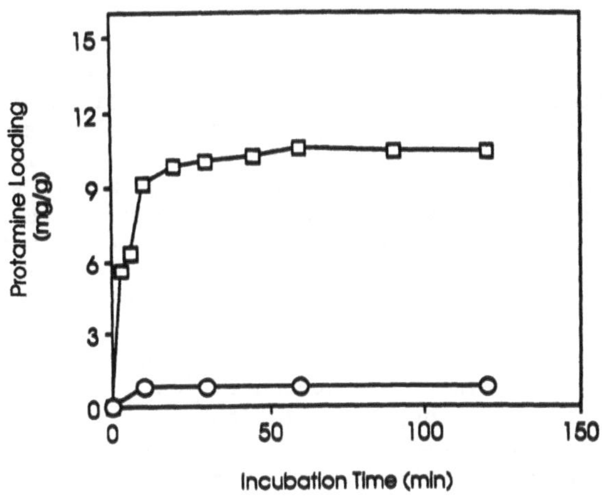


Figure 4. Effect of incubation time on protamine loading on the cellulose fibers. Ground fiber particles were used for these studies. Experimental procedures were described in detail in the Methods. For each gram of the ground fibers, 20 mL of the CNBr solution (0.1 g/mL) were used during the activation process, and 5 mL of the protamine solution (10 mg/mL) were used during the coupling process. Ground fibers which were not treated with CNBr were used as the control (circles) to identify the amount of protamine adsorbed on the fiber through a nonspecific, physical adsorption.

studies were therefore conducted with incubation times longer than 30 minutes to ensure the completion of the immobilization process. Control experiments using untreated fibers displayed a protamine adsorption of about 0.8 mg of protamine per gram of the untreated fibers, indicating the presence of certain nonspecific protamine adsorption on the fibers.

The protamine-bound hollow fiber bundles were tested for their efficiency in removing heparin, using heparinized normal saline solutions. When the heparin solution was recirculated through the protamine fiber, via a peristatic pump at a flow rate of 50 mL/min, a quick change in heparin concentration in the reservoir was observed. Figure 5 shows the typical adsorption versus time curves of the protamine filter. All the curves followed a similar pattern: a rapid heparin adsorption that occurred within the first ten minutes of recirculation, followed by an asymptotic approach of the heparin concentration to an equilibrium value. Table 2 demonstrates that heparin adsorption on the protamine filter was dependent upon the protamine loading on the filter increasing at higher protamine loadings. At a heparin concentration of 1 mg/mL, filters containing 5.0, 12.2, and 20.3 mg protamine/g fiber (Filter No 5, 6, and 7, respectively) removed 1.0, 1.8, and 3.3 mg heparin/g fiber, respectively. In addition, heparin adsorption on the protamine filter was also strongly dependent on the initial heparin concentration in the solution. As shown in Table 2, Filter No. 4, 6, and 8 which contained a similar protamine loading (about 13 mg/g fiber) removed 0.4, 1.8, and 12.4 mg of heparin/g fiber at a heparin concentration of 0.1, 1, and 10 mg/mL, respectively. It is interesting to note that the amount of heparin adsorbed on the filter through nonspecific, physical adsorption was also dependent on the initial heparin concentration. Bundles that were not treated with protamine adsorbed 0.1, 0.3, and 2.0 mg heparin/g fiber at heparin concentrations of 0.1, 1.0, and 10.0 mg/mL, respectively.

Table 2. Effect of protamine loading and heparin concentration on heparin adsorption on the cellulose fibers. Intact hollow fiber bundles containing immobilized protamine (i.e., the protamine filters) were used for these studies. Heparin solution was recirculated through the protamine filter using a peristatic pump at flow rate of 50 mL/min for 1 hour. Experimental procedures were described in detail in the Methods.

| Filter Number (#) | Protamine Loading (mg/g filter) | Initial Heparin Concentration (mg/mL) | Amount of Heparin Adsorbed (mg/g fiber) |
|---|---|---|---|
| Plain Filters (i.e., Filters were not treated with CNBr) | | | |
| 1 | - | 0.1 | 0.1 |
| 2 | - | 1.0 | 0.3 |
| 3 | - | 10.0 | 2.0 |
| Protamine Filters | | | |
| 4 | 13.5 | 0.1 | 0.4 |
| 5 | 5.0 | 1.0 | 1.0 |
| 6 | 12.2 | 1.0 | 1.8 |
| 7 | 20.3 | 1.0 | 3.3 |
| 8 | 13.0 | 10.0 | 12.4 |

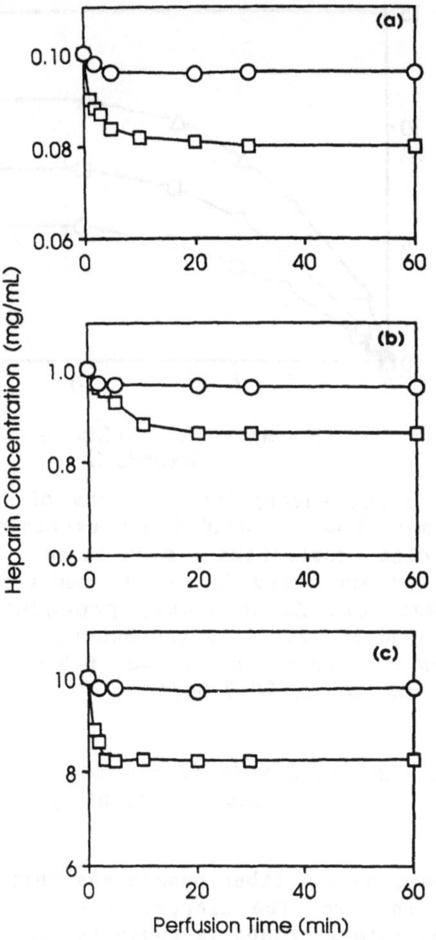


**Figure 5.** Transient change of heparin concentration in the reservoir. Intact hollow fiber bundles containing immobilized protamine (i.e., the protamine filters) were used for these studies. Experimental procedures were described in detail in the Methods Section.

Fine fiber particles, obtained by grinding the intact, protamine-bound hollow fibers, were used for the adsorption isotherm studies. Heparin adsorption on the protamine-bound fibers, as shown in Figure 6 by plotting the amount of heparin adsorbed (Q) versus the equilibrium heparin concentration (C) in the solution, appeared to follow the Langmuir adsorption equation, $Q = KQ_s/(1 + KC)$,[35] where K and $Q_s$ are the adsorption constant (mL/mg) and saturation capacity (mg/g fiber), respectively. At the low heparin concentration region (i.e., 0.01 - 1.0 mg/mL). the amount of heparin adsorbed was proportional to the equilibrium heparin concentration. At the high heparin concentration region, however, the amount of heparin adsorbed reached to a plateau which represented the saturation capacity ($Q_s$) of the protamine-bound fibers and was independent of the equilibrium heparin concentrations. By applying the linear regression analysis, the average K value was estimated to be 0.23 ± 0.06 mL/mg (i.e., $3.11 \times 10^6$ $M^{-1}$, assuming an average molecular weight of 13,500 daltons for heparin).[19] The saturation capacities were found dependent on the protamine loading on the adsorbent. The calculated $Q_s$

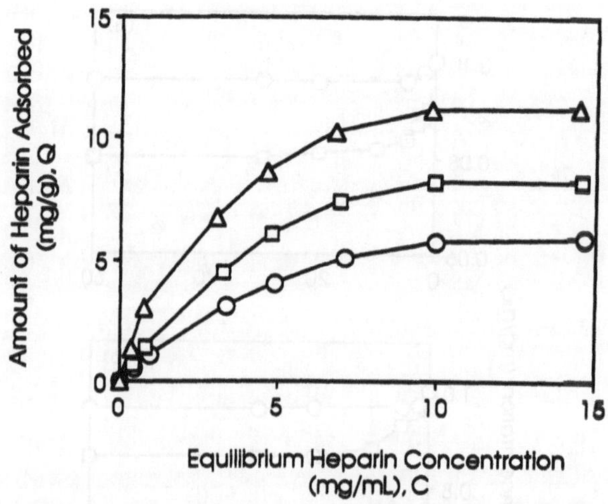


Figure 6. Equilibrium adsorption isotherm of heparin on the protamine-bound ground fiber particles. Fine fiber particles containing immobilized protamine were prepared according to the procedures described in the Methods. Experimental procedures were also described in detail in the Methods. Protamine loading on the fiber was : o——o, 6.5 mg/g; □——□, 12.7 mg/g; and Δ——Δ, 20.7 mg/g.

values were 5.8, 8.1, and 11.2 mg/g fibers for the adsorbent containing 6.5, 12.7, 20.7 mg of immobilized protamine per gram of the fibers, respectively.

The protamine-bound hollow fiber bundle was tested for its efficiency in removing heparin *in vivo*. The diappearance of heparin anticoagulant activity in the three animal groups is shown in Figure 7. Animals receiving intravenous protamine administration displayed a rapid and systemic heparin neutralization. Three minutes following protamine administration, the blood heparin level as measured by the ΔPTT assay, was already dropped to the minimum. Although the ACT assay indicated a full reversal of heparin anticoagulant activity (the ACT values were back to the baseline), a more sensitive ΔPTT assay nevertheless revealed that nearly 10% of the initial heparin activity remained. On the other hand, heparin removed by the protamine filter appeared to follow a slower regional type, two compartment kinetic model. A rapid removal, presumably due to the immediate binding of heparin to the filter, occurred within 5 minutes following the opening of the protamine filter. This rapid removal was followed by a second compartment which exhibited a much slower but relatively linear removal rate over a period of 60 minutes. At the 60 minute mark, the remaining heparin level had declined to a value close to that observed in animals receiving intravenous protamine administration. It should be noted that part of the heparin disappearance during the use of the protamine filter was attributed to the natural metabolism of heparin, as this was reflected by the time-dependent decay in heparin activity in the control animals who did not receive any protamine treatment (Figure 7). Blood continued to flow in an unrestricted fashion in the protamine filter during the course of the experiment (60 minutes). However, the flow seemed to slow down slightly near the end of the experiment (about 75 minutes). Microthrombi were seen in almost all the filters of animals receiving intravenous protamine, but only in one third of the protamine filters after the blood had been washed out from the used filters with

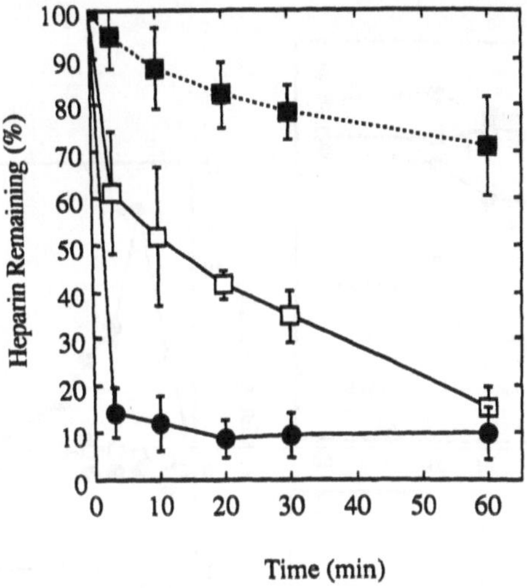


Figure 7. *In vivo* heparin removal. The experimental procedures were described in the Methods section. Blood heparin level was determined by the aPTT heparin clotting assay. The data were normalized to 0% activity before heparin administration and 100% activity after heparin administration. (Solid squares), (open squares), and (circles) represent Group I (i.e., the control dogs that received no protamine treatment), Group II (i.e., dogs attached with the protamine filter), and Group III (i.e. dogs that received intravenous administration of protamine) animals. The data were presented as the mean ± S.D.

the saline solution. Thrombus deposition on the filter was also found much less serious in the protamine filter.

Protamine-induced complications were monitored by measuring the hemodynamic responses in the dog. As shown in Figure 8, the control dogs with untreated fiber bundles (indicated by the dotted lines) did not display any clinical significant hemodynamic changes with time. None of the changes were statistically significant, when the data obtained at each time interval was compared to that at time zero. On the other hand, dogs receiving intravenous protamine administration demonstrated very important but typical protamine responses. Marked transitory changes were observed in all of the hemodynamic indices measured. A profound fall in systemic arterial blood pressure (-40 mmHg, Figure 8A), cardiac output (-1.6 L/min, Figure 8B), and mixed venous oxygen saturation (-8%, Figure 8C), in conjunction with a concomitant progressive increase in pulmonary artery systolic (+13 mmHg, Figure 8D) and diastolic pressures (+10 mmHg, Figure 8E) were noted for dogs receiving intravenous protamine.

In contrast, protamine filter did not elicit any protamine-related hemodynamic responses (Figure 8). None of the changes observed at any particular time interval was statistically significant, when the data was compared to that at time zero. There were also essentially no statistically significant differences in all of the parameters measured when the changes were compared with those in the control animals. The only change which might be statistically significant ($0.05 < p < 0.1$) was that in the cardiac output at 20 minutes. However, this low statistical $p$ value may

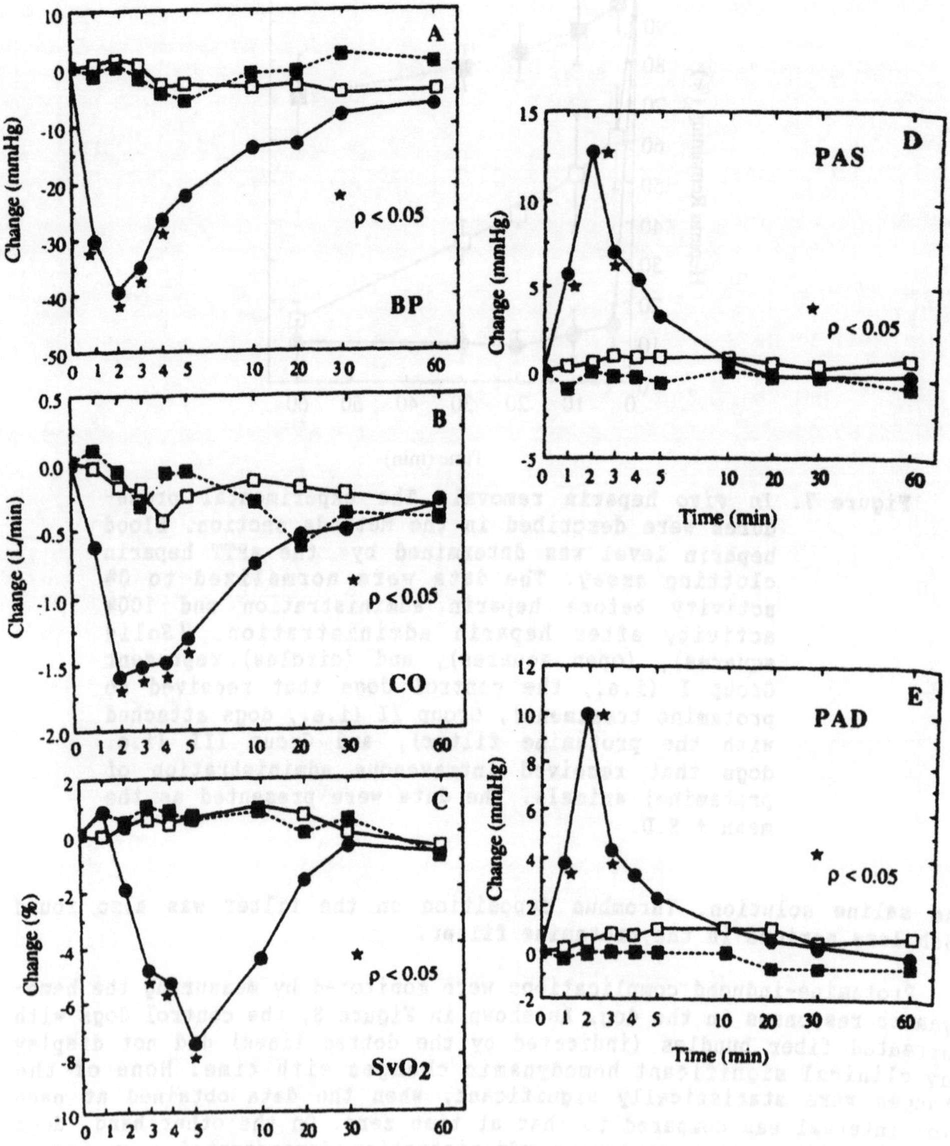


**Figure 8.** Hemodynamic changes in dogs. Hemodynamic parameters measured include (A) systemic arterial blood pressure (BP), (B) cardiac output (CO), (C) mixed venous oxygen saturation ($SvO_2$), (D) pulmonary artery systolic pressure (PAS), and (E) pulmonary artery diastolic pressure (PAD). The experimental procedures were described in detail in the Methods section. (Solid squares), (open squares), and (circles) represent Group I (i.e., the control dogs that received no protamine treatment), Group II (i.e., dogs attached with the protamine filter), and Group III (i.e., dogs that received intravenous administration of protamine) animals, respectively. Data were analyzed according to the statistical procedures described in the Methods.

be the consequence of the relatively large drop in cardiac output (-0.5 L/min) in the control animals, which could be attributed to an additional, supplement administration of pentobarbital to certain dogs at the 20 min interval.

The effects of the protamine filter on red blood cell counts, white blood cell counts, platelet counts, hematocrit, and total blood hemoglobin were examined. As shown in Figure 9A, the red blood cell counts in all animals, including both the control and experimental dogs, fluctuated around 100% over the entire period of the experiment. The change in red blood cell counts among different animal groups was not statistically significant, nor was the change with time statistically significant.

The white blood cell counts in animals receiving intravenous protamine administration dropped rapidly within the first 10 minutes following protamine administration, and after which the counts began to rebound (Figure 9B). The transient decrease in white blood cell counts was statistically significant ($p < 0.01$). The white blood cell counts dropped to an average value of 36 ± 14% of baseline at 3 minutes and returned to 97 ± 7% of baseline at 30 minutes. On the other hand, animals with the protamine filter showed only a slight drop in the white blood cell counts. The white blood cell counts were almost unchanged at 3 minutes, reduced to a value of 88 ± 8% of baseline at 10 minutes, returned to a value of 95 ± 9% of baseline at 20 minutes, and then remained constant thereafter. These changes were statistically significant, when the data were compared to those obtained from the control animals at the same time intervals.

The decline in platelet counts appeared to follow a similar pattern (Figure 9C). For animals receiving protamine, the platelet counts declined to a value of 32 ± 8% at 3 minutes, and then gradually rebounded to a value of 67 ± 7% of baseline at 30 minutes. However, for animals attached to the protamine filter, the maximum drop in platelet counts, which occurred at 10 minutes, was only 84 ± 5% of baseline. The control animals showed similar changes. None of changes was statistically significant, when the data were compared with those obtained from the control animals at the same time intervals.

The total hemoglobin remained constant at approximately 100% for both the control and the experimental animals (Figure 10A). The normalized hematocrit also remained unchanged over the entire course of the experiment for all of the animals tested (Figure 10B).

The effect of the protamine filter on the complement system was examined. As shown in Figure 11, the plasma complement level in the control animals, as measured by the $CH_{50}$ hemolytic assay, declined continuously over a period of 20 minutes until it reached a residual level of 80 ± 2% of baseline, and then remained constant thereafter. The changes were statistically significant ($p < 0.05$), when the data were compared to that at time zero. For animals receiving intravenous protamine treatment, however, the complement level was reduced in a much more rapid fashion. The residual complement level dropped to a value of 80 ± 4% of baseline at 3 minutes, and then remained at the same level thereafter. These changes from the baseline were also statistically significant ($p < 0.05$). Conversely, for animals attached to the protamine filter, the maximum drop in plasma complement, which occurred at about 20 minutes, was only to a value of 90 ± 3% of baseline. This value was statistically different ($p < 0.05$) from those observed in the control group and the group receiving intravenous protamine, indicating that the use of protamine-bound hollow fibers had lessened the induction of complement activity by both protamine and the hollow fibers.

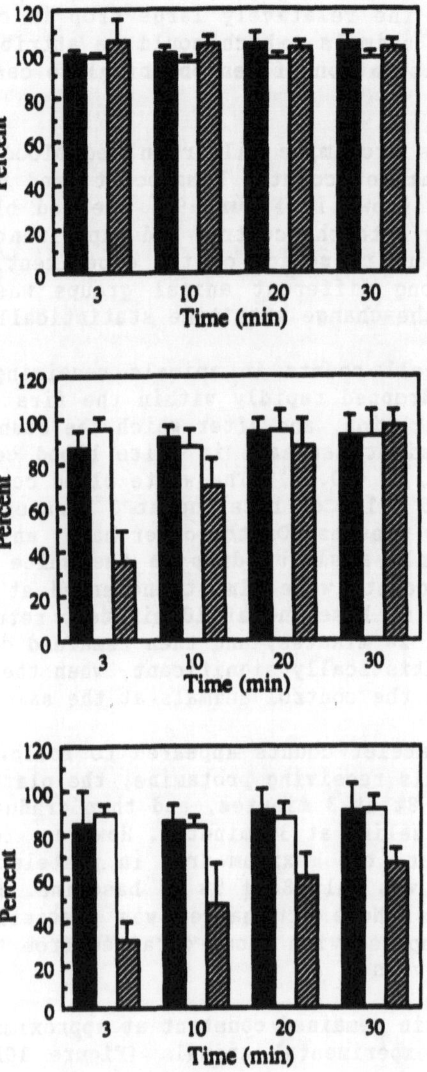


Figure 9. The effect of the filter on the (A) red blood cell counts, (B) white blood cell counts, and (C) platelet counts. All data are plotted as percent of baseline before opening the fiber bundle. The data were presented as mean ± S.D. (Solid bar), (open bar), and (hatched bar) represent Group I (i.e., the control dogs that received no protamine treatment), Group II (i.e., dogs attached with the protamine filter), and Group III (i.e., dogs that received intravenous administration of protamine) animals, respectively.

## DISCUSSION

To develop a protamine filter for extracorporeal blood heparin removal, protamine sulfate was immobilized onto regenerated cellulose hollow fibers. A detailed study of the parameters affecting the immobilization process was conducted in order to establish the most optimal protamine

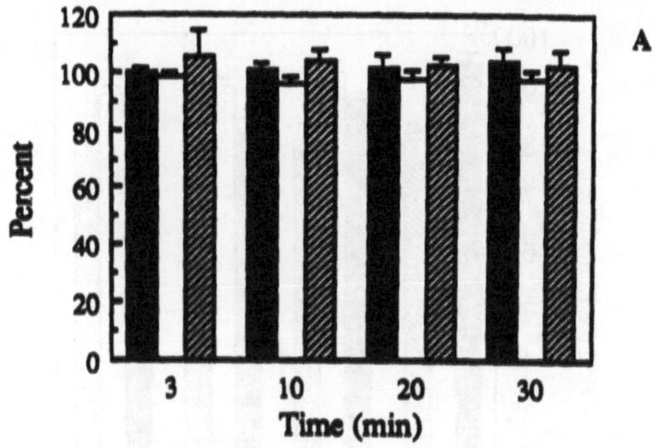


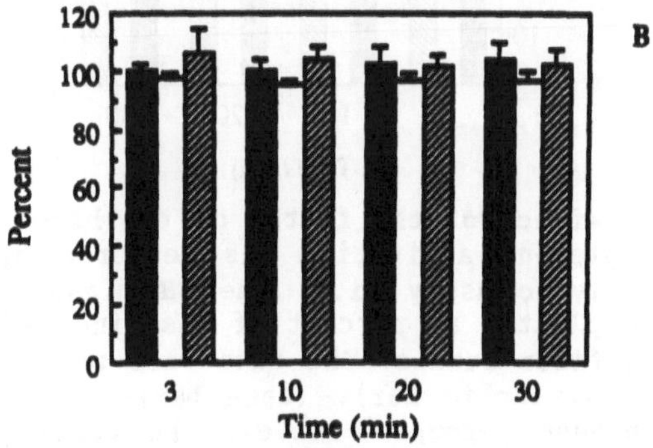


Figure 10. The effect of the filter on (A) total blood hemoglobin, and (B) hematocrit. All data are plotted as percent of baseline before opening the fiber bundle. The data were presented as mean ± S.D. (Solid bar), (open bar) and (hatched bar) represent Group I (i.e., the control dogs that received no protamine treatment), Group II (i.e., dogs attached with the protamine filter), and Group III (i.e., dogs that received intravenous administration of protamine) animals, respectively.

immobilization process. Ground fiber particles were selected for the immobilization studies for two reasons. First, ground fibers would offer the opportunity for a uniform cyanogen bromide activation and protamine immobilization, which in a sense was the crucial element for accurate and consistent evaluation of the immobilization process. Second, by using ground fibers and vigorous agitation, the intrinsic parameters would be measured in the absence of mass transfer limitation that could occur to macromolecular species such as protamine.

Protamine immobilization onto the CNBr-activated fibers followed a bimolecular reaction mechanism, and was dependent on both the cyanate ester concentration on the fibers and the protamine concentration in the coupling solution. Increasing the concentration of either of these two reactants resulted in an increase of the protamine loading on the fibers. In addition, protamine loading on the fibers was also dependent on the duration of the coupling step, increasing with the prolongation of the incubation time. At room temperature, and using sufficient reactant concentrations, the immobilization process was completed within an incubation period of about 30 minutes.

A linear relationship was observed when the protamine loading on the fibers was plotted against the cyanate ester concentration (Figure 3B).

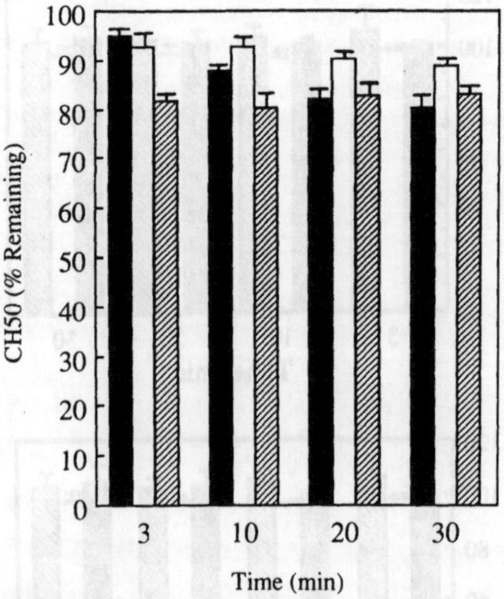


Figure 11. The effect of the filter on the blood complement. Complement activation was measured by the total hemolytic assay (i.e. the $CH_{50}$ assay). All data are plotted as percent of baseline before opening the fiber bundle. The data were presented as mean ± S.D. (Solid bar), (open bar) and (hatched bar) represent Groups I (i.e., the control dogs that received no protamine treatment), Group II (i.e., dogs attached with the protamine filter), and Group III (i.e., dogs that received intravenous administration of protamine) animal, respectively.

The slope of the plot yielded a ratio of 1.7 (mg/µmole), suggesting that one µmole of cyanate esters would bind 1.7 mg of protamine. Assuming an average molecular weight of 4,000 daltons for protamine,[37] a molal stoichiometric ratio of 0.43 (mole-protamine/mole-cyanate ester) was estimated for the reaction between protamine and the cyanate ester groups. Thus, in spite of the presence of abundant amino (i.e., arginine) groups on the protamine molecule, each protamine molecule would be linked to the membrane surface by only two or three isourea linkages. Since a certain amount of cyanate ester groups might have already been leached off the support by hydrolysis[38] before protamine could be attached to them, the molal stoichiometric ratio of 0.43 (mole-protamine/mole-cyanate ester) was an apparent value and only represented the lower limit.

Previous studies involving the use of CNBr-activated materials for protein immobilization have addressed the potential instability of the cyanate ester derivatives due to hydrolysis of cyanate ester.[38-40] Results obtained in current study appeared to confirm their findings. As shown in Table 1, the protamine loading on the fibers increased with increasing the protamine concentration in the coupling solution. If hydrolysis had not occurred, then the amount of available cyanate esters on the fibers would have remained unchanged. Under conditions where protamine was in large excess in comparison with the cyanate ester groups, and where the incubation period was carried out long enough to assure the completion of protamine immobilization (e.g., 2 hours of incubation), the protamine loading on the fibers should have been independent, rather than dependent on the protamine concentration in the coupling solution. These results suggested that the cyanate ester derivatives were not stable under the experimental conditions employed, and were continuously inactivated during the coupling process. Because of this continuous inactivation, protamine loading on the fibers was governed by the competition of two reaction kinetics; the rate of coupling of protamine to the cyanate ester groups, and the kinetics of hydrolysis of the cyanate ester groups. For fibers possessing the same cyanate ester content, increasing the protamine concentration in the coupling solution would increase the kinetics of protamine coupling and, as a consequence, would increase the protamine loading on the fibers.

Protamine loading on the fibers is thus deemed to be controlled by three independent parameters: the CNBr concentration during the activation step, the duration of the coupling step, and the protamine concentration in the coupling solution. The CNBr concentration is normally below 0.1 g/mL, because concentration beyond this limit can, at times, induce structural deterioration of the fibers and weaken their mechanical strength. The incubation period is generally controlled to one hour or slightly longer to assure that the immobilization process is completed. It therefore becomes evident that altering the protamine concentration in the coupling solution is the easiest and most convenient means to control the protamine loading on the fibers.

A recirculating system consisting of a heparin reservoir and a pump was adopted to characterize heparin adsorption for the purpose of simulating the *in vivo* blood circulation situation where the protamine filter would eventually be applied. All the heparin adsorption versus time curves appeared to follow a similar pattern: a rapid drop of heparin concentration in the reservoir with time, with an asymptotic approach to a steady state value. During the initial stage of operation, heparin in the feed solution was almost completely taken up by the protamine filter. The recycling of the heparin-depleted solution to the reservoir resulted in the rapid drop of heparin concentration in the reservoir. As the protamine filter gradually became equilibrated or saturated, the net heparin removal by the filter was reduced to zero and the heparin concentration in the reservoir gradually approached the steady state.

The adsorption isotherm was characterized using the protamine-bound fiber particles obtained by grinding the intact protamine filter. As previously discussed, the Cuprophane membrane employed for the construction of the hollow fiber has dense, isotropic structures with a nominal molecular weight cut-off of 10,000 daltons.[28] Diffusion of heparin, which has an average molecular weight of 13,000 daltons, is therefore hindered due to the size exclusion effect of the membrane pores. On the other hand, protamine molecules with an average molecular weight of 4,000 daltons can diffuse relatively easily through the membrane pores during the immobilization process. As a consequence, only a fraction of protamine which locates at the surfaces of the fibers are subject to heparin binding. Thus, the use of ground, protamine-bound fiber particles for the adsorption isotherm studies was to expose all the protamine molecules, both on the fiber surface and within the fiber matrix, to heparin binding.

*In vitro* results (Figure 6) showed that heparin adsorption on the protamine-bound fiber membrane followed the Langmuir adsorption model.[36] At the low heparin concentration region (0.01 - 1.0 mg/mL), adsorption was first order and the amount of heparin adsorbed was linearly proportional to the equilibrium heparin concentration. The adsorption in this case was determined by the equilibrium between heparin in the bulk solution and heparin adsorbed on the solid phase, which could be expressed by $Q = KQ_sC$ from the Langmuir equation. As the heparin concentration was increased from 1 to 10 mg/mL, the amount of heparin adsorbed on the fibers leveled off and the adsorption was mixed order. When the heparin concentration was further increased to above 10 mg/mL, the amount of heparin adsorbed reached a plateau which represented the saturation capacity ($Q = Q_s$) of the protamine-bound fibers. In this case, the adsorption was zero order and essentially independent of the substrate concentration.

Figure 7 shows that *in vivo* heparin removal by the protamine filter follows a two compartment kinetic model in which an early, bonding of the heparin to the device is followed by a second, slower phase of ongoing heparin removal. This kinetic model seems ideally suited to the accomplishment of both adequate extracorporeal heparinization as well as safe limitation of systemic anticoagulation. In some clinical settings, the heparin level can not be increased sufficiently to prevent clotting of the devices because of the increased risk of severe hemorrhagic complications.[41] The use of the protamine filter in series with these extracorporeal devices may allow the safe usage of adequate and higher doses of heparin in the circuit.

Based on the *in vivo* results, 1 mg of immobilized protamine was estimated to neutralize approximately 60 USP units of heparin, which is considerably lower than that reported in the literature for free protamine.[27,42] Clinically, one mg of free protamine is generally known to neutralize 80-120 USP units of heparin.[27,42] The decrease in the effectiveness in heparin neutralization for immobilized protamine may be due to the shielding effect resulting from heparin-protamine binding. Binding of heparin to the immobilized protamine would provide a shielding effect over the neighboring unoccupied protamine molecules, causing these molecules to become sterically inaccessible to other heparin molecules.

Unlike intravenous protamine administration which provides a systemic heparin reversal, the use of protamine filter allows for only a regional heparin removal. Thus the protamine filter can never accomplish a full systemic removal of heparin anticoagulation due to the mixing of the filter-treated blood with the untreated blood arriving from other branches of the vascular system. For instance, Figure 7 shows that approximately 15% of the initial heparin activity still remained after the use of the protamine filter. The remaining heparin level, however, can be reduced to below a certain level (e.g., < 0.2 unit/mL blood) by designing a filter with required surface area and protamine loading.

Protamine-induced hypotensive responses were examined in dogs. The dog was selected as the animal model because it is known to magnify human responses noted with protamine reversal of heparin.[37,43] Among all the species studied, the canine vascular system was the most vulnerable to protamine.[43] Conahan and co-workers reported that the cardiovascular effects of protamine sulfate appeared to be more benign in man than in dogs.[44] Thus, if the protamine filter does not cause complications in dogs, it is unlikely that filter would pose any adverse responses in patients. Consistent with the findings reported by other investigators,[27,37,43] dogs with intravenous administration of protamine displayed noticeable hemodynamic responses. Systemic hypotension, pulmonary artery hypertension, decreases in cardiac output, and declines in mixed venous oxygen saturation occurred between 1 to 10 min after the injection of protamine. These responses were transitory, and 20 min after the injection, the hemodynamic parameters had returned to their baseline levels. On the other hand, the use of the protamine filter did not elicit any statistically significant adverse hemodynamic response over the entire course of the experiment. It is known that the majority of protamine reaction may result from the direct contact of protamine with cells in certain organs.[27,43] As previously discussed, the use of a protamine filter would essentially eliminate any possible contact between protamine and various organ systems. In addition, our results also indicated that the use of protamine filter significantly reduced the activation of the complement system (Figure 11). The lack of hemodynamic response to the protamine filter may therefore be accounted for in terms of absence of these two mechanisms.

The use of immobilized protamine significantly attenuated the protamine-induced thrombocytopenic response. Peripheral platelet counts were reduced by 68% in animals receiving intravenous protamine, but only 16% in animals with the protamine filter. The precise nature of how the immobilized protamine suppresses the thrombocytopenic response caused by free protamine remains unknown. Jacques indicated that when thrombocytopenia was experimentally induced by glycogen, subsequent protamine administration did not decrease the animals systemic blood pressure,[43] suggesting that platelets are important components for this hypotensive response. Recent studies have also suggested a correlation between thrombocytopenia and the development of hypotension.[37,45] The diminution in the degree of thrombocytopenia may therefore be accounted for in terms of the lack of the related hypotension response to the protamine filter.

Numerous investigators have noted that transient granulocytopenia is manifested in patients undergoing dialysis[46] or protamine treatment.[47,48] This phenomenon has been linked to the reversible pulmonary leukosequestration resulting from the activation of granulocytes by complement-derived fragments.[46-48] Our results seem to agree with this mechanism. Animals that received intravenous protamine administration and demonstrated the highest degree of complement activation were also those which displayed the most significant drop in white blood cell counts (i.e., a drop of 64%). On the other hand, animals that were attached to the protamine filter and experienced the least degree of complement activation showed a drop of only 12% in white blood cell counts. Thus, the use of the protamine filter could also significantly diminish the protamine induced granulocytopenic response.

No hemolysis occurred during the operation of the protamine filter. The red blood cell counts, hematocrit, and the blood hemoglobin levels remained statistically unchanged over the one hour experiment. The difference in values of these hemotologic parameters between the untreated filter and the protamine filter groups was also negligible, suggesting that the protamine filter caused no further blood damages than found with the untreated filter.

Like antibody-antigen complexes, Gewurz and co-workers found that heparin-protamine complexes also activated the complement system.[49,50] Consistent with their *in vitro* findings, our results showed that 3 minutes following protamine administration, the animal's plasma complement level rapidly declined by 20%. However, a reduction of only 10% of the $CH_{50}$ values was observed in dogs treated with the protamine filter. Fiedel, et al,[50] reported that when heparin and protamine was combined in multiple proportions, the proportion which resulted in optimal precipitation was also the one that induced maximal complement consumption. It was therefore suggested that the heparin-protamine complex must achieve a critical size in order to activate Clq.[50,51] For the protamine filter, since protamine is covalently immobilized on the fiber surface, it would presumably only allow adsorption of a single layer of heparin. Thus, the reduction in complement activation by the protamine filter may be due to the lack of sufficient size for the heparin-protamine complex on the filter to bind Clq.

It is interesting to note that the protamine filter induced even less complement activation (or consumption) than the untreated fiber bundle.

The control dogs with the untreated fiber bundle displayed a drop in the complement level that was twice as great as that observed in the animals with the protamine filter. Several reports have indicated that the presence of hydroxyl groups on the cellulose membrane would result in the activation of the complement system through the alternative pathway.[52,53] The activation most likely results from the interaction of the thioester groups on the complement components (C3 or C4) with the -OH groups on the cellulose membrane.[52,53] In the protamine filter, since the -OH groups on the cellulose fibers have been converted into isourea linkages (i.e., the linkage between the protamine and the fiber), they may no longer be available for binding of the complement proteins. The decrease in complement activation by the protamine filter may also suggest that a protamine-coated cellulose membrane is more compatible with blood than the original untreated membrane.

The most severe and persistent response to protamine appears to occur in patients who have been previously exposed to protamine.[20,37] This enhanced protamine response is believed to be related to an immune-mediated mechanism.[44,54-56] In certain patients, IgE or IgG reaginic antibodies are produced in response to protamine.[44,54-56] These antiprotamine antibodies are attached to mast cells which are present in the liver, lungs, and other organ tissues. The second exposure to protamine (i.e., protamine secondary) results in a reaction of protamine with its antibodies on the cell surface. Once the antigen-antibody complex is formed, it triggers the degranulation of the mast cell, releasing histamine and other vasoactive amines.[27,37,43] Although the use of immobilized protamine would not prevent protamine from interacting with antibodies which are fixed to circulating basophiles, it would however block the conjugation of protamine to the antibodies which are fixed to the mast cells. Since the basophiles are present in relatively low numbers compared to that of the tissue mast cells, the use of the protamine filter theoretically would significantly reduce the antibody-mediated anaphylactic response resulting from protamine secondary. Studies related to this area and involving the use of protamine-sensitized animals will be conducted in our laboratory.

The protamine filter thus provides an unique means to control simultaneously both heparin and protamine induced complications. This control would significantly improve the safety of ECBC procedures. With nearly twenty million such procedures performed each year, the benefits of the filter would be widespread. In addition, the improved blood compatibility and reduced complement-activating ability of the protamine filter may enable it to replace current hemofilters such as those employed in hemodialysis or hemofiltration procedures.

From a practical standpoint, the protamine filter also possess two major advantages. One advantage lies in the simplicity and flexibility of operating the device. The filter can be easily interfaced with current ECBC devices without any additional apparatus or invasive procedures. The other advantage is the potential acceptance of the filter for clinical use. The filter is designed with a blood compatible material (i.e., cellulose hollow fibers obtained from a clinical hemodialyzer) and a clinically accepted drug (i.e., protamine), and at the same time significantly minimizes the toxic effects of the drug. Its acceptance for clinical use, particularly in patients with increased risk of bleeding complications from required heparin use, would, therefore, be highly anticipated. In a more general view, the protamine filter may serve as a prototype for other selective drug removal systems. It may also open up the possibility for the use of other immobilized chemicals for extracorporeal blood treatment, particularly when these chemicals have potential toxic side effects and can not be administered safely into the vascular system.

## ACKNOWLEDGMENTS

The authors thank Dr. T. W. Wakefield and Dr. G. O. Till for their assistance and valuable advice. The study was supported by the NIH Grant HL 38353 and by the Whitaker Foundation Biomedical Engineering Grant.

## REFERENCES

1. H. E. Kambic, S. Murabayashi, and Y. Nose, C&EN News, April 14, 31 (1986).
2. A. J. Wing, F. B. Brunner, H. O. A. Bryynger, C. Jacobs, and P. Kramer, in: "*Replacement of Renal Function by Dialysis*," W. Drukker, F. M. Parson, and J. F. Maher, Eds., Martinus Nijhoff, Boston, 1983, p. 850.
3. R. G. Mason, H. Y. K. Chuang, and S. F. Mohammad, in: "*Replacement of Renal Function by Dialysis*," W. Drukker, F. M. Parson, and J. F. Maher, Eds., Martinus Nijhoff, Boston, 1983 p. 186.
4. A. S. Gervin, Surg. Gynecol. Obstet., **140**, 789 (1975).
5. J. R. Fletcher, A. E. McKee, M. Mills, K. C. Suyden, and C. M. Herman, Surgery, **80**, 214 (1976).
6. R. D. Swartz, and F. K. Port, Kidney Int., **16**, 513 (1979).
7. M. A. Galen, S. M. Steinberg, and E. G. Lourie, Ann. Intern. Med., **82**, 359 (1975).
8. N. E. Tolkoff-Rubin, J. Nardini, R. N. Leslie, S. T. Fang, R. H. Rubin, Dialy. Transplant, **15**, 125 (1986).
9. J. Porter, and J. Jick, J. Am. Med. Assoc., **237**, 9 (1977).
10. K. Atsumi, I. Fujimasa, K. Imachi, M. Nakajima, S. Tsukagoshi, K. Mabushi, K. Motoruma, A. Kouno, T. Ono, A. Miyamoto, N. Takido, and N. Inou, AIAIO J., **8**, 155 (1985).
11. M. C. Smith, D. Kowit, J. W. Crow, A. E. Cato, G. D. Park, A. Hassid, and M. J. Dunn, Am. J. Med., **73**, 669 (1982).
12. A. Laned, I. R. McGregor, R. Michalski, and V. V. Kakkar, Thromb. Res., **12**, 256 (1978).
13. R. D. Swartz, Nephron., **28**, 65 (1981).
14. R. Pinnik, T. Wiegmann, and D. Diederich, N. Engl. J. Med., **308**, 258 (1983).
15. H. Miyama, N. Harumiya, Y. Mori, and H. Tanzawa, J. Biomed. Mater. Res., **11**, 251 (1977).
16. R. Langer, R. L. Linhardt, P. M. Galliher, M. M. Flanagan, C. L. Cooney, and M. D. Klein, Science, **217**, 261 (1982).
17. H. Bernstein, V. C. Yang, D. Lund, M. Randhawa, W. Harmon, and R. Langer, Kidney Int., **32**, 452 (1987).
18. V. C. Yang, H. Bernstein, and R. Langer, in: "*Bioreactor Immobilized Enzymes and Cells: Fundementals and Applications*," Moo-Young, Ed., Elsevier, U.K., 1988, p. 83.
19. V. C. Yang, H. Bernstein, C. L. Cooney, J. Kadam, and R. Langer, Thromb. Res., **44**, 599 (1986).
20. J. M. Weiler, P. Freiman, M. D. Sharath, W. J. Matzger, J. M. Smith, H. B. Richerson, Z. K. Ballas, P. C. Halverson, D. J. Shulan, S. Matsuo, and R. L. Wilson, J. Allergy Clin. Immunol., **75**, 297 (1985).
21. N. M. Katz, Y. D. Kim, R. Siegelman, S. A. Ved, S. W. Ahmed, and R. B. Wallace, J. Thorac. Cardiovasc. Surg., **94**, 881 (1987).
22. J. K. Kirklin, D. E. Chenoweth, D. C. Naftel, E. H. Blackstone, J. W. Kirklin, D. D. Bitran, J. G. Reves, and P. N. Samuelson, Ann. Thorac. Surg., **41**, 193 (1986).
23. G. N. Olinger, R. M. Becker, and L. I. Bonchek, Ann. Thorac. Surg., **29**, 20 (1980).
24. C. A. Cobb, III., and D. L. Fung, Surg. Neurol., **17**, 245 (1982).

25. C. Kurusz, 6th Annual Meeting of Pathophsiology and Extracorporeal Technology, San Diego, CA (1986).
26. V. C. Yang, US Patent 4,800,016.
27. L.B. Jaques, Can. Med. Assoc. J., 108, 1291 (1973).
28. W. Bendel, "*Developments in the Fields of Cuprophane Membranes and Hollow Fibers*," Enka Ag, West Germany, 1975.
29. J. Kohn, and M. Wilchek, Biochem. Biophys. Res. Commun., 84, 7 (1978).
30. M. M. Bradford, Anal. Biochem., 72, 248 (1976).
31. L. B. Jaques, Science, 206, 528 (19790.
32. J. M. Walenga, J. Fareed, D. Hoppensteadt, R. M. Emanuele, CRC Crit. Rev. Clin. Lab. Sci., 22, 361 (1986).
33. P. H. Schur, Hum. Pathol., 14, 338 (1983).
34. S. C. March, I. Parikh, P. Cuatrecasas, Anal. Biochem., 60, 149 (1974).
35. J. Zar, "*Biostatistical Analysis*," Prentice Hall Inc., Englewood, 1974.
36. D. M. Ruthven, "*Principles of Adsorption and Adsorption Processes*," Wiley & Sons, New York (1984).
37. J. C. Horrow, Anesth. Analg., 64, 348 (1985).
38. V. C. Yang, and C. L. Teng, Polym. Mater. Eng. Sci., 58, 116 (1988).
39. M. Wilchek, T. Oka, and Y. J. Topper, Proc. Nat. Acad. Sci., 72, 1055 (1975).
40. H. Bernstein, V. C. Yang, and R. Langer, Biotechnol. Bioeng., 30, 196 (1987).
41. R. D. Swartz, and F. K. Port, Kidney Int., 16, 513 (1979).
42. E. F. Reynolds, Ed., "*Martindale: The Extrapharmacopoeia*," The Pharmaceutical Press, London, 1982, p. 390.
43. L. B. Jaques, Brit. J. Pharmacol., 4, 135 (1949).
44. T. J. Conahan, III., R. W. Andrews, and H. MacVaugh, III., Anesth. Analg., 60, 33 (1981).
45. T. W. Wakefield, C. B. Hantler, B. Lindblad, W. M. Whitehouse, Jr., and J. C. Stanley, J. Vasc. Surg., 3, 885 (1986).
46. K. M. Skubitz, and P. R. Craddock, J. Clin. Invest., 67, 1383 (1981).
47. T. J. Flahertz, P. R. Craddock, D. E. Hammersmidt, and C. F. Moldow, Blood, 51, 731 (1978).
48. H. S. Jacob, P. R. Craddock, D. E. Hammersmidt, and C. F. Moldow, N. Eng. J. Med., 302, 789 (1980).
49. R. Rent, N. Ertel, R. Eisenstein, and H. Gewurz, J. Immunol., 114, 120 (1975).
50. B. A. Fiedel, R. Rent, R. Myhrman, and H. Gewurz, Immunology, 30, 161 (1976).
51. N. Cavarocchi, H. V. Schaff, T. A. Orszulak, H. A. Homburger, W. A. Schnell, Jr., and J. R. Pluth, Surgery, 98, 525 (1985).
52. R. M. Hakim, J. Breillatt, J. M. Lazarus, and R. Port, N. Eng. J. Med., 311, 878 (1984).
53. D. E. Chenoweth, A. K. Cheung, and L. W. Henderson, Kidney Int., 24, 764 (1983).
54. D. R. Jackson, Angiology, 21, 295 (1970).
55. J. D. Lakin, T. J. Blocher, D. M. Strong, and M. W. Yocum, J. Allergy Clin. Innumol., 61, 102 (1978).
56 M. E. Weiss, D. Nyhan, Z. Peng, J. C. Horrow, E. Lowenstein, C. Hirshman, and N. F. Adkinson, N. Eng. J. Med., 320, 886 (1989).

# POLYMERIC SYSTEMS FOR DIAGNOSTIC ULTRASOUND CONTRAST AGENTS

Beth Schrope,[a] Peng Shen [b] and Margaret Wheatley[a,b]

(a) Institute of Biomedical Engineering and Science
(b) Department of Chemical Engineering
Drexel University, Philadelphia, PA 19104

The use of contrast agents has been widely accepted in the area of diagnostic radiography for many years and the original concept of an ultrasound contrast agent was first proposed as early as 1968 by Gramiak, with the injection of indocyanine green into the ascending aorta and heart. However, despite this early beginning, no universally accepted contrast agent for ultrasound now exists. This is primarily due to problems with stability, size and/or toxicity of the agents which have been investigated. Development of an effective ultrasound contrast agent would be highly significant for the health care industry, since it would greatly expand the scope of ultrasound (a noninvasive and safe procedure) as a diagnostic technique. Amongst the best systems that demonstrate effective ultrasound contrast (also termed high echogenicity) are agents which present a gas/liquid interface to the impinging ultrasound beam. These have included hand agitated solutions and microencapsulated gas bubbles. The latter systems have the advantage of stability, but are often too large to be injectable. Capsules larger than 10 µm become entrapped in the capillary bed of the lung. We describe here the use of ionotropic gelation of the naturally occurring polysaccharide alginate, for the microencapsulation of air. Two procedures have been investigated. A novel jet head has been developed which allows co-extrusion of a solution of sodium alginate and air to produce nascent micro-encapsulated air bubbles which fall into a hardening solution of calcium ions. A second method employs ultrasound to introduce cavitation-induced bubbles into the sodium alginate prior to capsule formation by spraying. Power spectra of these preparations demonstrate strong resonant frequencies at characteristic wavelengths, with resonant peaks that are a function of capsule size and wall characteristics. B-mode sector scans indicate that the capsules are highly echogenic.

## INTRODUCTION

One of the major advantages of ultrasound as a diagnostic technique is that it is both noninvasive and safe. Advances in ultrasound imaging have so far involved improvements in the signal processing methods, such as the advent of grey scale imaging and color doppler. Despite these advances, ultrasound still has major deficiencies in particular areas. These include the imaging of certain metastatic lesions, for example in the kidney and liver; imaging of vessels and body cavities such as the heart; specific organ targeting; real time blood flow analysis; and dynamic measurements for specific uptake studies. It is in these areas that ultrasound contrast agents may be expected to have the greatest impact.

The major requirements of a contrast agent are that it must be safe, it must be amenable to parenteral administration, and it must be strongly echogenic, that it must have acoustic properties that generate a strong echo (backscatter) from the impinging ultrasound beam, since it is this echo returning to the transducer that is translated into the ultrasound image. Safety involves the toxicity of the agent and the fate of the agent in the body. Parenteral administration requires that the agent be less than 10 µm in size, and preferably below 5 µm, so that it can pass unimpeded through the capillary bed of the lung. Finally, the strength of the backscatter depends on both the experimental conditions and upon differences in the physical properties of the scatterer and the suspending medium, in this case blood. The strength of the enhancement of backscatter is quantified by a parameter known as the scattering cross section of the agent. The primary aim then, is to develop a highly echogenic agent that is both safe and less than 10 µm in size.

The concept of an ultrasound contrast agent to enhance ultrasound backscatter dates from the work of Gramiak, et. al., in 1968, in which indocyanine green was injected into the ascending aorta and heart.[1] Many subsequent investigations of cardiac ultrasound contrast agents have involved injection of a liquid vehicle that may or may not include pre-existing gas bubbles. Most of the early reports of contrast enhancement have involved the heart.[2-4] It has been shown that even rapid injection of a solution devoid of pre-existing gas bubbles, (e. g., saline)[5] into the venous system generates enough bubbles by cavitation at the injection site to produce a cloud of echoes. Hand agitated solutions of compounds such as indocyanine green,[6] Renografin[7] and sugar solutions[8] have also been described. These latter solutions have greater viscosity than either saline or sugar solutions, which renders some stability to the bubbles. However, the bubble size is directly proportional to the solution viscosity, which also appears to have a deleterious transient effect on coronary flow. This effect may also be due to the high osmolarity of these solutions, but even in the absence of these effects these agents are limited by size, for example sonicated Renografin[8] produces large bubbles of about 10 µm. In addition to agitated/sonicated solutions, investigators have explored other approaches which include development of high reflectivity microspheres of collagen,[9] and gelatin capsules containing entrapped nitrogen microbubbles.[10-12] Colloidal suspensions of perfluorochemicals have also been used as an ultrasound contrast agent for enhancement of the liver and spleen.[13,14] Perfluoro-chemicals give relatively little immediate contrast but are taken up by Kupffer cells in the liver and by the spleen, where they enhance echogenicity. Similar effects were found with 1 µm particles of Iodipamide ethyl ester (IDE), where it was shown that 95% of the injected particles accumulated in the liver, and 5% in the spleen.[15]

Encapsulated gas bubbles offer particular advantages in stability over hand agitated systems, but frequently present problems with size. Capsules larger than 10 µm become entrapped in the capillary bed of the lung. One agent which has been used is a sugar molecule matrix composed of galactose (SHU-454),[16-18] consisting of sugar crystals of 10 µm or



Cosmetic and Pharmaceutical Applications of Polymers
Edited by C.G. Gebelein et al., Plenum Press, New York, 1991

more in size with gas bubbles trapped within the crystalline matrix. Bubbles are generated just prior to use by dissolving the crystalline matrix in a sterile solution, liberating the entrapped bubbles and allowing injection of the mixture. The next generation preparation, SHU-508, produces bubbles with a size range of 2-8 µm, and has been demonstrated to pass the pulmonary bed and to opacify the left side of the heart in normal human volunteers.[19] This is a major advance since it is hard to achieve good contrast in the high pressure environment of the left ventricle, where encapsulated bubbles tend to collapse. Another promising agent is formed by microencapsulation of air bubbles in human albumin. Albunex$^R$ consists of albumin microspheres with a size range of 0.5-10 µm, produced by sonication of a 5% human serum albumin solution.[20-23] There is some confusion in the literature as to whether or not Albunex$^R$ does survive in the left ventricle.

## 1. Theoretical Considerations that Influence the Choice of Systems to be Used as a Diagnostic Ultrasound Contrast Agent

Contrast agents could potentially impact favorably on three acoustic parameters; enhancement of backscatter (also known as echogenicity), increase in the rate of attenuation and alteration of the speed of sound propagation. Of these, enhancement of backscatter is by far the predominant parameter that would be influenced, since ultrasound image generation relies on detecting and analyzing the ultrasound waves that are scattered from the site of interest back to the emitting transducer. This being the case, it is important to analyze what parameters affect the backscatter of ultrasound in order to set criteria for the design of the optimal system for use as a contrast agent.

The power that is backscattered to the receiving transducer is proportional to a parameter known as the scattering cross-section, σ, which has a strong dependence on both the radius, r, of the scatterer, and the density and compressibility differences between the scatterer and the continuum.[24] This is represented by Equation (1).

$$\sigma = \left[\frac{4}{9}\pi r^2 (kr)^4\right]\left[\left(\frac{\kappa_s - \kappa}{\kappa}\right)^2 + \frac{1}{3}\left(\frac{3(\rho_s - \rho)}{2\rho_s - \rho}\right)^2\right] \qquad \text{(Equation 1)}$$

$\sigma$ = scattering cross section
$r$ = radius of scatterer $\ll \lambda$
$k$ = wave number
$\kappa_s$ = adiabatic compressibility of scatterer
$\kappa$ = adiabatic compressibility of embedding medium
$\rho_s$ = density of scatterer
$\rho$ = density of embedding medium

For a given particle size (less than the ultrasound wavelength), Lubbers[25] has calculated that solid particles produce one order of magnitude more scatter than red cell aggregates (which, in turn, are similar to soft tissue). Gas bubbles however, are predicted to produce four orders of magnitude more scatter than solid particles. There is also a direct relation between the effective scattering cross section of a small cloud of scatterers and the total number of bodies in the cloud, so that a pulse of scatterers can produce a significant increase in scattering directly proportional to the concentration of scatterers in the cloud.

Free gas bubbles present an additional phenomenon when stimulated by ultrasound radiation, since they undergo oscillatory motion such that the bubble acts like a harmonic oscillator with a characteristic resonant frequency given by the highly simplified Equation 2.[26]

$$f_o r = \sqrt{\frac{3\gamma P_o}{4\pi^2 \rho_o}} \qquad \text{(Equation 2)}$$

f = resonant frequency
$r^o$ = radius
$\gamma$ = adiabatic ideal gas constant
$P_o$ = ambient fluid presure
$\rho_o$ = ambient fluid density

This can result in a scattering cross sectional area three orders of magnitude greater than the physical size. For bubbles of less than 10 µm, the resonant frequency of these bubbles lies within the medical imaging range of 1-10 MHz. However, in practice free gas bubbles of less than 10 µm, generated in an aqueous medium, collapse instantaneously (less than a tenth of a second) under the force of their own surface tension. Microencapsulation increases stability, but imparts stiffness to the bubbles, which results in resonance damping.

## 2. The Choice of System for This Study

We describe here our investigations of the ionotropic gelation of sodium alginate to produce a flexible polymer coat for microbubbles of air. We are interested in microencapsulation of a gas in order to produce a highly echogenic agent. It is hoped that alginate will prove to be a good choice for encapsulation because it forms a flexible hydrogel, and could both deform to pass through very small capillaries (similar to the manner in which erythrocytes pass through such spaces) and withstand the high pressures of the left side of the heart without rupture. Our long term goal is to produce microcapsules that are below 10 µm in size, and which remain stable in the circulation for at least 10 minutes. It is hoped that the stability and size would allow for intravenous injection, free passage through the pulmonary bed, survival in the high pressure environment of the left ventricle of the heart, and persistence in the circulation for more than one pass.

# MATERIALS AND METHODS

## 1. Microcapsule Formation

The system described here consists of microbubbles of gas entrapped in a crosslinked polymeric matrix of calcium alginate. The microcapsules are prepared by forcing a 1.5 wt% aqueous solution of sodium alginate (Keltone HV or LV, Merck and Co., Inc., Chicago, IL) from a 10 mL syringe (mounted in a Razel syringe pump, model A-99J), through a blunt ended, 26G syringe needle, around which flows a coaxial stream of air which detaches the drops of alginate as they emerge from the needle tip (Figure 1). The nascent capsules fall into a solution of 1.2 wt% calcium chlor-

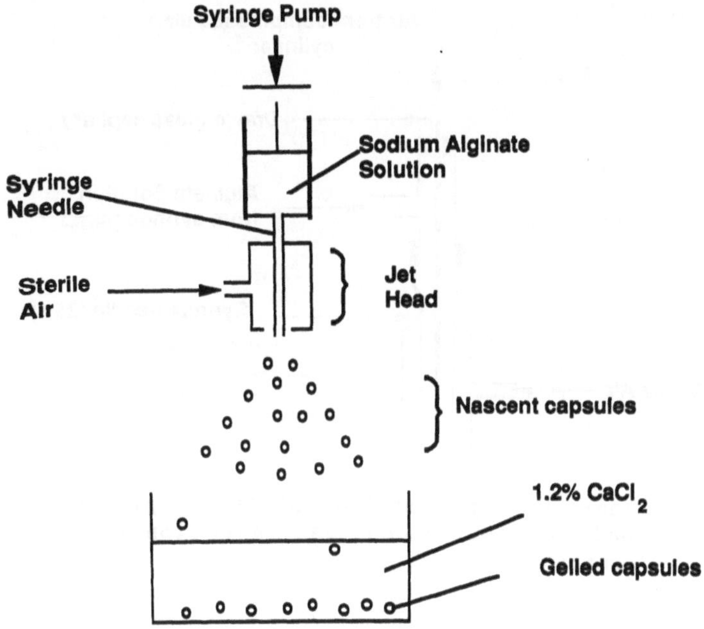


Figure 1. Apparatus for ionotropic gelatin of alginate to produce microcapsules.

ide, gel upon contact with the divalent calcium ions and harden with time. Most capsules prepared by this method, described in the literature to date are in excess of 500 µm diameter.[27-33] Various parameters affecting capsule size were investigated, including viscosity of the sodium alginate solution, speed of extrusion of the alginate, and air flow rate around the needle tip. Microcapsule size was measured against a calibrated grid mounted in the eye piece of an Olympus IMT2 Inverted microscope, equipped with photographic capability.

### 1A. Microencapsulation of Gas Bubbles by Co-extrusion

We developed a new concept of forming a microcapsule around a gas bubble, which resulted in the design and fabrication of a prototype triple-barreled jet head, for simultaneous development of an alginate drop and injection of a gas bubble inside the drop. Gas bubbles were introduced into the nascent capsules through a glass capillary which was hand drawn into a micron sized tip (close to 5 µm) in our laboratory and inserted through the center of the syringe needle. The air for microencapsulation was fed to the capillary from a compressed air cylinder. A coaxial stream of air was used to detach the alginate drop from the needle tip as before (Figure 2).

### 1B. Gas Entrapment in the Sodium Alginate Solution

Various other methods of gas entrapment were explored, in which gas was entrapped in the sodium alginate solution before extrusion through the syringe needle and prior to capsule formation (which was again by spraying). Hand agitation of the sodium alginate (either by use of a vortex mixer, or repeated introduction and expulsion of a solution through a 20 mL syringe via an 18G needle), processing of the solution through a tissue homogenizer and sonication of the sodium alginate solu-

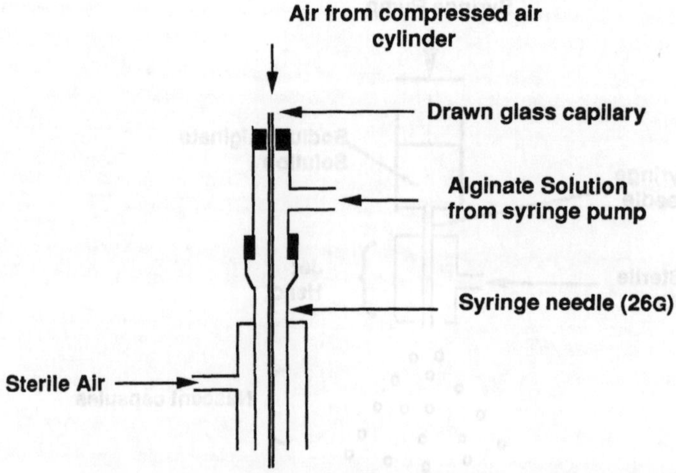


Figure 2. Schematic representation of the triple-barrel jet head for simultaneous gas injection and capsule formation.

tion (bath sonicator [Laboratory Supplies], and probe sonicator [Heat Systems W-385 with 1/2" tapped probe]) were all investigated. Surfactant (Tween 20 [Sigma, St. Louis, MS] at 0.01 wt%) was added to the sodium alginate to stabilize the gassed solution.

## 2. Tests for Echogenicity

An air/liquid or air/solid interface will reflect ultrasound upon impact of an incident ultrasonic wave, creating an echo in all directions. Many objects also exhibit resonance scattering, that is the object has a characteristic frequency associated with its echoes. Strong echoes (defined as echoes with amplitudes greater then the noise of the sensing device) are created by large bubbles. We however, are limited to sizes below 10 µm, since we are interested in developing a system that will pass unimpeded through the capillary bed of the lung. It is therefore imperative to test the echogenicity of our microcapsule system *in vitro*, and to follow the impact upon this echogenicity of both bubble size and size distribution. This is done by investigating both the resonance scattering of the agent and the intensity of backscatter that it can produce.

### 2A. Pulse Echo Studies: Generation of a Power Spectrum

A chamber was constructed in which to generate a power spectrum (defined as the ratio of the spectra generated from a chamber containing microcapsules and an identical chamber containing no capsules). The experimental setup is outlined in Figure 3, and consists of a glass water tank containing two glass sample chambers. A transducer, linked to a Panametric ultrasonic analyzer model 5052UA, emits ultrasound at all frequencies and receives the returning echo which is analyzed on a 3585A Hewlett Packard spectrum analyzer.

### 2B. Backscattering Studies: B-Mode Sector Image Generation

In order to demonstrate that the microcapsules did produce a strong

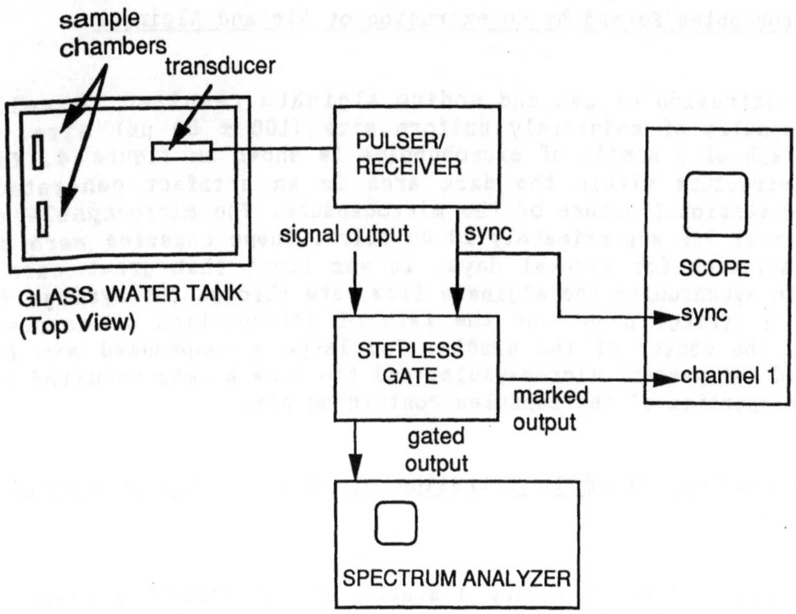


**Figure 3.** Apparatus for generation of a power spectrum.

echo under ultrasound stimulation, they were tested in a chamber that was setup to generate a B-mode sector image. This is a mode of display that is frequently used in medical imaging, and gives a two dimensional image of a slice through the scanned object. The capsules were suspended in buffer in a capped plastic tube, which was placed in a water bath below a 1/2 mHz focused transducer. The transducer acted as a transmitter and also received ultrasound radiation scattered back from the tube in the bath. The signal was analyzed on an ATL medical doppler flow analyzer. The resulting image was compared with a control consisting of an image from a tube that contained buffer alone.

## RESULTS AND DISCUSSION

### 1. Parameters that Impact on Capsule Size

Careful analysis of the microencapsulation system identified the critical factor controlling the microcapsule size to be the speed of air passing around the syringe needle, although the type of alginate that was used (high or low viscosity) and the concentration of the sodium alginate solution also had a small effect. Shear effects were noted, which produced a sub-population of very small solid alginate spheres that were ripped off the outside of the nascent capsules at high air speeds. By adjusting these parameters, we achieved solid alginate capsules as small as 30 µm in diameter. However, with use of the redesigned, triple-barrel jet head to introduce air into the system the size of the capsule prepared under identical conditions, increased to 100 ± 50 µm. When using the coaxial air stream to detach alginate drops from the needle tip, a maximum flow rate was reached, beyond which the vortex that formed in the chloride solution beneath the spray assembly became so large that capsule formation was impossible.

## 2. Microbubbles Formed by Co-extrusion of Air and Alginate

Co-extrusion of gas and sodium alginate resulted in well formed microcapsules of moderately uniform size (100 ± 50 μm) size. A photomicrograph of a sample of microbubbles is shown in Figure 4. The multilayer structure within the dark area is an artifact generated by the three dimensional nature of the microcapsule. The microcapsule wall (the light area) was approximately 10 μm thick. These capsules were stable in $CaCl_2$ solution for several days. It was found that great care must be taken to synchronize the alginate flow rate through the syringe (controlled by a syringe pump) and the rate of introduction of compressed air through the center of the needle. Too large a compressed air flow rate ruptured the nascent microcapsule, and too slow a rate resulted in a very small proportion of the capsules containing air.

## 3. Microbubbles Formed by Extrusion of Sodium Alginate Containing Entrapped Gas

The best method of making a stable, gassed sodium alginate solution was sonication of a 1.0 wt% alginate solution containing 0.01 wt% Tween 20. Sonication used a 1/2 inch probe, at an output power of 100W for 30 seconds. The foam generated by this procedure (mostly second generation bubbles from cavitation in the solution), was sprayed at once through a 26G syringe needle into a rapidly stirred $CaCl_2$ solution, as outlined in Figure 1. The resulting microencapsulated bubbles floated to the surface of the solution. Under microscopic examination, the bubbles were found to be individually encapsulated and 35 ± 5 μm in size, Figure 5. A great deal of alginate debris was generated by this method of capsule preparation and can be seen on the micrograph as translucent strands and empty capsules. These empty capsules may have either burst shortly after entering the chloride solution, or they may have been formed from alginate that contained no air bubble. The capsule walls are very thin, around 3-6 μm. As with the co-extruded capsules, the multilayer appearance of the dark center is an artifact generated by the 3-dimensional nature of the capsule. These bubbles were stable for around twelve hours, after which time the solid alginate strands settled to the bottom of the container, and many large gas bubbles (0.5 mm compared to 40 μm capsules) were visible trapped amongst alginate strands at the liquid surface. The bubbles were dislodged to the atmosphere when the tube was shaken. This behavior suggested that the gas bubbles were released by capsule rupture after they were sprayed into the $CaCl_2$ and coalesced to form larger bubbles which were then caught in the debris from the alginate capsule walls. If escape was by slow diffusion across the capsule membrane, it is unlikely that large gas bubbles would form in this manner at the liquid surface.

## 4. Power Spectra of Alginate-microencapsulated Air

### 4A. Microbubbles formed by co-extrusion

The power spectrum for microbubbles formed by co-extrusion of air and alginate is shown in Figure 6, and indicates that the microbubbles are acting as a harmonic oscillator. A strong resonant frequency was registered at 226.8 KHz. Treating the microbubbles as a harmonic oscillator with the properties of free air bubbles, this resonance frequency would correspond to a bubble diameter of 14 μm. The measured diameter was 100-150 μm, an order of magnitude larger. This is consistent with damping of

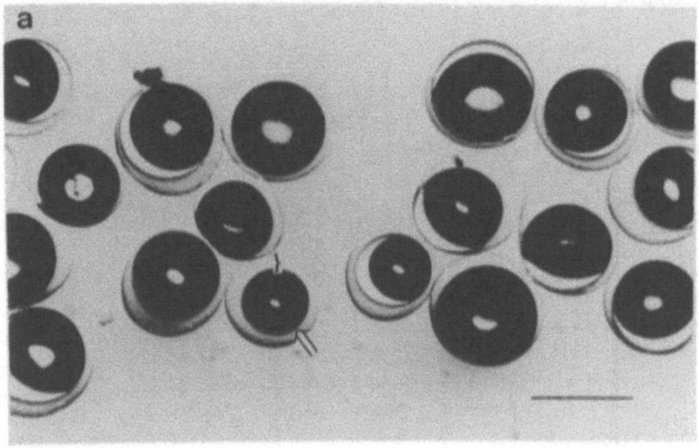

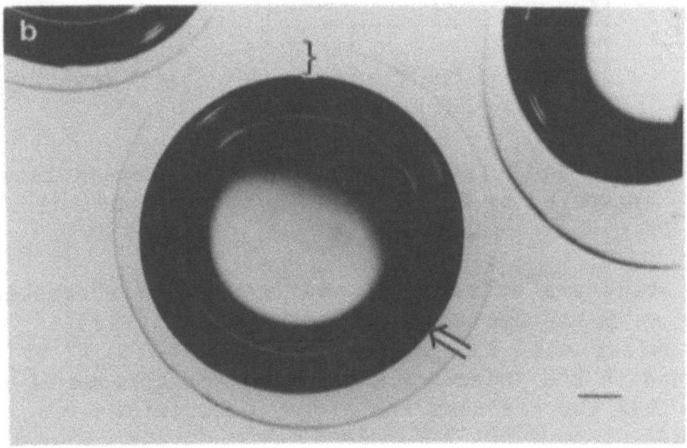

Figure 4. Photomicrographs of alginate microbubbles prepared by the co-extrusion method. Magnification (a) x 190 (bar = 100 µm), (b) x 700 (bar = 10 µm). The } = capsule wall, ← Entrapped gas.

the resonance by the capsule wall and is qualitatively similar to results obtained in the literature for vinylidene dichloride-acetonitrile copolymer microcapsules with a diameter of 30-40 µm.[34,35]

## 4B. Microbubbles Formed by Extrusion of Alginate Containing Entrapped Gas

Microbubbles prepared by spraying sonicated sodium alginate also displayed resonance behavior (Figure 7), however the spectrum indicated that the sample had a more heterogeneous size distribution than that found with the microbubbles formed by coextrusion. The major resonance peek was at 820 KHz, which would correspond to free air bubbles of 3.9 µm diameter. The measured diameter was 30-40 µm, again an order of magnitude greater in size than that predicted by the resonant frequency. As with the larger capsules, this indicates that the capsules are not acting as a free resonating gas bubbles, but are exhibiting damping of the resonance, most probably due to the presence of the capsule wall. Since resonance normally increases the effective backscatter cross section by a factor of

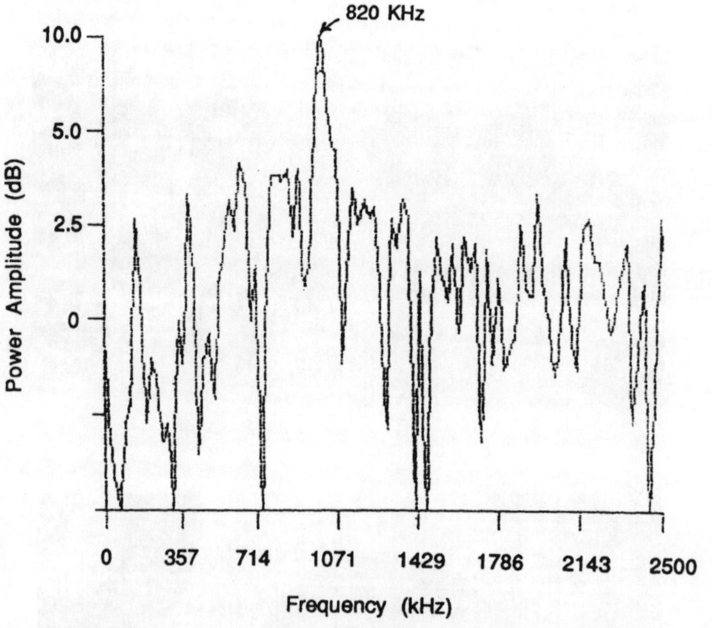


Figure 7. Power spectrum for alginate-microencapsulated air (prepared by the sonication/spraying method).

specks in the tank are echoes returned from bubbles created by disturbances in the water of the tank. When the tube is filled with capsules suspended in buffer (8b, 150 µm capsules made by coextrusion) a strong echo is returned to the transducer, giving a clear image of the tube, and demonstrating that the capsules are highly echogenic.

## CONCLUSIONS

We are extremely encouraged by our preliminary study of polymer-coated air bubbles for use in diagnostic ultrasound. We have developed two methods of preparing alginate-coated microbubbles, both of which produce echogenic species. The resonant peaks obtained from these microbubbles indicates some damping of oscillation compared to free air bubbles. This is most probably due to the relative rigidity of the capsule wall. B-mode sector images indicate that the capsules made by both methods are strongly echogenic. Current and future research will address two issues: the size of the microcapsule must be reduced to below 10 µm, and the stability must be increased.

## ACKNOWLEDGMENTS

We gratefully acknowledge the expertise of Drs. Barry Goldberg and Peter Burns, at the division of diagnostic ultrasound, Thomas Jefferson University Hospital, PA, without whom this project would not have started. This work was supported in part by a grant from the Commonwealth of Pennsylvania's Ben Franklin Partnership through the Ben Franklin Technical center of South East Pennsylvania, an affiliate of the University City Science Center.

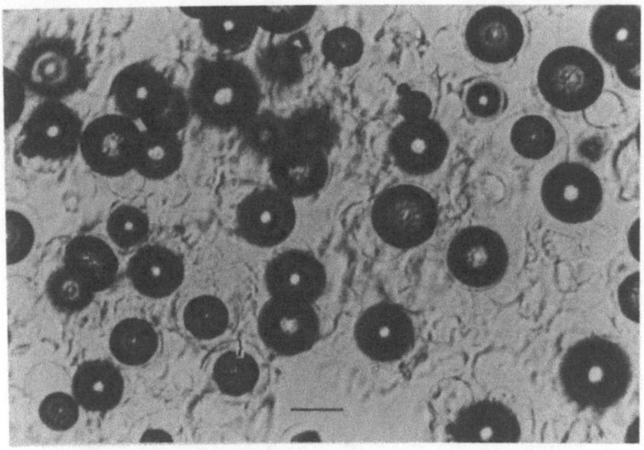

Figure 5. Light micrograph of calcium alginate microbubbles formed by the sonication/spraying technique. } = microcapsule wall. Magnification x 325, bar = 30 µm.

a thousand compared to the physical size, this small damping should not detract too severely from the advantages of a microencapsulated gaseous contrast agent over those of either a solid sphere or free bubbles.

## 5. B-mode Sector Images

B-mode sector images taken from a medical doppler flow analyzer are shown in Figure 8. When the tube beneath the transducer is filled with buffer alone (8a), reflections of the tube/water interface and from the bottom of the tank are seen, but no echo is returned from the inside of the tube. The position of the tube is outlined for clarity. The bright

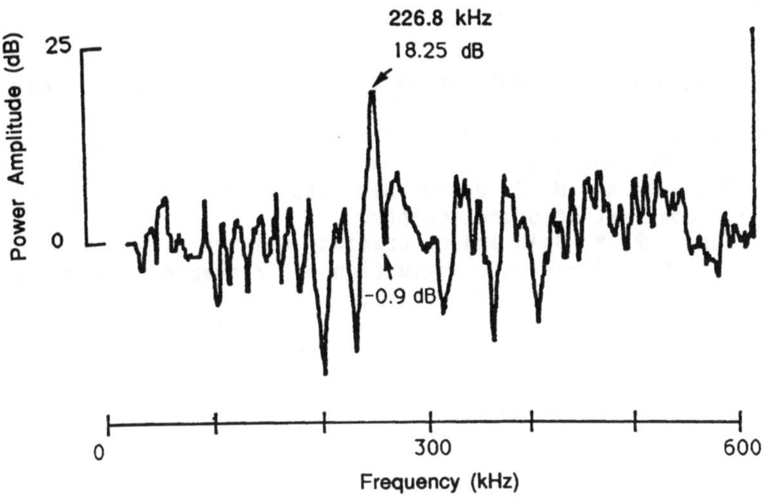


Figure 6. Power spectrum for alginate-microencapsulated air (prepared by the multiple-barrel method).

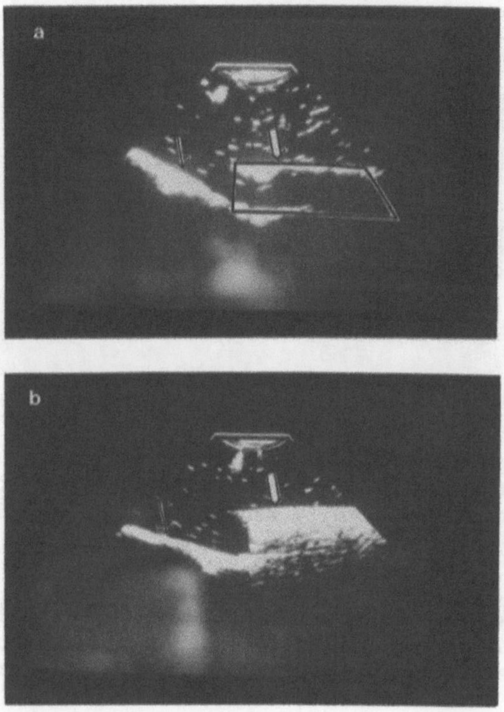


Figure 8. B-mode sector images of (a) tube filled with buffer, (b) tube filled with 150 μm capsules prepared by the coextrusion technique and suspended in buffer ⇓ tube/water interface, ↓ reflection from base of tank, position of the tube ▬.


## REFERENCES

1. R. Gramiak and P. M. Shah, Invest. Radiol., 3, 356-358 (1968). "Echocardiography of the aorta root."
2. N. C. Nanda, S. Stewart and R. Gramiak, Invest. Radiol., 3, 356-366 (1968). "Echocardiography of the aorta root."
3. R. L. Rothbald and N. C. Nanda, Seminars in Ultrasound, 2, 1130-1135 (1981). "Contrast Echocardiography."
4. N. C. Nanda, P. M. Shah and R. Gramiak, Clinical Res., 24, 233A, (1976). "Echocardiographic evaluation of tricuspid valve incompetence by contrast injections. (Abstract)."
5. F. W. Kremkau and E. Carstensen, Am. J. Roentgenol., 3, 159 (1968). "Ultrasonic detection of cavitation at the catheter tips."
6. H. Feigenbaum, J. Stone, D. Lee, W. Nasser and S. Chang, Circulation, 41, 615-621 (1970). "Identification of ultrasound echoes from the left ventricle by use of intracardiac injections of indocyanine green."

7. M. C. Ziskin, A. Bonakdapour, D. P. Weinstein and P. R. Lynch, Invest. Radiol., 7, 500-505 (1972). "Contrast agents for diagnostic ultrasound."
8. S. B. Feinstein, F. J. T. Cate, Z. Werner, K. Ong, G. Maurer, C. Tei, P. M. Shah, S. Meerbaum and E. Corday, J. Am. Coll. Cardiol., 3, 14-20 (1984). "Two dimensional contrast echocardiography, I, *In vitro* development and quantitative analysis of echo contrast agents."
9. J. Ophir, A. Gobuty, R. E. McWhirt and M. F. Makland, Ultrasonic Imaging, 2 67-77 (1980). "Ultrasonic backscatter from contrast-producing collagen microspheres."
10. J. Ophir, N. F. Makland, A. Gobuty and R. E. McWhirt, Proc. Ultrasonic Imaging Conference, Cannes, France (1980). "Progress in the development of sonographic contrast agents. (Abstract)"
11. B. A. Carroll, R. J. Turner, E. G. Tickner, D. B. Boyle and S. W. Young, Invest. Radiol., 15(3), 260-266 (1980). "Gelatin encapsulated microbubbles as ultrasonic contrast agents."
12. R. S. Meltzer, V. King and L. E. Teichholz, J. Am. Coll. Cardiol., 5, 978-982 (1985). "Generating precision microbubbles for use as an echocardiographic contrast agent."
13. R. F. Mattrey, F. W. Scheible, B. B. Gosink, G. R. Leopold, D. M. Long and C. B. Higgins, Radiology, 145(30), 759-762 (1982). "Perfluorocetylbromide: Liver/spleen-specific and tumor-imaging ultrasound contrast material."
14. R. F. Mattrey, G. Strich, R. E. Shelton, et al., Radiol., 163, 339-343 (1987). "Perfluorochemicals as ultrasound contrast agents for tumor imaging and hepatosplenography; preliminary clinical results."
15. M. R. Violante, K. Mare and H. W. Fisher, Invest. Radiol., 40, 16 (1981). "Biodistribution of a particulate hepatolienographic CT contrast agent; a study of iodipamide ethyl ester in the rat."
16. M. D. Smith, O. L. Kwan, H. J. Reiser and A. N. DeMaria, J. Am. Coll. Cardiol., 3(4), 992-998 (1984). "Superior intensity and reproducibility of SHU-454, a new right heart contrast agent."
17. M. D. Smith, J. L. Elion and R. R. McLure, J. Am. Coll. Cardiol., 13, 1622-1623 (1989). "Left heart opacification with peripheral venous injection of a new saccharide echo contrast agent in dogs."
18. H. Becher and R. Schlief, Am. J. Cardiol., 64, 374-377 (1989). "Improved sensitivity of color Doppler by SHU-454."
19. R. Schlief, T. Staks, M. Mahler, et al., Echocardiography, 7, 1-4 (1990). "Successful opacification of the left heart chambers on echocardiographic examination after intravenous injection of a new saccharide based contrast agent."
20. S. B. Feinstein, J. Am. Coll. Cardiol., 8, 251-253 (1986). "Myocardial perfusion imaging: contrast echocardiography today and tomorrow."
21. M. W. Keller, S. B. Feinstein, D. D. Watson, et al., Am. Heart J., 114, 570-575 (1987). "Successful left ventricular opacification following periferal venous injection of a sonicated contrast agent: an experimental evaluation."
22. S. B. Feinstein, P. A. Heidenreich, C. D. Dick, et al., Circulation, 78 (Suppl II), 11-565 (1988). "'Albunex': a new intravascular ultrasound contrast agent: preliminary safety and efficacy results. (Abstract)"
23. S. B. Feinstein, R. M. Lang, C. Dick, et al., J. Am. Coll. Cardiol., 11, 59-65 (1988). "Contrast echocardiography during coronary artiography in humans, Perfusion and anatomic studies."
24. P. M. Morse and K. U. Ingartd, "*Theoretical Acoustics*," McGraw Hill, New York, 1968, p. 427.
25. J. Lubbers and J. W. VanDenBerg, Ultras. in Med. and Biol., 2, 301 (1976). "An ultrasonic detector for micro gas emboli in a blood flow line."

26. K. Kinsler and A. R. Frey, "*Fundamentals of Acoustics*," 2nd Ed., Wiley and Sons, New York, 1962, Chapter 10.
27. M. Kierstan and C. Bucke, Biotech. Bioeng., 9, 387-397 (1977). "Immobilization of microbial cells subcellular organelles and enzymes in calcium alginate gels."
28. P. Scheurich, H. Schnabl, U. Zimmermann and J. Klein, BBA, 598, 645-651 (1980). "Immobilization and mechanical support of individual protoplasts."
29. G. Pilwat, P. Washausen, J. Klein and U. Zimmermann, Z. Naturforsch., 35C, 352-356 (1980). "Immobilization of human red blood cells."
30. F. Lim and A. M. Sun, Science, 210, 908-910 (1980). "Microencapsulated Islets as Bioartificial Endocrine Pancreas."
31. G. M. O'Shea, M. F. A. Goosen and A. M. Sun, BBA, 804, 133-136 (1984). "Prolonged survival of transplanted islets of langerhans encapsulated in a biocompatible membrane."
32. M. A. Wheatley, H. Eisen and R. Langer, Proc. 12th Int. Symp. Control. Rel. Bioactive Mat., N. A. Peppas and R. J. Haluska, Eds., Contr. Rel. Soc. Inc., 1985, p. 245. "Alginate/Poly-L-Lysine Microcapsules as a Vehicle for Controlled Release of Macromolecules."
33. M. F. Goosen, G. M. O'Shea, G. M. Gharapetian, H. Chou and A. M. Sun, Biotechnol. Bioeng., 27, 146 (1985). "Optimization of microcapsule parameters: semipermeable microcapsules as bioartificial pancreas."
34. K. Ishihara, A. Kitabake, J. Tanouchi, K. Fujii, M. Uematsu, Y. Yoshida, T. Kamada, T. Tamura, K. Chichara and K. Shirae, J. Appl. Phys., 27, Supl. 27-1, 125-127 (1987). "New Approach to Noninvasive Manometry Based on Pressure dependent Resonant Shift of Elastic Microcapsules in Ultrasonic Frequency Characteristics."
35. T. Tamura, K. Chihara, K. Shirae, K. Ishihara, T. Nagakura, J. Tanouchi and A. Kitabake, J. Appl. Phys., 28, Supl. 28-1, p. 211-213 (1988). "Dynamic Pressure Measurement Using Two-Frequency Ultrasound."

# FACILE ACTIVATION OF TRISACRYL GELS WITH 2-FLUORO-1-METHYLPYRIDINIUM SALT (FMP): APPLICATIONS IN AFFINITY CHROMATOGRAPHY AND ENZYME IMMOBILIZATION

T. T. Ngo

Bioprobe International, Inc.
14272 Walnut Avenue
Tustin, CA 92680 USA

2-Fluoro-1-methylpyridinium-toluene-4-sulfonate was used to rapidly activate, under mild conditions, hydroxyl groups of Trisacryl GF-2000. Trisacryl GF-2000 is a synthetic, non-ionic, hydrophilic synthetic gel derived from copolymerization of N-acryloyl-2-amino-2-hydroxymethyl-1,3-propanediol and a bifunctional crosslinker. The activated groups have long half-life in aqueous medium. The activated gel can be used to couple a variety of amino or thiol carrying ligands. A thiol group containing enzyme, B-galactosidase was facilely coupled to the activated gel with good retention of activity. Antigens such as human IgG can also be directly coupled to the activated gel for use in immunoaffinity chromatography. Amino, carboxy, thiol and activated thiol gels can be prepared from FMP-activated gel. Low molecular weight ligands, such as gentamicin, were coupled to carboxy-terminal gels and they were used for affinity supports for purifying anti-gentamicin antibodies.

## INTRODUCTION

In attempts to obtain a chromatographic gel with its mechanical, structural and chemical properties superior to those made of natural polymers, such as dextran, agarose or cellulose, a number of synthetic polymeric materials have been prepared. Examples of synthetic polymeric materials include Eupergit C, which is a beaded polymeric gel prepared from a copolymerization of methacrylamide, allylglycidyl-ether and N-methylene-bis-methacrylamide;[1-4] Fractogel TSK, a semi-rigid gel which is made from a copolymer of oligo-ethylene glycol, glycidyl methacrylate and pentaerythrol-dimethacrylate;[5-7] HEMA gel is a macroporous hydrophilic copolymer of 2-hydroxylethyl methacrylate and ethylenedimethcrylate[8] and Trisacryl GF-2000 is a non-ionic gel prepared by copolymerization of N-acryloyl-2-amino-2-hydroxymethyl-1,3-propane-diol and a bifunctional crosslinker.[9] As a chromatographic gel, Trisacryl GF-2000, appears to have a number of desirable characteristics, such as the high degree of hydrophilicity, high porosity, the chemical stability toward organic solvents and denaturing and chaotropic reagents, chemically defined structure and good pressure-flow rate profile. However, Miron and

*Cosmetic and Pharmaceutical Applications of Polymers*
Edited by C.G. Gebelein *et al.*, Plenum Press, New York, 1991

Wilchek[10] pointed out that the hydroxyl groups in Trisacryl GF-2000 are not as reactive as those of polysaccharides, due perhaps to the higher pK of the hydroxyl groups. It is also possible that the trihydroxyl groups of the gel are not too accessible due to steric crowding.

In this communication, I shall describe the activation of the hydroxyl groups of Trisacryl GF-2000 by 2-fluoro-1-methylpyridinium-toluene-4-sulfonate (FMP), a versatile reagent which has been successfully used to activate Sepharose and Fractogel.[11-13] The properties of FMP-activated gel will also be compared to the same gels activated by other methods.[10,14]

## MATERIALS AND METHODS

### 1. Materials

Trisacryl GF-2000 was purchased from LKB Instruments, Inc. High purity 2-fluoro-1-methylpyridinium-toluene-4-sulfonate (FMP) was obtained from BioProbe International, Inc. $I^{125}$-labeled bovine serum albumin was purchased from ICN Biomedicals, Inc. Sodium dodecyl sulfate, tris (hydroxymethyl) aminomethane (Tris), porcine pepsin, o-nitrophenyl-B-D-galactopyranoside (ONPG), gentamicin, 1-ethyl-3-(3-dimethylaminopropyl)-carbodiimide·HCl (EDC), 3-(2-pyridyldithiol)-propionic acid N-hydroxysuccinimide (SPDP), sodium EDTA, L-cysteine 5,5'-dithiobis-2-nitro-benzoic acid, dithiothreitol and bovine serum albumin were purchased from Sigma Chemical Co. Triethylamine 2,2'-dithiopyridine, purchased from Fluka Chemical Corporation. High purity solvents, "Baker Resi-Analyzed" grade acetone, "Photrex" grade N, N-dimethylformamide (DMF) and ethyl alcohol were purchased from J. T. Baker Chemical Company. Human IgG was purchased from Miles Laboratories, Inc. Affinity purified goat anti-human IgG was purchased from Tago, Inc. N,N'-Bis-(3-aminopropyl)-ethylenediamine, and succinic anhydride and 6-aminocaproic acid, cystamine dihydrochloride were purchased from Aldrich Chemical Company. *E. coli* B-galactosidase was purchased from Boehringer Mannheim Biochemicals.

### 2. Activation of Trisacryl GF-2000 with 2-fluoro-1-methylpyridium toluene-4-sulfonate (FMP)

Prior to the reaction with FMP, Trisacryl GF-2000 gel (1200 mL.) was washed successively with: 5 x 1600 mL distilled water; 3 x 1600 mL ethyl alcohol; and 1600 mL dry acetone. After removing excess acetone by vacuum filtration, the gel was suspended in 1600 mL dry acetone and stirred for 10 minutes. The gel was washed one more time with 1600 mL dry acetone and excess solvent was removed by vacuum filtration. The gel was then placed in a vacuum drying oven at 80°C for 14 hours. The dry gel was suspended in 500 mL, dry acetone containing 2.6 mL dry triethylamine, mixed with 50 g FMP and 75 mL TEA in 750 mL dry N,N-dimethylformamide, and stirred at room temperature for 0.5-3 hours. This reaction procedure is outlined in Figure 1.

After filtrating off the unreacted FMP, the gel was washed with 1 L dry DMF and suspended in 1 L dry DMF with stirring for 10 minutes. The FMP-activated gel was then washed successively with: 2 x 1000 mL dry acetone; 1000 mL acetone. The washed FMP-activated gel was dried at room temperature under vacuum for 14 hours. The activated gel is stored in a

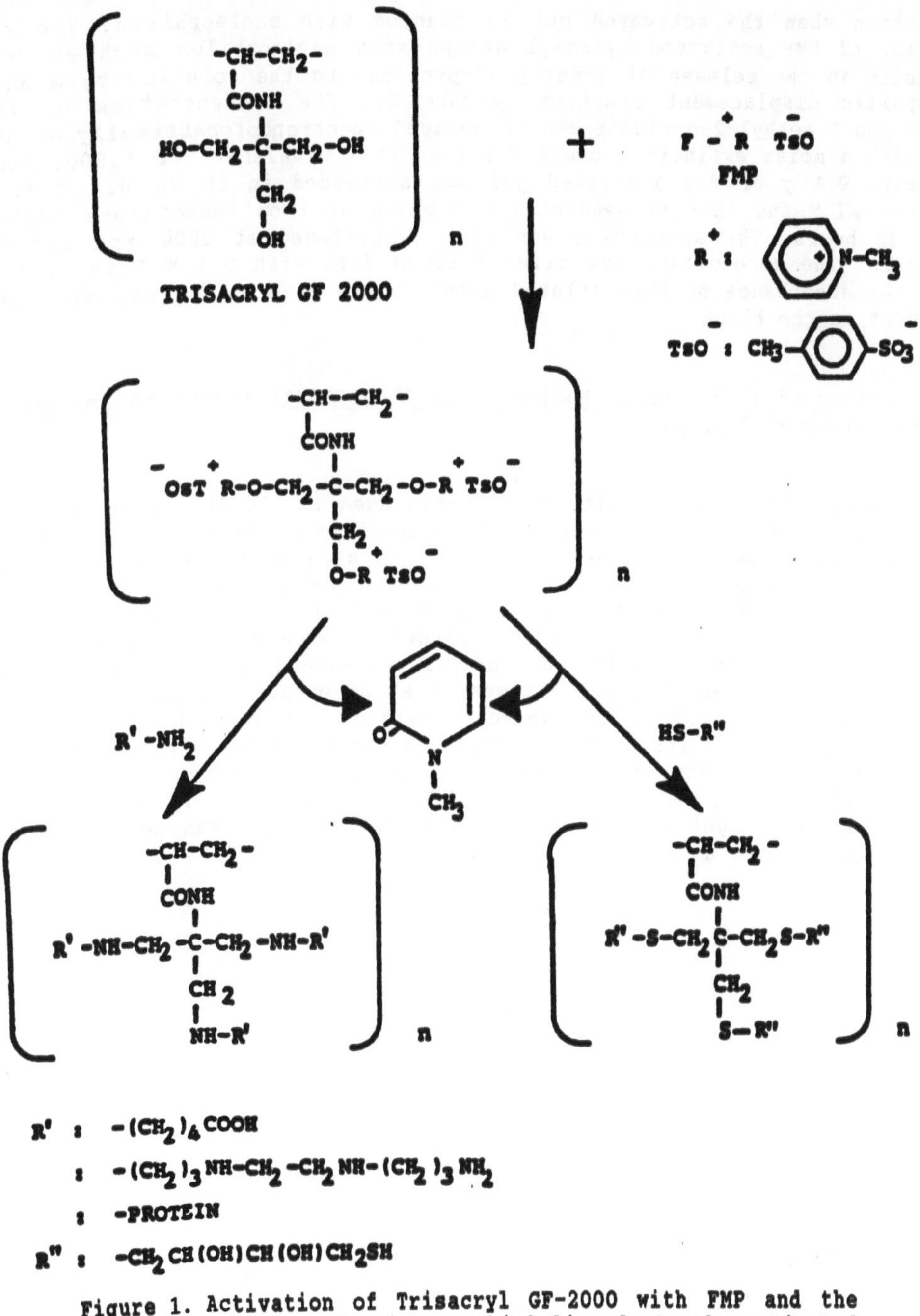


Figure 1. Activation of Trisacryl GF-2000 with FMP and the coupling of amino or thiol ligands to the activated gel.

desiccator and kept at 4°C. Under this condition the gel is stable for months.

## 3. Determination of the Density of FMP Activated Hydroxyl Groups of Trisacryl GF-2000

The number of hydroxyl groups of trisacryl GF-2000 activated by FMP can be quantitatively determined by the amount of FMP released into the solution when the activated gel is reacted with nucleophiles. The reaction of FMP activated hydroxyl groups with nucleophiles such as Tris results in the release of 1-methyl-2-pyridone to the solution by a nucleophilic displacement reaction (Figure 2). The concentration of the released 1-methyl-2-pyridone can be assayed spectrophotometrically at 297 nm with a molar extinction coefficient at this wavelength of 5,900. Routinely, 0.1 g of dry activated gel was suspended in 10 mL 0.1 M Tris buffer pH 9 and this suspension was tumbled at room temperature (22°C) for 20 hours. The suspension was then centrifuged at 2000 rpm for 10 minutes. The supernatant was diluted 10-50 fold with 0.1 M Tris pH 9.0 and the absorbance of this diluted solution was read at 297 nm using the diluent as the blank.

## 4. Coupling of $I^{125}$-Labeled Bovine Serum Albumin ($I^{125}$-BSA) to FMP-Activated Trisacryl GF-2000

The specific radioactivity of $I^{125}$-BSA used in the following experiments was 20,600 cpm/mg protein. Two experimental protocols for bind $I^{125}$-BSA were used. Procedure (a): FMP activated Trisacryl GF-2000 was washed with distilled water, 1 g of wet activated gel was added to 2 mL $I^{125}$-BSA in 0.1 M $NaHCO_3$ (5 mg/mL). The suspension was tumbled overnight. The next day the suspension was centrifuged at 2000 rpm for 15 minutes. The supernatant was discarded and the gel was washed with 4 mL distilled water, 3 times, 4 mL 1M NaCl, 3 times, 4 mL 8M urea, 3 times, 4 mL 1.5N KCN, 3 times, 4 mL 10% sodium dodecyl sulfate, 3 times. The washed gel was then counted for its radioactivity in a gamma counter. Procedure (b): One hundred milligrams of dry FMP-activated Trisacryl GF-2000 was added to 1 mL $I^{125}$-BSA in 0.05 M $NaHCO_3$, pH 8.5 (40 mg BSA/mL). The suspension was tumbled overnight at room temperature. Then the gel was centrifuged, washed and counted as described above.

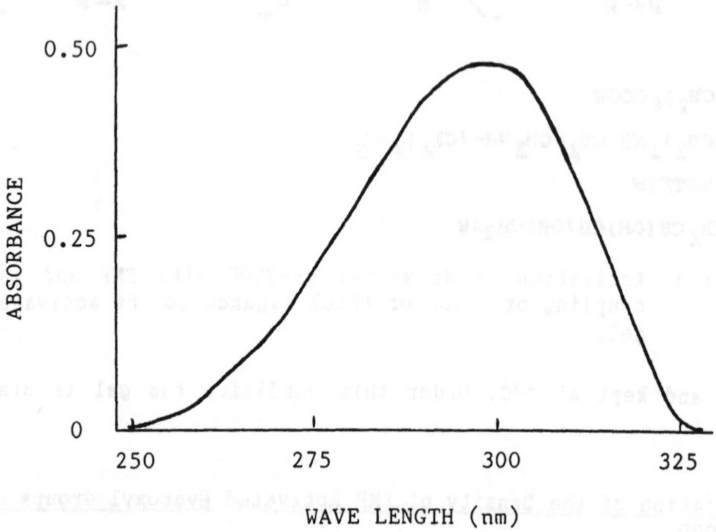


**Figure 2. Absorption spectrum of 1-methyl-2-pyridone.**

## 5. Coupling of *E. coli* B-galactosidase to FMP-Activated Trisacryl GF-2000

*E. coli* B-galactosidase, free of nucleophilic substances such as Tris. sulfhydryl agents, was dissolved in 0.1 M sodium phosphate, pH 8.5 at concentration ranges of 0.2-1 mg/mL. To one mL of enzyme solution was added 5-50 mg of dry FMP activated Trisacryl GF-2000. The suspension was tumbled at room temperature for 45-120 minutes.

The suspension was washed in a funnel with 5 x 10 mL 0.1 M sodium phosphate, pH 8.5, 5 x 10 mL 0.1 M sodium phosphate, pH 8.0, containing 0.5 M NaCl and 5 x 10 mL 0.1 M sodium phosphate pH 8.

## 6. Coupling of Pepsin to FMP-Activated Trisacryl GF-2000

Pepsin powder (1.5 g) was dissolved in 100 mL 0.02 M sodium phosphate at pH 5.5. To this pepsin solution was added 15 g of dry FMP-activated Trisacryl GF-2000. After adjusting the pH to 5.5, the suspension was tumbled at room temperature for 30 minutes and then at 4°C for 16 hours. The gel was washed with 200 mL 0.5 M NaCl in 1 mM HCl, 5 times, and then suspended in 500 mL 0.5 M NaCl in 1 mM HCl for 15 minutes. After washing with 200 mL 1 mM HCl containing 1 mM NaCl, 0.05% Na $N_3$ and 50% glycerol. The immobilized pepsin was stored in the same solution at 4°C.

## 7. Coupling of Human IgG to FMP-Activated Trisacryl GF-2000

Purified human IgG (100 mg) was dissolved in 25 mL 0.05 M sodium bicarbonate buffer, pH 8.7. The dry powder of FMP-activated Trisacryl GF-2000 (5 g) was added to the solution of human IgG. The suspension was stirred for 15 minutes and then tumbled for 24 hours at room temperature. The gel was then washed with 500 mL bicarbonate buffer and resuspended in 50 mL 0.1 M ethanolamine in 0.1 M Tris buffer at pH 8.5 and tumbled at room temperature for 12 hours. The immobilized human IgG gel was extensively washed in a sequential way with 125 mL of 0.01 M sodium acetate pH 4, 0.01 M sodium acetate pH 4 containing 0.5 M NaCl, 0.01 M Tris pH 9, 0.5 M NaCl, 0.01 M Tris pH 9 containing 0.5 M NaCl and finally with phosphate buffered saline. The immobilized human IgG gel was stored in the last buffer solution at 4°C.

## 8. Coupling of N, N-bis (3-aminopropyl) ethylene diamine to FMP-Activated Trisacryl Gel: Preparation Amino-Terminal Gel

FMP-activated Trisacryl GF-2000 (130 g) was suspended in dry DMF. To this suspension was added 17 mL N, N-bis (3-amino propyl) ethylene diamine. The suspension was vigorously agitated at room temperature. After 24 hours, the gel was washed successively with 1 L distilled water, five times, and with 1 L 0.5 M NaCl solution twice. The gel was then suspended in 2 L 1N NaCl and agitated for 30 minutes. Then the gel was washed again with 1 L 0.5M NaCl twice, with 1 L distilled water 3 times and with 1 L ethanol twice. The washed amino-terminal gel was stored at 4°C in 0.1 M sodium phosphate, pH 7.4, containing 0.15 M NaCl and 0.1% sodium azide.

9. Coupling of 6-Aminocaproic acid to FMP-Activated Trisacryl Gel: Preparation of "Short Arm" Carboxy-Terminal Gel

FMP-activated Trisacryl GF-2000 (15 g) was suspended 100 mL 0.1 M $NaHCO_3$ containing 10 mmoles aminocaproic acid and 1 mL triethylamine and the suspension was agitated at room temperature for 25 hours. The reacted gel was washed with 300 mL 0.5 N NaCl in 0.1 M phosphate buffer pH 7.5 three times, then with 300 mL distilled water twice. The washed gel was resuspended in 200 mL 0.2 M Tris buffer, pH 9, containing 0.1 M L-crysteine and was agitated at room temperature for 24 hours. Finally the gel was washed with 300 mL 1M NaCl three times, then with 200 mL distilled water twice. The gel was stored in distilled water at 4°C.

10. Coupling of Succinate to Amino-Terminal Gel: Preparation of "Long Arm" Carboxy-Terminal Gel

Amino-terminal gel (65 g) was suspended in 500 mL DMF containing 3 mmoles 4-dimethylaminopyridine and 50 mmoles triethylamine. To the gel suspension was added 20 mmole of succinic anhydride. After 15 minutes stirring, the suspension was tumbled at room temperature for 24 hours. The succinylated gel was washed successively with 1 L distilled water 5 times, with 1 L 0.5M NaCl twice and again with 1 L distilled water 5 times. The resulting carboxy-terminal gel was stored in 0.1 M phosphate buffer, at pH 7.5, containing 0.1% sodium azide.

11. Coupling of Gentaminicin to "Short Arm" Carboxy-Terminal Gel

"Short arm" carboxy-terminal gel (30 mL) was washed with 300 mL distilled water and then suspended in 80 mL distilled water which contained 500 mg gentamicin. After adjusting the pH of the suspension to 4.0-4.2, 500 mg of 1-ethyl-3-(3-dimethylaminopropyl)-carbodiimide·HCl was added to the suspension. The gel suspension was agitated at room temperature for 24 hours. The gentamicin-coupled gel was washed with 300 mL distilled water 3 times, with 200 mL 1M NaCl 3 times, with 300 mL PBS 3 times. The gentamicin-coupled gel was stored in PBS at 4°C.

12. Coupling of Gentamicin to "Long Arm" Carboxy-Terminal Gel

"Long arm" carboxy-terminal gel (50 mL) was washed with 500 mL distilled water and resuspended in 50 mL distilled water. After adjusting the pH of the suspension to 4.0-4.2, 1.9 g of 1-ethyl-3-(3-dimethylamino-propyl)-carbodiimide·HCl was added and it was stirred for 2 minutes before adding 5 g gentamicin. The pH of the suspension was maintained at 4.0-4.2. The gel was continuously agitated for the next 12 hours. The gentamicin-coupled gel was washed with 250 mL distilled water 5 times, with 250 mL 0.5 N NaCl 5 times, with 250 mL distilled water 5 times and with PBS 5 times. The gel was stored in PBS at 4°C.

13. Coupling of Dithiothreitol to FMP-Activated Trisacryl gel: Preparation of Thiol and Activated Thiol gels

FMP-activated Trisacryl GF-2000 (33 g) was suspended in 100 mL methanol which contained 2.6 g dithiothreitol and 3.3 mL triethylamine. The suspension was tumbled at room temperature for 24 hours. The gel was washed with 200 mL distilled water 5 times and was resuspended in 250 mL 0.1M phosphate buffer pH 7 containing 0.1M L-cysteine and 0.5M NaCl and stirred at room temperature for 30 minutes. The gel was then washed with 200 mL 0.1M phosphate, pH 7.0 containing 0.5M NaCl and with 200 mL 0.025M sodium bicarbonate which contained 60% acetone and 1 mM EDTA. The gel was stored in this solution at 4°C.

The activated thiol gel was prepared by reacting the thiol gel with 2,2'-dithiopyridine.[15] Thus, 200 mM 2,2'-dithiopyridine was dissolved in a solution consisted of 40% 0.025M NaCO containing 1 mM EDTA and 60% acetone. This solution was filtered through 5 µm nylon filter. Thiol gel (100 mL) was suspended in 100 mL of the above solution of 2,2'-dithiopyridine and was tumbled at room temperature for 2 hours. The activated thiol gel was washed with 200 mL acetone twice, with 200 mL mixed solution of the $NaHCO_3$-acetone 5 times and finally the gel was stored in 0.025M NaHCO containing 1 mM EDTA at 4°C.

## 14. Determination of Thiol Group Density of the Thiol Gel

Approximately 0.5 mL thiol gel was added to a 15 mL graduated conical centrifuge tube. To this tube was added 0.025M NaHCO to 10 mL mark. The tube was centrifuged at 2000 rpm for 10 minutes and the exact volume of the packed gel was measured. The gel was then vigorously agitated and 3 mL of the suspension was removed and added to a test tube containing 40 µl 5,5'-dithiobis (2-nitrobenzoic acid) (4 mg/mL). Again the gel suspension, which was turning yellow, was vigorously mixed for 1 minute. After 5 minutes the suspension was centrifuged at 2000 rpm for 5 minutes. The supernatant liquid was diluted with equal volume of 0.025M $NaHCO_3$ and its absorbance at 412 nM was read in a spectrophotometer. The blank used 0.025M $NaHCO_3$ in place of the gel. The molar extinction coefficient of 13,600 was used in the calculation.

## 15. Affinity Purification of Antibody (General Procedure)

Affinity gel (10 to 15 mL) was poured into an 1 cm diameter column. The column was washed sequentially with the 5 elution buffers and finally with 5 column volumes of phosphate buffered saline again. Eluent was continuously monitored and fraction collected for further use.

## 16. Determination of Amino Group Density of Amino-Terminal Gel

The amino group density of the amino-terminal gel was determined by a spectrophotometric method recently developed by Ngo.[16] The method involves reacting the solid support with an excess of activated acylating agent, SPDP in the presence of 4-dimethylaminopyridine as the catalyst. After the reaction, the gel was washed and treated with dithiothreitol to quantitatively release pyridine-2-thione quantified reading its absorbance at 343 nm in a spectrophotometer.

## 17. Determination of Carboxy Group Density of Carboxy-Terminal Gel

The carboxy group density of the carboxy-terminal gel was measured by a colorimetric procedure recently developed by Ngo.[17] The procedure called for coupling of cystamine at pH 4-4.5 to the carboxy groups of the gel using 1-ethyl-3-(3-dimethyl aminopropyl)-carbodiimide. The gel bound disulfides was subsequently reduced to sulfhydryl group density of the gel which is equivalent to the density of the carboxy groups, was quantified by using the Ellman's reagent 5,5'-dithiobis-(2-nitrobenzoic acid).[18]

## 18. Determination of the Activity of Immobilized Pepsin Gel

Immobilized pepsin was assayed using hemoglobin as the substrate according to a previously described method.[19]

## 19. Determination of the Activity of Immobilized B-Galactosidase Gel

Immobilized B-galactosidase was assayed using o-nitrophenylgalactosidase as the substrate according to a previous method.[20]

# RESULTS AND DISCUSSION

The activation of hydroxyl groups of Trisacryl GF-2000 by most activating reagents proves to be difficult. Miron and Wilchek showed that chloroformates, unlike other well-known activating agents were able to facilely activate the hydroxyl groups of Trisacryl GF-200 and the activated gel react readily with amino ligand and formed a stable carbamate linkage.[10] In this paper, I show that FMP too can readily activate the hydroxyl groups of Trisacryl GF-2000 in about 30 minutes reaction time, at an ambient temperature of about 22°C. The activated gel readily reacts with both amino or thiol carrying ligands. The linkages between the gel and the amino or thiol ligands are, respectively, the stable secondary amine or thiol-ether bonds. The reaction scheme for FMP activation of Trisacryl gel and the subsequent coupling of the activated gel to amino or thiol ligands is shown in Figure 1. The displacement of 1-methylpyridyl groups from the activated gel upon reaction with nucleophiles gives rise to 1-methyl-2-pyridone which has a clear UV absorption spectrum with a wavelength of maximal absorption at 295 nm and a molar extinction coefficient of 5900 (Figure 2). The release of 1-methyl-2-pyridone provides a convenient means for determining the degree of FMP activation by simply treating the gel with a basic solution and measuring the absorbance of the solution at 295 nm. Approximately 50 µmoles of 1-methyl-2-pyridyl groups were commonly found in one gram FMP-activated Trisacryl GF-2000. Using $I^{125}$-labeled bovine serum albumin, FMP-activated Trisacryl GF-2000 were able to bind greater than 2 mg of the protein per mL of the gel. The chloroformate activated Trisacryl gel bound about one mg protein per wet gram of the gel.[10]

*E. coli* B-galactosidase was used as an example of ligand carrying both amino and thiol reactive groups for coupling to the FMP-activated Trisacryl gel. Figure 3 shows the extent of enzyme binding to the activated gel. Under this experimental condition, FMP-activated gel was able to bind 7 mg of enzyme per mL of the gel. The time course of binding

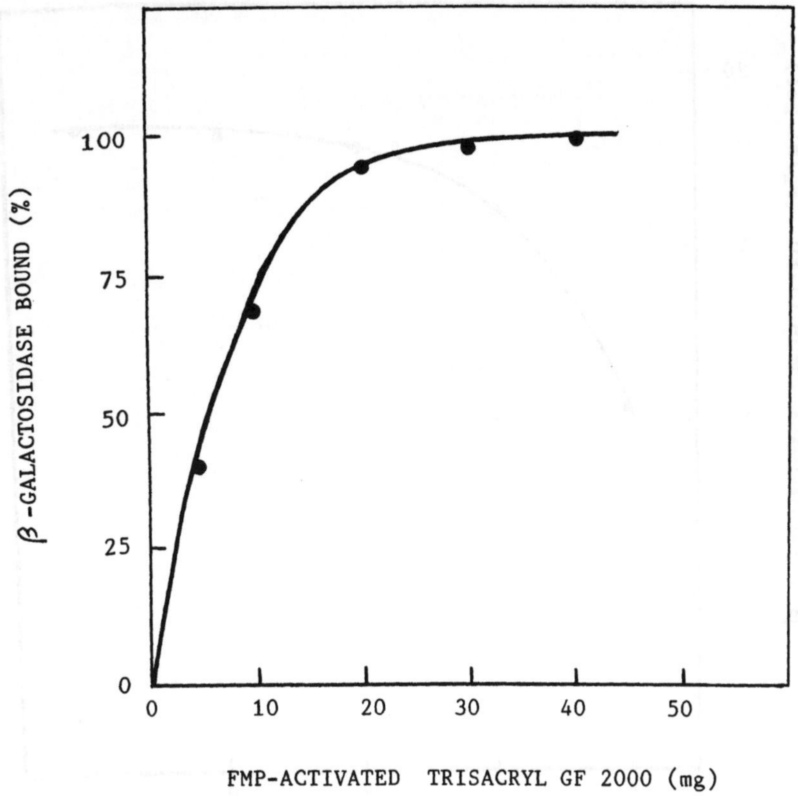


**Figure 3. Binding of B-galactosidase to FMP-activated Trisacryl GF-2000.**

showed that the reaction proceeds rapidly with most of the reaction occurring in the first 30 minutes (Figure 4). The effect of coupling pH on the rate of coupling was studied from pH 6 to 10 (Figure 5). The amount of enzyme coupled increases with increasing pH and reaches maximum after pH 8.

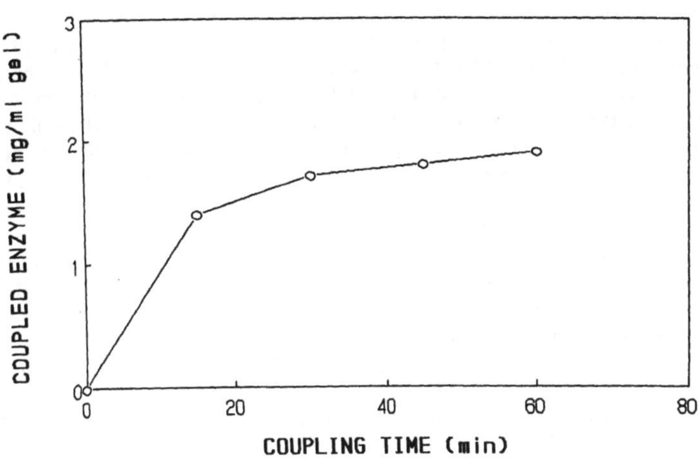


**Figure 4. Time course of B-galactosidase binding to FMP-activated Trisacryl GF-2000.**

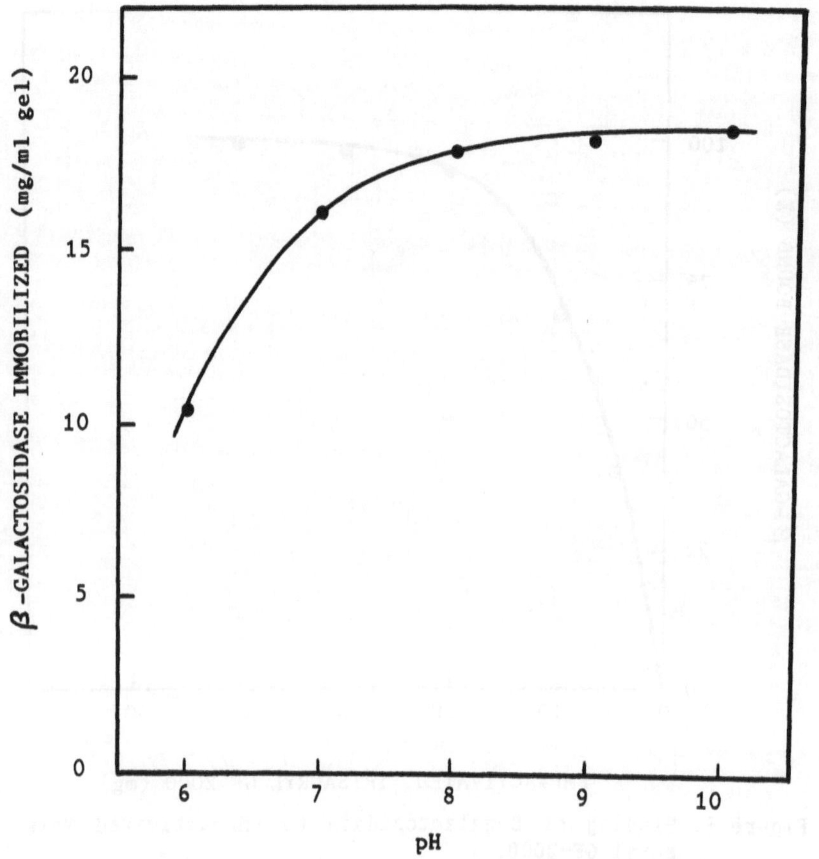


Figure 5. Effects of coupling pH on the amount of B-galactosidase coupled to FMP-activated Trisacryl GF-2000.

The half-life of the activated group in aqueous solution was relatively long. For example, at pH 6 the half-life was greater than 290 hours and at pH 10 it was greater than 10 hours (Figure 6).

The immobilization of pepsin to solid phase has been of considerable interest because of the utility of immobilized pepsin in the production of F(ab')2 fragments from intact IgG and the challenge of immobilizing a protein that is stable only in acidic pH. A systematic study on the immobilization of pepsin on FMP-activated Trisacryl GF-2000 showed that the efficiency of coupling using 3 mg pepsin and varying amount of activated gels (5 to 50 mg dry gel) ranges from 20 with 5 mg gel to 99% with 50 mg gel. Immobilized pepsin gel prepared by using FMP-activated Trisacryl GF-2000 as described here exhibited a very high specific activity of 2358 units per mL gel. This is in contrast to pepsin immobilized on the same gel, Trisacryl GF-2000, pre-activated with carbonyl diimidazole which gave a specific activity of only 2.2 units per mL gel. Thus immobilized pepsin prepared from FMP-activated gel has a specific activity 1072 fold higher than that prepared by using carbonyl diimidazole activated gel.[14]

Coupling of human IgG to FMP-activated gel resulted in an affinity gel having 2.9 mg human IgG covalently bound to the gel. This IgG gel was shown to be a good affinity sorbent for purifying anti human IgG (Figure

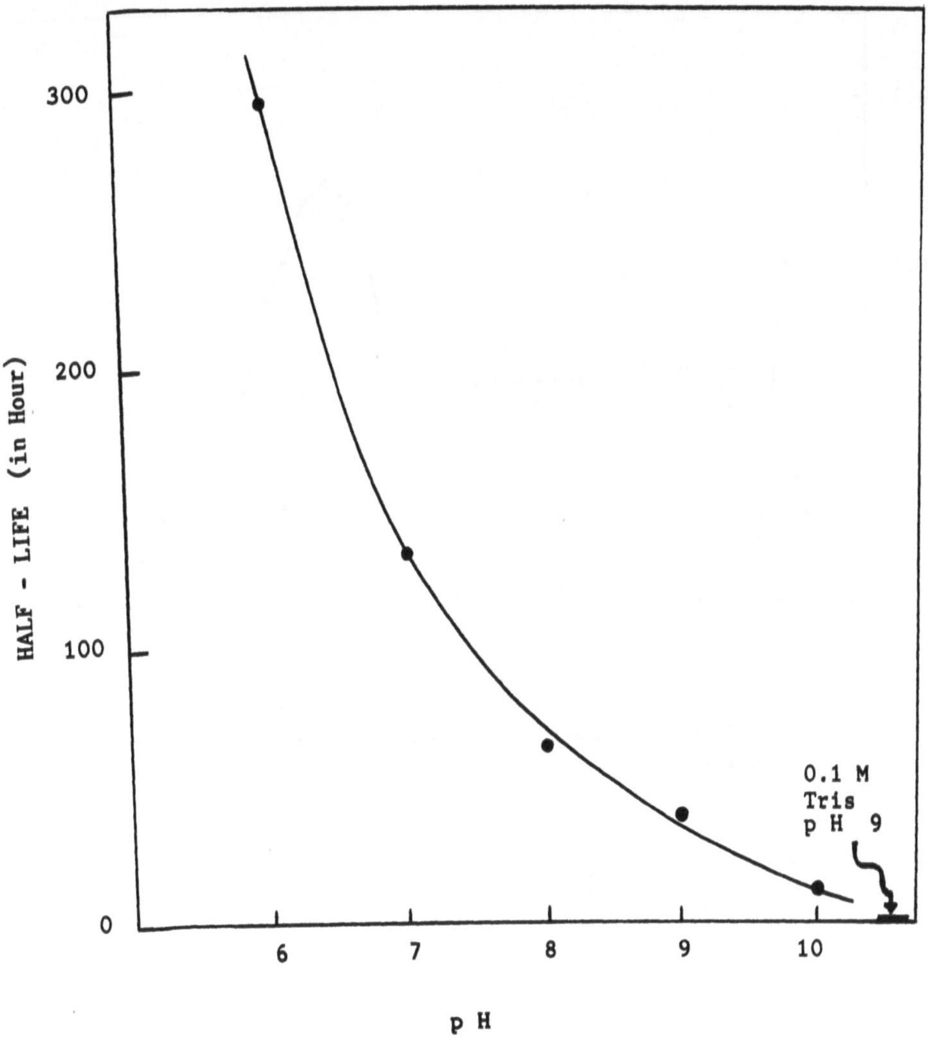


Figure 6. Half-life of the activated group of FMP-activated Trisacryl GF-2000 at various pH.

7). Using an affinity-purified IgG to test the efficiency of the affinity gel, the recovery of purified IgG was 94%.

Affinity purification of rabbit anti-gentamicin antibodies was initially carried out by using gentamicin covalently attached to a short carboxy-terminal gel. This carboxyl-terminal gel was prepared by reacting the FMP-activated gel with aminocaproic acid. Gentamicin was attached to the carboxylic groups via a reaction promoted by 1-ethyl-3,3-(dimethylaminopropyl)-carbodiimide. The efficiency of this affinity gel for purifying anti-gentamicin antibodies was rather low; only about 0.24 mg of antibody can be isolated per mL affinity gel (Figure 8). On the other hand, when gentamicin was coupled to "long arm" carboxy-terminal gel, using much less antiserum (one seventh) and a much smaller column (12 mL rather than 30 mL gel), a higher efficiency was obtained (0.37 mg antibody was isolated per mL affinity gel) (see Figure 9). The "long arm" carboxy-terminal gel was prepared by first reacting the FMP-activated gel

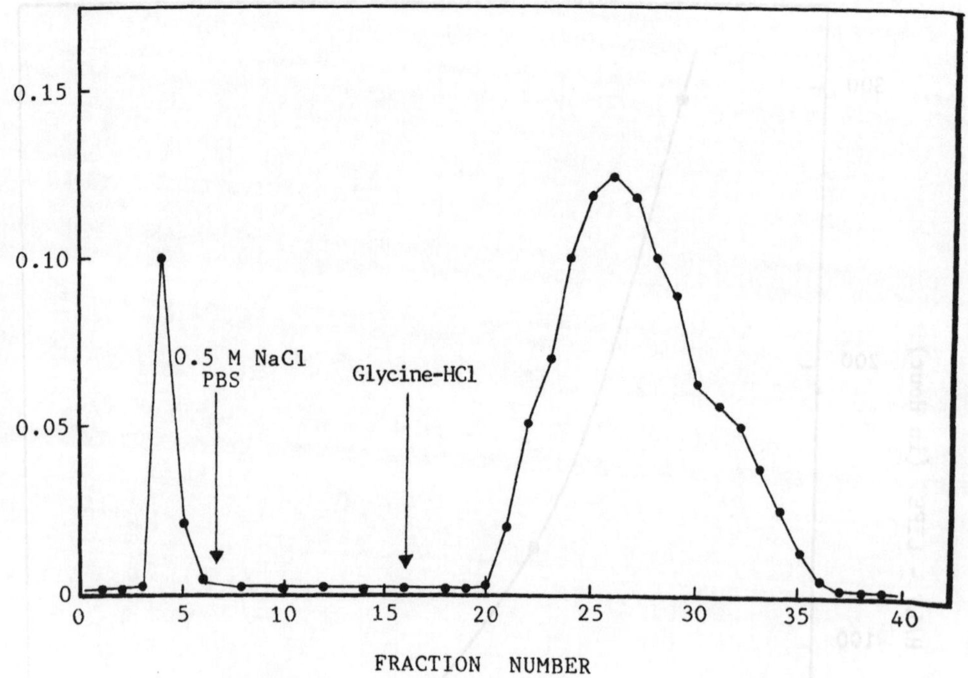


**Figure 7. Immunoaffinity purification of rabbit anti-human IgG using human IgG immobilized on FMP-activated Trisacryl GF-2000.**

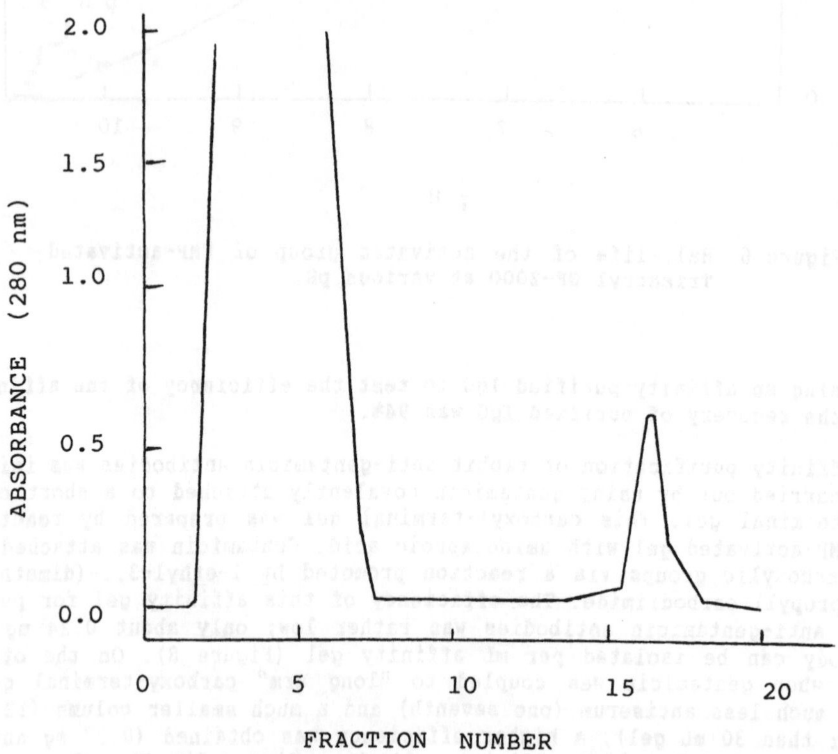


**Figure 8. Immunoaffinity purification of rabbit anti-gentamicin using gentamicin coupled to "short arm" carboxy-terminal gel.**

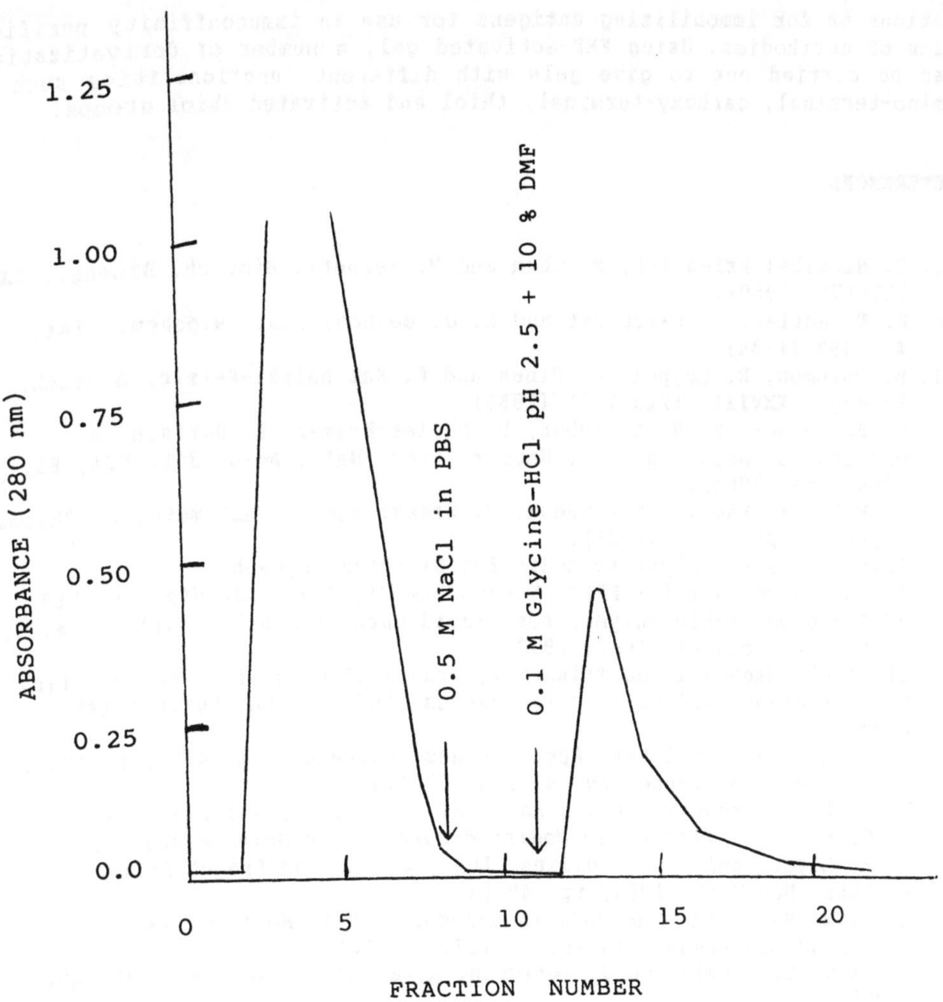


Figure 9. Immunoaffinity purification of rabbit anti-gentamicin using gentamicin coupled to "long arm" carboxy-terminal gel.

with N,N-bis (3-aminopropyl) ethylene diamine and then reacting the amino-terminal gel with succinic anhydride to generate a 16-atom spacer group.

The number of sulfhydryls was determined to be 9 µmoles per g dry gel. The activated thiol gel was shown to bind 3 mg B-galactosidase per g dry activated thiol gel.

## CONCLUSION

FMP was shown to be an effective reagent to activate the hydroxyl groups Trisacryl GF-2000 gel. The activated groups showed excellent half-lives ranging from hours to days depending on the pH of the solution. Ligand coupling can be performed at pH 6 to 10. Ligand with either free amino or free thiol groups can be coupled directly. The FMP-activated gel can be used directly for immobilization of enzymes for bioreactor applications or for immobilizing antigens for use in immunoaffinity purification of antibodies. Using FMP-activated gel, a number of derivatizations can be carried out to give gels with different functionalities such as amino-terminal, carboxy-terminal, thiol and activated thiol groups.

## REFERENCES


1. O. Hannibal-Friedrich, M. Chun and M. Sernetz, Biotech. Bioeng., XXII, 157-175 (1980).
2. P. E. Butler, D. Fairhurst and R. J. Beynon, Anal. Biochem., 141, 494-498 (1984).
3. B. Solomon, R. Koppel, G. Pines and E. Katchalski-Katzir, Biotech. Bioeng., XXVIII, 1213-1221 (1986).
4. U. Bamberger, P. H. Scheuber, B. Sailer-Kramer, K. Bartsch, A. Hartman, G. Beck and D. K. Hammer, Proc. Natl. Acad. Sci. USA, 83, 7054-7058 (1986).
5. Y. Kato, K. Kamiya, T. Iwaeda, H. Sasaki and T. Hashimato, J. Chromatography, 205, 185 (1981).
6. Fractogel TSK, Polymeric Media for Biochromatography.
7. TSK-Gel Toyopearl for FFLC, TechnoChem 21, Toyo Soda Mfg. Co. Ltd.
8. HEMA Biocompatible Polymer for Liquid Chromatography, Bulletin #179, Alltech Associates, Inc., 1990.
9. Ultrogel, Magnogel and Trisacryl, Practical Guide for Use in Affinity Chromatography and Related Techniques, 2nd edition, Reactif IBF, France.
10. T. Miron and M. Wilchek, App. Biochem. Biotech., 11, 445-456 (1985).
11. T. T. Ngo, Bio/Technology, 4, 134-137 (1986).
12. T. T. Ngo, Makromol. Chem., Macromol. Symp., 17, 229-239 91988).
13. T. T. Ngo in: "*Chemically Modified Surfaces in Science and Industry*," D. E. Leyden and W. T. Collins, Eds., Gordon and Breach Science Publishers, New York, 1988, pp. 49-66.
14. M. T. W. Hearn in: "*Methods in Enzymology*," K. Mosbach, Ed., Volume 135, Academic Press, New York, 1987, pp. 102-117.
15. R. Axen, H. Drevin and J. Cerlsson, Acta Chem. Scan., B29, 471-474 (1975).
16. T. T. Ngo, App. Biochem. Biotech., 13, 213-219 (1986).
17. T. T. Ngo, App. Biochem. Biotech., 13, 207-212.
18. G. L. Ellman, Arch. Biochem. Biophys., 82, 70-77 (1959).
19. Worthington, Enzymes, Enzyme Reagents, Related Biochemicals, Worthington Biochemical Corp., New Jersey, p. 239-242 (1977).
20. T. T. Ngo, D. Narinesingh and K. Laidler, Biotech. Bioeng., 18, 119 (1976).

**CONTRIBUTORS**

G. Blümel, 263
Institut für Experimentelle Chirurgie,
Technische Universität München,
Ismaninger Str. 22, D-8000,
München 80, Germany

Steven F. Bolling, 231
University of Michigan,
R5014 Kresge II,
Ann Arbor, MI 48109-0576

George L. Brode, 105 117, 129
Union Carbide Chemicals and Plastics Company Inc.,
P. O. Box 670,
Bound Brook, NJ 08805

Cynthia W. Butler, 255
Department of Chemistry,
Florida Atlantic University,
Boca Raton, FL. 33431

Charles E. Carraher, Jr., 255, 269
Department of Chemistry,
Florida Atlantic University,
Boca Raton, FL. 33431

Weiliam Chen, 329
College of Pharmacy,
The University of Michigan,
Ann Arbor, MI 48109 -1065

Tai Cheng, 1, 169
Advanced Cardiovascular Systems, Inc.,
3200 Lakeside Drive,
Santa Clara, CA 95052-8167

Jerry R. Conklin, 15
Larkin Laboratory,
The Dow Chemical Company,
Midland, MI 48640

Victoria Dole, 73
Johnson & Johnson Consumer Products, Inc.,
Grandview Road,
Skillman, NJ 08558

Parvin Dorostkar, 231
University of Michigan,
R5014 Kresge II,
Ann Arbor, MI 48109-0576

K. Egbaria, 215
College of Pharmacy,
University of Michigan,
Ann Arbor, MI 48109

Robert Eury, 169
Advanced Polymer Systems, Inc.,
Redwood City, CA 94063

Y. L. Fan, 311
Specialty Chemicals Division,
Union Carbide Corporation,
Bound Brook, NJ 08805

D. R. Flanagan, 239
Division of Medicinal and Natural Products Chemistry and
Division of Pharmaceutics, College of Pharmacy,
University of Iowa,
Iowa City, Iowa 52242

David Fleisher, 201
College of Pharmacy,
University of Michigan,
Ann Arbor, MI 48109

David T. Floyd, 49
Goldschmidt Chemical Corporation,
914 E. Randolph Road,
Hopewell, VA 23860

You-Yin Fu, 347
College of Pharmacy,
University of Michigan,
Ann Arbor, Michigan, 48109-1065

Koichiro Fukuda, 225
Faculty of Pharmaceutical Sciences,
Science University of Tokyo,
Shinjuku-ku, Tokyo 162, Japan

Charles G. Gebelein, 1
Dept. of Chemistry,
Youngstown State University,
Youngstown, OH 44555-0001

E. D. Goddard, 117, 129
Union Carbide Corp.,
Bound Brook, NJ 08805

Gershon Golomb, 231
Hebrew University of Jerusalem,
Jerusalem, Israel, 91120

W. C. Harris, 117, 129
Union Carbide Corp.,
Bound Brook, NJ 08805

Ken-ichiro Hayashi, 95
Department of Agricultural Biochemistry and Biotechnology,
Tottori University,
Tottori 680, Japan

Shigehiro Hirano, 95
Department of Agricultural Biochemistry and Biotechnology,
Tottori University,
Tottori 680, Japan

Keiji Hirochi, 95
Department of Agricultural Biochemistry and Biotechnology,
Tottori University,
Tottori 680, Japan

Thomas P. Johnston, 231
University of Illinois at Chicago,
833 S. Wood Street,
Chicago, IL 60612

Alan Kadish, 231
Northwestern University Medical Center,
250 E. Superior,
Chicago, IL 60611

Atsushi Kato, 225
Faculty of Pharmaceutical Sciences,
Science University of Tokyo,
Shinjuku-ku, Tokyo 162, Japan

Cherng-Ju Kim, 209
Faculty of Pharmacy,
University of Toronto,
Toronto, Ontario, Canada M5S 2S2

Jae-Seung Kim, 347
College of Pharmacy,
University of Michigan,
Ann Arbor, Michigan, 48109-1065

Sung Wan Kim, 193
Department of Pharmaceuticals and
Center for Controlled Chemical Delivery,
University of Utah,
Salt Lake City, Utah 84112

Joachim Kohn, 321
Department of Chemistry,
Rutgers - The State University of New Jersey,
New Brunswick, New Jersey 08903

Akihiro Kondo, 33
Wakayama Res. Lab.,
Kao Corp.,
1334, Minato, Wakayama

Tamotsu Kondo, 225
Faculty of Pharmaceutical Sciences and
Research Institute for Biosciences,
Science University of Tokyo,
Shinjuku-ku, Tokyo 162, Japan

S. L. Kopolow, 39, 137
GAF Chemicals Corporation,
1361 Alps Road,
Wayne, New Jersey 07470

Jim H. Kou, 201
Syntex Research, 3401 Hillview Ave.,
Palo Alto, CA 94304

Nancy E. Larsen, 147
Biomatrix, Inc.,
Ridgefield, New Jersey

Ping I. Lee, 209
Faculty of Pharmacy and
Dept. of Chemical Engineering & Applied Chemistry,
University of Toronto,
Toronto, Ontario, Canada M5S 2S2

E. Leshchiner, 129, 147
Biomatrix Inc.,
65 Railroad Ave.,
Ridgefield, NJ 07657

P. S. Leung, 129
Union Carbide Corp.,
Bound Brook, NJ 08805

Robert J. Levy, 231
University of Michigan,
R5014 Kresge II,
Ann Arbor, MI 48109-0576

R. J. Linhardt, 239
Division of Medicinal and Natural Products Chemistry and
Division of Pharmaceutics, College of Pharmacy,
University of Iowa,
Iowa City, Iowa 52242

Jue-Chen Liu, 73
Johnson & Johnson Consumer Products, Inc.,
Grandview Road,
Skillman, NJ 08558

R. Login, 39, 137
GAF Chemicals Corporation,
1361 Alps Road,
Wayne, New Jersey 07470

Karen Longe, 169
Advanced Polymer Systems, Inc.,
Redwood City, CA 94063

Gary Martino, 25
National Starch and Chemical Corporation,
Bridgewater, New Jersey 08807

Toshiya Mikami, 95
Department of Agricultural Biochemistry and Biotechnology,
Tottori University,
Tottori 680, Japan

Sergio Nacht, 83, 169
Research & Development,
Advanced Polymer Systems, Inc.,
3696 Haven Ave.,
Redwood City, CA 94063

Peng Shen, 371
Department of Chemical Engineering,
Drexel University,
Philadelphia, PA 19104

Rivka Siden, 231
University of Michigan,
R5014 Kresge II,
Ann Arbor, MI 48109-0576

Amnon Sintov, 231
Perio-Products Ltd.,
Box 23950,
Jerusalem, Israel, 91232

A. Stemberger, 263
Institut für Experimentelle Chirurgie,
Technische Universität München,
Ismaninger Str. 22, D-8000,
München 80, Germany

Hisashi Tachibana, 95
Department of Agricultural Biochemistry and Biotechnology,
Tottori University,
Tottori 680, Japan

M. Tazi, 39, 137
GAF Chemicals Corporation,
1361 Alps Road,
Wayne, New Jersey 07470

M. R. Thomas, 9
Dow Chemical U. S. A.,
2020 Dow Center,
Midland, MI 48674

Avinash G. Thombre, 159
Pfizer Central Research,
Groton, CT 06340

Chung-ye Tseng, 73
Johnson & Johnson Consumer Products, Inc.,
Grandview Road,
Skillman, NJ 08558

Rikio Tsushima, 33
Wakayama Res. Lab.,
Kao Corp.,
1334, Minato, Wakayama

M. Unkauf, 263
Institut für Experimentelle Chirurgie,
Technische Universität München,
Ismaninger Str. 22, D-8000,
München 80, Germany

Dan W. Urry, Ph.D., 181
Bioelastics Research, Ltd.,
1075 South Thirteenth Street,
Birmingham, Alabama 35205

# SUBJECT INDEX

MIX
Papier aus verantwortungsvollen Quellen
Paper from responsible sources
FSC® C105338

If you have any concerns about our products,
you can contact us on
**ProductSafety@springernature.com**

In case Publisher is established outside the EU,
the EU authorized representative is:
**Springer Nature Customer Service Center GmbH**
**Europaplatz 3, 69115 Heidelberg, Germany**

Printed by Libri Plureos GmbH
in Hamburg, Germany